Turbulent Shear Flows 9

Selected Papers from the Ninth International Symposium on Turbulent Shear Flows, Kyoto, Japan, August 16 - 18, 1993

Editors:

F. Durst N. Kasagi B. E. Launder
F. W. Schmidt K. Suzuki J. H. Whitelaw

With 325 Figures

Springer-Verlag
Berlin Heidelberg New York
London Paris Tokyo
Hong Kong Barcelona Budapest

Franz Durst
Lehrstuhl für Strömungsmechanik, Friedrich-Alexander-Universität,
Cauerstraße 4, D-91058 Erlangen, Germany

Nobuhide Kasagi
Department of Mechanical Engineering, University of Tokyo,
7-3-1 Hongo, Bunkyo-Ku, Tokyo 113, Japan

Brian E. Launder
Department of Mechanical Engineering, University of Manchester,
Institute of Science and Technology,
Manchester, M60 1QD, U.K.

Frank W. Schmidt
Department of Mechanical Engineering, The Pennsylvania State University,
University Park, PA 16802, USA

Kenjiro Suzuki
Department of Mechanical Engineering, Kyoto University,
Kyoto 606-01, Japan

James H. Whitelaw
Department of Mechanical Engineering, Imperial College,
London, SW7 2BX, U.K.

ISBN-13:978-3-642-78825-3 e-ISBN-13:978-3-642-78823-9
DOI: 10.1007/978-3-642-78823-9

CIP data applied for

Softcover reprint of the hardcover 1st edition 1995

Production: PRODUserv Springer Produktions-Gesellschaft, Berlin
Data conversion: Hilmar Schlegel, Berlin
SPIN: 10081579 55/3020 - 5 4 3 2 1 0 Printed on acid-free paper

Preface

The inaugural Symposium on Turbulent Shear Flows was held at The Pennsylvania State University in 1977. Thereafter the locations for the biennial symposium have alternated between the USA and Europe. However, the ninth Symposium on Turbulent Shear Flows was awarded to Japan in recognition of the strong support researchers of the Pacific Rim countries have given previous symposia. The University of Kyoto was the host institution and the meeting was held in the International Conference Hall. The Local Arrangements Committee did a superb job scheduling traditional Japanese dinners and arranging visits to the many cultural treasures in the Kyoto region.

The meeting attracted more than 260 offers of papers. Thirty-three sessions were scheduled to accommodate the 138 papers accepted for oral presentation. In addition a poster session was scheduled on each of the three days to accommodate a total of 42 poster presentations.

From the presentations at the symposium 24 have been selected for inclusion in this volume. The authors of these papers have revised them taking into consideration comments made during their oral presentation and recommendations made by the Editors. Four subject areas are identified, namely closures and fundamentals, free flows, wall flows, and combustion and recirculating flows. Eminent authorities have prepared introductory articles for each topic to put the individual contributions in context with each other and with related research.

The symposium was organized in cooperation with the American Society of Mechanical Engineers, Architectural Institute of Japan, Combustion Society of Japan, the Heat Transfer Society of Japan, the Gas Turbine Society of Japan, the Japan Society for Aeronautical and Space Sciences, the Japan Society of Civil Engineers, Japan Society of Fluid Mechanics, the Japan Society of Mechanical Engineers, the Japan Society of Multiphase Flow, and the Society of Chemical Engineers, Japan.

Financial support for the ninth symposium was provided by 62 Japanese industries and institutions.

The technical program of the symposium was the responsibility of the Papers Committee comprised of K. Suzuki (Chairman), F. Durst, N. Kasagi and B. E. Launder. This committee was guided by at least two reviews of each extended abstract provided by members of the Advisory Committee. The Advisory Committe consisted of the following members:

R. J. Adrian
H. Alfredsson
R. A. Antonia
D. W. Bechert
G. Bergeles
R. W. Bilger
R. Blackwelder
P. Bradshaw
K. Bremhorst
I. P. Castro
M. K. Chung
M. Coantic
J. Cousteix
R. Friedrich
I. S. Gartshore
V. W. Goldschmidt
E. Gutmark
H. Ha Minh
K. Hanjalic
T. J. Hanratty
J. A. C. Humphrey
F. Hussain
H. Kawamura
J. Kim
M. Kiya
T. Kobayashi
E. Krause
M. Lesieur
M. A. Leschziner
J. L. Lumley
Y. Nagano
I. Nakamura
I. Nezu
F. Ogino
M. Peric
S. B. Pope
W. C. Reynolds
W. Rodi
A. M. Savill
U. Schumann
G. X. Shen
R. L. Simpson
A. J. Smits
M. Sokolov
M. Sommerfeld
T. Takeno
A. Taylor
H. Ueda
D. Wang
M. Wolfshtein
J. C. Wyngaard
A. Yoshizawa

Members of the Advisory Committee also served as Session Chairmen at the symposium and have been continually valuable sources of advice and helpful criticism. The Editors wish to thank the staff of Springer-Verlag for their advice and considerable help in producing a volume in keeping with our aims.

University Park, PA, 1994 The Editors

Contents

Part I **Closures and Fundamentals**

Introductory Remarks
B. E. Launder 3

Two-Scale Double-Layer Model in Wall Bounded Turbulent Flow
E. Arad, M. Wolfshtein 7

Modelling of Intercomponent Transfer in Reynolds Stress Closures of Homogeneous Turbulence
M. Hallbäck, T. Sjögren, and A. V. Johansson 21

Closure of the Dissipation Tensor and the Pressure–Strain Tensor Based on the Two-Point Correlation Equation
M. Oberlack 33

Turbulence Statistics of Rotating Channel Flow
H. I. Andersson and R. Kristoffersen 53

Dynamic Modeling of Rotating Turbulence
K. D. Squires and U. Piomelli 71

Large Eddy Simulation and Second-Moment Closure Model of Particle Fluctuating Motion in Two-Phase Turbulent Shear Flows
O. Simonin, E. Deutsch, and M. Boivin 85

Part II **Free Turbulent Flows**

Introductory Remarks
M. Sokolov 119

Numerical Simulation of Low Re-Number Turbulence in Round Jets
J.H. Chen, J.J. Lienau, and W. Kollmann 123

Mixing in Turbulent Opposed Jet Flows
E. Mastorakos, A. M. K. P. Taylor, and J. H. Whitelaw 147

Thermal Characteristics of Turbulent Vortices in the Wake of a Circular Cylinder
J. Mi and R. A. Antonia 165

Second-Moment Closure Predictions of Buoyant Jets in Neutral and Stratified Cross-Flows
A. D. Gosman, R. Liu, and J. J. McGuirk 187

On Coherent Vortex Formation in Axisymmetric Compressible Shear Layers
K. Yu, E. Gutmark, and K. C. Schadow 207

Part III **Wall Flows**

Introductory Remarks
R. F. Blackwelder 221

Detailed Measurement of the Near-Wall Region of a Turbulent Pipe Flow
F. Durst, J. Jovanović, and J. Sender 225

Direct Numerical Simulation of Turbulent Plane Couette-Poiseuille Flows: Effect of Mean Shear Rate on the Near-Wall Turbulence Structures
A. Kuroda, N. Kasagi, and M. Hirata 241

Large Eddy Simulation of Turbulent Flow in Concentric Annuli with a Thin Inner Rod
S. Satake and H. Kawamura 259

Direct and Large-Eddy Simulations of Transition of a Supersonic Boundary Layer
F. Ducros, P. Comte, and M. Lesieur 283

Fine Structure of Reynolds Shear Stress in an Unsteady Turbulent Channel Flow
S. Tardu, M. Q. Feng, and G. Binder 301

Computation of Oscillating Turbulent Flows at Transitional Re-Numbers
K. Hanjalić, S. Jakirlić, and I. Hadžić 323

Structure of Heat Transfer in the Thermal Layer Growing in a Fully Developed Turbulent Flow
Y. Nagano, H. Sato, and M. Tagawa 343

Structure of the Turbulent Boundary Layer and Heat Transfer Downstream of a Vortex Generator Attached to a LEBU Plate
K. Inaoka and K. Suzuki 365

Part IV **Combustion and Recirculating Flows**

Introductory Remarks
D. Bradley 385

Transient Flame Growth in a Developing Shear Layer
R. P. Lindstedt and V. Sakthitharan 389

Spectral Measurement of Reactive and Conserved Scalars in a Turbulent Reactive-Scalar-Mixing Layer
J. D. Li, R. J. Brown, and R. W. Bilger 411

Turbulent Mixing and Non-gradient Diffusion in Baffle-stabilized Flames
P. Ferrão and M. V. Heitor 427

Turbulent Premixed Flame Propagation Revisited Results with a New Model
T. Mantel, R. Borghi, and A. Picart 439

A Description of Turbulent Premixed Flames Based on Flamelet Time Statistics
Ch. Mounaïm-Rousselle and I. Gökalp 459

Index of Contributors 473

Part I

Closures and Fundamentals

Introductory Remarks

B. E. Launder

UMIST, Manchester England

Of the papers forming this chapter, three are concerned with the provision of turbulence models that allow the mean flow to be computed by solving the Reynolds equations, two provide direct or large-eddy simulations of the flows under study while the last provides results from both approaches. This year marks the centenary of Osborne Reynolds' famous paper on decomposing turbulent motion into mean and fluctuating parts, a process leading to the appearance of effective turbulent (or, as we now call them, Reynolds) stresses in the averaged momentum equations. While, for fundamental work and certain particular fields of application, more elaborate approaches can now be employed, Reynolds-averaged schemes are destined to remain the principal route for computing practically interesting turbulent flows some way into the next century.

One of the factors affecting how much longer Reynolds-averaging remains the main approach to computing turbulent flows is the quality of the models developed for computing the Reynolds stresses. The papers represented here tackle three distinct and important problems in model refinement. The first by Arad & Wolfshtein addresses the question of turbulent time scales. At high Reynolds numbers nearly all approaches adopt just a single turbulent time scale, (k/ε). Arad and Wolfshtein develop a suggestion first made at TSF2 (Hanjalić *et al.* 1980) of employing a second turbulent time scale, thereby providing distinct (albeit interdependent) scales for the large scale interactions (τ) and the fine scale (k/ε). Although the authors apply their ideas within the context of an eddy viscosity approach its extension to use in a Reynolds-stress closure is self evident.

It is, indeed, at Reynolds-stress closure that the other two modelling papers are directed. The contribution of Hallbäck, Sjögren & Johansson tackles two modelling issues: that of characterizing the anisotropy of the dissipation tensor in terms of the anisotropy of the stress tensor; and that of devising a functional form for the Rotta coefficient in the slow part of the pressure-strain term. Curiously, while the return-to-isotropy coefficient in the slow pressure-strain term is shown to be a strong function of turbulence Reynolds number, the proposed relationship linking the dissipation and stress anisotropies contains no Reynolds number influences. The authors make recourse to their own hot-wire measurements (and DNS data) to help calibrate their model - a nice demonstration that even in the simple homogeneous flows considered the hot-wire has not been rendered obsolete by the computer.

The other modelling paper by Oberlack is on a grand scale and, indeed, has been significantly expanded since it was presented at the TSF meeting in Kyoto. Essentially, the two-point double-velocity-correlation equation is used as the basis for deriving a transport equation for the dissipation tensor which is arrived at by way of a tensor length-scale equation. Following this particular route ensures that the end equation is focused on the energy-containing range of turbulence rather than on the very fine scale responsible for the dissipation itself. The resultant form, eq (35) in the paper, exhibits, as it should, more varied interconnections between the stress, strain and dissipation fields than do existing semi-empirical equations for ε. It is interesting to note that when eq (35) is contracted (assuming local isotropy or, possibly, the relation suggested by Hallbäck *et al.* to express ε_{ij} in terms of ε) one obtains an equation for ε containing the production rate of turbulence energy as the primary generation but also secondary sources dependent upon the mean strain and the vorticity. The author would probably object strongly to the indicated simplification of his ε_{ij} equation but, for some years to come, it will scarcely be feasible to obtain *two* second-rank tensors from transport equations in practical CFD software - and one is inevitably that for the Reynolds stress tensor. As a finale to his paper Oberlack considers the modelling of the pressure-strain correlation and, without making additional closure assumptions, arrives at Rotta's model for the "slow" part and a new linear form for the "rapid" or mean-strain contribution (equations (41) and (39) respectively). For the record, it is noted that the model of $\Phi_{ij\,\mathrm{rapid}}$ appearing in his ref [10] was in fact first published and tested in Launder *et al.* (1972), thus pre-dating the cited publication of Naot *et al.* [14]. The latter, however, adopted a novel and powerful approach to obtaining the equation on which Oberlack's paper, itself, can be said to have built.

Andersson & Kristoffersen's paper, while derived from the DNS results obtained by the same authors (Kristoffersen & Andersson, 1993) is aimed at providing data to help the development and testing of second-moment closures. The flow considered is that in an infinite plane channel, $U(y)$, where rotation occurs about the z-axis. While the range of results will pose an interesting challenge for closure schemes, it is perhaps those at the highest Rossby number that are of most interest. The progressive collapse of turbulence on the suction side with increase in *Ro*, and the corresponding augmentation on the pressure side could be foreseen qualitatively but it is of great value to have the detailed results available showing precisely how this happens.

Squires and Piomelli contribute another paper on the simulation of rotating flows, this time employing a large-eddy simulation (LES), in which, locally, the Smagorinsky parameter, C, is determined from the resolved field, eq (6). This practice springs from developments in subgrid modelling proposed by Germano *et al.*. (1991) and Lilly (1992) which represent important steps towards making LES usable for practical flows; indeed the present contribution shows the application of the approach to physically complex flows. The simulations of both the decay of rotating grid turbulence and the rotating channel flow considered above achieve impressive agreement with the DNS results. For the latter flow, neither the DNS nor the LES show the very marked decrease in wall friction on the suction side that was exhibited by the earlier experimental studies, a discrepancy here attributed to the fact that the experiments were of necessity carried out in a finite-width channel. This seems an area where the LES scheme could be used to test this conjecture by changing the geometry considered from the infinite channel considered here to the finite aspect ratio channel examined in the experiment.

The final contribution is a composite LES and second-moment closure exploration by Simonin, Deutsch and Boivin of a turbulent flow containing discrete spherical particles in dilute concentration. Again this is a contribution that has been considerably expanded since its oral presentation at TSF9. Both approaches underline the fact that mean shear in the fluid stream has a large effect on the turbulent velocity of the particles, the most striking effect of inertia being to increase the anisotropy of a particle's fluctuations relative to that of the fluid in which it is immersed. The modelling approach in particular has many novelties especially for those, like the writer, who, apart from coping with inclement rain showers, spend their working lives in a single-phase environment.

References

Germano M, Piomelli U, Moin P and Cabot W H, 1991, *Phys. Fluids A*, **3**, 1760

Hanjalić K, Launder B E and Schiestel R, 1980 *Turbulent Shear Flows 2* pp. 36-49 (Ed L J S Bradbury *et al.*), Springer Verlag, Heidelberg

Kristoffersen R and Andersson H, 1993, *J Fluid Mech.* **256**, 163

Launder B E, Morse A P, Spalding D B and Rodi W, 1972 *Proc Conf on Free Turbulent Shear Flows*, NASA SP-321, pp. 361-426

Lilly D K, 1992, *Phys. Fluids A*, **4**, 633

Two-Scale Double-Layer Model in Wall Bounded Turbulent Flow

Eran Arad and Micha Wolfshtein

Faculty of Aerospace Engineering, Technion,
Israel Institute of Technology, Haifa 32000, Israel

Abstract

A two-scale turbulence model for the solution of wall bounded flows is presented. The turbulent diffusion and production of energy are related to the scale of the big eddies, while the decay term in the energy equation is obtained from a modified dissipation equation. The three-equation model utilizes partial differential equations in the external part of the boundary layer only, and is matched with a single equation model in the inner region. The model is compared with data of duct flow and attached adverse pressure gradient flows. Good agreement with experimental data and results of Direct Numerical Simulation is obtained.

1. Introduction

1.1 Background

Two equation turbulence models, and in particular the energy-dissipation model, have become a popular and widely used tool for the prediction of turbulent flows. However, the prediction of turbulent quantities in the vicinity of solid walls still remains a difficult task for practitioners of the models. Several authors reported deficiencies of such models: Incompatible low-Reynolds number damping functions[1,2], wrong asymptotic behavior of quantities like the Reynolds stresses or turbulent energy near walls[2] and lack of natural boundary condition for the ε equation on a wall[3]. Other models with simpler near wall boundary conditions, such as the $k-\omega$ model, yield asymptotically inconsistent solutions for the turbulent kinetic energy[3]. As a result, the prediction of the near-wall peaks of the turbulent kinetic energy and dissipation rate are not good[2,4]. This fact has become more and more obvious as Direct Numerical Simulations (DNS) and improved measurement techniques provided more detailed and accurate information on the thin and extremely nonhomogeneous region adjacent to solid walls.

The quest for better two equation models appear to be very difficult, as all two-equation models can be shown to differ from one another mainly in source terms[5]. Indeed, the deficiencies related to the ε equation near a wall occur also in other scale formulations. Therefore improvement of the applicability and universality of the models must extend beyond the scope of conventional two-equation models.

An additional deficiency of two equation models is the assumption of spectral similarity which is implied in all such models. There is no reason to expect that large and small eddies develop in the same way. Therefore the scale for the dissipation should not be computed from the scale for the viscosity.

In this paper we try to address these problems.

1.2 Present Contribution

The model proposed here still utilizes the turbulent energy, but with two scale equations: A time scale equation for the big, energy carrying eddies, and a dissipation equation for the small dissipative eddies. The turbulent diffusion and production of turbulent energy are related to the scale of the big eddies, (the time scale τ in this paper[6]), while the decay term in the energy equation is obtained from a dissipation equation, representing the high wave number scale. The three-equation model utilizes partial differential equations only in the external part of the flow. In the inner region we use a single equation model[7]. The two scales are calculated in the inner layer by algebraic relations and matched with the external solution on the interface. The algebraic scales are so devised as to ensure that they exhibit the correct asymptotic solution on the wall.

In this paper we consider only two-dimensional parabolic flows near solid walls. These cases are easier to compute, and the amount of experimental data for them is very large. Moreover, such flows constitute an important ingredient of most elliptic flow fields. Therefore they are very suitable for testing new models. Solutions were obtained using an efficient boundary layer solver, utilizing a fourth order OCI scheme[8].

2. Two-Scale Model

Single-point one-scale models are based on the assumption of spectral similarity, and thus use a single time scale. Yet, different turbulent interactions are associated with different parts of the spectrum, making this assumption highly simplistic[9]. Still, solution in the wave number space is very complicated and demands large resources. Therefore, Hanjalić, Launder and Schistel[9] suggested to consider two domains in the wave number space, and to calculate typical scale and energy for each domain. A four-equation model was defined that way.

The present model continues this line of thought, but with only three equations, for the two scales (for the low and the high wave number spectral region) and for the energy of the large eddies. The energy of the dissipative eddies is not computed as it is negligibly small. Thus the energy equation is written in the conventional form:

$$\varrho \frac{\mathrm{D}k}{\mathrm{D}t} = \frac{\partial}{\partial x_j}\left[\left(\mu + \frac{\mu_t}{\sigma_k}\right)\frac{\partial k}{\partial x_j}\right] + P_k - \varrho\varepsilon\,, \tag{1}$$

where P_k is the production term and depends on the mean strain. The turbulent diffusion and production are related to the low wave numbers scale (the time scale τ in this model), and the decay term is associated with the high wave number scale (the dissipation rate ε). Thus, the turbulent viscosity is given by:

$$\mu_t = C_\mu \varrho k \tau \,. \tag{2}$$

The time scale is an integral scale, and as such we consider it more suitable for an eddy viscosity estimation. The time scale equation[6] is derived from the auto-correlation equation. Its exact equation is much simpler than that of the dissipation rate, and requires a lower level of modelization. The model equation is the following:

$$\varrho \frac{\mathrm{D} k\tau}{\mathrm{D} t} = \frac{\partial}{\partial x_j}\left[\left(\mu + \frac{\mu_t}{\sigma_\tau}\right)\frac{\partial k\tau}{\partial x_j}\right] + C_{g_\tau}\tau P_k - \varrho C_{d_\tau} k \,. \tag{3}$$

Eq. (3) may actually be regarded as an eddy viscosity equation.

The dissipation rate, ε, is used as the high wave number scale. The governing equation is written in high Reynolds number form:

$$\varrho \frac{\mathrm{D}\varepsilon}{\mathrm{D} t} = \frac{\partial}{\partial x_j}\left[\left(\mu + \frac{\mu_t}{\sigma_\varepsilon}\right)\frac{\partial \varepsilon}{\partial x_j}\right] + \varrho\frac{\varepsilon_p \varepsilon}{k} - \varrho C_{d_\varepsilon}\frac{\varepsilon^2}{k} \,. \tag{4}$$

The source term for the dissipation does not depend on the mean strain field, but on ε_p, the energy transfer rate related to the large scales, which is defined as follows:

$$\varepsilon_p = C_{g_\varepsilon}\frac{k}{\tau} \,. \tag{5}$$

With this formulation the turbulent energy dissipation rate responds only slowly to the applied mean strain, giving a better representation of a rapidly changing turbulence field. The decay term of the dissipation, on the other hand, depends only on the high wave number region. Various authors have shown that irrotational strain affects the turbulence transport in decelerating flow[10]. This effect must be present in the model equations. However, as the dissipation is a small scale phenomenon, it may be postulated that it is not significantly affected by the irrotational strain. Of course, this argument is not valid for the conventional dissipation equation, which accounts for the large scales. However, in the present case, where the dissipation represents only small scales, and the dissipation equation is different from the conventional one, this argument is legitimate. Therefore, the contribution of irrotational strain is added only to the energy and time scale equations:

$$\begin{aligned} P_k^{\mathrm{IR}} &= -\left(\overline{u^2} - \overline{v^2}\right)\frac{\partial U}{\partial x} \,, \\ P_{k\tau}^{\mathrm{IR}} &= C_{g_\tau^{\mathrm{IR}}}\tau P_k^{\mathrm{IR}} \,. \end{aligned} \tag{6}$$

As any eddy viscosity model, the present model does not provide a good estimate of the values of the diagonal stresses. Therefore we chose to relate them to the turbulent kinetic energy by the following empirical coefficients:

$$\frac{\overline{u^2}}{k} = 1.1 \qquad \frac{\overline{v^2}}{k} = 0.25 \,. \tag{7}$$

Selection of the values of all constants has been done in a way which ensures consistency with previous models and experimental data, to give:

σ_k	C_{g_T}	C_{d_T}	σ_T	C_{g_ε}	C_{d_ε}	σ_ε	$C_{g_T^{IR}}$	C_μ
1	0.475	0.05	0.3	0.5	1.85	1.3	7.3	0.09

3. Near Wall Analysis

Turbulent flows in the vicinity of solid walls are different from other turbulent flows in two respects: Firstly, the level of turbulence diminishes as the wall is approached, and therefore viscous effects become dominant there. This effect is usually referred to as the low turbulence effect. The second effect is that of the strong anisotropy associated with the rapid decay of the velocity component normal to the wall (relative to the tangential components). This second factor makes low turbulence near walls quite different from low turbulence away from walls, and justifies the use of wall coordinates (the "+" system) for the modelling of the near wall region. Obviously, such a formulation is less convenient to use when non orthogonal coordinates are used. Still, it is difficult to describe wall effects without utilizing the distance from the wall.

3.1 Double Layer Concept

The wall region can be usually divided into some sub-regions: Viscous sub layer, buffer zone and fully turbulent zone. Various authors[11,12] have suggested lately the use of a special model in the immediate vicinity of the wall, based on the nearly universal nature of this layer. The solution of this model is then matched to the solution of an external model, on the interface between the inner and outer regions. The matching is performed in the region of $Y_+ \approx 50 - 150$. Low sensitivity of the solution to the location of the interface was observed.

Wolfshtein[7] suggested the use of two length scales L_μ and L_ε in the context of one equation turbulence model. The two scales are related to the eddy viscosity and the dissipation rate of the turbulent kinetic energy in the following manner

$$\mu_t = \varrho L_\mu k^{1/2} = \frac{-\varrho\overline{uv}}{\partial U/\partial y} ; \qquad \varepsilon = \frac{k^{3/2}}{L_\varepsilon} . \tag{8}$$

It may be noted here that examination of the scales obtained by DNS[13], showed that the distribution of the scale of the small eddies near the wall is not linear.

3.2 Scale Analysis Near The Wall

The distribution of the scales near the wall can be calculated by asymptotic analysis. The analysis is based on a power series expansion of the fluctuating velocities near the wall[2] in terms of the distance from the wall, x_2. In order to achieve universality we have to use the wall coordinates $y = x_2 u_\tau/\nu$ (often referred to as y_+). The expansion near the wall ($x_2 = 0$) is:

$$\begin{aligned} u_1 &= b_1 y + c_1 y^2 + d_1 y^3 + O(y^4) \\ u_2 &= c_2 y^2 + d_2 y^3 + O(y^4) \\ u_3 &= b_3 y + c_3 y^2 + d_3 y^3 + O(y^4) \\ p &= a_p + b_p y + c_p y^2 + d_p y^3 + O(y^4)\,, \end{aligned} \tag{9}$$

where b_1, b_2 etc. are functions of (x_1, x_3, t). Moreover, the exact solution for the mean velocity in a one-dimensional boundary layer near a wall is:

$$U = y - \frac{y^2}{2Re} + O(y^3)\,. \tag{10}$$

Using the series of (9), to evaluate L_μ (defined in 8), gives:

$$L_\mu = -\frac{\Gamma_1}{\sqrt{2\Gamma_3}} y^2 - \frac{\overline{\Lambda_1 \left(Re_\tau b_1 \Gamma_3 - b_1 b_3 c_3 + c_1 b_3^2\right) - 2\Gamma_2 \Gamma_3}}{\sqrt{2\Gamma_3}\Gamma_3} y^3 + O(y^4)\,, \tag{11}$$

where

$$\begin{aligned} \Gamma_1 &= \overline{b_1 \frac{\partial b_1}{\partial x_1}} + \overline{b_1 \frac{\partial b_3}{\partial x_3}} = \overline{b_1 \Lambda_1} \\ \Gamma_2 &= \overline{b_1 d_2} \\ \Gamma_3 &= \overline{b_1^2} + \overline{b_3^2} \\ \Gamma_4 &= \overline{b_1 c_1} + \overline{b_3 c_3}\,. \end{aligned} \tag{12}$$

The dissipation length scale, L_ε (defined in 8), can be evaluated by the series (9) as well:

$$L_\varepsilon = \sqrt{\frac{\Gamma_3}{8}} y^3 - \frac{1}{2} \frac{\Gamma_4}{\sqrt{2\Gamma_3}} y^4 + O(y^5)\,. \tag{13}$$

It is apparent that the distributions of the two scales near the wall are quite different from one another: L_μ varies as y^2 while L_ε varies as y^3. The model chosen reproduces the parabolic or cubic parabola near the wall, and matches the external distribution away from the wall. The algebraic form of the matching functions was determined by reference to DNS results. To illustrate the point the reader is referrd to figures 4 and 9 which show the parabolic distributions of the scales near the wall as obtained from DNS data. The large eddie's length scale varies linearly in the buffer layer, and then becomes nearly uniform in the outer part of the boundary layer. This distribution is represented by an exponential matching function quite well. The dissipation length scale shows a bulge in the buffer layer (in the region of $5 < y < 20$). Therefore a more complicated matching function is required for it, based on a hyperbolic tangent.

Thus the algebraic formulae used to calculate the length scales near the wall are:

$$\begin{aligned} L_\mu &= kC_\mu^{1/4} y \left[1 - \exp(-\alpha_\mu y)\right] \\ L_\varepsilon &= \left(\frac{L_\varepsilon}{y}\right)_m \left[1 - \exp(-\alpha_{\varepsilon_1} y)\right]^2 \tanh(\alpha_{\varepsilon_2} y). \end{aligned} \tag{14}$$

The value on the interface $\left(L_\phi/y\right)_{\mathrm{m}}$ is set by the solution of the external differential model and the coefficients α_μ, α_{ε_1} and α_{ε_2} depend on the pressure gradient.

4. The Numerical Scheme

The numerical scheme is described elsewhere and will not be discussed in detail here. The numerical problem is that of a system of four parabolic equations (for the velocity, the turbulent energy and the two scales). These equations are solved by a marching scheme, utilizing the Thomas algorithm for tri-diagonal systems. The differential equations are replaced by finite difference equations using an Operator Compact Implicit (OCI) scheme. The scheme is fourth order in the cross stream direction, but only of first order in the marching direction. The scheme has been thoroughly tested for accuracy and convergence.

5. Results and Discussion

Duct flow and flat plate boundary layer cases, for which DNS results are available, were computed, to gain better insight into the turbulent variables near the wall. Then the model was used for calculation of adverse pressure gradient cases, and compared with experimental results.

5.1 Turbulent Variables and Budgets in Duct Flow

DNS data for fully developed duct flow[2,14] was used for comparisons with the computations. DNS result are available for $Re_\tau = 180$ and $Re_\tau = 395$. In figures 1-3 the profiles of the turbulent energy, the shear stress and dissipation rate are presented for $Re_\tau = 180$. The results are compared with results of the Jones and Launder[15] two-equation model (designated by JL) and DNS results. The present turbulent energy and shear stress distribution are in very good agreement with the DNS data. The standard $k - \varepsilon$ model (JL) under-predicts the peak of the turbulent energy near the wall. This deficiency was already reported for other types of flow fields[5]. The distribution of the dissipation rate is not as good, showing a certain difference between the present and DNS results in the region of $Y_+ \approx 10 - 20$. Yet, it appears that this disagreement does not affect the other variables. Furthermore, the plateau which is drawn in that region by DNS data, does not appear in experimental dissipation profile[1], as pointed out by Rodi and Mansour[13]. In figure 4 the two length scales which are defined by the present model, L_μ and L_ε, are compared with the scales of the DNS results. The figure demonstrates the smooth transition between the internal zone, where algebraic formulation is used, to the outer three-equation region. The agreement with DNS result is good. Zooming in the vicinity of the wall (Fig. 5), shows correct behavior of the scales.

Profiles of k, $-\overline{u'v'}$, ε and the scales for $Re_\tau = 395$ are presented in figures 6-9, respectively. The good agreement with DNS results is generally conserved, with somewhat better agreement in ε, and worse in the energy peak. Comparisons are made with the results of Launder and Sharma[16], designated by LS, and the advantage of the present model is evident.

In figure 10, the budget of terms in the turbulent energy equation, at $Re_\tau = 180$ is plotted. It appears that the general behavior of all terms is well reproduced by the model. In the region next to the wall and in the external layer, the agreement is very good. Some discrepancies appear in the $Y_+ \approx 5-20$ region, in the turbulent diffusion and dissipation terms. The pressure diffusion term, which has a small but non-zero value near the wall, has no modelization in the present version.

5.2 Turbulent Variables in Flat Plat Boundary Layer

The DNS results of Spalart[17] for a flat plat boundary layer, were used as another test case. Results for this case are available at three Re_ϑ numbers: 300, 670 and 1410. The prediction of the profiles in the first low Reynolds number is not good, but high level of agreement with the DNS data is obtained in the two higher Reynolds numbers. In figures 11-12, the turbulent energy for the two above mentioned Reynolds numbers is plotted. Very good agreement with the DNS data is obtained. The comparison of the shear stress, presented in figures 13-14, yields good agreement as well. Another interesting aspect that is considered here, is the influence of the Reynolds number on the scales in the vicinity of the wall. As can be seen in figures 15-16, the Reynolds number influence was well reproduced by the model.

5.3 Adverse Pressure Gradient Boundary Layer

This type of parabolic flow is more difficult to predict, and the agreement of the computations with experiments is usually poorer. Two cases are reported here, namely the Nagano et. al.[18] experiment and the Samuel and Joubert[19] experiment.

The Nagano et al. Experiment. Nagano *et al.*[18] reported on experimental results for an attached boundary layer flow at moderate to strong adverse pressure gradient of $0.76 \leqslant \beta \leqslant 4.66$, where

$$\beta = \frac{\delta_\star}{\tau_W}\frac{\mathrm{d}\overline{P}}{\mathrm{d}x}\,.$$

Re_ϑ in this case is between 1290 to 3350. In figures 17-18 the comparison of the turbulent energy with experimental data is presented for the two cases, showing a high level of agreement. The predicted shear stress (figures 19-20) exhibits the same behavior. Once again the correct behavior of the scale near the wall is obtained by the model (figure 21), though the values of the scales are under-predicted in the region of $Y_+ \approx 20-40$.

The Samuel and Joubert Experiment. Samuel and Joubert[19] report on a two-dimensional unseparated diffuser flow. In this case both the first and second streamwise pressure derivatives are positive, as is the case in many practical cases. Due to the high pressure gradient the boundary layer approached separation at the end of the test section. Thus the current parabolic calculation may not be valid there. The results of the computations are shown in Fig. 22-23. In Fig. 22 the profiles of the turbulence energy ($2k/U_e^2$, where U_e is the local free stream velocity) are plotted at various stations along the boundary layer. Very good agreement is obtained for distances smaller than 2.5 m. Further downstream the agreement deteriorates, but this may be the consequence of the thickening of the boundary layer as it approaches separation. The Reynolds stress profiles are shown in Fig. 23, for five stations along the wall. Here the agreement is better although the quality of the agreement deteriorates along the wall.

6. Discussion and Conclusions

The three-equation double-layer model is based on general arguments, and satisfies the theoretical near wall distributions of the turbulent components. It appears to reproduce the low level turbulence of the wall region in two-dimensional boundary layers well, for both standard cases and high adverse pressure gradients. The agreement with DNS results, experimental data and asymptotic analysis is very good in most cases. The derivation of the model equations is general, and allows an easy extension to two and three-dimensional cases. However, one should take account of the non-local element near the wall due to the use of the normalized distance from the wall. We argue that this is necessary in order to distinguish between low level turbulence away from walls, and low level turbulence in the vicinity of solid walls. This non-local behaviour may impose certain difficulties when non-orthogonal coordinates are used. Still, we believe that the wall effect should be properly taken into account. The distance from the wall allows an easy estimate of the wall effect, and is therefore a practical way of accounting for the high anisotropy of the turbulence near solid walls.

Acknowledgments

The authors are grateful to Prof. P. Bradshaw for providing the Direct Numerical Simulation data which is used in this paper, and for his helpful suggestions.

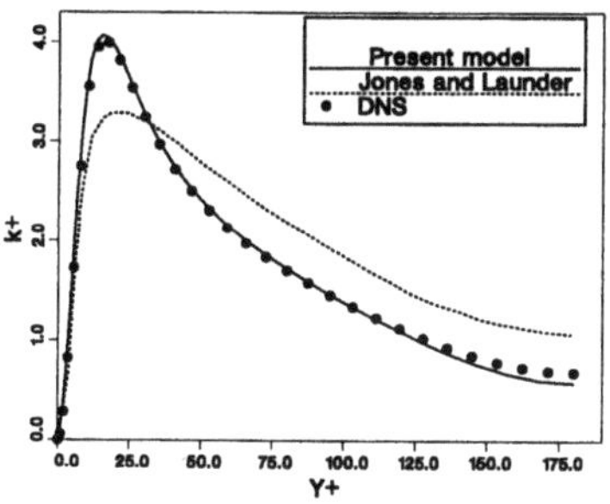

Fig. 1. Turbulent energy for duct flow, for $Re_\tau = 180$.

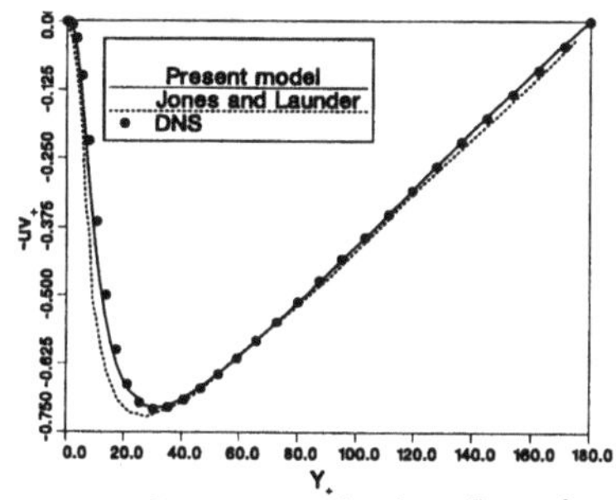

Fig. 2. Shear stress for duct flow, for $Re_\tau = 180$.

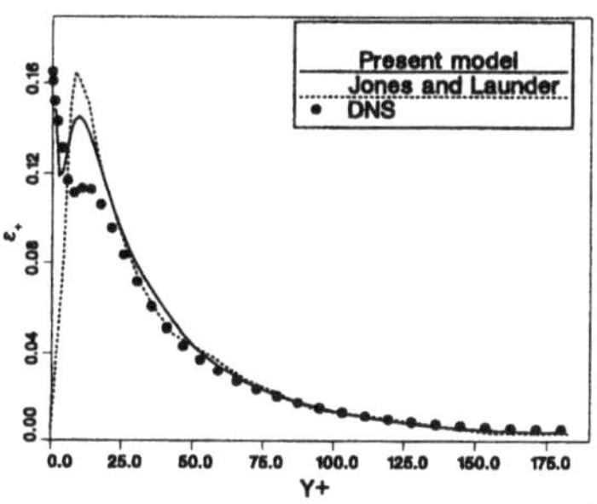

Fig. 3. Dissipation rate for duct flow, for $Re_\tau = 180$.

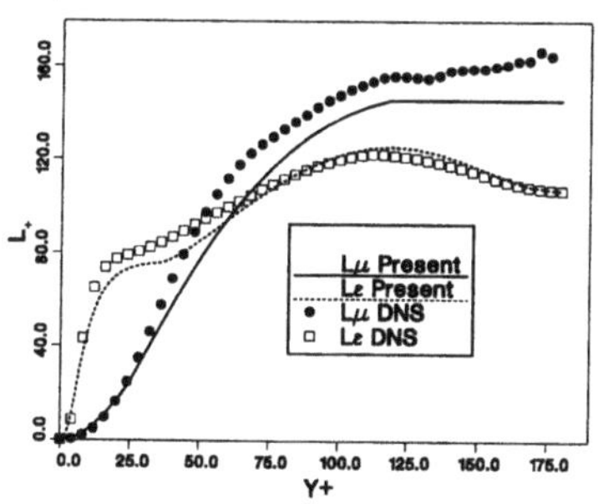

Fig. 4. Scales in duct flow, for $Re_\tau = 180$.

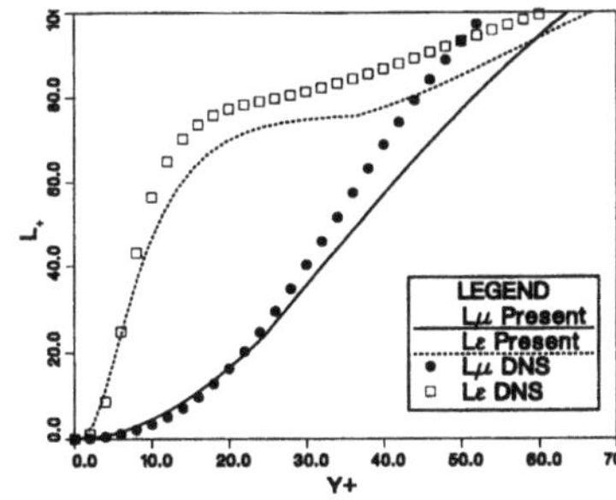

Fig. 5. Scales close to the wall,in duct flow, for $Re_\tau = 180$.

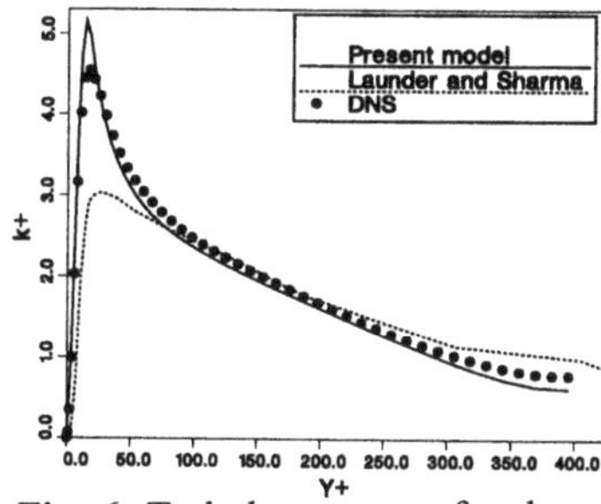

Fig. 6. Turbulent energy for duct flow, for $Re_\tau = 395$.

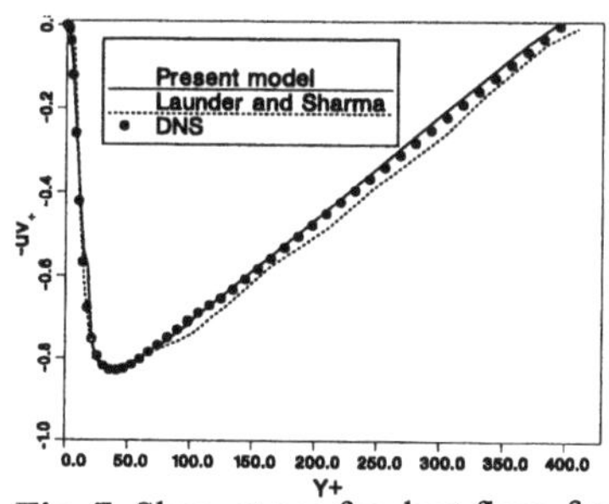

Fig. 7. Shear stress for duct flow, for $Re_\tau = 395$.

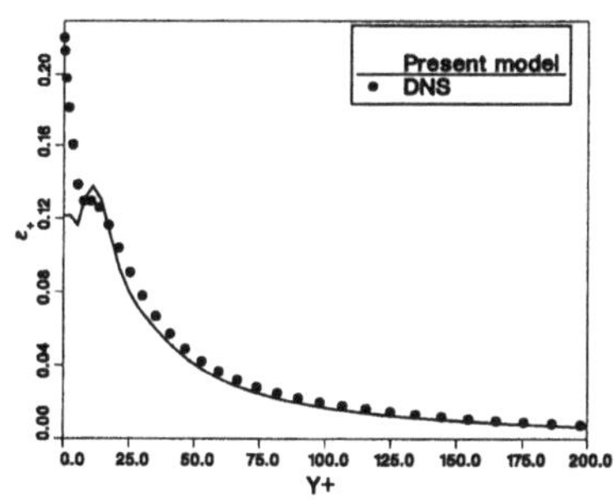

Fig. 8. Dissipation rate for duct flow, for $Re_\tau = 395$.

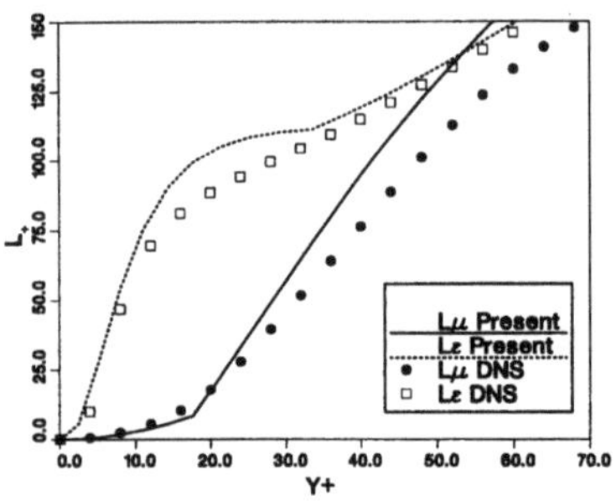

Fig. 9. Scales in duct flow, for $Re_\tau = 395$.

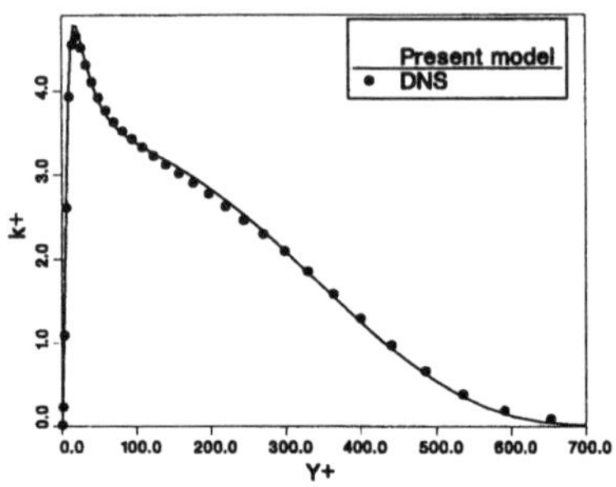

Fig. 12. Turbulent energy in boundary layer, ($Re_\vartheta = 1410$).

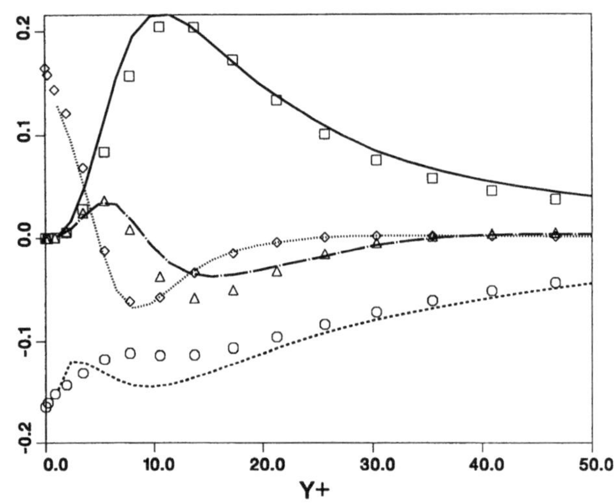

Fig. 10. Budget of terms in energy equation, for duct flow ($Re_\tau = 180$).

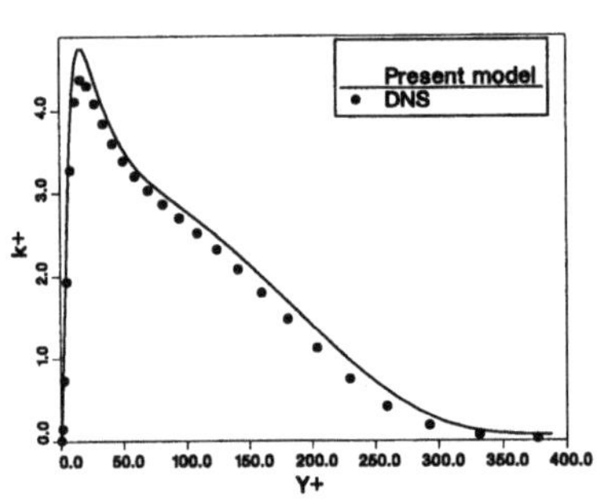

Fig. 11. Turbulent energy in boundary layer, ($Re_\vartheta = 670$).

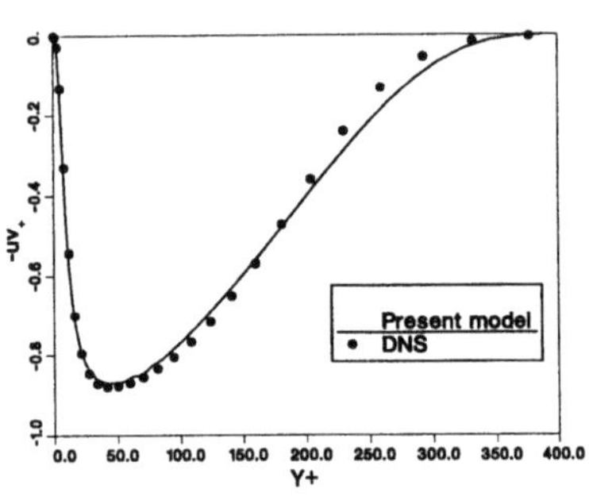

Fig. 13. Shear stress in boundary layer, ($Re_\vartheta = 670$).

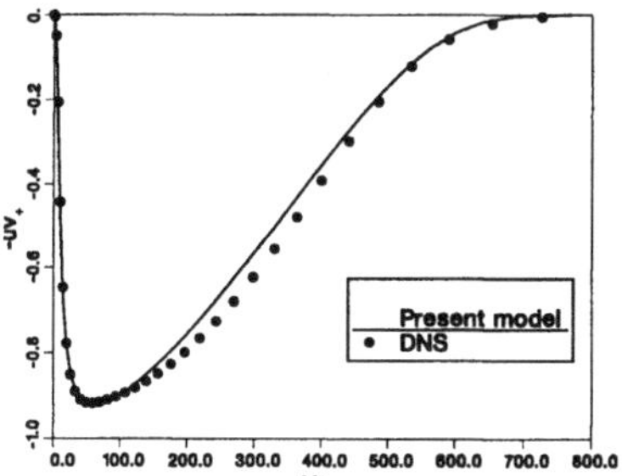

Fig. 14. Shear stress in boundary layer, (Re_ϑ = 1410).

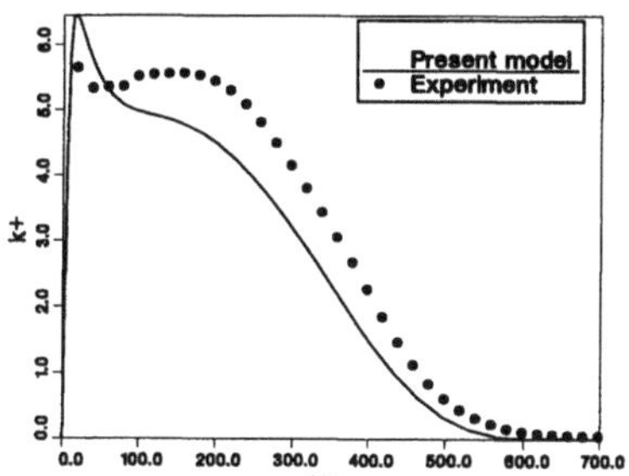

Fig. 18. Turbulent energy in adverse pressure gradient boundary layer, (β =2.05).

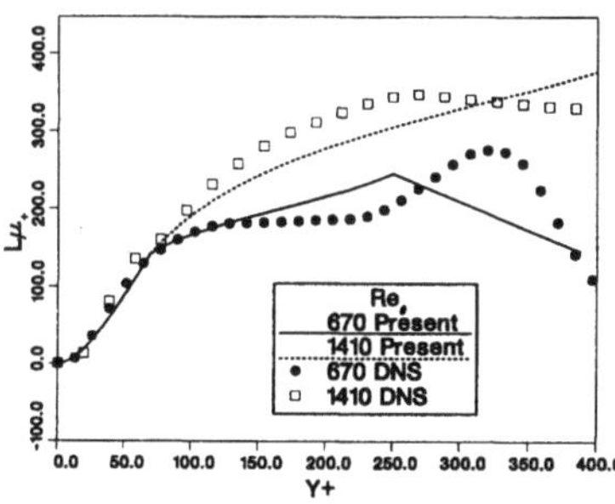

Fig. 15. Large eddies scales in boundary layer.

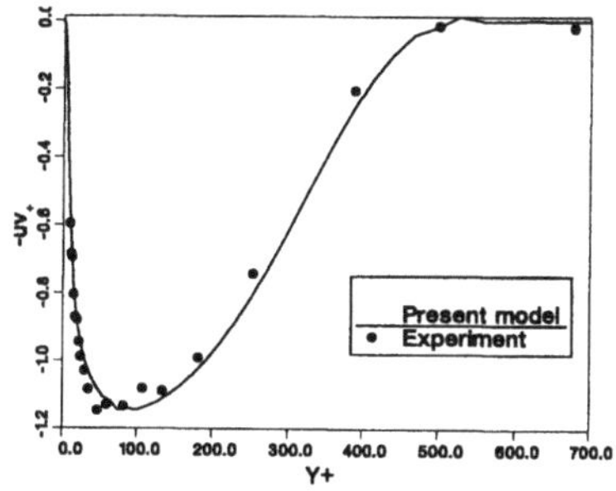

Fig. 19. Shear stress in adverse pressure gradient boundary layer, (β =0.76).

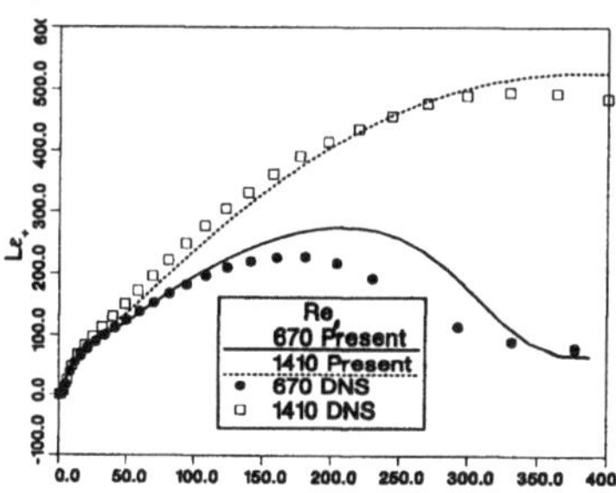

Fig. 16. Small eddies scales in boundary layer.

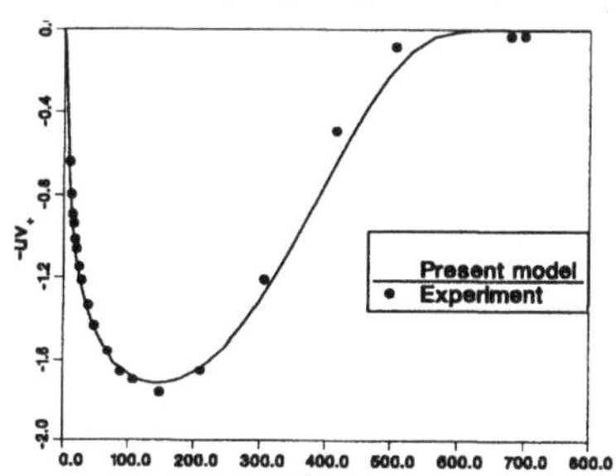

Fig. 20. Shear stress in adverse pressure gradient boundary layer, (β =2.05).

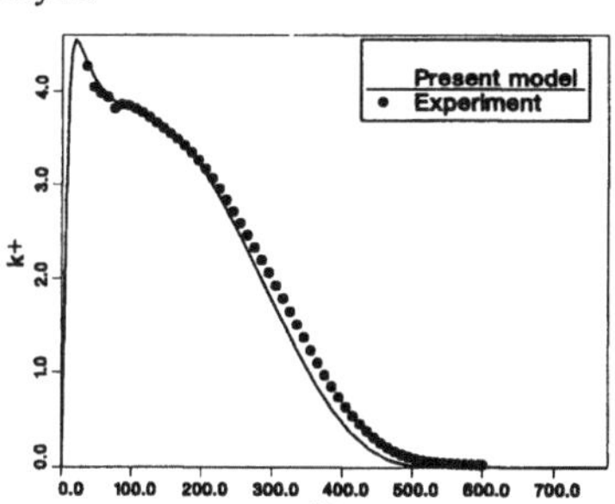

Fig. 17. Turbulent energy in adverse pressure gradient boundary layer, (β = 0.76).

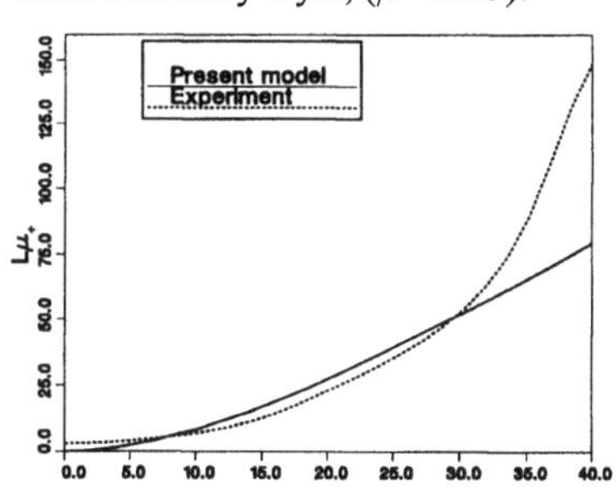

Fig. 21. Large eddies scales in adverse pressure gradient boundary layer, (β =0.76).

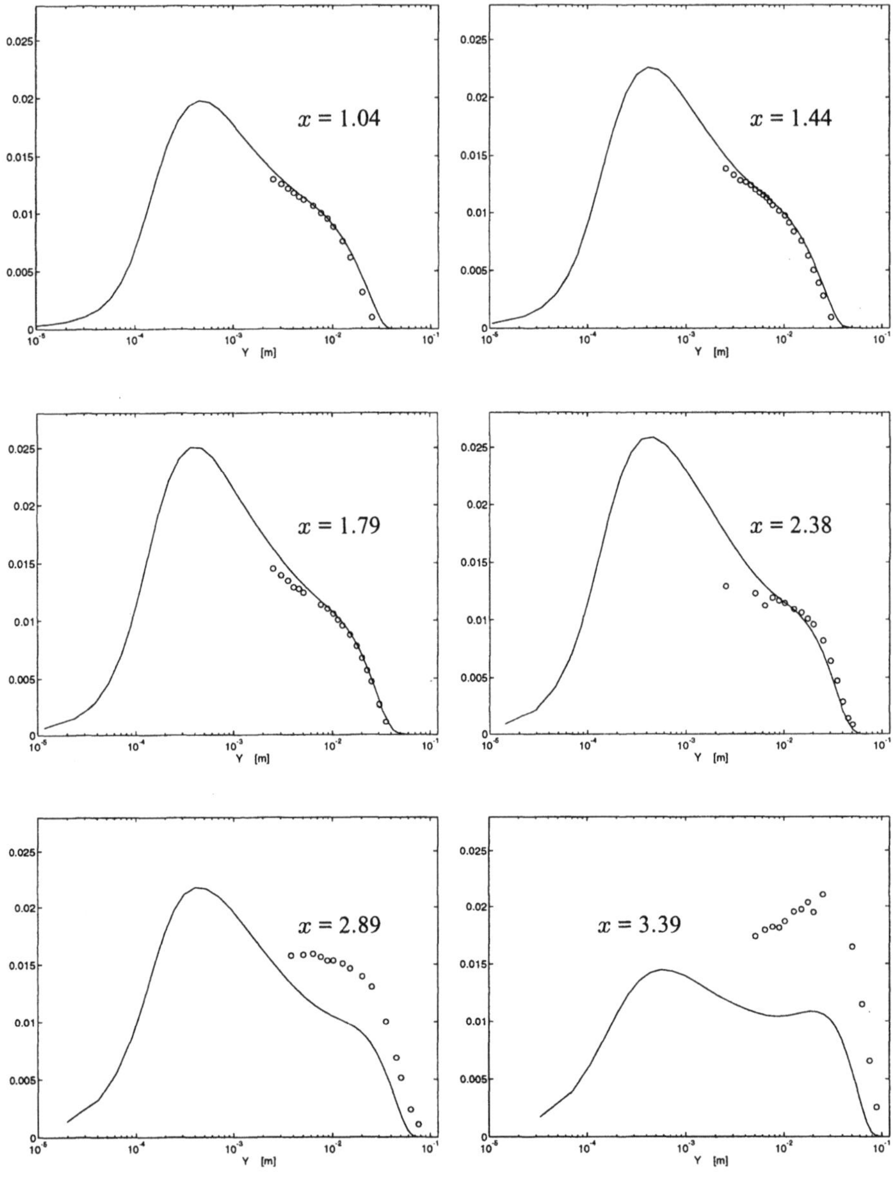

Fig. 22. Turbulent energy $(2k/U_e^2)$ profiles in various x locations along the wall. Solid line is present computations, open circles are experimental data[19].

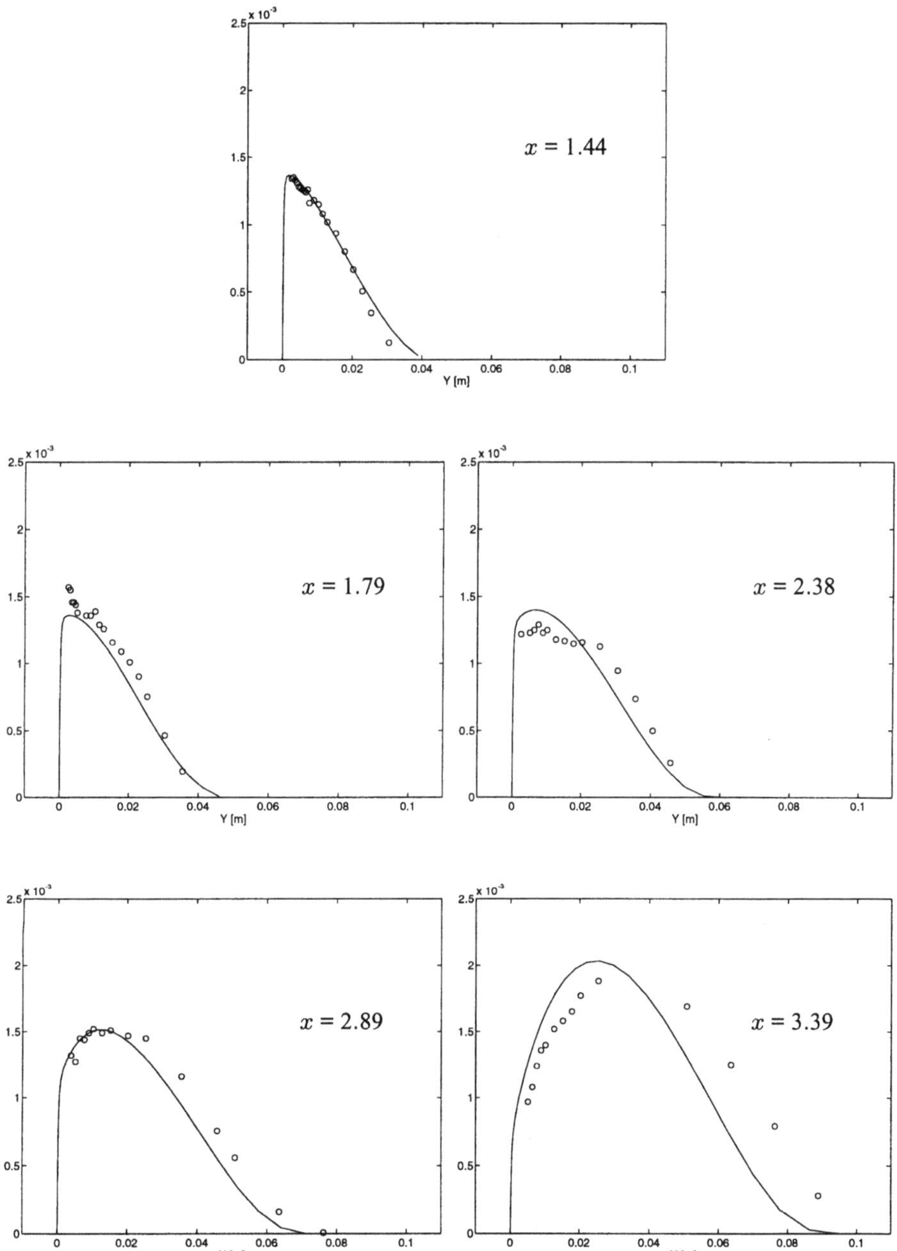

Fig. 23. The Reynolds stress $(\overline{u'v'}/U_e^2)$ profiles in various x locations along the wall. Solid line is present computations, open circles are experimental data[19].

References

1. Patel, V. C., Rodi, W. and Scheuerer, G., "Turbulence Models for Near Wall and Low Reynolds Number Flow: A Review", AIAA J., Vol. 33, № 9, 1985.
2. Mansour, M. N., Kim, J. and Moin, P., "Reynolds-Stress and Dissipation-Rate Budgets in a Turbulent Channel Flow", Journal of Fluid Mechanics, Vol. 194, pp. 15-44, 1988.
3. Speziale, C. G., Abid, R. and Anderson, E. C., "Critical Evaluation of Two-Equation Models for Near-Wall Turbulence", AIAA J., Vol. 30, № 2, Feb. 1992.
4. Kline, S. J., Cantwell, B.J. and Lilley, G.M. Ed., "The 1980-81 AFOSR-HTTM Stanford Conference on Complex Turbulent Flows", Stanford Univ., 1981.
5. Arad, E. and Wolfshtein, M., "A Comparative Study of Length Scale Equations in Turbulent Boundary Layers", Proc. TSF 7, 1989, pp. 5.4.1-5.4.6.
6. Zeierman, S. and Wolfshtein, M., "Turbulent Time Scale for Turbulent Flow Calculation", AIAA J., Vol. 24, № 10, 1986.
7. Wolfshtein, M., "The Velocity and Temperature Distribution of One Dimensional Flow with Turbulence Augmentation and Pressure Gradient", Int. J. of Heat and Mass Trans., Vol. 12, 1969.
8. Arad, E. and Wolfshtein, M., "Analysis of Fourth Order Accurate Solution for Wall Bounded Turbulent Flow ", to be published.
9. Hanjalić, K., Launder, B. E. and Schiestel, R., "Multiple Time Scale Concept in Turbulent Transport Modelling", TSF 2 Symp., London 1979, pp. 10.31-10.36.
10. Rodi, W. and Scheuerer, G., "Scrutinizing the $k - \varepsilon$ turbulence model under adverse pressure gradient conditions", ASME Trans. J. Fluid Eng., Vol. 108, pp. 174-179, June 1986.
11. Chen, H. C. and Patel, V. C., "Near Wall Turbulence Models for Complex Flows Including Separation", AIAA J. Vol. 26, № 6, June 1988.
12. Rodi, W.. "Experience With Two-Layer Models Combining The $k - \varepsilon$ Model With a One Equation Model Near The Wall", AIAA-91-0216, Reno 1991.
13. Rodi, W. and Mansour, N. N., "One Equation Near Wall Turbulence Modeling With The Aid of Direct Simulation Data", CTR Summer prog., Stanford 1990.
14. Bradshaw, P., private communication, 1992.
15. Jones, W. P. and Launder B. E., "The Calculation of Low Reynolds Number Phenomena With a Two-Equation Model of Turbulence", Int. J. Heat Mass trans., Vol. 16, 1973, pp. 1119-1130
16. Launder, B. E. and Sharma, B. I., "Application of The Energy Dissipation Model of Turbulence to Calculation of Flow near Spinning Disk", Letters in Heat and Mass Trans., Vol. 1, 1974, pp. 131-138.
17. Spalart, P., "Direct simulation of turbulent boundary layer up to $Re_\vartheta = 1410$", Journal of Fluid Mechanics, Vol. 187, 1988, pp. 61-98.
18. Nagano, Y., Tagawa, M. and Tsuji, T., "Effects of adverse pressure gradient on mean flows and turbulence statistics in a boundary layer", TSF 8, Springer Verlag 1991.
19. Samuel, A.G., and Joubert, P.N., "A boundary layer developing in an increasingly adverse pressure gradient", Journal of Fluid Mechanics, Vol. 66, Pt 3, 1974, pp. 481-505.

Modelling of Intercomponent Transfer in Reynolds Stress Closures of Homogeneous Turbulence

Magnus Hallbäck, Torbjörn Sjögren, and Arne V. Johansson

Department of Mechanics, KTH, 100 44 Stockholm, Sweden

Abstract

This work deals with the problem of modeling the *slow pressure-strain rate correlation* and the *dissipation rate anisotropy tensor* in the context of *classical* one-point turbulence closures based on the modeling of the Reynolds stress transport equations. Both quantities are significant in the intercomponent energy transfer in anisotropic turbulence. A model for the Reynolds number dependence of the slow pressure-strain rate correlation is presented together with comparisons with direct numerical simulation data. Experimental results from direct measurements of the dissipation rate tensor in axisymmetric turbulence are presented which show that the dissipation rate is strongly anisotropic even at a turbulence Reynolds number of nearly 5000.

Nomenclature

a_{ij}	$\equiv \frac{\overline{u_i u_j}}{K} - \frac{2}{3}\delta_{ij}$	stress anisotropy
e_{ij}	$\equiv \frac{\varepsilon_{ij}}{\varepsilon} - \frac{2}{3}\delta_{ij}$	dissipation anisotropy
II_a	$\equiv a_{ik}a_{ki}$	anisotropy invariants
III_a	$\equiv a_{ik}a_{kj}a_{ji}$	
K	$\equiv \frac{1}{2}\overline{u_i u_i}$	turbulent kinetic energy per unit mass
L_f	$\equiv \frac{1}{2K}\int \Phi_{\alpha\alpha}(re_\alpha)\mathrm{d}r$	longitudinal macro length scale
p		fluctuating pressure
Re_T	$\equiv \frac{4K^2}{\nu\varepsilon}$	turbulent Reynolds number
S_{ij}	$\equiv \frac{1}{2}(U_{i,j} + U_{j,i})$	mean flow strain rate tensor
$S^\star$	$\equiv \sqrt{\frac{1}{2}S_{ij}S_{ji}}\,\frac{2K}{\varepsilon}$	strain rate parameter
$S_0^\star$	$\equiv S^\star(t=0)$	strain rate at beginning of the strain
U_i		mean flow velocity
u_i		fluctuating velocity
ε_{ij}	$\equiv 2\nu\overline{u_{i,k}u_{j,k}}$	dissipation rate tensor
ε	$\equiv \frac{1}{2}\varepsilon_{ii}$	total dissipation rate
η	$\equiv (\nu^3/\varepsilon)^{1/4}$	Kolmogorov length scale

$\lambda^2_{\alpha\beta}$	$\equiv 2\overline{u^2_\alpha} \Big/ \overline{u^2_{\alpha,\beta}}$	Taylor micro length scales
ν		kinematic viscosity
Π_{ij}	$\equiv \overline{\frac{p}{\varrho}(u_{i,j} + u_{j,i})}$	pressure-strain rate
ϱ		fluid density
$\Phi_{ij}(\boldsymbol{r})$	$\equiv \overline{u_i(\boldsymbol{x}) u_j(\boldsymbol{x}+\boldsymbol{r})}$	two-point velocity correlation function

no summation over Greek indices

1. Introduction

A basic aim and general trend in the development of turbulence models is to improve the range of validity of the models. The price to be paid in order to achieve the increased generality is a greater complexity of the models. In complex flows we typically encounter effects of strong streamline curvature or body forces caused by, *e.g.*, system rotation. Since the kinetic energy equation is unaffected by system rotation, the lowest level of single-point closure at which such effects enter explicitly is that in which transport equations are formulated for the individual Reynolds stress components. In these, the Coriolis force gives rise to terms that directly will influence the intercomponent transfer. Other phenomena such as secondary flows in ducts with non-circular cross section clearly need modeling at, at least, this level, since the flow in the cross stream planes is driven by the anisotropic distribution of the Reynolds stresses.

In the present contribution Reynolds stress closures based on the modeling of the transport equations for $\overline{u_i u_j}$ and ε are considered. In closures of this type, which will be referred to as *classical*, the modeling of the terms involved follows a route where constitutive relations of continuum mechanics type are assumed. The present study is restricted to homogeneous turbulence. The investigated models have been tested in highly anisotropic turbulence generated by means of strong mean strains imposed on the turbulence, both in direct numerical simulations (DNS) and in laboratory experiments. The main effort has been directed to the study of the return-to-isotropy process in which the slow pressure-strain rate and the dissipation rate tensors are active. The tensorial evolution of these tensors become trivial in the case of statistically axisymmetric turbulence which simplifies the quantification of these tensors, and the reduced number of degrees of freedom also allows direct experimental determination of the velocity derivative moments constituting ε_{ij}.

2. Axisymmetric Turbulence

In axisymmetric turbulence it is feasible to measure the dissipation rate tensor ε_{ij} by nearly standard hot-wire techniques. The two-point correlation function $\Phi_{ij}(\boldsymbol{r})$ in axisymmetric turbulence with reflectional symmetries can be expressed in terms of four scalar invariant functions (see, *e.g.*, Batchelor[1]),

$$\begin{aligned}\Phi_{ij}(\boldsymbol{r}) &= A(\sigma,\mu) r_i r_j + B(\sigma,\mu)\delta_{ij} \\ &+ C(\sigma,\mu) n_i n_j + D(\sigma,\mu)(n_i r_j + n_j r_i)\end{aligned}$$

where $\boldsymbol{n}$ is the direction of symmetry and the scalar arguments are defined as $\sigma \equiv r^2$ and $\mu \equiv \boldsymbol{r} \cdot \boldsymbol{n}$. The scalar functions A, B, C and D are coupled via the continuity condition,

$$2\frac{\partial A}{\partial \sigma}\sigma + \frac{\partial A}{\partial \mu}\mu + 4A + 2\frac{\partial D}{\partial \sigma}\mu + \frac{\partial D}{\partial \mu} + 2\frac{\partial B}{\partial \sigma} = 0$$

$$2\frac{\partial C}{\partial \sigma}\mu + C\mu + 4D + 2\frac{\partial D}{\partial \sigma}\sigma + \frac{\partial D}{\partial \mu}\mu + \frac{\partial B}{\partial \mu} = 0$$

Using these relations one may show that all elements of ε_{ij} can be determined from the measured data of four independent instantaneous velocity derivative moments. These may be chosen as, with the x-axis parallel to the axis of symmetry,

$$\overline{\left(\frac{\partial u}{\partial x}\right)^2}, \quad \overline{\left(\frac{\partial u}{\partial y}\right)^2}, \quad \overline{\left(\frac{\partial v}{\partial x}\right)^2}, \quad \overline{\left(\frac{\partial v}{\partial y}\right)^2}$$

From these we may determine the other velocity derivative moments contributing to the elements of ε_{ij}.

$$\begin{aligned}
\overline{\left(\frac{\partial v}{\partial z}\right)^2} &= 3\overline{\left(\frac{\partial v}{\partial y}\right)^2} - \overline{\left(\frac{\partial u}{\partial x}\right)^2}, \\
\overline{\left(\frac{\partial w}{\partial y}\right)^2} &= \overline{\left(\frac{\partial v}{\partial z}\right)^2}, \quad \overline{\left(\frac{\partial w}{\partial z}\right)^2} = \overline{\left(\frac{\partial v}{\partial y}\right)^2}, \\
\overline{\left(\frac{\partial w}{\partial x}\right)^2} &= \overline{\left(\frac{\partial v}{\partial x}\right)^2}, \quad \overline{\left(\frac{\partial u}{\partial z}\right)^2} = \overline{\left(\frac{\partial u}{\partial y}\right)^2}
\end{aligned}$$

Furthermore all derivative correlations of the type $\overline{u_{\alpha,\gamma}u_{\beta,\gamma}}$, $\alpha \neq \beta$ are zero which implies that all the off-diagonal elements of ε_{ij} are zero. Thus the dissipation rate tensor may be quantified in terms of only two independent elements,

$$\begin{aligned}
\varepsilon_{11} &= 2\nu\left(\overline{\left(\frac{\partial u}{\partial x}\right)^2} + 2\overline{\left(\frac{\partial u}{\partial y}\right)^2}\right) \\
\varepsilon_{22} &= \varepsilon_{33} = 2\nu\left(\overline{\left(\frac{\partial v}{\partial x}\right)^2} - \overline{\left(\frac{\partial u}{\partial x}\right)^2} + 4\overline{\left(\frac{\partial v}{\partial y}\right)^2}\right)
\end{aligned}$$

These relations where derived in a slightly different manner by George & Hussein[2] who also used them for direct measurements of ε_{ij} in an axisymmetric turbulent jet. Since the anisotropy tensors have zero trace they contain only one independent element, for instance for the dissipation rate anisotropy

$$e_{11} = 2\left(\frac{1}{1 + 2\varepsilon_{22}/\varepsilon_{11}} - \frac{1}{3}\right) \tag{1}$$

and similarly for the stress anisotropy tensor. With a hot-wire probe configuration with two X-wires separated a small distance in the cross-stream direction and the frozen turbulence hypothesis of Taylor the necessary cross-stream and streamwise derivatives can be obtained. Of course the requirements of sufficient spatial resolution and low noise must be fulfilled.

3. Experimental Setup

The experiments have been carried out in a low turbulence, low speed wind tunnel. Turbulence was generated by a monoplane square rod grid with 10 cm mesh width positioned in the stagnation chamber about 20 mesh widths upstream of the contraction which has an area ratio of nine. The measurements presented in the present paper were all conducted at a free stream velocity in the test section of 21 m/s. The test section is 7 m long and has a rectangular cross-section of 0.8m $\times$ 1.2m. The variation of the mean velocity along the centreline of the test section is less than 1%. This means that the production term of the turbulence kinetic energy equation is less than 2% of the total dissipation rate term and effects due to mean distortion of the turbulence field in the test section are negligible.

In the contraction the turbulence is strongly distorted by the accelerating mean flow producing a turbulence state at the beginning of the test section that is highly anisotropic with a relative turbulence level just below 1%. The turbulence level in the streamwise direction is about 0.2%. It should here be mentioned that the free-stream streamwise turbulence level without the grid is as low as 0.02% (see Johansson[6]).

In the test section the turbulence is allowed to relax towards isotropy in the absence of any mean strain. The length of the test section corresponds to about 0.5 eddy-turnover timescales and the turbulence Reynolds number at 21 m/s was $Re_T \approx 4500$. A four channel constant temperature hot-wire anemometer system together with a double X-probe configuration was used for all the velocity and velocity derivative measurements. The hot-wire X-probes, which where homemade, had 2.5μm platinum wire sensors and the measurement volume of each X-probe was $(0.75\text{mm})^3$. The anemometer system was connected to a Macintosh computer with a four channel simultaneous sampling and AD-converter board. The Macintosh computer was also used to control the traversing mechanism and the speed of the wind tunnel. The system allowed automatization of the calibration and measurement procedures. An angular calibration procedure was used where third order polynomials were fitted to the calibration data over a mean velocity range of 20 to 22 m/s and $\pm 4.5°$. The sampling was carried out at a rate corresponding to about one 'sample' per turbulence macro time scale. Each 'sample' consisted of three consecutive data points per wire taken at a rate suitable for central difference approximations of the time derivatives, see below. At each downstream position 40 000 thousand 'samples' were taken for each lateral probe-separation. A typical experiment for the determination of ε_{ij} and related quantities along the centreline of the test section could typically take about 10 hours.

4. Direct Numerical Simulations

A pseudo-spectral direct simulation code (see *e.g.* Rogallo[7]) was used for detailed investigation of the above described flow case. Statistically isotropic fields, in conformity with a prescribed energy spectrum, consisting of random Fourier modes were allowed to relax into developed states and were then used as initial fields for imposed axisymmetric mean strain rate simulations producing anisotropic fields that were subsequently allowed to relax. The imposed axisymmetric strains corresponded to the distortion caused by a passage through a wind-tunnel contraction with an area ratio of three and the value of the strain rate parameter, $S_0^\star = 3$, is typical for a laboratory experiment.

Specific attention was paid to the problem of the Reynolds number dependence of the slow pressure-strain rate term which was studied in a set of runs where the Reynolds number was systematically varied over a decade and a half. These simulation were carried out using also a varying number of node points to obtain equal resolution of the small scales in all runs. The simulations where stopped when the half box two-point correlations started to depart from zero due to growing turbulence length scales. The simulations for the highest Reynolds number (≈ 400) involved 172^3 (dealiased) modes and was carried out on a 32k processor CM200. The simulations have been reported in greater detail in Hallbäck[4].

5. Error Estimate

Two-point central differences have been used to approximate the velocity derivatives. Since for practical reasons, such as finite probe-size and noise contamination of the signals, it is impossible to use infinitely small separations between the two X-probes it is important to study the effect on the measured value of the derivative from finite probe separation. Here, since we are also interested in determining the Taylor micro-scales, we have chosen to study how the measured value of $\lambda_{\alpha\beta}$

$$\lambda^2_{\alpha\beta_m} = 2\overline{u_\alpha u_\alpha} \bigg/ \overline{\left(\frac{u_\alpha(x + \frac{r}{2}e_\beta) - u_\alpha(x - \frac{r}{2}e_\beta)}{r} \right)^2}$$

depends on the probe separation re_β. By expanding the velocity field around x one may show that the measured value of $\lambda_{\alpha\beta}$ relates to the exact value through

$$\left(\frac{\lambda_{\alpha\beta_m}}{\lambda_{\alpha\beta}} \right)^2 = 1 + \mathrm{B}_{\alpha\beta} \left(\frac{r}{\lambda_{\alpha\beta}} \right)^2 + \dots$$

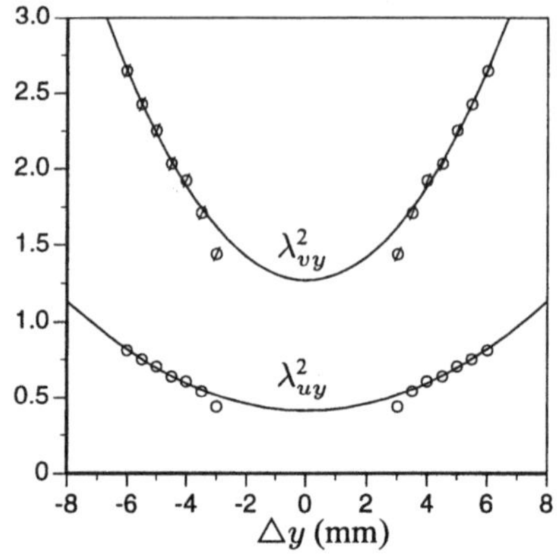

Fig. 1. The measure values of the micro-scales $\lambda^2_{\alpha y}$ (cm^2) versus probe separation Δy at x = 0.6m. Parabolas have been fitted to obtain the extrapolated value for zero separation.

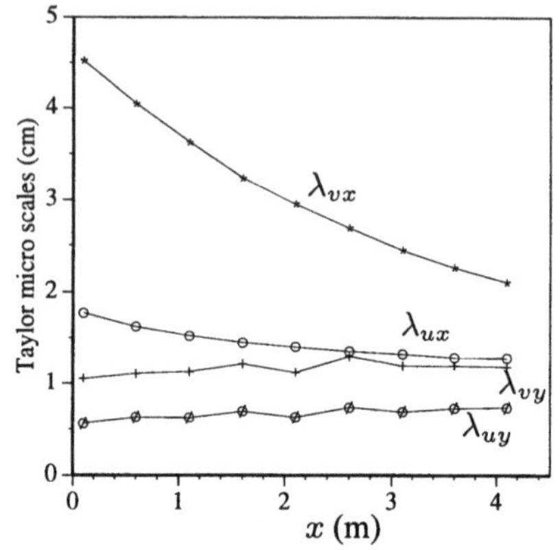

Fig. 2. The micro length scales $\lambda_{\alpha\beta}$ versus downstream distance x. Note that the length scales in the streamwise direction, *i.e.* $\lambda_{\alpha x}$, decrease during the relaxation.

In figure 1 the measured values of λ^2_{uy} and λ^2_{vy} are shown as a function of different separations Δy in the cross-stream direction. The overestimation of $\lambda_{\alpha y}$ with Δy is caused by an underestimation of the velocity derivative moments due to the fact that the turbulence scales in the range between the probe separation and the Kolmogorov micro-scale (η) are not resolved. In the present experiment $\eta \approx 0.6$mm. For the lowest values of the separation the contamination from noise becomes a problem and is reflected in figure (1) by a departure of the measured value of $\lambda_{\alpha y}$ from the expected parabola. The same type of plot can be established for time- (or stream-) wise separations. The time-wise derivations do not constitute any problem. These derivatives were determined using only two different separations. The $\lambda_{\alpha y}$ micro-scales were determined for seven different Δy separations. In figure 2 the downstream evolution of the four independent Taylor micro-scales are shown. Note that the streamwise micro-scales decrease during the relaxation in contrast to the situation in isotropic turbulence. This is due to the initial streamwise elongation of the turbulence scales.

6. The ε_{ij} Measurements

The above considerations and results imply that for moderate separations the measured velocity derivative relates to the exact as

$$\overline{\left(\frac{\partial u_\alpha}{\partial x_\beta}\right)^2_m}(re_\beta) = \frac{\overline{u_{\alpha,\beta}u_{\alpha,\beta}}}{(1+\mathrm{B}_{\alpha\beta}(r/\lambda_{\alpha\beta})^2+\cdots)}$$

If this expression is inserted into the expressions for ε_{ij} and subsequently in the expression (1) for the anisotropy measure e_{11} we see that

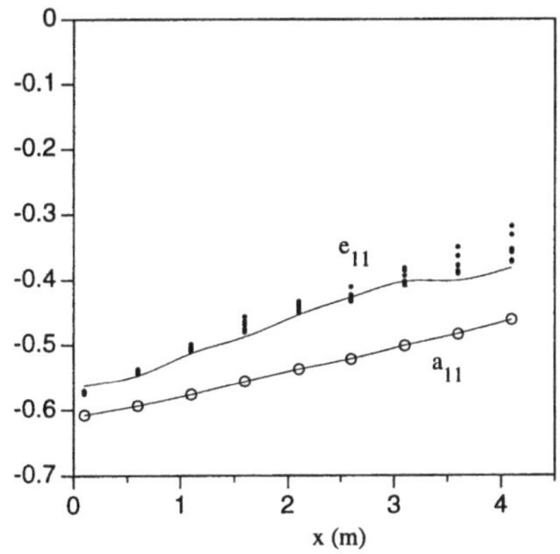

Fig. 3. e_{11} and a_{11} versus downstream distance x. Each point represents the measured value of e_{11} for a certain separation $\triangle y$. The line through the points represents the value of e_{11} obtained through extrapolation to zero separation.

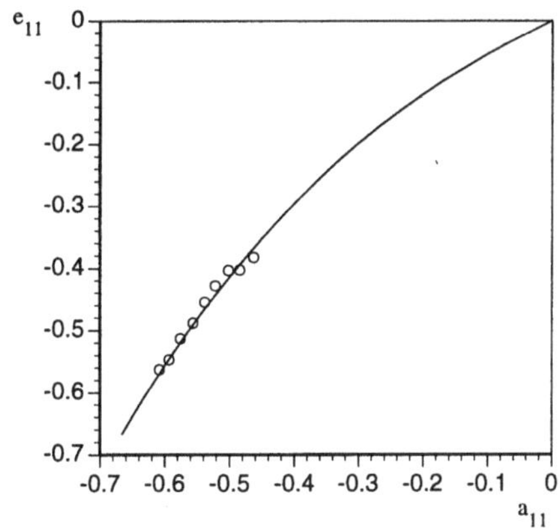

Fig. 4. e_{11} vs a_{11} from measurements (based on extrapolated data) at $Re_T \approx 4500$ and according to the Hallbäck *et al.* model.

$$e_{11,\text{meas}} = e_{11}\left(1 + k_y \triangle y^2 + O(\triangle y^4)\right)$$

where the error coefficient k_y is a function of quantities that are all determined in the measurements

$$k_y = k_y\left(\overline{\left(\frac{\partial u}{\partial x}\right)^2}, \overline{\left(\frac{\partial u}{\partial y}\right)^2}, \overline{\left(\frac{\partial v}{\partial x}\right)^2}, \overline{\left(\frac{\partial v}{\partial y}\right)^2}, \mathrm{B}_{uy}, \mathrm{B}_{vy}, \overline{u^2}, \overline{v^2}\right).$$

In the present measurements $k_y \in [-0.026; 0.015]\text{mm}^{-2}$ depending on downstream position. The corresponding error coefficient for x has a range of $k_x \in [0.000\,4; 0.006]\text{mm}^{-2}$. The micro-scales were determined for seven different $\triangle y$-separations and these data were then extrapolated down to zero separation. In figure 3 the Reynolds stress anisotropy a_{11} and the dissipation rate anisotropy e_{11} are shown. Each individual e_{11}-value corresponds to a certain value of $\triangle y$. We may note that the measured value of e_{11} is essentially insensitive to the separation used in the derivative approximation. Hence, the double X-probe configuration gives a very well-conditioned problem for determination of e_{11}, which otherwise is a quantity quite difficult to determine accurately.

The measured total dissipation rate, based on extrapolated data, was compared to the dissipation rate obtained through balancing of the terms of the kinetic energy equation. The directly measured dissipation was in fact 50% larger than that obtained from the kinetic energy equation. This indicates that the contamination from system noise may be a severe problem.

The measured data have been compared with an algebraic model for the relation between e_{ij} and a_{ij} suggested by Hallbäck *et al.*[3].

$$e_{ij} = \left[1 + \alpha\left(\frac{1}{2}II_a - \frac{2}{3}\right)\right] a_{ij} - \alpha\left(a_{ik}a_{kj} - \frac{1}{3}II_a\delta_{ij}\right) \tag{2}$$

with $\alpha = 3/4$. In figure 4 the experimental data, where the velocity derivative terms were determined from measured Taylor micro-scales that were extrapolated to zero separation (figure 1), shown together with the model (2).

7. The Slow Pressure-Strain Rate

Although not rigorously shown all experiences indicate that the slow pressure-strain rate term plays the role of redistributing energy among velocity components towards isotropy. The simplest possible way to model this tendency of isotropization mathematically in terms of one-point measures is to apply dimensional analysis and assume that the pressure-strain rate is proportional to the deviation from isotropy as was suggested by Rotta[8];

$$\overline{\frac{p^{(s)}}{\varrho}\left(\frac{\partial u_i}{\partial x_j} + \frac{\partial u_j}{\partial x_i}\right)} = -C_R\frac{K^{3/2}}{L_f}\left(\frac{\overline{u_iu_j}}{K} - \frac{2}{3}\delta_{ij}\right) \tag{3}$$

where L_f is the longitudinal turbulence macro length scale and C_R is a model coefficient. In isotropic turbulence the pressure has, indeed, been shown to scale on the kinetic energy $\sqrt{\langle(p/\varrho)^2\rangle} \sim K$ (Batchelor[1] §8.3). In homogeneous turbulence L_f is normally taken as $L_f \sim K^{3/2}/\varepsilon$, which gives

$$\Pi_{ij}^{(s)} = -C_1\varepsilon a_{ij} \tag{4}$$

where C_1 is known as the "Rotta constant", often assigned a value of about 1.5 to 1.8. This type of modeling has been adopted by most modellers although C_1 is in general not a constant but rather a function of the turbulence state. A strong dependence on the Reynolds number was inferred by Weinstock[11,12] with $C_1 \rightarrow 3.6$ as $Re_T \rightarrow \infty$. This type of behaviour was also found by Hallbäck[5] in DNS with Re_T ranging over one and a half decade. In these simulations developed, isotropic turbulence was subjected to an axisymmetric mean distortion from which it was allowed to relax towards isotropy. The relaxation phases of runs with different turbulent Reynolds number (Re_T: 10 → 400) were compared and the Rotta constant C_1 was determined as a function of Re_T (figure 5).

Actually, the Re_T trend observed in the DNS data can be captured if the finite-Reynolds-number corrected relation (5) between ε and L_f suggested by Rotta[8]

$$\varepsilon = \nu B_1\frac{K}{L_f^2} + B_2\frac{K^{3/2}}{L_f} \tag{5}$$

which may be rewritten on the form

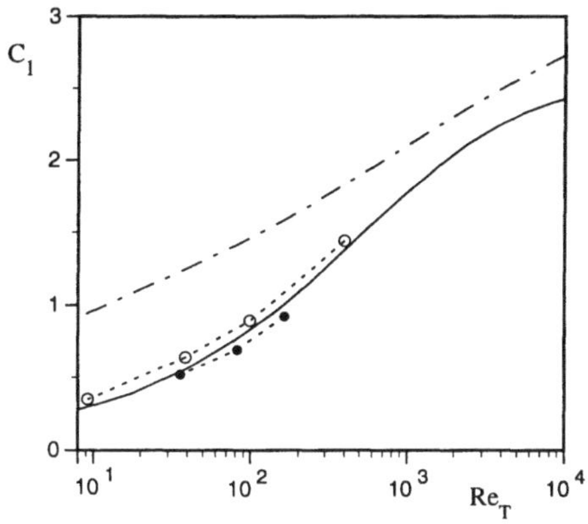

Fig. 5. The Rotta parameter C_1 from DNS of axisymmetric relaxation ({•} and {∘}), according to Weinstock {chain-dashed line} and according to the present proposal {solid line}. The anisotropies in the DNS were $a_{11} = -0.38$ and $a_{11} = -0.30$ respectively.

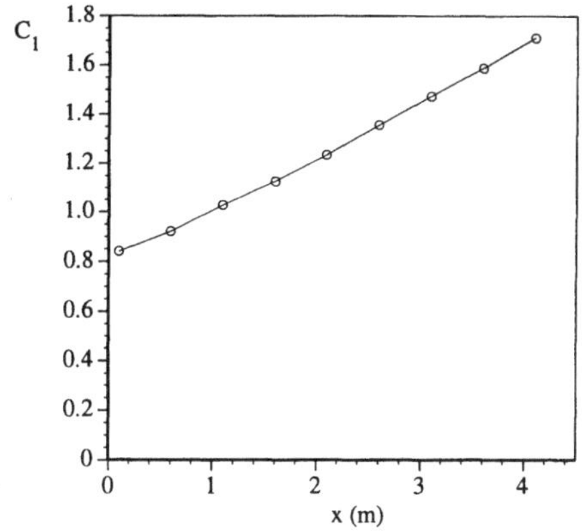

Fig. 6. Experimentally determined values of C_1 from relaxation of anisotropic turbulence, based on direct measurements of the dissipation rate anisotropy.

$$\varepsilon = \frac{1}{f(Re_T)} \frac{K^{3/2}}{L_f} \tag{6}$$

is used to express the original Rotta proposal (3) on the form (4). If C_R is assumed constant then the "Rotta constant" C_1 varies with Re_T as

$$C_1 = C_R f(Re_T) \tag{7}$$

where $f(Re_T)$ is given by

$$f(Re_T) = \sqrt{\frac{Re_T}{4B_1} + \left(\frac{B_2}{8B_1} Re_T\right)^2} - \frac{B_2}{8B_1} Re_T \tag{8}$$

We may obtain an estimate of B_1 by approximating the correlation function as $\Phi_{\alpha\alpha}(re_\alpha) = \overline{u_\alpha u_\alpha} \exp\left(-r^2/2\lambda_{\alpha\alpha}^2\right)$ (cf. Schumann & Patterson[9]) in the low Re_T limit which yields $B_1 = 5\pi$. Assuming a $-5/3$-power law energy spectrum representing turbulence at high Reynolds numbers one gets that $B_2 \to 0.31$ as $Re_T \to \infty$. Good agreement with DNS is obtained for $C_R = 0.8$, figure 5. For this choice of C_R the "Rotta constant" C_1 approaches $C_R/B_2 = 2.58$ as $Re_T \to \infty$. The DNS data also showed that, indeed, C_R exhibits a significantly weaker variation with the Reynolds number than C_1. This is in agreement with the results by Schumann & Patterson[10] who find that C_R varies less than C_1 with both Re_T and different initial conditions.

From the experimental data of e_{11} and a_{11} one may determine also the $\Pi_{ij}^{(s)}$-term by use of the transport equation for a_{ij}, which in the absence of a mean strain field reads

$$\frac{da_{ij}}{dt} = \frac{1}{K}\Pi_{ij}^{(s)} - \frac{\varepsilon}{K}(e_{ij} - a_{ij})$$

In figure 6 the 'Rotta constant' is shown during the relaxation. Direct comparison with the simulation data is not possible due to the stronger degree of anisotropy and the fact that the value of C_1 is continuously developing during the relatively short period of time that is covered by the experiment (about 0.5 eddy turnover times).

8. Conclusions

The possibilities of experimental determination of velocity derivative moments contributing to the elements of the dissipation rate tensor have been explored. It has been found that the procedure using two X-probes for determining normalized quantities, such as e_{11}, is well conditioned.

The anisotropy of the dissipation rate tensor was found to be of comparable magnitude to that of the Reynolds stress tensor in the present flow case of relaxation of axisymmetrically strained turbulence. This result is in good agreement with the algebraic one-point model by Hallbäck *et al.*.

From DNS of homogeneous turbulence the numerical value of the so called Rotta constant was found to exhibit a strong Reynolds number dependence. A simple model, based on Rotta's original proposal for the slow pressure-strain rate model, was suggested in order to account for this Reynolds number dependence. The model was shown to agree well with the DNS data.

The direct measurements of the dissipation rate anisotropy were also used to determine a numerical value of the Rotta constant from the experimental data of the relaxation process. In the experiment, however, the experimental conditions were such that this value did not reach an equilibrium within the test section. Further work is planned to improve this situation to enable direct comparisons between experiments and DNS.

Acknowledgement

Financial support from NUTEK is gratefully acknowledged. CRAY-time has been granted by the Swedish National Supercomputer Center. CM200 computer time was granted through a contract between the US Army Research Office and the University of Minnesota for the Army High Performance Computing Research Center.

References

[1] Batchelor, G.K. 1953 *The theory of homogeneous turbulence*, Cambridge.

[2] George, W.K. & Hussein, H.J. 1991 "Locally axisymmetric turbulence". *J. Fluid Mech.* **233**, 1-23.

[3] Hallbäck, M. Groth, J. & Johansson, A.V. 1990 "An algebraic model for nonisotropic turbulent dissipation rate in Reynolds stress closures". *Phys. Fluids A* **2**, 1859-1866.

[4] Hallbäck, M. 1993 "Development and critical evaluation of turbulence models through direct numerical simulations of homogeneous turbulence". Trita-Mek-93-2, Dept. of Mech., Royal Institute of Technology, Stockholm, Sweden.

[5] Hallbäck, M. 1993 "Development of Reynolds Stress Closures of Homogeneous Turbulence". PhD Thesis, Dept. of Mech., Royal Institute of Technology, Stockholm, Sweden.

[6] Johansson, A.V. 1992 "A low-speed wind-tunnel with extreme flow quality – design and tests". In *Proc. of the 18th ICAS congress*, Paper ICAS-92-3.8.1.

[7] Rogallo, R.S. 1981 "Numerical Experiments in Homogeneous Turbulence". NASA-TM81315, Ames Research Center, USA.

[8] Rotta, J. 1951 "Statistische Theorie nichthomogener Turbulenz I". *Zeitschrift für Physik* **129**, 547-572.

[9] Schumann, U. & Patterson, G.S. 1978 "Numerical study of pressure and velocity fluctuations in nearly isotropic turbulence". *J. Fluid Mech.* **88**, 685-709.

[10] Schumann, U. & Patterson, G.S. 1978 "Numerical study of the return of axisymmetric turbulence to isotropy". *J. Fluid Mech.* **88**, 711-735.

[11] Weinstock, J. 1981 "Theory of pressure-strain-rate correlation for Reynolds-stress turbulence closures. Part 1. Off-diagonal elements". *J. Fluid Mech.* **105**, 369-396.

[12] Weinstock, J. 1982 "Theory of the pressure-strain rate. Part 2. Diagonal elements". *J. Fluid Mech.* **116**, 1-29.

Closure of the Dissipation Tensor and the Pressure–Strain Tensor Based on the Two-Point Correlation Equation

Martin Oberlack

Institut für Technische Mechanik, RWTH Aachen, 52056 Aachen, Germany

Abstract

On the basis of the two-point velocity correlation equation a dissipation tensor equation and the pressure–strain correlation are derived by means of asymptotic analysis and frame invariance considerations. The new dissipation tensor equation can account for nonisotropy effects as well as for strong streamline curvature. The pressure–strain correlation is expressed as a functional of the two-point correlation, leading to an exact formulation of the rapid part of the pressure–strain correlation. In this turbulence modelling approach the only term which still needs ad hoc closure assumptions is the triple correlation of the fluctuating velocities. Hence, a consistent formulation of the return term in the pressure–strain correlation and the dissipation tensor equation is achieved.

1. Introduction

Current second moment closure models still suffer from three unresolved problems: the closure of the velocity pressure gradient correlation Ψ_{ij}, the triple correlation ${}^{\mathrm{t}}D_{ij}$ and the dissipation tensor ε_{ij}. The velocity pressure gradient correlation is split into the pressure–strain correlation (redistribution) Φ_{ij} and the pressure diffusion ${}^{\mathrm{p}}D_{ij}$ which is modelled together with the triple correlation as a gradient type diffusion term. In nearly all common models the pressure–strain term is written as a linear (or more recently a nonlinear) algebraic function of local turbulence parameters, in particular the anisotropy tensor and its invariants, the mean flow gradient and a scalar length scale [10], [13], [17].

Beside the model of Tagawa et al. [23], in which he proposed a closure for all the seven unknown terms in the ε-tensor equation in an ad hoc manner, all common dissipation function models are based on a scalar evolution equation. Most formulations are variations of the scalar dissipation equation first proposed by Hanjalić and Launder [7]. The tensorial character of the dissipation function is modelled either as an isotropic tensor $\varepsilon_{ij} = (\delta_{ij}/3)\varepsilon$ or anisotropy of the dissipation tensor is supposed to be a function of the Reynolds stress anisotropy tensor b_{ij}.

The deficiencies of the dissipation models are mainly related to the isotropic ε-equation, the reasons being manifold: Firstly, a closed ε-equation cannot be derived from basic principles and is mainly an empirical equation, because in the exact ε-equation a very large number of complicated terms would have to be modelled. Secondly, recent direct simulation data of turbulent channel flows have revealed the strong non-isotropic nature of the dissipation process [1], [9], [11]. However, the main disadvantage is the fact that the isotropic dissipation equation cannot account for streamline curvature and all proposed ad hoc models for that problem lead to spurious results in flow situations, where the models have not been adjusted to [22].

On the other hand, Rotta [17] and later Donaldson and Sandri [5], Sandri [18] have shown in a series of papers that a fundamental length scale equation can be obtained from the balance equation for the two-point velocity correlation. Unfortunately in all of the papers [5], [17] and [18] a large number of terms have been modelled, even where no modelling is needed because exact equations can be derived.

Here we will follow the route initiated by Rotta and try to extract new information from the two-point correlation tensor equations. By introducing a model for the two-point triple correlation function only, we will derive expressions for the dissipation tensor and the pressure–strain correlation.

2. Governing Equations

Splitting the flow velocity and the pressure into their means $\bar{u}_i, \bar{p}$ and their fluctuations u_i, p and defining the two-point, one-time correlation tensors

$$
\begin{aligned}
R_{ij}(\boldsymbol{x},\boldsymbol{r},t) &= \overline{u_i(\boldsymbol{x},t)\, u_j(\boldsymbol{x}^{(1)},t)} \ , \\
R_{(ik)j}(\boldsymbol{x},\boldsymbol{r},t) &= \overline{u_i(\boldsymbol{x},t)\, u_k(\boldsymbol{x},t)\, u_j(\boldsymbol{x}^{(1)},t)} \ , \\
\overline{pu_j}(\boldsymbol{x},\boldsymbol{r},t) &= \overline{p(\boldsymbol{x},t)\, u_j(\boldsymbol{x}^{(1)},t)} \ ,
\end{aligned}
\tag{1}
$$

which are functions of the physical and the correlation space coordinates $\boldsymbol{x}$ and $\boldsymbol{r} = \boldsymbol{x}^{(1)} - \boldsymbol{x}$ respectively, the two-point correlation equation for incompressible fluids in a rotating frame reads as follows:

$$
\begin{aligned}
\frac{\mathrm{D}R_{ij}}{\mathrm{D}t} = & -R_{kj}\frac{\partial \bar{u}_i(\boldsymbol{x},t)}{\partial x_k} - R_{ik}\frac{\partial \bar{u}_j(\boldsymbol{x}+\boldsymbol{r},t)}{\partial x_k} \\
& - \left[\bar{u}_k(\boldsymbol{x}+\boldsymbol{r},t) - \bar{u}_k(\boldsymbol{x},t)\right]\frac{\partial R_{ij}}{\partial r_k} \\
& - \frac{1}{\varrho}\left[\frac{\partial \overline{pu_j}}{\partial x_i} - \frac{\partial \overline{pu_j}}{\partial r_i} + \frac{\partial \overline{u_i p}}{\partial r_j}\right] \\
& + \nu\left[\frac{\partial^2 R_{ij}}{\partial x_k \partial x_k} - 2\frac{\partial^2 R_{ij}}{\partial x_k \partial r_k} + 2\frac{\partial^2 R_{ij}}{\partial r_k \partial r_k}\right] \\
& - \frac{\partial R_{(ik)j}}{\partial x_k} + \frac{\partial}{\partial r_k}\left[R_{(ik)j} - R_{i(jk)}\right] \\
& - 2\,\Omega_k\left[\epsilon_{kli}R_{lj} + \epsilon_{klj}R_{il}\right] \ .
\end{aligned}
\tag{2}
$$

$\mathsf{D}/\mathsf{D}t$ is the substantial derivative and Ω_k is the angular velocity of the coordinate system relative to an inertial frame. The divergence of (2) leads to a Poisson equation for $\overline{pu}_j$ which reads

$$\begin{aligned}
&\frac{1}{\varrho}\left[\frac{\partial^2\overline{pu_j}}{\partial x_k\partial x_k} - 2\frac{\partial^2\overline{pu_j}}{\partial r_k\partial x_k} + \frac{\partial^2\overline{pu_j}}{\partial r_k\partial r_k}\right] = \\
&- \quad 2\left[\frac{\partial\bar{u}_k}{\partial x_l} + \epsilon_{mlk}\Omega_m\right]\left[\frac{\partial R_{lj}}{\partial x_k} - \frac{\partial R_{lj}}{\partial r_k}\right] \\
&- \quad \left[\frac{\partial^2 R_{(kl)j}}{\partial x_k\partial x_l} - 2\frac{\partial^2 R_{(kl)j}}{\partial x_k\partial r_l} + \frac{\partial^2 R_{(kl)j}}{\partial r_k\partial r_l}\right] .
\end{aligned} \tag{3}$$

All of the dependent variables in (1) have to satisfy the continuity conditions

$$\frac{\partial R_{ij}}{\partial x_i} - \frac{\partial R_{ij}}{\partial r_i} = 0, \qquad \frac{\partial R_{ij}}{\partial r_j} = 0 \ , \tag{4}$$

$$\frac{\partial R_{(ik)j}}{\partial r_j} = 0 \quad \text{and} \quad \frac{\partial\overline{pu_i}}{\partial r_i} = 0 \ . \tag{5}$$

It is worth mentioning that the only remaining term to be modelled in the whole system (1) – (5) is the triple correlation $R_{(ik)j}$, to be discussed in Sec. (3.2).

Clearly, (2) can be reduced to the Reynolds stress equations by introducing the limit $\boldsymbol{r} \to 0$:

$$\frac{\mathsf{D}\overline{u_iu_j}}{\mathsf{D}t} = P_{ij} + \Phi_{ij} - 2\,\varepsilon_{ij} + {}^{\mathrm{t}}D_{ij} + {}^{\mathrm{p}}D_{ij} + {}^{\nu}D_{ij} + C_{ij} \ , \tag{6}$$

where

$$\begin{aligned}
P_{ij} &= -\left[\overline{u_iu_k}\frac{\partial\bar{u}_j}{\partial x_k} + \overline{u_ku_j}\frac{\partial\bar{u}_i}{\partial x_k}\right] , \\
\Phi_{ij} &= \overline{\frac{p}{\varrho}\left[\frac{\partial u_j}{\partial x_i} + \frac{\partial u_i}{\partial x_j}\right]} , \\
\varepsilon_{ij} &= \nu\overline{\frac{\partial u_i}{\partial x_k}\frac{\partial u_j}{\partial x_k}} , \\
{}^{\mathrm{t}}D_{ij} &= -\frac{\partial\overline{u_iu_ju_k}}{\partial x_k} , \\
{}^{\mathrm{p}}D_{ij} &= -\frac{\partial}{\partial x_k}\left[\overline{p/\varrho(u_i\delta_{jk} + u_j\delta_{ik})}\right] , \\
{}^{\nu}D_{ij} &= \nu\frac{\partial^2\overline{u_iu_j}}{\partial x_k\partial x_k} , \\
C_{ij} &= -2\,\Omega_k\left[\epsilon_{kli}\overline{u_ju_l} + \epsilon_{klj}\overline{u_iu_l}\right] .
\end{aligned} \tag{7}$$

The Reynolds stress equations obviously contain more functions that need to be modelled than the two-point correlation equation. The information that is needed for the closure is contained in the two-point correlation equations.

3. Derivation of the Tensor–Length Scale Equation

In order to compute the major unknown terms in the Reynolds stress equation, ε_{ij} and Φ_{ij}, without solving the system (1)–(5) in its total complexity of six dimensions, we will consider an integral form of these equations.

The continuity equations (4) will be satisfied by introducing a tensor potential V_{mn} which is related to the correlation function R_{ij} by

$$R_{ij} = \epsilon_{ikm}\,\epsilon_{jln}\left[\frac{\partial}{\partial x_k} - \frac{\partial}{\partial r_k}\right]\frac{\partial V_{mn}}{\partial r_l}\,, \tag{8}$$

where ϵ_{ijk} is the alternating tensor. For example, if isotropic turbulence is considered, the ansatz

$$V_{mn} = -\frac{1}{2}\,\bar{u}^2\,\delta_{mn}\int_0^r f(r)\,r\,\mathrm{d}r \quad \text{with} \quad r = |\boldsymbol{r}| \tag{9}$$

leads with (8) to the known result of the isotropic correlation function

$$R_{ij} = \bar{u}^2\left[\delta_{ij}\left(f + \frac{r}{2}f'\right) - \frac{r_i\,r_j}{2r}f'\right]\,, \tag{10}$$

first developed by Kármán and Howarth [8].

To preserve the tensor character of the correlation function and to account for nonisotropic effects in R_{ij}, invariant theory [20] may be used to derive a generalization of (9). To avoid unclosed terms in the length scale tensor equation as well as in the pressure–strain correlation, the ansatz for V_{mn} should obey the following restrictions

- V_{mn} should be only a function of the correlation vector r_k and a symmetric tensor $\int_0^r F_{mn}\,r\,\mathrm{d}r$,
- V_{mn} should be linear in the tensor $\int_0^r F_{mn}\,r\,\mathrm{d}r$,
- V_{mn} should have the dimension $\left[r^2 F_{mn}\right]$, where F_{mn} has the dimension of the Reynolds stress tensor.

The most general formulation contains eight independent tensor functions where three of them become singular or drop out completely in the subsequent calculations. The remaining five are

$$\begin{aligned} V_{mn} \;=\; & \alpha_1 \int_0^r F_{mn}\,r\,\mathrm{d}r + \alpha_2\,\delta_{mn}\int_0^r F_{kk}\,r\,\mathrm{d}r \\ & +\alpha_3\,\delta_{mn}\,\frac{r_k\,r_l}{r^2}\int_0^r F_{kl}\,r\,\mathrm{d}r + \alpha_4\,\frac{r_m\,r_n}{r^2}\int_0^r F_{kk}\,r\,\mathrm{d}r \\ & +\alpha_5\left[\frac{r_m\,r_k}{r^2}\int_0^r F_{nk}\,r\,\mathrm{d}r + \frac{r_n\,r_k}{r^2}\int_0^r F_{mk}\,r\,\mathrm{d}r\right]. \end{aligned} \tag{11}$$

The five coefficients α_1–α_5 are not all independent of each other, because the tensor F_{ij} should obey the one-point limit

$$\overline{u_i u_j} = R_{ij}(\boldsymbol{x}, \boldsymbol{r} = 0, t) = F_{ij}(\boldsymbol{x}, \boldsymbol{r} = 0, t) \ , \tag{12}$$

which yields

$$\alpha_1 = 1 - \alpha_3 + \alpha_5 \quad \text{and} \quad \alpha_2 = \frac{1}{2}\left[-1 + \alpha_4\right] \ . \tag{13}$$

In anticipation of the result in Sect. 5, the system (11)–(13) covers the result of Naot [14] for the pressure–strain correlation but contains three additional free parameters. They do not contribute to the pressure–strain correlation but have particular relevance for the tensor length scale equation.

3.1 Asymptotic Expansion of the Two-Point Correlation Equation

At this stage we will introduce the first of two basic asymptotic limits which are essential for the following analysis. We postulate each element of F_{ij} to be, to the leading order, solely a function of the absolute value of the correlation distance $\boldsymbol{r}$, but still of the spatial coordinate vector $\boldsymbol{x}$ and the time t

$$F_{ij} = F_{ij}(\boldsymbol{x}, |\boldsymbol{r}|, t) \ . \tag{14}$$

Introducing the generalized ansatz for V_{mn} (11) into (8) by considering restriction (14), it follows that R_{ij} can be expressed in terms of the new function F_{ij}.

In turbulent flows not close to solid walls the variation in correlation space dominates the variations in physical space. In this regime we may introduce a second asymptotic expansion for the variables in the equations (1) – (5) using the parameter γ, which is the ratio between the integral length scale l and a characteristic mean flow length scale L, as a small perturbation parameter

$$\frac{l}{L} = \gamma \ll 1 \ . \tag{15}$$

This assumption is justified by extensive measurements in a large amount of turbulent flow fields as well as by direct numerical simulation [2]. Numerical values for γ are of the order of 0.1 at most. However, (15) loses its validity close to solid walls, because here the variation in correlation and in physical space are of the same order of magnitude.

A consistent asymptotic expansion for the flow variables in the two-point correlation system (1) – (5) is the following

Mean flow characteristics:

$$x_k = O(1) \ , \qquad \bar{u}_i = O(1) \ . \tag{16}$$

Turbulence characteristics:

$$\begin{aligned} l = O(\gamma) \quad &\Rightarrow \quad r_k = \xi_k \gamma \ , \quad r = |\boldsymbol{r}| = \xi \gamma \ , \\ u_k = O(\gamma) \quad &\Rightarrow \quad R_{ij} = O(\gamma^2) \ , \quad R_{(ij)k} = O(\gamma^3) \ , \quad \overline{p u_j} = O(\gamma^3) \ . \end{aligned} \tag{17}$$

The leading order terms $(\cdot)^{(0)}$ of (2) and (3) obey the equations

$$\begin{aligned}\frac{\mathrm{D}R_{ij}^{(0)}}{\mathrm{D}t} &= -R_{kj}^{(0)}\frac{\partial \bar{u}_i(\boldsymbol{x})}{\partial x_k} - R_{ik}^{(0)}\frac{\partial \bar{u}_j(\boldsymbol{x})}{\partial x_k} \\ &\quad -\frac{\partial \bar{u}_k(\boldsymbol{x})}{\partial x_l}\xi_l\frac{\partial R_{ij}^{(0)}}{\partial \xi_k} + \frac{1}{\varrho}\left[\frac{\partial \overline{pu_j}^{(0)}}{\partial \xi_i} - \frac{\partial \overline{u_ip}^{(0)}}{\partial \xi_j}\right] \\ &\quad +\frac{\partial}{\partial \xi_k}\left[R_{(ik)j}^{(0)} - R_{i(jk)}^{(0)}\right] \\ &\quad -2\Omega_k\left[\epsilon_{kli}R_{lj}^{(0)} + \epsilon_{klj}R_{il}^{(0)}\right]\end{aligned} \tag{18}$$

and

$$\frac{1}{\varrho}\frac{\partial^2 \overline{pu_j}^{(0)}}{\partial \xi_k \partial \xi_k} = 2\left[\frac{\partial \bar{u}_k}{\partial x_l} + \epsilon_{mlk}\,\Omega_m\right]\frac{\partial R_{lj}^{(0)}}{\partial \xi_k} - \frac{\partial^2 R_{(kl)j}^{(0)}}{\partial \xi_k \partial \xi_l} \ . \tag{19}$$

Applying the scale split (16) and (17) to (8) it follows that

$$R_{ij}^{(0)} = -\epsilon_{ikm}\,\epsilon_{jln}\frac{\partial^2 V_{mn}^{(0)}}{\partial \xi_k \partial \xi_l} \ . \tag{20}$$

The terms in (11) have all the same order of magnitude, so they all contribute to $V_{mn}^{(0)}$ in (20). In the system (18), (19) and (20) the triple correlation function $R_{(ik)j}^{(0)}$ is the only term to be modelled. In the following the superscript $^{(0)}$ will be omitted.

3.2 Closure of the Triple Correlation

Any model for $R_{(ik)j}$ has to comply with (5). Thus we express the triple correlation as a tensor potential $H_{(ik)m}$

$$R_{(ik)j} = \epsilon_{jlm}\frac{\partial H_{(ik)m}}{\partial r_l} \ , \tag{21}$$

where $H_{(ik)m}$ as well as $R_{(ik)j}$ have to be symmetric in the first two indices.

For simplicity we assume $H_{(ik)m}$ to be solely a functional of the correlation distance r_k and the symmetric tensor F_{ij}. From this we can build a minimal integrity basis to get a frame invariant formulation for $H_{(ik)m}$. In general this leads to eight terms by restricting to linear invariants of F_{ij} with the dimension $\left[r^2 F_{ij}\right]$ divided by an integral time scale [20]. The eight terms are not all linearly independent so we restrict ourselves to two terms plus a nonlinear isotropic turbulence model developed and tested previously [15], [16].

The isotropic model

$$H_{(ik)m}^{[1]} = \left[\epsilon_{mil}\,r_k + \epsilon_{mkl}\,r_i\right]\frac{r_l}{4\,r}K \tag{22}$$

leads to the known result of Kármán and Howarth [8], for which we [15] proposed a model for K

$$K = \frac{\sqrt{2}}{5(3\,\mathfrak{C})^{3/2}} \sqrt{2\,k - F_{jj}} \frac{\partial F_{nn}}{\partial r} \quad . \tag{23}$$

Here, $\mathfrak{C}$ is the Kolmogorov constant and k is the turbulent kinetic energy.

The new nonisotropic model is defined by

$$\begin{aligned} H^{[2]}_{(ik)m} &= \beta_1\, \epsilon_{mnl} \left[F_{in}\, r_k + F_{kn}\, r_i\right] \frac{r_l\, k^{3/2}}{\mathfrak{L}_{jj}} \\ &+ \beta_2 \left[\epsilon_{ilm}\, F_{lk} + \epsilon_{klm}\, F_{li}\right] \frac{r^2\, k^{3/2}}{\mathfrak{L}_{jj}} \quad . \end{aligned} \tag{24}$$

The expression $\mathfrak{L}_{jj}$ is an integral operator applied to F_{jj}, as defined by (25) in the next section. The final model consists of the sum of the two terms $H^{[1]}_{(ik)m}$ and $H^{[2]}_{(ik)m}$.

In Sect. 5 we will find that the nonisotropic model (24) leads to Rotta's return-to-isotropy term of the pressure–strain correlation [17], while the isotropic model, of course, vanishes completely.

3.3 The Tensor Length Scale Equation

To derive a tensor length scale equation, the Batchelor integral operator $\frac{1}{4\pi}\int_{V_\xi}(\cdot)\,\frac{\mathrm{d}^3\xi}{\xi^2}$ will be applied to (18) which eliminates the correlation coordinate ξ_k. To make the quantity

$$\mathfrak{L}_{ij} = \frac{1}{4\pi} \int_{V_\xi} F_{ij}\, \frac{\mathrm{d}^3\xi}{\xi^2} \tag{25}$$

the new dependent transport variable the remaining coefficients α_3–α_5 will be rescaled to

$$a_1 = \frac{\alpha_3}{7}\, \frac{24}{5 + 3\alpha_3 - 5\alpha_5} \quad \text{and} \quad a_2 = \frac{1}{5}\, \frac{5 + 3\alpha_3 - 5\alpha_5}{1 + 3\alpha_4 + 2\alpha_5} \quad , \tag{26}$$

because only two of the parameters α_3–α_5 are linearly independent.

Inserting now (11) and (20) together with the model for the triple correlation (22) and (24) into (18) and (19), solving (19) for $\overline{pu_j}$ and applying the Batchelor integral to (18) we obtain after some rearrangements an equation for the tensor $\mathfrak{L}_{ij}$

$$\frac{\mathrm{D}\mathfrak{L}_{ij}}{\mathrm{D}t} = {}^{1\mathfrak{L}}P_{ij} + {}^{2\mathfrak{L}}P_{ij} + {}^{\mathfrak{L}}\Phi_{ij} - {}^{\mathfrak{L}}E_{ij} + {}^{\mathfrak{L}}C_{ij} \quad , \tag{27}$$

where

$$\begin{aligned} {}^{1\mathfrak{L}}P_{ij} &= -\left[\mathfrak{L}_{ik}\, \frac{\partial \bar{u}_j}{\partial x_k} + \mathfrak{L}_{jk}\, \frac{\partial \bar{u}_i}{\partial x_k}\right] \quad , \\ {}^{2\mathfrak{L}}P_{ij} &= \left[\frac{2}{3} - \frac{28}{15}\, a_1\, a_2 - \frac{8}{15}\, a_2\right] \delta_{ij}\, \frac{\partial \bar{u}_k}{\partial x_l}\, \mathfrak{L}_{kl} \quad , \\ {}^{\mathfrak{L}}\Phi_{ij} &= \left[\frac{2}{15} - \frac{1}{3a_2} + \frac{32}{15}\, a_1\right] \bar{S}_{ij}\, \mathfrak{L}_{kk} \end{aligned}$$

$$
\begin{aligned}
&+\left[\frac{2}{5}+\frac{56}{15}a_1\right]\left[\overline{W}_{ik}\,\mathfrak{L}_{jk}+\overline{W}_{jk}\,\mathfrak{L}_{ik}\right]\\
&+\left[\frac{4}{5}-\frac{16}{5}a_1\right]\left[\bar{S}_{ik}\,\mathfrak{L}_{jk}+\bar{S}_{jk}\,\mathfrak{L}_{ik}-\frac{2}{3}\,\delta_{ij}\,\bar{S}_{kl}\,\mathfrak{L}_{kl}\right] \qquad (28)\\
&-c_{\mathrm{R}}^{\mathfrak{L}}\left[\frac{\mathfrak{L}_{ij}}{\mathfrak{L}_{kk}}-\frac{\delta_{ij}}{3}\right]k^{3/2}\ ,
\end{aligned}
$$

$$
\begin{aligned}
{}^{\mathfrak{L}}E_{ij} &= c_{\mathfrak{L}}\,k^{3/2}\,\frac{\mathfrak{L}_{ij}}{\mathfrak{L}_{kk}}\ ,\\
{}^{\mathfrak{L}}C_{ij} &= -2\,\Omega_k\left[\epsilon_{kli}\,\mathfrak{L}_{jl}+\epsilon_{klj}\,\mathfrak{L}_{il}\right]\ .
\end{aligned}
$$

The quantity $\mathfrak{L}_{ij}$ divided by the kinetic energy has the dimension of a length scale. Equation (27) plays the role of a new length scale equation in nonisotropic turbulence modelling and $\mathfrak{L}_{ij}$ will therefore be referred to as a length scale tensor function. The coefficients $c_{\mathrm{R}}^{\mathfrak{L}}$ and $c_{\mathfrak{L}}$ are functions of the triple correlation parameter $\mathfrak{C}$, β_1 and β_2.

In (28) $\bar{S}_{ij}$ is the mean rate-of-strain term defined by the symmetric part of the mean flow gradient tensor

$$
\frac{\partial \bar{u}_i}{\partial x_j} = \bar{S}_{ij} + \bar{\omega}_{ij}\ . \qquad (29)
$$

Here $\bar{\omega}_{ij}$ is the mean vorticity tensor and $\overline{W}_{ij}$ is the intrinsic or absolute mean vorticity tensor defined by

$$
\overline{W}_{ij} = \bar{\omega}_{ij} + \epsilon_{mji}\,\Omega_m\ . \qquad (30)
$$

Let us trace back the origin of each single term in (28). The first production term ${}^{1\mathfrak{L}}P_{ij}$ and the Coriolis term ${}^{\mathfrak{L}}C_{ij}$ have their clear counterpart in the first and last line of (2). The second production term ${}^{2\mathfrak{L}}P_{ij}$ and one part of the redistribution term ${}^{\mathfrak{L}}\Phi_{ij}$ comes up due to the term in line two of (2) which has no correspondence in the Reynolds stress equation, because it vanishes in the limit of $r \to 0$. The major part of the redistribution term ${}^{\mathfrak{L}}\Phi_{ij}$ which has vanishing trace, leads back to the pressure-velocity correlation in line three of (2). The latter and the dissipation term ${}^{\mathfrak{L}}E_{ij}$ are the only terms that contain the model for the triple correlation.

For second moment closure we need a frame invariant and dimensionally correct relation between the length scale tensor function $\mathfrak{L}_{ij}$ and the dissipation tensor ε_{ij}.

4. The Dissipation Tensor Equation

4.1 Invariant Modelling

In turbulence modelling at the level of two-equation models an ε- or an l-equation are synonymous formulations for the same physical process. A condensation of this fact is the well known scalar relation $\varepsilon \sim k^{3/2}/l$. A generalization of this relation to tensor notation should be an invariant formulation that connects the length scale tensor function with the dissipation tensor.

If one assumes the length scale tensor function to be only a function of the dissipation and the Reynolds stress tensor the Cayley Hamilton theorem yields (see [20])

$$\mathcal{L}_{ij} = \frac{1}{2}\frac{k^{5/2}}{\varepsilon}\sum_{\lambda,\mu=0}^{2}\varphi_{(\lambda,\mu)}\left[b^{\lambda}_{ik}\,d^{\mu}_{kj} + d^{\mu}_{ik}\,b^{\lambda}_{kj}\right] , \tag{31}$$

where b_{ij} and d_{ij} are the anisotropy part of the Reynolds stress and the dissipation tensor respectively

$$b_{ij} = \frac{\overline{u_i u_j}}{2k} - \frac{\delta_{ij}}{3} \quad \text{and} \quad d_{ij} = \frac{\varepsilon_{ij}}{\varepsilon} - \frac{\delta_{ij}}{3} . \tag{32}$$

In general, each $\varphi(\lambda, \mu)$ depends on the eight nonlinear tensor invariants

$$\begin{aligned} &II_b = b^2_{kk} \;,\; III_b = b^3_{kk} \;,\; II_d = d^2_{kk} \;,\; III_d = d^3_{kk} \;, \\ &I_{bd} = b_{ij}d_{ji} \;,\; II_{bd} = b^2_{ij}d_{ji} \;,\; III_{bd} = b_{ij}d^2_{ji} \;\text{ and }\; IV_{bd} = b^2_{ij}d^2_{ji} \;. \end{aligned} \tag{33}$$

To drop the huge number of unknown function $\varphi(\lambda, \mu)$ for the further analysis we restrict $\mathcal{L}_{ij}$ to be only a linear function of b_{ij} and d_{ij}. Thus all nonlinear tensor products in (31) vanish and the remaining coefficients $\varphi(0,0)$, $\varphi(1,0)$ and $\varphi(0,1)$ are constant. The second constant can be fixed to 0 by a procedure similar to the realizability conditions developed by Schumann [19].

With the aid of (31) one can convert the tensor length scale equation (27) to a dissipation tensor equation given below.

4.2 Derivation of the Dissipation Tensor Equation

Taking the substantial derivative $\mathsf{D}/\mathsf{D}t$ of (31) one can derive a new tensor dissipation equation by replacing the substantial derivatives of $\mathcal{L}_{ij}$ and $\overline{u_i u_j}$ by the right hand side of the Reynolds stress and the length scale tensor (6) and (27) respectively

$$\frac{\mathsf{D}\varepsilon_{ij}}{\mathsf{D}t} = {}^{1\varepsilon}P_{ij} + {}^{2\varepsilon}P_{ij} + {}^{\varepsilon}\Phi_{ij} - {}^{\varepsilon}E_{ij} + {}^{\varepsilon}C_{ij} . \tag{34}$$

Equation (34) contains four arbitrary constants: $a_0 = \frac{\varphi(0,0)}{\varphi(0,1)a_2}$, a_1, $c^{\varepsilon}_{\mathrm{R}}$ and c_{ε_2}. c_{ε_2} and $c^{\varepsilon}_{\mathrm{R}}$ are functions of the triple correlation model coefficients, where c_{ε_2} has been set to 1.92 to fulfill the limit of isotropic turbulence. The remaining three should be adjusted to some reference flows as can be seen in Sect. 6.

The final form for each term in (34) then follows as

$$\begin{aligned} {}^{1\varepsilon}P_{ij} &= -\varepsilon_{ik}\frac{\partial \bar{u}_j}{\partial x_k} - \varepsilon_{jk}\frac{\partial \bar{u}_i}{\partial x_k} , \\ {}^{2\varepsilon}P_{ij} &= -5\,\varepsilon_{ij}\frac{\overline{u_l u_k}}{2k}\frac{\partial \bar{u}_l}{\partial x_k} \\ &\quad + \frac{2}{15\,a_0}\left[\left(2\frac{\varepsilon_{ij}}{\varepsilon} - \frac{1}{3}\,\delta_{ij}\right)(4 + 14\,a_1) + 5\,a_0\,\delta_{ij}\right]\varepsilon_{kl}\frac{\partial \bar{u}_k}{\partial x_l} , \end{aligned}$$

$$
\begin{aligned}
{}^{\varepsilon}\Phi_{ij} &= \left[\frac{2}{15} - a_0 + \frac{32}{15}\,a_1\right] \varepsilon\,\bar{S}_{ij} \\
&+ \left[\frac{4}{5} - \frac{16}{5}\,a_1\right] \left[\bar{S}_{ik}\varepsilon_{jk} + \bar{S}_{jk}\varepsilon_{ik} - \frac{2}{3}\,\delta_{ij}\bar{S}_{kl}\varepsilon_{kl}\right] \\
&+ \left[\frac{2}{5} + \frac{56}{15}\,a_1\right] \left[\overline{W}_{ik}\varepsilon_{jk} + \overline{W}_{jk}\varepsilon_{ik}\right] \\
&- c_{\mathrm{R}}^{\varepsilon}\,\frac{\varepsilon^2}{k} d_{ij} \ , \\
{}^{\varepsilon}E_{ij} &= c_{\varepsilon_2}\,\frac{\varepsilon}{k}\varepsilon_{ij} \ , \\
{}^{\varepsilon}C_{ij} &= -2\,\Omega_k\left[\epsilon_{kli}\,\varepsilon_{jl} + \epsilon_{klj}\,\varepsilon_{il}\right] \ .
\end{aligned}
\tag{35}
$$

The production term ${}^{1\varepsilon}P_{ij}$ and the Coriolis term ${}^{\varepsilon}C_{ij}$ are consistent with the exact dissipation equation as they should be, because they are the only terms requiring no closure.

The final result in (34) and (35) obviously no longer has the main disadvantages pointed out in the Introduction. It was derived by successive approximations employing asymptotic limits and tensor invariance considerations. Only the triple correlation has to be modelled. As indicated by the tensor character of (34) the dissipation model is nonisotropic. One can easily see that (34) and (35) can also account for strong streamline curvature.

Besides the dissipation tensor there is a second unknown term in the Reynolds stress equation which has the same order of magnitude as ε_{ij}: the redistribution or pressure–strain correlation Φ_{ij}. We will show that this term can be tackled in the frame of our theory without additional modelling assumptions.

5. The Pressure–Strain Correlation

It was first shown by Chou [3] that the pressure–strain correlation Φ_{ij} can be expressed as an integral of the two-point correlation function

$$
\Phi_{ij} = \overline{\frac{p}{\varrho}\left[\frac{\partial u_j}{\partial x_i} + \frac{\partial u_i}{\partial x_j}\right]} = \Phi_{ij_{\text{rapid}}} + \Phi_{ij_{\text{return}}} + \Phi_{ij_{\text{surf}}} \ , \tag{36}
$$

where

$$
\begin{aligned}
\overline{p\frac{\partial u_i}{\partial x_j}} &= \frac{1}{4\pi}\int\limits_{V_r}\left[2\left(\frac{\partial \bar{u}_k}{\partial x_l}(\boldsymbol{x}+\boldsymbol{r}) + \epsilon_{mlk}\,\Omega_m\right)\left(\frac{\partial^2 R_{il}}{\partial x_j \partial r_k} - \frac{\partial^2 R_{il}}{\partial r_j \partial r_k}\right)\right. \\
&\left. + \left(\frac{\partial}{\partial x_j} - \frac{\partial}{\partial r_j}\right)\frac{\partial^2 R_{i(kl)}}{\partial r_k \partial r_l}\right]\frac{\mathrm{d}^3 r}{|\boldsymbol{r}|} + \left.\overline{p\frac{\partial u_i}{\partial x_j}}\right|_{\text{surf}} \ .
\end{aligned}
\tag{37}
$$

The first and the second term of (37) are called the rapid and the return term of the pressure–strain correlation respectively.

The surface integral $\Phi_{ij_{\text{surf}}}$ is only relevant near solid boundaries and thus has been neglected in our analysis.

With the scale split (16) and (17), (37) reads to leading order

$$\overline{p\frac{\partial u_i}{\partial x_j}} = -\frac{1}{4\pi}\int\limits_{V_\xi}\left[2\left(\frac{\partial \bar{u}_k}{\partial x_l}(\boldsymbol{x}) + \epsilon_{mlk}\,\Omega_m\right)\frac{\partial^2 R_{il}^{(0)}}{\partial\xi_j\partial\xi_k} + \frac{\partial^3 R_{i(kl)}^{(0)}}{\partial\xi_j\partial\xi_k\partial\xi_l}\right]\frac{\mathrm{d}^3\xi}{|\boldsymbol{\xi}|}\ . \tag{38}$$

Equation (38) does not necessarily lead to an algebraic function for the pressure–strain correlation. Without any restriction for the correlation functions one has to solve a transport equation for (38) as is shown in Appendix A.

Introducing now (20) into (38) and considering (11)–(14), the rapid term integral in (38) can be calculated exactly:

$$\begin{aligned}\Phi_{ij_{\text{rapid}}} &= c_1\,\bar{S}_{ij}\,k + c_2\,\left[\overline{W}_{ik}b_{jk} + \overline{W}_{jk}b_{ik}\right]\,k \\ &\quad + c_3\left[\bar{S}_{jk}b_{ik} + \bar{S}_{ik}b_{jk} - \frac{2}{3}\delta_{ij}\bar{S}_{kl}b_{kl}\right]\,k\ ,\end{aligned} \tag{39}$$

where

$$c_1 = \frac{4}{5}\ ,\quad c_2 = \frac{4}{5} + 7\,a_3\ ,\quad c_3 = \frac{12}{5} - 9\,a_3\quad\text{and}\quad a_3 = \frac{35}{16}\alpha_3\ . \tag{40}$$

The rapid-term model is similar to the Launder, Reece and Rodi model [10] (LRR) and was first developed by Naot [14].

It should be emphasized that we could calculate the coefficients for our model *without* introducing any particular form of the two-point correlation function.

For comparison, we have included the coefficients of the LRR and the Gibson and Launder model (GL) [6] in Table 1. Beside the coefficient c_1 which has been calculated by Crow [4] using rapid distortion theory the other constants in Table 1 differ significantly from our model where we have assumed $\alpha_3 = 0$.

We think that the reason for this disagreement is twofold: Firstly, most models are proposed in combination with a scalar dissipation equation and anisotropy effects of the dissipation tensor are modelled within the pressure–strain correlation. Secondly, the model constants are usually fixed due to experiments and thus other turbulent effects which are not separable from the pressure–strain correlation are included in the constants.

Finally, on introducing the model for the triple correlation (24) together with (21) into (38), Rotta's well-known model [17] for the return term of the pressure–strain correlation

$$\Phi_{ij_{\text{return}}} = -c_{\text{R}}\,\varepsilon\,b_{ij} \tag{41}$$

is retained as mentioned in Sect. 3.2. The isotropic part (22) and (23) vanishes in (38) and thus does not contribute to (41).

Table 1. Constants of the pressure–strain models.

	c_1	c_2	c_3
LRR	0.8	1.31	1.75
GL	0.8	1.2	1.2
present model	0.8	0.8	2.4

It should be emphasized that the basic assumption in the theory is the limit in (14). It leads to an algebraic function for the pressure–strain correlation as was first suggested by Lumley [13] and which has been followed by nearly all authors. In addition (14) avoids further unclosed terms in the tensor length scale equation beside the triple correlation function.

6. Results

We have tested our new dissipation tensor model in four independent homogeneous turbulent flows. For clarity, we will summarize the model constants in (35), (40) and (41) that have been calculated by realizability and numerical optimization

$$a_0 = 0.05\ ,\quad a_1 = 0\ ,\quad c_R^\varepsilon = 0.8\ ,\quad c_{\varepsilon_2} = 1.92\ ,\quad a_3 = 0 \quad \text{and} \quad c_R = 4\ . \tag{42}$$

The initial conditions at $t = 0$ for all test cases below correspond to a state of isotropic turbulence where

$$b_{ij} = 0\ ,\qquad k = k_0 \quad \text{and} \quad \varepsilon = \varepsilon_0\ . \tag{43}$$

Each test case will be compared with the large eddy simulation (LES) of Bardina et al. [1] or with the direct numerical simulation (DNS) data of Lee and Reynolds [11]. In addition we have included the results for the wide spread LRR model.

It can be shown that any solution for homogeneous shear flow depends only on the parameter $\varepsilon_0/(S\,k_0)$ and Ω/S where S is the mean shear field parameter defined in (45), (46) and (47) for each particular flow to be considered. Ω is the rotation rate defined in (45).

The simple dependence on one single initial parameter is a direct consequence of our having introduced the nondimensional variables

$$k^* = \frac{k}{k_0}\ ,\qquad \varepsilon^* = \frac{\varepsilon}{\varepsilon_0} \quad \text{and} \quad t^* = S\,t\ . \tag{44}$$

This notation will be adopted for all figures to follow.

The problem of homogeneous shear flow in a rotation frame of reference will be considered first. The mean velocity gradient and the rotation rate are given by

$$\frac{\partial \bar{u}_i}{\partial x_j} = \begin{pmatrix} 0 & S & 0 \\ 0 & 0 & 0 \\ 0 & 0 & 0 \end{pmatrix} \quad \text{and} \quad \Omega_i = (0, 0, \Omega) \ . \tag{45}$$

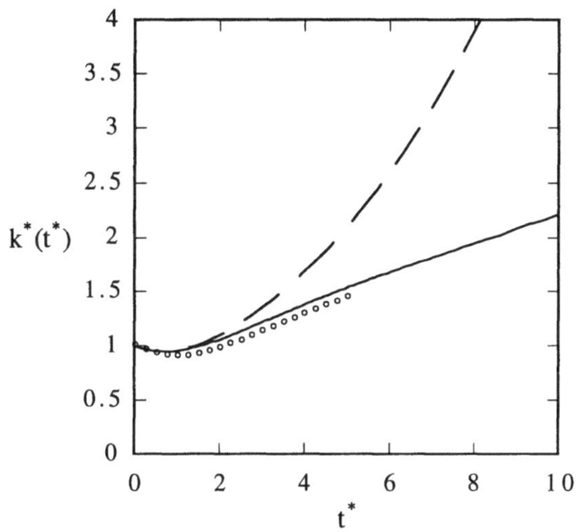

Fig. 1. Evolution of the turbulent kinetic energy in a rotating shear flow with the isotropic initial condition $\varepsilon_0/(S\,k_0) = 0.18$ and $\Omega/S = 0$. Comparison of the dissipation tensor model (———) with the LRR model (– – –) and the large eddy simulation of Bardina et al. (1983) (∘).

In Figs. 1–3 we have compared the new dissipation tensor model with the LES data of Bardina et al. [1] and the LRR model for three different values of the frame rotation parameter Ω.

From Fig. 1 it is clear that the dissipation tensor model is well in line with the LES data while the LRR model cannot capture the weak increase of turbulence energy. In Figs. 2 and 3 the new dissipation model gives comparable results to the LRR model while neither can capture the weak increase of turbulent energy for $\Omega/S = 0.5$.

The second test case considers the evolution of the turbulent energy and the Reynolds stress anisotropy tensor for homogeneous plain strain in a non-rotating frame. The mean velocity gradient is given by

$$\frac{\partial \bar{u}_i}{\partial x_j} = \begin{pmatrix} S & 0 & 0 \\ 0 & -S & 0 \\ 0 & 0 & 0 \end{pmatrix} \ . \tag{46}$$

From Fig. 4 we find excellent agreement for the LRR model as well as for the dissipation tensor model with the DNS data of Lee and Reynolds [11]. On the other the hand the corresponding non-zero components of the Reynolds stress anisotropy

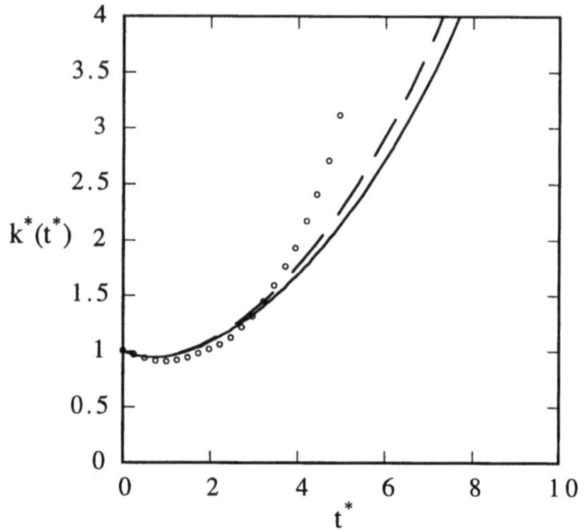

Fig. 2. Evolution of the turbulent kinetic energy in a rotating shear flow with the isotropic initial condition $\varepsilon_0/(S\,k_0) = 0.18$ and $\Omega/S = 0.25$. Comparison of the dissipation tensor model (———) with the LRR model (– – –) and the large eddy simulation of Bardina et al. (1983) (∘).

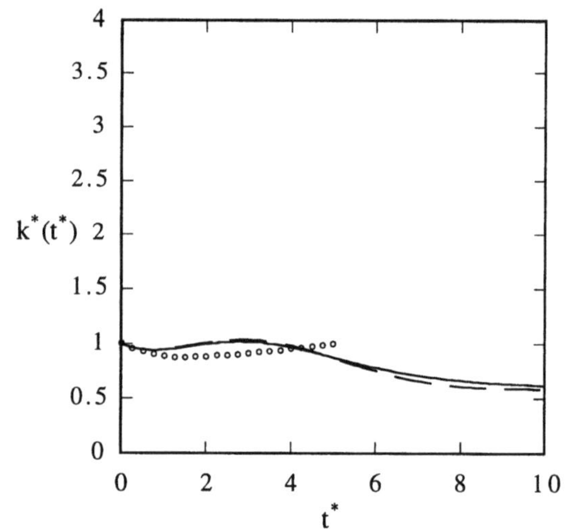

Fig. 3. Evolution of the turbulent kinetic energy in a rotating shear flow with the isotropic initial condition $\varepsilon_0/(S\,k_0) = 0.18$ and $\Omega/S = 0.5$. Comparison of the dissipation tensor model (———) with the LRR model (– – –) and the large eddy simulation of Bardina et al. (1983) (∘).

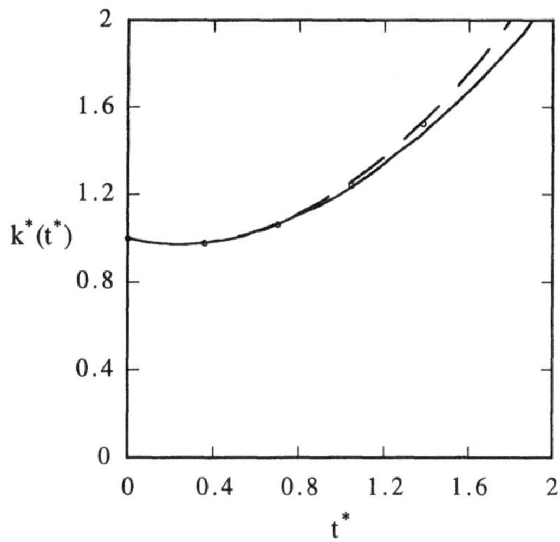

Fig. 4. Evolution of the turbulent kinetic energy in homogeneous plain strain with the isotropic initial condition $\varepsilon_0/(S\,k_0) = 0.25$. Comparison of the dissipation tensor model (——) with the LRR model (– – –) and the direct numerical simulation of Lee and Reynolds (1985) (○).

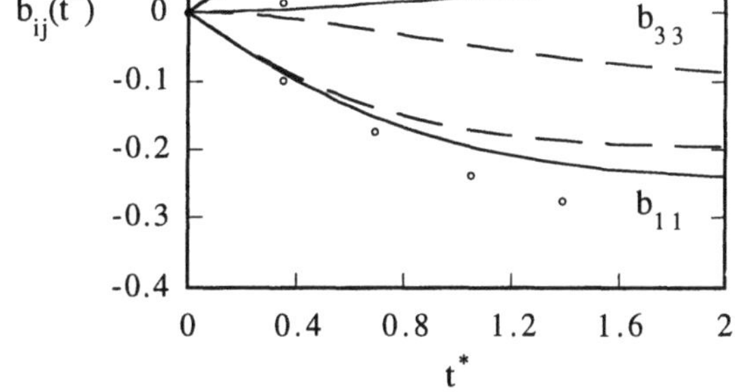

Fig. 5. Evolution of the Reynolds stress anisotropy tensor in homogeneous plain strain with the isotropic initial condition $\varepsilon_0/(S\,k_0) = 0.25$. Comparison of the dissipation tensor model (——) with the LRR model (– – –) and the direct numerical simulation of Lee and Reynolds (1985) (○).

tensors in Fig. 5 are significantly better for the dissipation tensor model. In particular the tensor component b_{33} in the LRR model decreases, while the dissipation tensor model captures the trend of a weakly increasing b_{33}.

Finally we will examine the performance of the dissipation tensor model for axisymmetric expansion and contraction. Both flows are three dimensional in contrast to the flows considered above. The mean velocity gradients for the expansion and the contraction flow are defined as

$$\frac{\partial \bar{u}_i}{\partial x_j} = \begin{pmatrix} -S & 0 & 0 \\ 0 & \frac{1}{2}S & 0 \\ 0 & 0 & \frac{1}{2}S \end{pmatrix} \quad \text{and} \quad \frac{\partial \bar{u}_i}{\partial x_j} = \begin{pmatrix} S & 0 & 0 \\ 0 & -\frac{1}{2}S & 0 \\ 0 & 0 & -\frac{1}{2}S \end{pmatrix} \tag{47}$$

respectively.

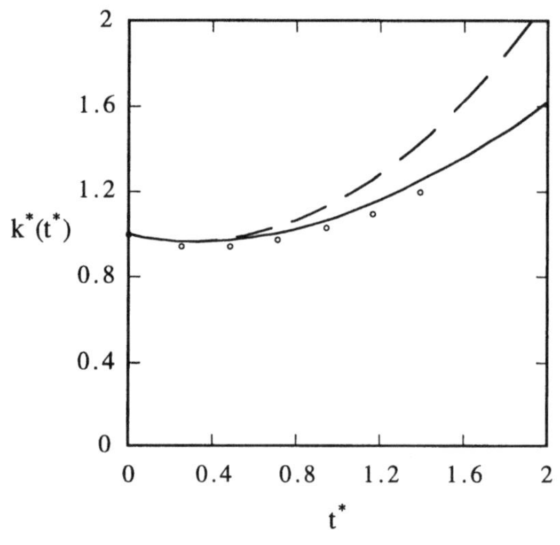

Fig. 6. Evolution of the turbulent kinetic energy in the axisymmetric expansion with the isotropic initial condition $\varepsilon_0/(S\,k_0) = 0.245$. Comparison of the dissipation tensor model (———) with the LRR model (– – –) and the direct numerical simulation of Lee and Reynolds (1985) (∘).

Again we have compared both flows with the LRR model and the DNS data of Lee and Reynolds [11]. From Figs. 6–9 it is clear that the dissipation tensor model performs much better than the LRR model and gives nearly identical results with the DNS data. It is interesting to note that the dissipation tensor model gives slightly better results for the three-dimensional test cases than for the much simpler two-dimensional homogeneous flows.

As a whole the dissipation tensor model yields significant improvements for a variety of homogeneous turbulent flows compared to the widely used LRR model. In

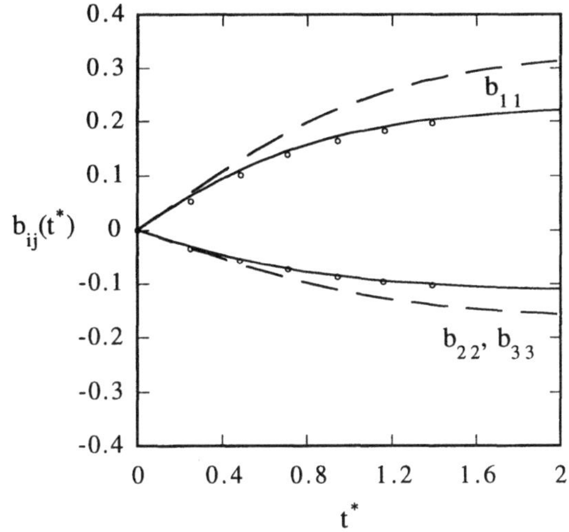

Fig. 7. Evolution of the Reynolds stress anisotropy tensor in the axisymmetric expansion with the isotropic initial condition $\varepsilon_0/(S\,k_0) = 0.245$. Comparison of the dissipation tensor model (———) with the LRR model (– – –) and the direct numerical simulation of Lee and Reynolds (1985) (∘).

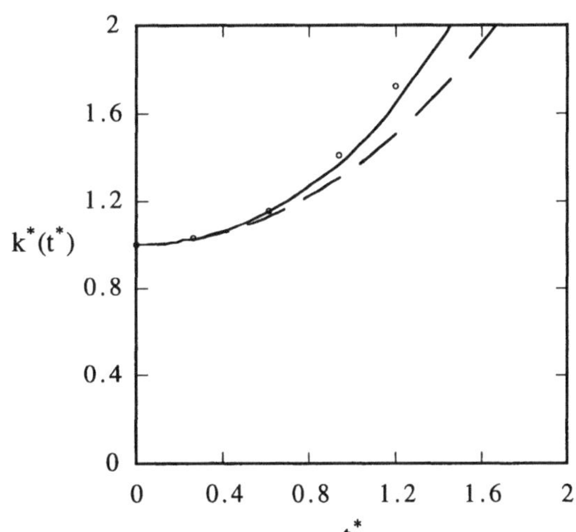

Fig. 8. Evolution of the turbulent kinetic energy in the axisymmetric contraction with the isotropic initial condition $\varepsilon_0/(S\,k_0) = 0.0179$. Comparison of the dissipation tensor model (———) with the LRR model (– – –) and the direct numerical simulation of Lee and Reynolds (1985) (∘).

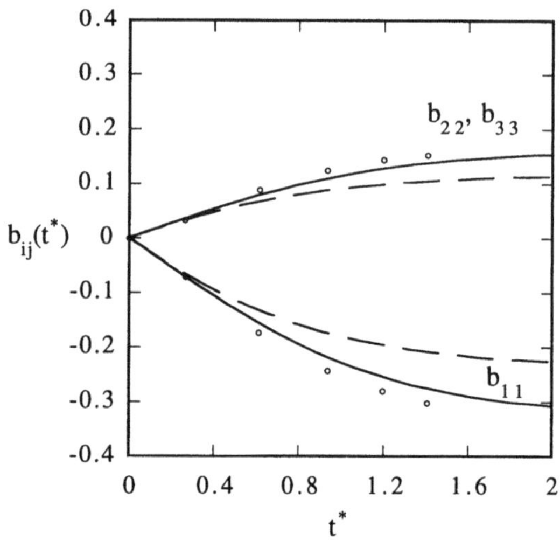

Fig. 9. Evolution of the Reynolds stress anisotropy tensor in the axisymmetric contraction with the isotropic initial condition $\varepsilon_0/(S\,k_0) = 0.0179$. Comparison of the dissipation tensor model (———) with the LRR model (– – –) and the direct numerical simulation of Lee and Reynolds (1985) (∘).

addition we could calculate the model on a less empirical basis and therefore the dissipation tensor model promises to perform better in other turbulent test cases as well. Further improvement should concern the model of the two-point triple correlation.

7. Summary

Using two asymptotic limits it has been possible to derive a dissipation tensor equation and the pressure–strain correlation with the help the two-point velocity correlation function.

The stronger of the two restrictions was to postulate each component of a two-point correlation tensor to be only a function of the absolute value of the correlation distance. The second asymptotic limit describes a scale split between the integral length scale and the mean flow length scale. Using perturbation methods in conjunction with the latter assumptions, it was possible to transform the two-point correlation equation to a tensor length scale equation by applying the Batchelor integral to the first order of the two-point correlation equation.

The resulting length scale equation has been converted into a dissipation equation by using tensor invariant considerations. In contrast to common dissipation equations the final model can account for nonisotropic as well as for curvature effects.

Introducing the two asymptotic limits into the integral formulation of the pressure–strain correlation, one finds an exact formulation for the rapid term.

Acknowledgement

The author would like to thank Prof. Dr.-Ing. N. Peters for many helpful discussions and his great interest in the progress of the work.

A. Transport Equation for the Rapid Term

The rapid term integral

$$X_{jk}^{il} = -\frac{1}{4\pi}\int\limits_{V_\xi} \frac{\partial^2 R_{il}^{(0)}}{\partial\xi_j\partial\xi_k}\frac{\mathrm{d}^3\xi}{|\boldsymbol{\xi}|} \tag{48}$$

in (38) leads with (18) and (19) to

$$\frac{\mathsf{D}X_{jk}^{il}}{\mathsf{D}t} = {}^{1\mathrm{X}}P_{jk}^{il} + {}^{2\mathrm{X}}P_{jk}^{il} + {}^{\mathrm{X}}\Phi_{jk}^{il} - {}^{\mathrm{X}}E_{jk}^{il} + {}^{\mathrm{X}}C_{jk}^{il}\ , \tag{49}$$

where

$$\begin{aligned}
{}^{1\mathrm{X}}P_{jk}^{il} &= -X_{jk}^{ml}\frac{\partial\bar{u}_i}{\partial x_m} - X_{jk}^{im}\frac{\partial\bar{u}_l}{\partial x_m}\ ,\\
{}^{2\mathrm{X}}P_{jk}^{il} &= X_{mj}^{il}\frac{\partial\bar{u}_m}{\partial x_k} + X_{mk}^{il}\frac{\partial\bar{u}_m}{\partial x_j} + X_{jkmn}^{il}\frac{\partial\bar{u}_m}{\partial x_n}\ ,\\
{}^{\mathrm{X}}\Phi_{jk}^{il} &= \left[\bar{S}_{mn} + \overline{W}_{mn}\right]\left[X_{imjk}^{nl} + X_{lmjk}^{ni}\right] + \frac{1}{2}\left[Y_{imjk}^{ml} + Y_{lmjk}^{mi}\right],\\
{}^{\mathrm{X}}E_{jk}^{il} &= Y_{jk}^{il} + Y_{jk}^{li}\ ,\\
{}^{\mathrm{X}}C_{jk}^{il} &= -2\,\Omega_m\left[\epsilon_{mni}X_{jk}^{nl} + \epsilon_{mnl}X_{jk}^{in}\right]\ .
\end{aligned} \tag{50}$$

The unclosed tensors X_{jkmn}^{il}, Y_{jkmn}^{il} and Y_{jk}^{il} are defined as

$$X_{jkmn}^{il} = -\frac{1}{4\pi}\int\limits_{V_\xi}\frac{\xi_m\xi_n}{\xi^2}\frac{\partial^2 R_{il}^{(0)}}{\partial\xi_j\partial\xi_k}\frac{\mathrm{d}^3\xi}{|\boldsymbol{\xi}|}\ , \tag{51}$$

$$Y_{jkmn}^{il} = -\frac{1}{4\pi}\int\limits_{V_\xi}\frac{\xi_m\xi_n}{\xi^2}\frac{\partial^3 R_{(io)l}^{(0)}}{\partial\xi_j\partial\xi_k\partial\xi_o}\frac{\mathrm{d}^3\xi}{|\boldsymbol{\xi}|} \quad\text{and}\quad Y_{jk}^{il} = Y_{jkmm}^{il}\ . \tag{52}$$

References

1. Bardina J., Ferziger J. H., Reynolds W. C.: Improved Turbulence Models Based on Large-Eddy Simulation of Homogeneous, Incompressible Turbulent Flows *Stanford University Tech. Rep.* **TF-19** (1983)
2. Bradshaw P., Mansour N.N., Piomelli U.: On Local Approximations of the Pressure–Strain Term in Turbulence Models *Center Turb. Res. Proc. Summer Progr.* (1987) 159-164
3. Chou P. Y.: On Velocity Correlations and the Solutions of the Equations of Turbulent Fluctuations *Quart. App. Math.* **3** (1945) 38-54
4. Crow S. C.: Viscoelastic Properties of Fine-Grained Incompressible Turbulence *J. Fluid Mech.* **33** (1968) 1-20
5. Donaldson C. duP., Sandri G.: On the Inclusion of Information on Eddy Structure in Second-Order-Closure Models of Turbulent Flows *AGARD Rep.* **CP** (1981) 25.1-25.14
6. Gibson M.M., Launder B. E.: Ground Effects on Pressure Fluctuations in the Atmospheric Boundary Layer *J. Fluid Mech.* **86** (1978) 491
7. Hanjalić K., Launder B. E.: A Reynolds Stress Model of Turbulence and its Application to thin Shear Flows *J. Fluid Mech.* **52** (1972) 609-638
8. von Kármán T., Howarth L.: On the Statistical Theory of Isotropic Turbulence *Proc. Roy. Soc.* **A 164** (1938) 192-215
9. Kim J., Moin P., Moser R.: Turbulence Statistics in Fully Developed Channel Flow at Low Reynolds Number *J. Fluid Mech.* **177** (1987) 133-166
10. Launder B. E., Reece G. J., Rodi W.: Progress in the Development of a Reynolds-Stress Turbulence Closure *J. Fluid Mech.* **68** (1975) 537-566
11. Lee M. J., Reynolds W. C.: Numerical Experiments on the Structure of Homogeneous Turbulence *Stanford University Tech. Rep.* **TF-24** (1985)
12. Lumley J. L., Newman G. R.: The Return to Isotropy of Homogeneous Turbulence *J. Fluid Mech.* **82** (1977) 161-178
13. Lumley J. L.: Computational Modeling of Turbulent Flows *Adv. Appl. Mech.* **18** (1978) 123-176
14. Naot D., Shavit A., Wolfshtein M.: Two-Point Correlation Model and the Redistribution of Reynolds Stress *Phys. Fluids* **16** (1973) 738-743
15. Oberlack M., Peters N., Kivotides D.: Solutions of the von Kármán-Howarth Equation by Gradient Flux Approximations *Eighth Symp. Turb. Shear Flows* (1991) **III-12-1**
16. Oberlack M., Peters N.: Closure of the Two-Point Correlation Equation as a Basis of Reynolds Stress Models *Appl. Sci. Res.* **51** (1992) 533-538
17. Rotta J.C.: Statistische Theorie Nichthomogener Turbulenz *Z. Phys.* **129** (1951) 547-572
18. Sandri G.: Fundamental Research in Turbulent Modeling *ARAP Rep.* **438** (1981)
19. Schumann U.: Realizability of Reynolds-Stress Turbulence Models *Phys. Fluids* **20** (1977) 721-725
20. Spencer A. J. M: Theory of Invariants *Continuum Physics* (ed. A. C. Eringen) **1** (1971) 239-353
21. Speziale C. G.: Turbulence Modeling: Present and Future, Comment 2. *Whither Turbul. Workshop, Ithaca, N.Y. Lect. Notes Phys.*, Vol. 357, ed. Lumley, J. L. (1990) 490-512, Springer Verlag
22. Speziale C. G.: Analytical Methods for the Development of Reynolds-Stress Closures in Turbulence *Annu. Rev. Fluid Mech.* **23** (1991) 107-157
23. Tagawa M., Nagano Y., Tsuji T.: Turbulence Model for the Dissipation Components of Reynolds Stresses *Eighth Symp. Turb. Shear Flows* (1991) **29-3-1**

Turbulence Statistics of Rotating Channel Flow

H. I. Andersson and R. Kristoffersen

Department of Applied Mechanics, Faculty of Mechanical Engineering,
The Norwegian Institute of Technology, N-7034 Trondheim, Norway

Abstract

Turbulence statistics of rotating channel flow are considered, including budget data for the individual non-zero components of the Reynolds stress tensor. The statistics have been compiled from direct numerical simulations of Kristoffersen & Andersson (1993) at six different rotation numbers $Ro_{\mathrm{m}} = 2h|\Omega|/U_{\mathrm{m}}$ in the range from 0.01 to 0.50. Complete Reynolds stress budgets are provided for the particular cases $Ro_{\mathrm{m}} = 0.15$ and $Ro_{\mathrm{m}} = 0.50$, while the variation with Ro_{m} of some important flow characteristics demonstrates the significant alterations in the flow field induced by the Coriolis force.

Nomenclature

C_f	skin-friction coefficient, $\tau_{\mathrm{w}}/\varrho U_{\mathrm{m}}^2$
D_{ij}^P	turbulent diffusion term due to pressure fluctuations
D_{ij}^T	turbulent diffusion term due to velocity fluctuations
D_{ij}^ν	viscous diffusion
G_{ij}	stress production due to system rotation
h	channel half-width
k	mean turbulent kinetic energy, $\overline{u_i u_i}/2$
p, P	fluctuating and mean pressure
P_{eff}	effective mean static pressure, $P - \varrho\Omega^2 r^2/2$
P_k	production of k, $P_{ii}/2$
P_{ij}	stress production due to mean shear
r	distance from axis of rotation
Re	Reynolds number, $U_\tau h/\nu$
Re_{m}	Reynolds number, $ReU_{\mathrm{m}}/U_\tau = U_{\mathrm{m}}h/\nu$
Ro	rotation number, $2h|\Omega|/U_\tau$
Ro_{m}	rotation number, $RoU_\tau/U_{\mathrm{m}} = 2h|\Omega|/U_{\mathrm{m}}$
S	vorticity ratio, $-2\Omega/(\mathrm{d}U/\mathrm{d}y)$
u_i, U_i	fluctuating and mean velocity components in x_i direction
u, v, w	fluctuating velocity components

U	streamwise mean velocity
U_{m}	bulk mean velocity
U_{τ}	global wall friction velocity, $\sqrt{-h\,\mathrm{d}(P_{\mathrm{eff}}/\varrho)\,/\mathrm{d}x}$
x_i	Cartesian coordinate in i-direction
Δx_i	coordinate increment in x_i-direction
x, y, z	streamwise, wall-normal and spanwise coordinates
δ_{ij}	Kronecker delta
ε	dissipation rate of k, $\varepsilon_{ii}/2$
ε_{ij}	dissipation rate
ϵ_{ijk}	Levi-Civita symbol
η	Kolmogorov's microscale for length, $\left(\nu^3/\varepsilon\right)^{1/4}$
ν	kinematic viscosity
Π_{ij}	velocity pressure-gradient correlation, $\Phi_{ij} + D_{ij}^{P}$
ϱ	density
τ	total shear stress
Φ_{ij}	pressure-strain correlation
Ω	angular velocity
$\overline{(\,)}$	Reynolds-averaged value
$(\,)^+$	inner variable
$(\,)_{\mathrm{m}}$	value based on U_{m}
$(\,)_{\mathrm{max}}$	value corresponding to maximum mean velocity
$(\,)_0$	value corresponding to $\overline{uv} = 0$
$(\,)_{\mathrm{s}}$	suction side value
$(\,)_{\mathrm{p}}$	pressure side value
$(\,)_{\mathrm{w}}$	wall value

1. Background and Motivation

Turbulent flows in rotating channels are of obvious engineering interest with applications associated with rotating devices such as turbines, pumps, compressors and cyclone separators. It is well known that the Coriolis body-force arising from the imposed system rotation has important implications, not only on the mean flow pattern, but also on the turbulence structure. The Coriolis force field may, for example, enhance the turbulence along the pressure side of a turbine blade and correspondingly reduce the turbulent activity along the suction side. The latter suppression of the turbulent agitation may eventually lead to laminarization of the flow, thereby promoting the tendency of the boundary layer to separate from the surface of the blade.

Motivated by the crucial role played by the Coriolis force in fluid machinery, a number of experimental investigations on rotating channel flow have been carried out during some 25 years, of which the experiments conducted by Johnston, Halleen & Lezius (1972) and Johnston (1973) are among the pioneering ones. Very recently Nakabayashi *et al.* (1993) made available mean velocity data from a similar study. However, laboratory investigations of turbulence in rotating systems are more difficult to accomplish than most other experiments, simply because the flow apparatus has to be mounted on a rotating turntable. Accordingly, complete and accurate turbulence statistics of rotating channel flow have not been reported so far.

While experiments in rotating systems are complicated to perform, laboratory studies of fully-developed channel flows are even more difficult to accomplish in a rotating environment. Fully-developed flows are, on the other hand, particularly attractive objects for large-eddy simulations (LES) and direct numerical simulations (DNS). The LES approach to rotating channel flow was adopted by Kim (1983), Miyake & Kajishima (1986), and more recently by Tafti & Vanka (1991) and Squires & Piomelli (1993), whereas Kristoffersen & Andersson (1993) undertook a direct simulation of fully developed rotating plane Poiseuille flow.

The objective of this paper is to present statistical data of the numerically simulated flow fields of Kristoffersen & Andersson (1993) which are intended to be helpful in the development, testing and validation of turbulence closure models. While complete budgets of the non-zero Reynolds stress components at a fairly low rotation rate were presented by Kristoffersen *et al.* (1990), corresponding budget data at intermediate and high rotation are now made available.

2. Model Problem and Governing Equations

Consider fully developed and turbulent pressure-driven (Poiseuille) flow between two infinite parallel walls rotating in orthogonal mode with constant angular velocity Ω about the z-axis, as indicated in Fig. 1. The resulting flow, which is assumed incompressible, is characterized by two independent dimensionless parameters, namely the *Reynolds number* $Re = U_\tau h/\nu$ and the *rotation number* $Ro = 2h|\Omega|/U_\tau$, where $2h$ is the wall distance and the *global* wall friction velocity U_τ is defined in terms of the imposed mean pressure gradient. These parameters may alternatively be expressed in terms of the bulk mean velocity U_m, i.e. $Re_\mathrm{m} = U_\mathrm{m}h/\nu$ and $Ro_\mathrm{m} = 2h|\Omega|/U_\mathrm{m}$.

It may be interesting to notice that the two independent parameters Re and Ro can be interpreted as the ratios of the geometrical length scale h to the viscous length scale ν/U_τ and to half of the Coriolis length scale U_τ/Ω, respectively. Moreover, following Kitoh & Nakabayashi (1992), the ratio of the viscous length scale to the Coriolis length scale, i.e. $\Omega\nu/U_\tau^2$ or $Ro/2Re$, becomes an essential parameter for wall-bounded flows subject to system rotation.

Since the mean flow under consideration is fully developed in the streamwise direction, the streamlines are parallel with the walls and the turbulence field may be treated as being homogeneous in xz-planes. After decomposing the instantaneous flow variables into mean (U_i, P) and fluctuating (u_i, p) parts, the Reynolds-averaged Navier-Stokes equations governing the transport of mean momentum reduce to an ordinary differential equation in y, which can be integrated once to give a linear distribution of the total (i.e. viscous and turbulent) kinematic mean shear stress

$$\frac{\tau}{\varrho} = \nu\frac{\mathrm{d}U}{\mathrm{d}y} - \overline{uv} = -U_\tau^2\frac{y}{h} + const \tag{1}$$

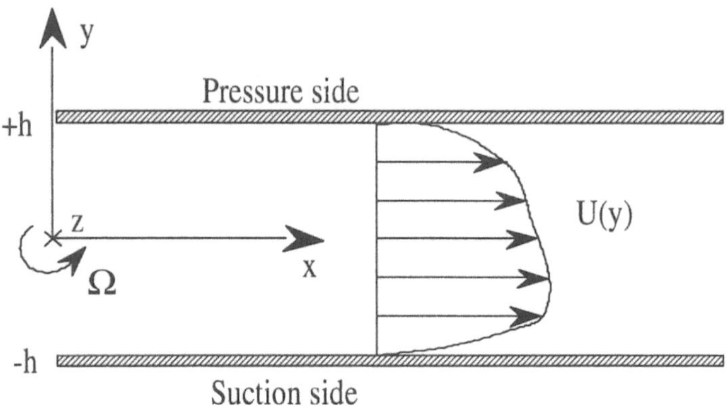

Fig. 1. Flow configuration and coordinate system. The mean velocity profile $U(y)$ is indicated for clockwise rotation $\Omega < 0$.

where the constant of integration depends on the imposed rotation and vanishes if $\Omega = 0$.

Following Launder *et al.* (1987), the transport equations for the individual Reynolds stress components can be expressed in Cartesian tensor notation for a rotating frame-of-reference as:

$$\frac{\mathsf{D}\overline{u_i u_j}}{\mathsf{D}t} = P_{ij} + G_{ij} + D_{ij}^{\nu} + D_{ij}^{T} + D_{ij}^{P} + \Phi_{ij} - \varepsilon_{ij} \tag{2}$$

where the different contributions on the right-hand side are defined as:

$$P_{ij} = -\left(\overline{u_i u_k}\frac{\partial U_j}{\partial x_k} + \overline{u_j u_k}\frac{\partial U_i}{\partial x_k}\right) \tag{3}$$

$$G_{ij} = -2\Omega\delta_{3k}\left(\overline{u_j u_{\mathrm{m}}}\epsilon_{ikm} + \overline{u_i u_{\mathrm{m}}}\epsilon_{jkm}\right) \tag{4}$$

$$D_{ij}^{\nu} = \nu\left(\frac{\partial^2 \overline{u_i u_j}}{\partial x_k \partial x_k}\right) \tag{5}$$

$$D_{ij}^{T} = -\frac{\partial}{\partial x_k}\left(\overline{u_i u_j u_k}\right) \tag{6}$$

$$D_{ij}^{P} = -\frac{1}{\varrho}\frac{\partial}{\partial x_k}\left(\overline{p u_i}\delta_{jk} + \overline{p u_j}\delta_{ik}\right) \tag{7}$$

$$\Phi_{ij} = \overline{\frac{p}{\varrho}\left(\frac{\partial u_i}{\partial x_j} + \frac{\partial u_j}{\partial x_i}\right)} \tag{8}$$

$$\varepsilon_{ij} = 2\nu\left(\overline{\frac{\partial u_i}{\partial x_k}\frac{\partial u_j}{\partial x_k}}\right) \tag{9}$$

The various terms in Eqs. (3) - (9) represent different physical mechanisms tending to change $\overline{u_i u_j}$, namely production due to mean shear (P_{ij}) and rotation (G_{ij}), viscous diffusion (D_{ij}^{ν}) and turbulent diffusion associated with velocity (D_{ij}^{T}) and pressure (D_{ij}^{P}) fluctuations, pressure-strain processes (Φ_{ij}), and viscous dissipation (ε_{ij}).

Due to the unidirectional mean flow and the assumed homogeneity in planes parallel with the walls, the Reynolds stress equations (2) reduce to a coupled set of 4 ordinary differential equations for the non-zero stress components $\overline{u^2}$, $\overline{v^2}$, $\overline{w^2}$ and $\overline{uv}$. Interestingly, some of the generation and diffusion terms vanish under these assumptions; see Table 1.

Table 1. Selected budget terms for unidirectional mean flow $U = U(y)$.

	i=1, j=1	i=2, j=2	i=3, j=3	i=1, j=2
P_{ij}	$-2\overline{uv}\,\mathrm{d}U/\mathrm{d}y$	0	0	$-\overline{v^2}\,\mathrm{d}U/\mathrm{d}y$
G_{ij}	$4\Omega\,\overline{uv}$	$-4\Omega\,\overline{uv}$	0	$-2\Omega\left(\overline{u^2}-\overline{v^2}\right)$
D_{ij}^{P}	0	$-2\varrho^{-1}\,\mathrm{d}\left(\overline{pv}\right)/\mathrm{d}y$	0	$-\varrho^{-1}\,\mathrm{d}\left(\overline{pu}\right)/\mathrm{d}y$

3. Numerically Generated Rotating Channel Flow

The three-dimensional time-dependent Navier-Stokes equations for incompressible flow in a rotating Cartesian frame-of-reference were solved numerically by Kristoffersen & Andersson (1993). The Reynolds number Re was prescribed as 194, whereas Re_m was about 2900, which is close to the Reynolds number of the accurate reference simulation of non-rotating channel flow by Kim *et al.* (1987). Here, however, simulations were carried out also for six non-zero rotation numbers Ro_m in the range from 0.01 to 0.50; see Table 2.

Table 2. Some characteristics of the rotating channel flow.

Ro_m	$U_{\tau p}/U_\tau$	$U_{\tau s}/U_\tau$	η_p^+	η_s^+	$-y_{max}/h$	$-y_0/h$
0	1.002	1.004	1.50	1.48	-0.010	0.002
0.01	1.023	0.970	1.44	1.55	-0.004	0.061
0.05	1.140	0.805	1.27	1.96	0.094	0.248
0.10	1.185	0.760	1.22	2.24	0.286	0.421
0.15	1.213	0.754	1.18	2.21	0.365	0.440
0.20	1.217	0.707	1.16	2.74	0.422	0.496
0.50	1.207	0.679	1.07	3.84	0.429	0.539

The computer code used for this purpose was an adapted version of the second-order accurate finite-difference code ECCLES (see Gavrilakis *et al.* (1986)), which was run without any sub-grid-scale model. The computational domain with dimensions $4\pi h \times 2h \times 2\pi h$ was divided into $128 \times 128 \times 128$ computational cells to assure that the grid was sufficiently fine to resolve all essential scales of the turbulence.

Because of the possible existence of large-scale vortical structures induced by the Coriolis force, the hypothesized homogeneity in xz-planes may be questionable. Johnston *et al.* (1972) did, in fact, observe a spanwise array of longitudinal roll-cells, which they interpreted as counter-rotating Taylor-Görtler vortices analogous to those arising from the centrifugal instability mechanism. Similar flow structures were detected in the large-eddy simulations of Kim (1983), Miyake & Kajishima (1986), and Tafti & Vanka (1991), as well as in the direct simulations of Kristoffersen & Andersson (1993). Interestingly, however, a persistent roll-cell pattern was neither observed in the laboratory apparatus nor in the numerically simulated flow fields, except for the particular case $Ro_m = 0.15$ considered by Kristoffersen & Andersson. Obviously, these vortical flow structures induce a spanwise variation of the flow variables, but the unsteady roll-cells can unfortunately not be isolated from the underlying turbulence. The various statistical correlations presented in the following section are therefore obtained by averaging over xz-planes and sampling in time. The sampling periods, over which statistics were computed, were typically in the range from $3h/U_\tau$ to $9h/U_\tau$.

4. Presentation and Discussion of Results

Statistics of the non-rotating case compared reasonably well with the most reliable experimental data available (see Andersson & Kristoffersen 1992). Some primary statistics of the rotating flow simulations (e.g. mean velocity profiles, Reynolds stress distributions and two-point velocity correlations) have been published elsewhere (Kristoffersen & Andersson 1993), whereas statistics of particular relevance for turbulence modelling are presented herein. It should be emphasized that the channel was rotated with *negative* angular velocity ($\Omega < 0$) so that the two channel walls at $y = -h$ and $y = +h$ became the suction and pressure sides, respectively. In accordance with the experimental observations of Johnston *et al.* (1972), the computed mean velocity profiles $U(y)$ shown in Fig. 2 became increasingly asymmetric with increasing rotation and ultimately exhibited an appreciable region with slope $\mathrm{d}U/\mathrm{d}y \approx 2\Omega$ in the centre of the channel. Due to the gradually increasing asymmetry of the flow field due to rotation, the wall-adjacent layer tends to become thinner at the pressure side while it thickens at the suction side. Correspondingly, Kolmogorov's microscale for length becomes asymmetric due to the imposed rotation. In global wall units, for instance, the Kolmogorov length scale becomes

$$\eta^+ \equiv \frac{\eta U_\tau}{\nu} = Re^{1/4}\left(\frac{\varepsilon h}{U_\tau^3}\right)^{-1/4} \tag{10}$$

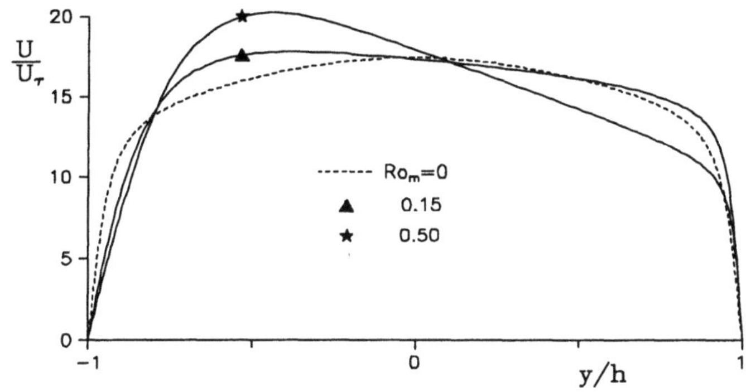

Fig. 2. Mean streamwise velocity profiles for some different rotation numbers.

where the local dissipation rate ε varies with $Ro_{\rm m}$. An estimate of η^+ in the near-wall regions can be obtained from Eq. (10) with actual values of ε at the respective channel walls. The resulting η^+-values, as given in Table 2, can thus be compared with the minimum grid spacing $\triangle y^+ \approx 0.5$ and the average grid width $(\triangle x^+ \triangle y^+ \triangle z^+)^{1/3} \approx 4.6$ adjacent to the walls. Since the non-uniform distribution of grid points is symmetrical about $y = 0$, the actual grid resolution improves at $y = -h$ and deteriorates at $y = +h$ with increasing angular velocity.

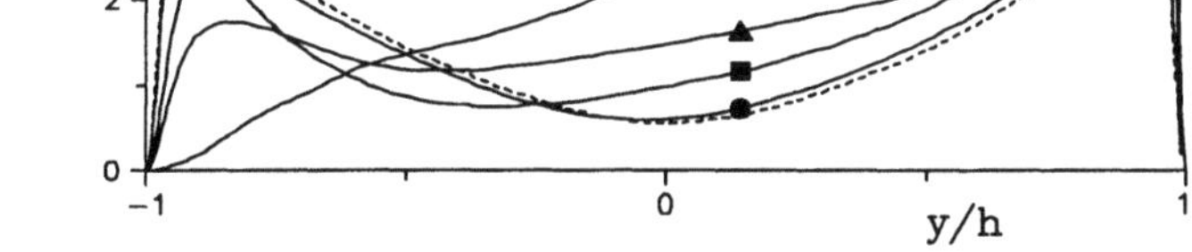

Fig. 3. Profiles of mean turbulent kinetic energy across the channel. The dotted line denotes non-rotating flow.

The variation of the mean turbulent kinetic energy $k = \overline{u_i u_i}/2$ and the structural parameter $\overline{uv}/k$ across the channel are shown for different rotation numbers in Figs. 3 and 4, respectively. At the lowest non-zero rotation rate, $Ro_{\rm m} = 0.01$, the profiles are only weakly modified by the system rotation, and opposite effects are observed

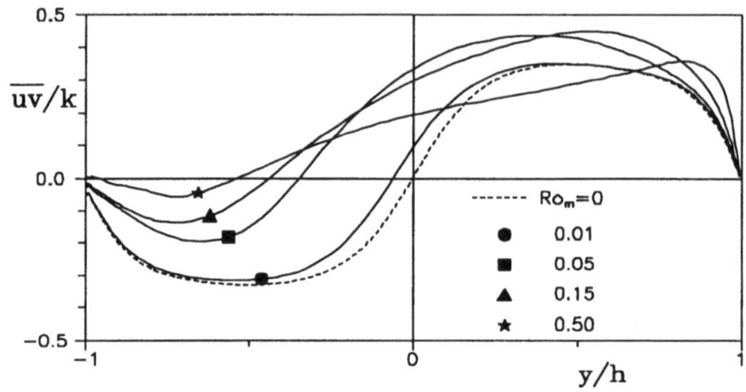

Fig. 4. Profiles of the structural parameter $\overline{uv}/k$ across the channel. The dotted line denotes non-rotating flow.

along the two sides of the channel. At this low rotation rate, the rotational stress generation G_{ij} in Eq. (4) is negligible compared to the production P_{ij} due to mean shear in Eq. (3) except for the wall-normal component $i = j = 2$. Since $P_{22} = 0$, the change-of-sign of $G_{22} = -4\Omega\overline{uv}$ from the pressure side to the suction side makes the rotational effect on the budgets antisymmetric about the centre. More specifically, the negative rotation rate makes G_{22} positive (negative) near the pressure (suction) side and the Coriolis contribution therefore tends to increase (reduce) the transverse stress component $\overline{v^2}$. Since $G_{11} = -G_{22}$ opposite effects of rotation may be anticipated on the streamwise component $\overline{u^2}$, and, consequently, $G_{ii} = 0$ and no Coriolis influence on the turbulence energy k should be expected. Interestingly, however, the total production of the streamwise component

$$P_{11} + G_{11} = (1 + S)P_{11} \text{ where } S = \frac{G_{11}}{P_{11}} = -\frac{2\Omega}{\mathrm{d}U/\mathrm{d}y} \tag{11}$$

is dominated by the changes in the mean shear production $P_{11} = -2\overline{uv}\mathrm{d}U/\mathrm{d}y$ resulting indirectly from the rotational-induced changes in $\overline{v^2}$ via the primary shear stress production $P_{12} = -\overline{v^2}\mathrm{d}U/\mathrm{d}y$. In conclusion, the overall effect of rotation is completely dominated by the Coriolis production of fluctuations normal to the walls, i.e. in the direction of inhomogeneity.

With increasing rotation rates, the turbulence level is gradually suppressed near the suction side, whereas the turbulent agitation is enhanced at the pressure side for Ro_m up to 0.10 and then apparently levels off until a distinct drop is observed for the highest rotation number $Ro_m = 0.50$. This suggests that other mechanisms come into play, which can be revealed only by a closer examination of the various contributions to the transport equation (2) for the individual second-moments.

Complete computed near-wall budgets for the non-zero Reynolds stress components are shown in Fig. 5 for $Ro_m = 0.15$ and in Fig. 6 for $Ro_m = 0.50$. The various budget terms defined in Eqs. (3)-(9) have been computed individually, except the dissipation rate ε_{12} in the shear-stress budget which has been obtained by difference. The computed terms have been normalized by U_τ^3/h, and the vertical scales in Figs. 5 and 6 are therefore different for the two channel halves in order to make the suction side data more visible. The separation between the tick marks on the abscissas (i.e. $\triangle y = 0.05h$) corresponds to $\triangle y^+ = \triangle y U_\tau/\nu = 9.7$ in global wall units. It is readily observed that the rotational stress generation G_{22} still plays a most significant role in the $\overline{v^2}$ budgets, except in the vicinity of the suction side in Fig. 6. On the other hand, G_{11} is practically negligible in the near-wall regions for Ro_m = 0.15 and remains small even at the highest rotation rate 0.50. Further from the pressure side, however, the dimensionless parameter S attains appreciable negative values, as shown in Fig. 7. This suggests a reduction in the total production $P_{11} + G_{11}$ in Eq. (11), which obviously contributes to the observed saturation of the kinetic energy for $Ro_\mathrm{m} > 0.10$. In fact, the flow apparently strives to make S equal to -1 over a significant portion of the cross-section. Here, it should be recalled that the particular parameter value $S = -1$ separates the destabilizing region from restabilization, the latter which may occur for $S < -1$; see Johnston *et al.*(1972) and Tritton (1992) for details.

Interestingly, the production of $\overline{uv}$ associated with the Coriolis force becomes comparable in magnitude with P_{12} for the rotation number Ro_m = 0.15 considered in Fig. 5. In typical shear flow situations $\overline{u^2}$ exceeds $\overline{v^2}$ and the shear stress generation $G_{12} = -2\Omega\left(\overline{u^2} - \overline{v^2}\right)$ will thus be of the same sign across the entire channel, whereas the primary shear production P_{12} inevitably attains the opposite sign of $\mathrm{d}U/\mathrm{d}y$ and thus changes its sign across the channel. The rotational production G_{12} has therefore the same effect on the channel flow as G_{22} as long as $\overline{u^2} > \overline{v^2}$. However, the augmention of $\overline{v^2}$ and the saturation and eventual suppression of $\overline{u^2}$ on the unstable pressure side tend to reduce the conventional stress anisotropy. The profiles of $\overline{v^2}/\overline{u^2}$ in Fig. 8 show the increasing isotropization of the flow adjacent to the pressure side, whereas a reversal of the anisotropy occurs in an appreciable outer region when $Ro_\mathrm{m} \geqslant 0.10$. Isotropization of the normal stresses was also observed in the large-eddy simulation by Kim (1983) for Ro_m = 0.068, while Koyama & Ohuchi (1985) reported a significant anisotropy reversal $\left(\overline{v^2} > 2\overline{u^2}\right)$ in the developing boundary layer on the pressure side of their rotating flow facility. An obvious consequence of this anomalous stress anisotropy is a change-of-sign of G_{12} (see for example Fig. 6) and therefore a tendency of $\overline{uv}$ to decay, as can be observed from Fig. 4.

Adjacent to the suction side, the usual stress anisotropy increases with Ro_m and the rotational shear stress generation G_{12} outweighs the conventional production P_{12} for Ro_m = 0.15, thereby leaving essentially a balance between pressure transport D_{12}^P and pressure-strain interactions Φ_{12}. This balance is even more pronounced in the suction side budget of $\overline{v^2}$ in Fig. 6, in which pressure diffusion D_{22}^P continues to play an essential role also further away from the wall. Mansour *et al.* (1988) introduced a velocity pressure-gradient term defined as $\Pi_{ij} = \Phi_{ij} + D_{ij}^P$, and observed that the complexity of the near-wall variation is substantially emphasized when Π_{ij} is split into these two parts. According to Taylor-series expansions about the wall, Φ_{12} balances with D_{12}^P, which both attain non-vanishing values at the walls, whereas the

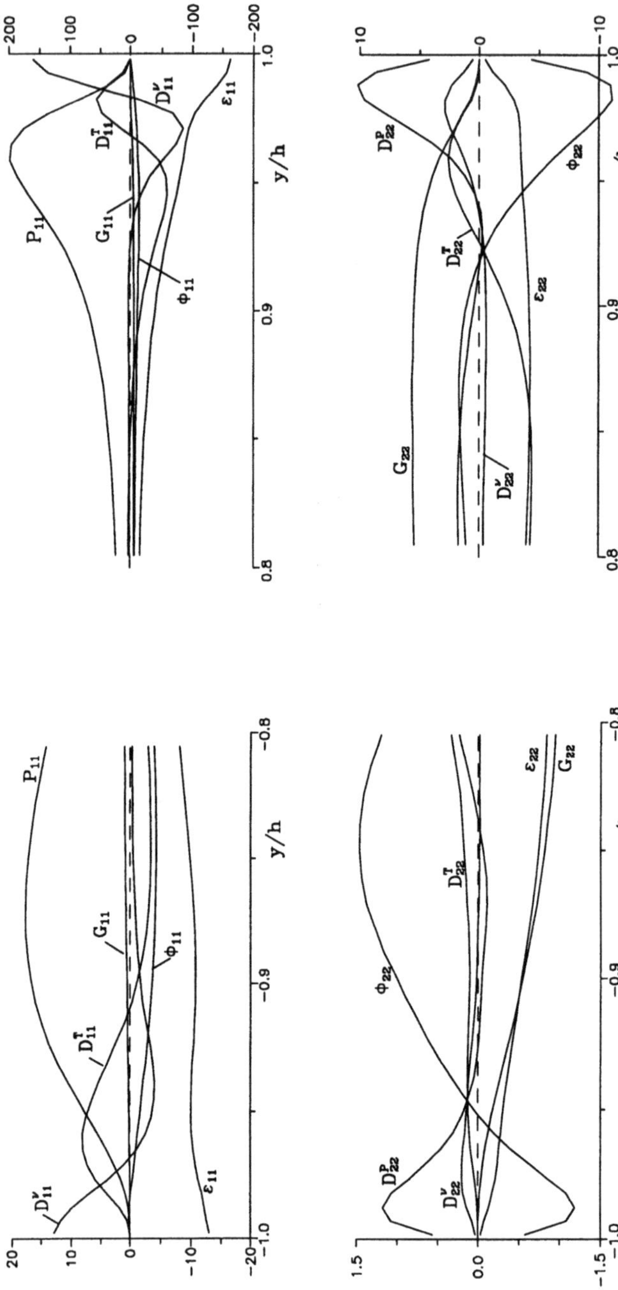

For caption see other side.

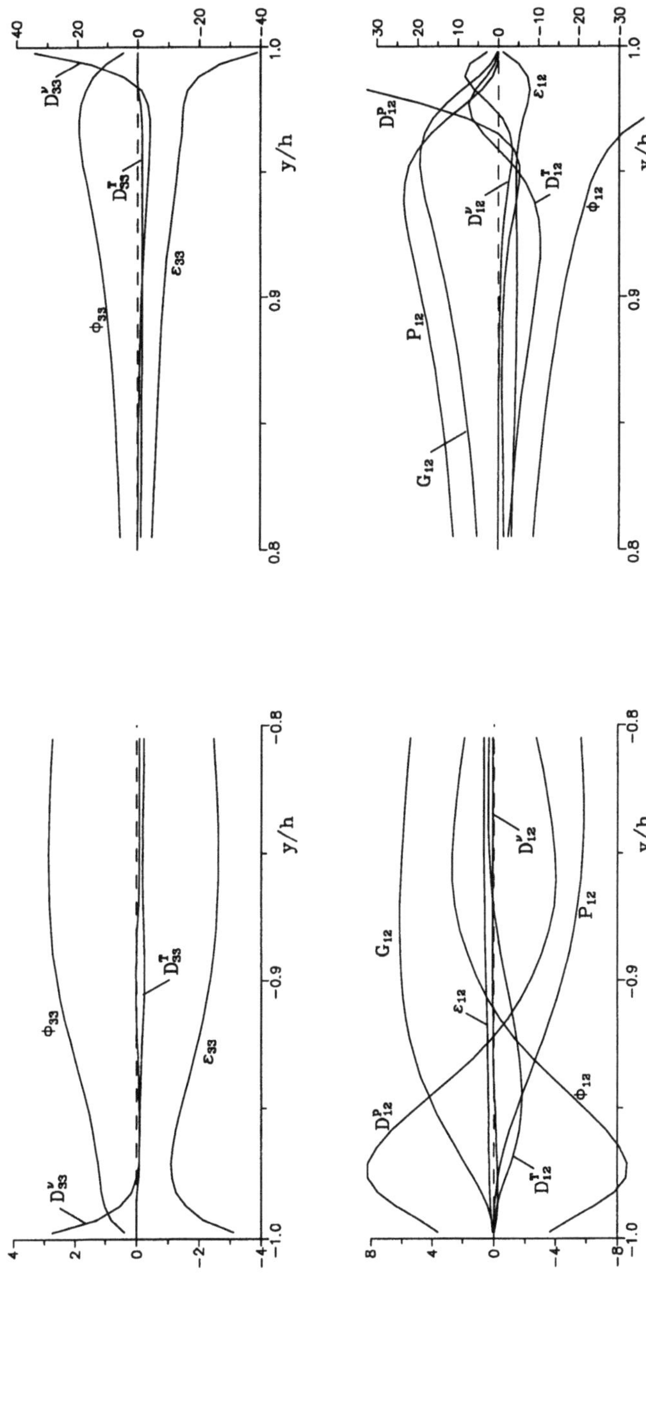

Fig. 5. Terms in the budgets of the non-zero Reynolds stress components in the near-wall regions for $Ro_{\mathrm{m}} = 0.15$.

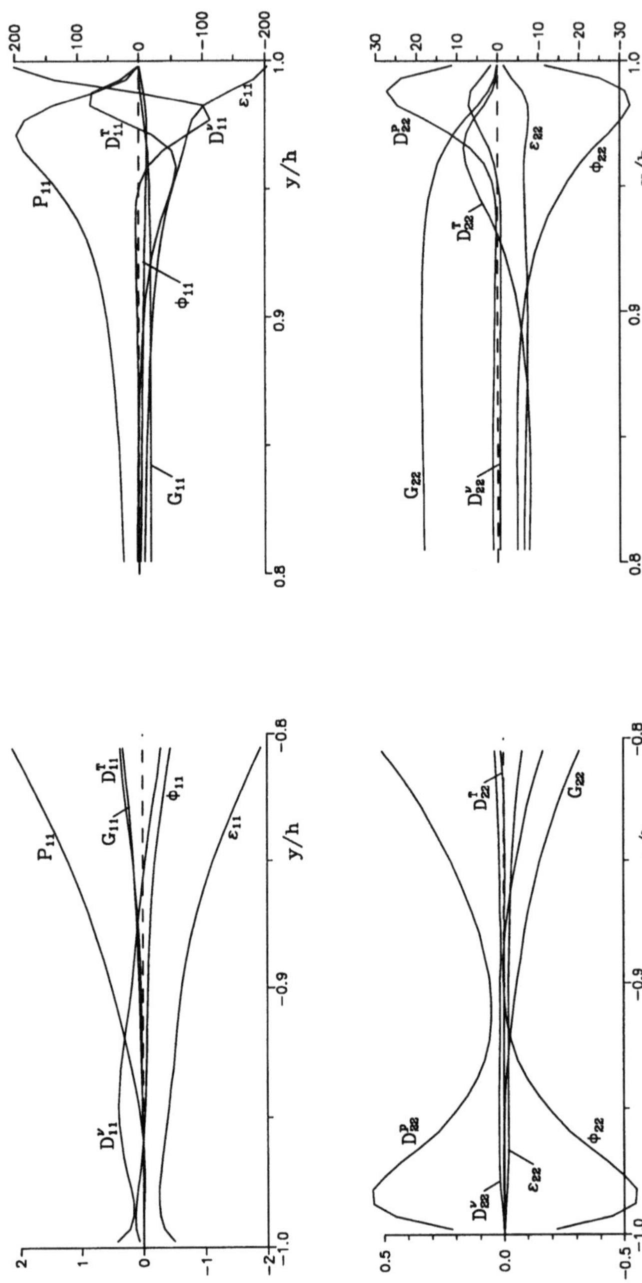

For caption see other side.

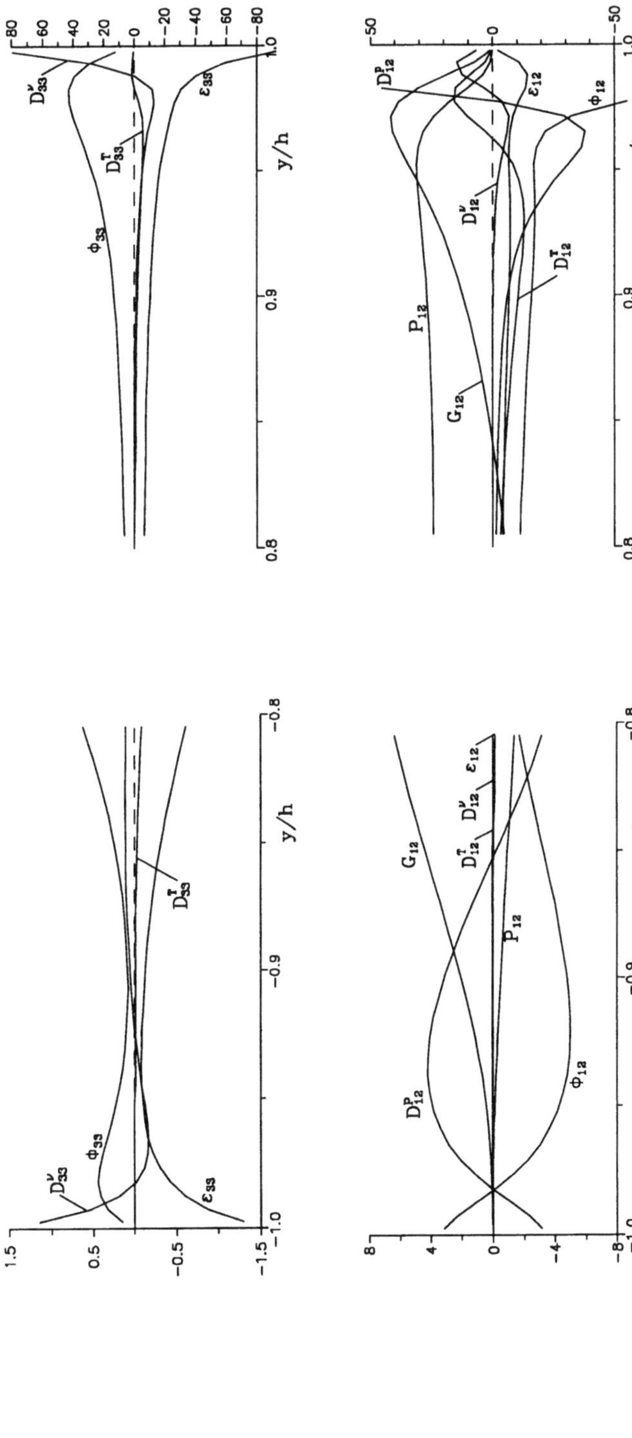

Fig. 6. Terms in the budgets of the non-zero Reynolds stress components in the near-wall regions for $Ro_m = 0.50$.

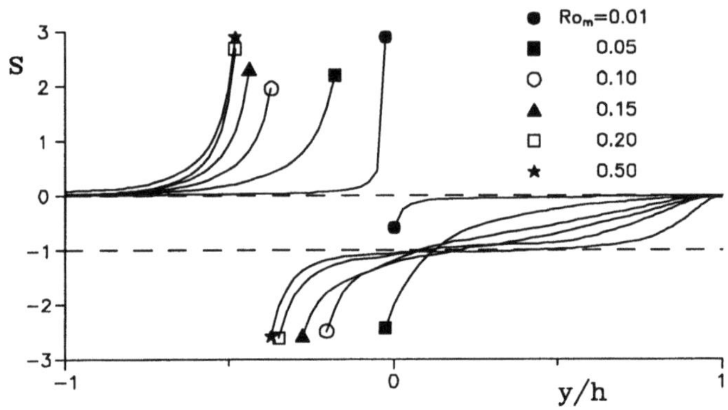

Fig. 7. Variation of $S = -2\Omega/(\mathrm{d}U/\mathrm{d}y)$ across the channel.

diagonal terms of Φ_{ij} tend to zero. In fact, very close to the pressure side the off-diagonal terms are out-of-scale in Figs. 5 and 6. A noticeable feature of the budgets at the suction side is the behaviour of the pressure diffusion D^P_{12}. Without rotation, the nondimensionalized D^P_{12} attains a value of about 10 at the wall (Andersson & Kristoffersen 1992), whereas corresponding wall-values of +3 and −3 were observed for rotation numbers of 0.15 and 0.50, respectively. These remarkable variations can be attributed to other than conventional mechanisms which come into play instead of stress production at high Ro_m, cf. Kristoffersen & Andersson (1993).

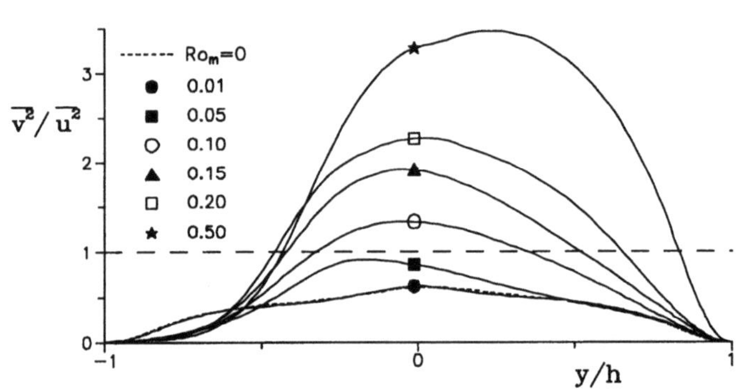

Fig. 8. Profiles of $\overline{v^2}/\overline{u^2}$ across the channel. The dotted line denotes non-rotating flow.

A number of engineering turbulence models involve the turbulent kinetic energy k and its dissipation rate ε as primary variables. Important modelling arguments, for example the well-established *wall-function approach*, (see, e.g. Launder & Spalding 1974), are often based on the assumption of local equilibrium, i.e. production and

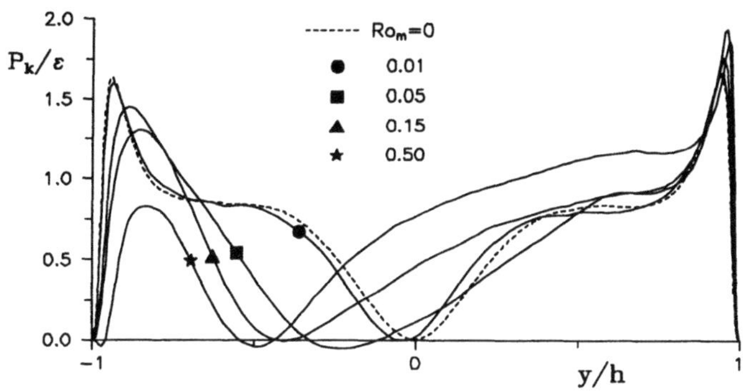

Fig. 9. Variation of P_k/ε across the channel. The dotted line denotes non-rotating flow.

dissipation of turbulent energy are in balance. It is therefore interesting to note in Fig. 9 that the assumption $P_k \approx \varepsilon$ holds reasonably well for all rotation numbers in the range $0.25 < y/h < 0.85$ on the pressure side. On the contrary, a plateau with $P_k \approx \varepsilon$ does not exist at the suction side for $Ro_m \geqslant 0.05$, thereby suggesting that the wall-function approach should be abandoned in rotating flow computations.

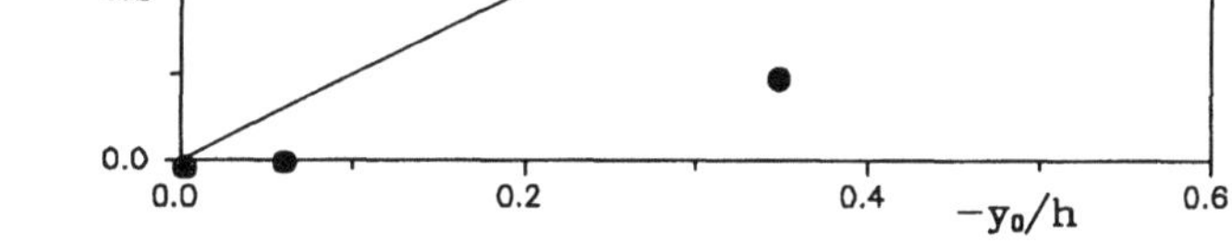

Fig. 10. Position y_{max} of maximum mean velocity versus position y_0 of zero shear stress $\overline{uv}$ for different rotation rates. The solid line denotes $y_{max} = y_0$.

Closer examination of the profiles displayed in Fig. 9 reveals localized areas with negative energy production, i.e. $P_k < 0$. The existence of such zones in certain flows with asymmetric mean velocity distribution was addressed by Eskinazi & Erian (1969) and Hinze (1970), and observed in rotating channel flow by Johnston *et al.*(1972). In the fully developed flow under consideration here, the phenomenon of negative energy production $P_k = -\overline{uv}\mathrm{d}U/\mathrm{d}y < 0$ arises when the position y_0

of zero turbulent shear stress no longer coincides with the point y_{max} of maximum mean velocity (or vanishing mean shear rate dU/dy). The separation of the points in Fig. 10 from the straight line $y_{max} = y_0$ signifies the existence of regions with negative production. Due to the imposed system rotation the positions of $\overline{uv} = 0$ and $dU/dy = 0$ shift monotonically towards the suction side as the rotation rate is being raised. Interestingly, however, the position of zero shear stress is displaced further than the position of maximum velocity, thereby forming localized zones with $P_k < 0$. Evidently, the presence of these zones of energy reversal imply that mean flow energy is extracted from the turbulence. In this connection it may be worthwhile to recall that the well-known inability of the $k-\varepsilon$ model to displace y_0 from y_{max} is an inherent shortcoming of Boussinesq-type eddy-viscosity models.

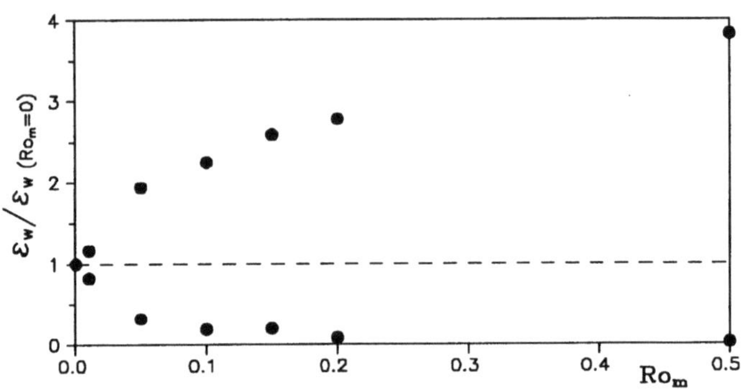

Fig. 11. Dissipation rates at the walls for different rotation numbers.

The relative importance of the various contributions to the Reynolds stress budgets are substantially altered by the imposed system rotation. Interestingly, however, the level of the dissipation rate of the turbulent kinetic energy at the walls seems to represent a useful measure of the order of the other processes in the immediate vicinity of the walls. Actually, ε_w balances the viscous diffusion of k towards the walls. The variation of ε_w with Ro_m, as displayed in Fig. 11, therefore effectively summarizes the noticeable changes in the near-wall budgets caused by the rotation.

Another, and more common, measure of the near-wall turbulence is the wall-friction τ_w. Based on the experimental data of Johnston *et al.* (1972), the best-fit formula for low Ro_m:

$$C_f/C_f(Ro_m = 0) = (1 \pm 1.75Ro_m)^2 \tag{12}$$

was proposed by Johnston (1973). Here, the positive and negative signs refer to the pressure and suction sides, respectively. More recently, Kitoh & Nakabayashi (1992) argued that $\Omega\nu/U_\tau^2$, i.e. $Ro/2Re$, would be a more appropriate scaling parameter since the wall-friction is supposed to depend not only on Ro but also on Re. In fact,

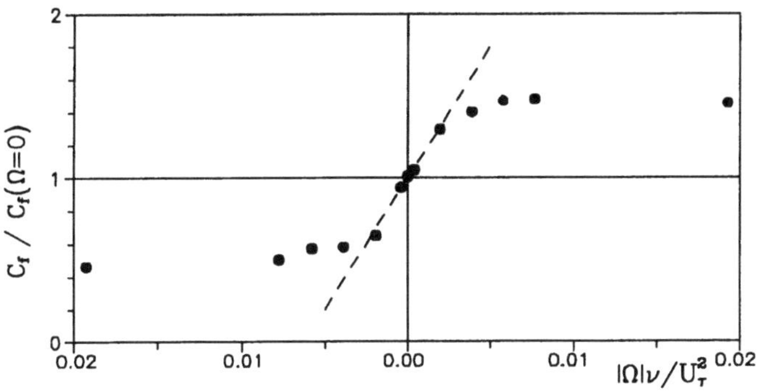

Fig. 12. Skin-friction coefficient for different rotation rates. The broken line represents the formulae in Eq. (13).

the experimental data of Nakabayashi *et al.* (1993) in the range $1850 \leqslant Re_m \leqslant 5000$ correlated with the formula:

$$C_f/C_f(\Omega = 0) = 1 \pm 160|\Omega|\nu/U_\tau^2 \tag{13}$$

It is therefore interesting to observe from Fig. 12 that the present DNS data are well represented by Eq. (13) for $|\Omega|\nu/U_\tau^2$ up to about 0.003, whereas the rotational effects are appreciably reduced as compared with (13) for higher rotation rates.

5. Concluding Remarks

Complete budget data for the individual non-zero Reynolds stress components have been provided for $Ro_m = 0.15$ and $Ro_m = 0.50$. The variation of several flow characteristics with Ro_m in the range from 0.01 to 0.50 has also been presented and compared with corresponding data for non-rotating channel flow. With increasing rotation rates, significant augmention and suppression of the turbulence were observed near the pressure and suction sides, respectively. The budget terms turned out to be even more affected by rotation than the Reynolds stresses themselves, and the dissipation rate of the turbulent kinetic energy at the walls become a representative measure of the magnitude of the dominant terms.

The presented turbulence statistics for the wide range of rotation numbers represent a useful reference for the assessment of turbulence closure models, and budget data for the particular case $Ro_m = 0.05$ have already been used for model validation by Kristoffersen *et al.* (1990) and Launder & Tselepidakis (1994). It has been inferred from the present results that the rotational stress generation in the $\overline{v^2}$-budget plays a crucial role for the variation of all the non-zero Reynolds stress components with Ro_m. From a turbulence modeller's point of view, it is therefore a challenge trying to reproduce the rather delicate balance between the various processes present in the $\overline{v^2}$-budget.

References

Andersson, H.I. & Kristoffersen, R. (1992): Statistics of numerically generated turbulence. *Acta Appl. Math.* **26**, 293–314.

Eskinazi, S. & Erian, F.F. (1969): Energy reversal in turbulent flows. *Phys. Fluids* **12**, 1988–1998.

Gavrilakis, S., Tsai, H.M., Voke, P.R. & Leslie, D.C. (1986): Large-eddy simulation of low Reynolds number channel flow by spectral and finite difference methods. In *Notes on Numerical Fluid Mechanics* (U. Schumann, R. Friedrich, eds), Vieweg, Vol. 15, pp. 105–118.

Hinze, J.O. (1970): Turbulent flow regions with shear stress and mean velocity gradient of opposite sign. *Appl. Sci. Res.* **22**, 163–175.

Johnston, J.P. (1973): The suppression of shear layer turbulence in rotating systems. *ASME J. Fluids Eng.* **95**, 229–236.

Johnston, J.P., Halleen, R.M. & Lezius, D.K. (1972): Effects of spanwise rotation on the structure of fully developed turbulent channel flow. *J. Fluid Mech.* **56**, 533–557.

Kim, J. (1983): The effect of rotation on turbulence structure. In *Proc. 4th Symposium on Turbulent Shear Flows*, Karlsruhe, pp. 6.14–6.19.

Kim, J., Moin, P. & Moser, R. (1987): Turbulence statistics in fully developed channel flow at low Reynolds number. *J. Fluid Mech.* **177**, 133–166.

Kitoh, O. & Nakabayashi, K. (1992): Analytical studies of two-dimensional channel turbulent flow subjected to Coriolis force (in Japanese). *Trans. JSME B* **58**, 2138–2142.

Koyama, H.S. & Ohuchi, M. (1985): Effects of Coriolis force on boundary layer development. In *Proc. 5th Symposium on Turbulent Shear Flows*, Ithaca, pp. 21.19–21.24.

Kristoffersen, R. & Andersson, H.I. (1993): Direct simulations of low Reynolds number turbulent flow in a rotating channel. *J. Fluid Mech.* **256**, 163–197.

Kristoffersen, R., Nilsen, P.J. & Andersson, H.I. (1990): Validation of Reynolds stress closures for rotating channel flows by means of direct numerical simulations. In *Engineering Turbulence Modelling and Measurements* (W. Rodi, E.N. Ganic, eds), Elsevier, pp. 55–64.

Launder, B.E. & Spalding, D.B. (1974): The numerical computation of turbulent flows. *Comp. Meth. Appl. Mech. Engng* **3**, 269–289.

Launder, B.E. & Tselepidakis, D.P. (1994): Application of a new second-moment closure to turbulent channel flow rotating in orthogonal mode. *Int. J. Heat Fluid Flow* **15**, 2–10.

Launder, B.E., Tselepidakis, D.P. & Younis, B.A. (1987): A second-moment closure study of rotating channel flow. *J. Fluid Mech.* **183**, 63–75.

Mansour, N.N., Kim, J. & Moin, P. (1988): Reynolds-stress and dissipation-rate budgets in a turbulent channel flow. *J. Fluid Mech.* **194**, 15–44.

Miyake, Y. & Kajishima, T. (1986): Numerical simulation of the effects of Coriolis force on the structure of turbulence. *Bull. JSME* **29**, 3341–3346 and 3347–3351.

Nakabayashi, K., Kitoh, O., Yamazoe, A. & Tsutsumi, Y. (1993): Velocity distributions of fully developed two-dimensional turbulent channel flow under effects of low-Reynolds number and Coriolis force (in Japanese). *Trans. JSME B* **59**, 389–396.

Squires, K.D. & Piomelli, U. (1993): Large-eddy simulation of rotating turbulence using the dynamic model. In *Proc. 9th Symposium on Turbulent Shear Flows*, Kyoto, pp. 17.3.1–17.3.6.

Tafti, D.K. & Vanka, S.P. (1991): A numerical study of the effects of spanwise rotation on turbulent channel flow. *Phys. Fluids A* **3**, 642–656.

Tritton, D.J. (1992): Stabilization and destabilization of turbulent shear flow in a rotating fluid. *J. Fluid Mech.* **241**, 503–523.

Dynamic Modeling of Rotating Turbulence

Kyle D. Squires[1] and Ugo Piomelli[2]

[1] Dept. of Mechanical Engineering, University of Vermont, Burlington, VT 05405, U.S.A.
[2] Dept. of Mechanical Engineering, University of Maryland, College Park, MD 20742, U.S.A.

Abstract

The dynamic subgrid-scale eddy viscosity model has been applied to the computation of two building-block rotating flows: initially-isotropic turbulence subjected to uniform system rotation, and turbulent channel flow with rotation about the spanwise axis. In each of these flows identical formulations of the model are able to account correctly for the effect of rotation on the subgrid scales, *e.g.*, inhibited energy transfer in rotating isotropic turbulence and the stabilizing/destabilizing effect of rotation on turbulent channel flow. LES predictions of rotating isotropic turbulence are in good agreement with the DNS results obtained by Mansour *et al.* (1991). First- and second-order statistics from the LES of rotating channel flow are also found to be in good agreement with DNS results.

1. Introduction and Background

Accurate prediction of turbulence in rotating reference frames is required for a wide variety of technologically important areas, such as the flow in turbomachines, or turbulence in the atmosphere. Thus, turbulence undergoing uniform system rotation has been the subject of a large number of theoretical, experimental, and computational studies (*e.g.*, Greenspan 1968, Bardina *et al.* 1985, Jacquin *et al.* 1990), that have demonstrated that the effect of system rotation on turbulence is subtle and remains exceedingly difficult to predict.

For turbulence that is initially isotropic, one of the principal effects of rotation is to inhibit energy transfer from large to small scales. Both experiments and direct numerical simulations (DNS) have shown that this leads to a reduction in turbulence dissipation and an associated decrease in the decay rate of turbulence energy (Veeravalli 1991, Mansour *et al.* 1991, 1992). Computations and experiments have also shown an increase in turbulence length scales along the rotation axis relative to those in non-rotating turbulence (Bardina *et al.* 1985, Jacquin *et al.* 1990). The presence of mean shear normal to the axis of rotation may have either a stabilizing or a destabilizing effect, depending on whether the angular velocity and mean shear have the same or opposite signs. In turbulent channel flow, for example, system rotation acts to both stabilize and destabilize the flow. Experiments (Johnston *et al.* 1972) and computations (Kim 1983, Tafti & Vanka 1991) have shown that on the unstable side Coriolis forces resulting from system rotation enhance turbulence-producing events, leading to an increase in turbulence levels. On the stable side Coriolis forces inhibit turbulence production and decrease turbulence levels. The increase in the component energies, however, is dependent on the rotation rate: at sufficiently high rotation rates streamwise fluctuations on the unstable channel wall are suppressed relative to the non-rotating case.

Prediction of these and other phenomena by K–ε models is difficult, since the effect of rotation does not appear *directly* in either of the transport equations for K or ε. As a result, various *ad hoc* modifications of the model have been proposed for computation of rotating flows. These modifications are typically the incorporation of additional terms which explicitly represent irrotational strains. However, Speziale *et al.* (1992) point out that in order to describe rotating turbulence accurately, some directional or two-point information needs to be incorporated into the model; an aspect difficult to incorporate into single point closures such as K–ε.

An alternative technique for the study of turbulent flows is the large-eddy simulation (LES) of the Navier-Stokes equations; in LES only the subgrid-scale motions are modeled. This permits the use of simpler parameterizations of the subgrid-scale (SGS) stresses. Nearly all previous LES computations of rotating turbulence utilized the Smagorinsky SGS model without any modification to account for system rotation (Bardina *et al.* 1985, Tafti & Vanka 1991). In rotating channel flow this may give rise to significant inaccuracies on the stable side, where the Smagorinsky model overestimates the SGS contribution to the shear stresses, and cannot predict a fully relaminarized state. The dynamic eddy-viscosity model developed by Germano *et al.* (1991), however, has been successfully applied to transitional (Germano *et al.* 1991) and relaminarizing (Esmaili & Piomelli 1993) flows, as well as to a number of turbulent flows (Germano *et al.* 1991, Moin *et al.* 1991). Its most attractive characteristic is that the model coefficient is calculated rather than specified *a priori*; this feature allows the SGS stresses to vanish in laminar flows and near solid walls, and also to respond more rapidly to the changes in the spectral content of the resolved scales.

The considerations made above regarding the inability of K–ε models to account for system rotation also apply to eddy-viscosity SGS stress models. Furthermore, Smith and Yakhot (1993) have shown that the short-time behavior of the Smagorinsky model is incorrect if the time scale of the external perturbation (in this case, the rotation) is smaller than the subgrid-scale eddy turnover time. In this respect, however, the dynamic model could be more accurate than models with a constant determined *a priori* because of the close coupling between the model coefficient and the spectrum of the resolved scales. The coefficient, for instance, is expected to decrease if the spectra of the resolved-scales decay rapidly as a result of rotation. Whether this behavior will take place on the correct time-scale is, however, unclear.

The primary objective of this study is to investigate the accuracy of the dynamic model in rotating turbulence. The main emphasis of this investigation will be placed on the response of the model coefficient to the changes in the large-scale velocity fluctuations (both in intensity and in spectral distribution) due to system rotation. The flow fields examined in this study are: (1) initially isotropic turbulence subjected to uniform rotation, and (2) fully-developed turbulent channel flow with axis of rotation in the spanwise direction.

2. Simulation Overview

Contained in §2.1 is an overview of LES and dynamic modeling. The reader is also referred to Germano *et al.* (1991) and Piomelli (1993) for a more complete discussion. The parameters for the simulations of rotating isotropic turbulence and fully-developed channel flow are presented in §2.2 and §2.3, respectively.

2.1 Filtered Equations of Motion

The governing equations for the filtered velocity and pressure in a rotating reference frame are

$$\frac{\partial \overline{u}_i}{\partial t} + \frac{\partial \overline{u}_i \overline{u}_j}{\partial x_j} = -\frac{1}{\varrho}\frac{\partial \overline{p}}{\partial x_i} + \nu \frac{\partial^2 \overline{u}_i}{\partial x_j \partial x_j} - \frac{\partial \tau_{ij}}{\partial x_j} + 2\epsilon_{ij3}\Omega \overline{u}_j, \tag{1}$$

and

$$\frac{\partial \overline{u}_j}{\partial x_j} = 0, \tag{2}$$

where ϵ_{ijk} is Levi-Civita's alternating tensor, $\overline{u}_i$ the filtered velocity, $\overline{p}$ the filtered pressure, and Ω the rotation rate. The effect of system rotation appears through the Coriolis term in (1). For the simulations performed in this study the axis of rotation is in the positive z (or x_3) direction.

The subgrid-scale (SGS) Reynolds stress, $\tau_{ij} = \overline{u_i u_j} - \overline{u}_i \overline{u}_j$, appearing in (2) is parameterized by an eddy viscosity model

$$\tau_{ij} - \frac{\delta_{ij}}{3}\tau_{kk} = -2\nu_{\mathrm{T}}\overline{S}_{ij} = -2C\overline{\Delta}^2 |\overline{S}|\overline{S}_{ij}\,, \tag{3}$$

in which δ_{ij} is the Kronecker delta and $|\overline{S}| = \sqrt{2\overline{S}_{ij}\overline{S}_{ij}}$ is the magnitude of the large-scale strain rate tensor

$$\overline{S}_{ij} = \frac{1}{2}\left(\frac{\partial \overline{u}_i}{\partial x_j} + \frac{\partial \overline{u}_j}{\partial x_i}\right) . \tag{4}$$

The trace of the subgrid-scale stress, τ_{kk}, is added to the pressure term. Closure of the SGS stress τ_{ij} is obtained through specification of the model coefficient C appearing in (3).

Following Germano *et al.* (1991), a second filter, the test filter, (denoted throughout this work by $\widehat{\cdot}$) is used to derive an expression for the model coefficient C. The key feature of this procedure is use of the identity

$$\mathcal{L}_{ij} = T_{ij} - \widehat{\tau_{ij}}\,, \tag{5}$$

which relates the resolved turbulent stresses, $\mathcal{L}_{ij} = \widehat{\overline{u}_i\overline{u}_j} - \widehat{\overline{u}}_i\widehat{\overline{u}}_j$, the subgrid-scale stresses τ_{ij}, and the subtest-scale stresses $T_{ij} = \widehat{\overline{u_iu_j}} - \widehat{\overline{u}}_i\widehat{\overline{u}}_j$. Assuming that the subtest stresses can be parameterized by an expression similar to that used to model τ_{ij} it is possible to obtain an expression for the model coefficient that reflects the local properties of the flow (Lilly 1992):

$$C = -\frac{1}{2}\frac{\mathcal{L}_{ij}M_{ij}}{M_{kl}M_{kl}}\,. \tag{6}$$

where

$$M_{ij} = \widehat{\overline{\Delta}}^2\,|\widehat{\overline{S}}|\widehat{\overline{S}}_{ij} - \overline{\Delta}^2\,\widehat{|\overline{S}|\overline{S}_{ij}}\,. \tag{7}$$

The filter widths at the subgrid and subtest levels are denoted by $\overline{\Delta}$ and $\widehat{\overline{\Delta}}$, respectively; for an eddy viscosity formulation the only parameter required for the model is the ratio $\widehat{\overline{\Delta}}/\overline{\Delta}$. The model coefficient given by (6) is, in principle, a function of the three spatial coordinates and time. In practice, however, local spatial averaging is required in order to eliminate the ill-conditioning which arises when the denominator in (6) is very small. In the present study the model coefficient is calculated through averaging of the numerator and denominator in (6) over homogeneous directions. Therefore, it is time-dependent in simulations of rotating isotropic turbulence, and a function of the normal coordinate and time in the simulations of fully-developed channel flow.

2.2 Parameters for Rotating Isotropic Turbulence

LES of rotating isotropic turbulence was performed using the pseudo-spectral method developed by Rogallo (1981). In these simulations 48^3 collocation points were used to resolve the large-scale field. LES predictions were compared to the (filtered) DNS results obtained by Mansour *et al.* (1991). Mansour *et al.* performed an extensive study of the effect of system rotation on isotropic turbulence using DNS resolutions of both 128^3 and 256^3. In their study a velocity field was allowed to develop under no rotation for long enough integration times such that energy transfer from large to small scales was well established (as evidenced by an equilibrium value of the velocity derivative skewness around -0.48) and the velocity field was undergoing a power-law decay. The developed velocity field was then used as the initial condition for simulations with non-zero rotation rates.

In this study LES results were compared to the DNS database of Mansour *et al.* obtained at the 128^3 resolution. To generate the initial conditions for the LES calculations a Fourier cutoff filter was applied to the developed DNS velocity field. Simulations using six non-zero Rotation numbers were compared to the DNS database. The Rotation number was defined as $Ro_\lambda = \Omega\lambda_g/u'$ where λ_g and u' are the Taylor microscale and rms velocity fluctuation in the initial (*i.e.*, developed) velocity field. The values of Ro_λ used in this study were 0, 1, 2, 3, 10, 20, and 200.

2.3 Parameters for Rotating Channel Flow

Simulations of fully-developed channel flow with rotation about the spanwise axis were performed using a Fourier-Chebychev collocation scheme (Zang & Hussaini 1986). Calculations were performed at a Reynolds number $Re_b = U_b(2\delta)/\nu = 5,730$ (based on the channel width, 2δ and bulk velocity, U_b). The Reynolds number based on friction velocity, u_τ, and channel half-width was $Re_\tau = u_\tau \delta/\nu \simeq 180$. Three non-zero rotation rates were examined: $Ro_b = \Omega(2\delta)/U_b = 0.069$, 0.144 and 0.21. The computational domain was $4\pi\delta \times 2\delta \times 4\pi\delta/3$ (in the streamwise, wall-normal and spanwise directions, respectively), and $48\times65\times64$ grid points were used. The initial conditions for the cases with non-zero rotation rates were obtained from equilibrium cases at $\Omega = 0$, and were in very good agreement with experimental and DNS data (Piomelli 1993). After rotation was applied the simulations were integrated forward in time to a new steady state, statistics being obtained by averaging over at least 4 dimensionless time units tu_τ/δ. Experimental measurements obtained by Johnston *et al.* (1972) exist, although at higher Reynolds numbers. A direct simulation of the intermediate ($Ro_b = 0.144$) case was carried out to compare with the LES results. The DNS used $128\times129\times128$ grid points, and was in very good agreement with the DNS results of Kristoffersen and Andersson (1993), which will be referred to as KA.

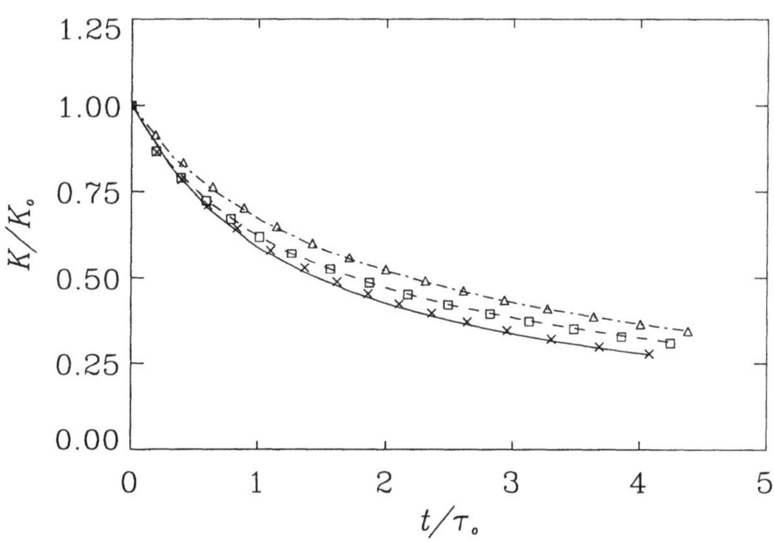

Fig. 1. Evolution of resolved-scale turbulence kinetic energy in rotating isotropic turbulence; K_0 and τ_0 are the values of the turbulence kinetic energy and eddy turnover time in the initial field, respectively.
DNS: × $Ro_\lambda = 1$, □ $Ro_\lambda = 3$, △ $Ro_\lambda = 20$;
LES: ——— $Ro_\lambda = 1$, - - - $Ro_\lambda = 3$, —·— $Ro_\lambda = 20$.

3. Results

3.1 Rotating Isotropic Turbulence

Comparison of the resolved-scale kinetic energy in rotating isotropic turbulence is shown in Fig. 1. This figure illustrates one of the characteristic effects of solid-body rotation on isotropic turbulence: the reduction in the non-linear transfer of energy from large to small scales, which results in reduced decay rates of turbulence kinetic energy with increasing Ro_λ. LES predictions are in excellent agreement with the DNS data of Mansour *et al.* (1991), the largest differences occurring at the smaller rotation rates.

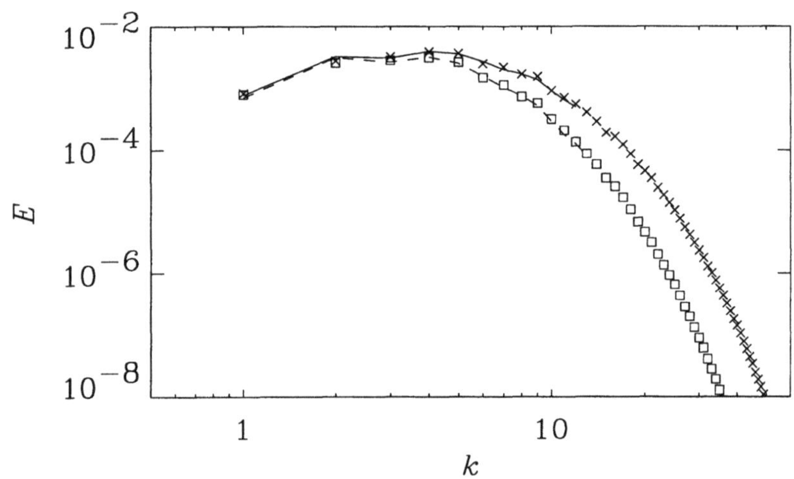

Fig. 2. Three-dimensional energy spectra of rotating isotropic turbulence, $Ro_\lambda = 3$. DNS: □ $t/\tau_0 = 1.26$, × $t/\tau_0 = 3.13$; LES: —— $t/\tau_0 = 1.26$, - - - $t/\tau_0 = 3.13$.

The three-dimensional energy spectra from simulations with $Ro_\lambda = 3$ are compared in Fig. 2 with the DNS data at two times in the evolution of the flow. The spectra shown in the figure are representative of those obtained at other Rotation numbers. It may be observed that not only is the total kinetic energy well predicted using the dynamic model, but the distribution of energy with wavenumber is also accurately captured. Agreement between LES results and DNS data is very good all the way to the cutoff wavenumber.

As was observed before, the effect of solid-body rotation is to reduce the non-linear cascade of energy from large to small scales, and, consequently, the decay rate of kinetic energy. One measure of energy transfer from large to small scales is the velocity derivative skewness. Since overall energy transfer is reduced by increasing rotation rates the derivative skewness should be decreased in rotating flows. Mansour *et al.* (1991) have shown that this quantity is indeed suppressed in rotating isotropic turbulence, and the time scale over which this occurs is of the same order as the rotation time scale, Ω. Shown in Fig. 3 is the time development of the model coefficient from simulations of rotating isotropic turbulence. Consistent with the effect of increasing rotation rate reducing the non-linear energy transfer the model coefficient is reduced for larger Ro_λ. More importantly, the time scale over which the reduction occurs is also of the same order as the rotation time scale. Thus, the dynamic model not only accounts for the reduction in downscale energy transfer but also reflects the proper time scale over which this occurs in rotating turbulence.

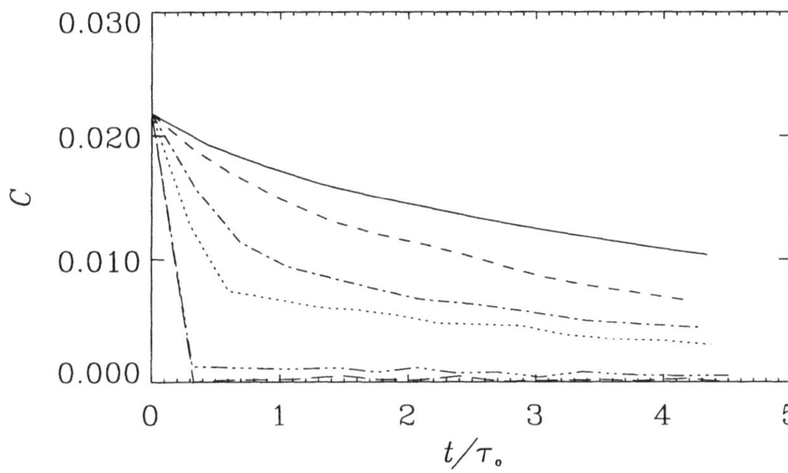

Fig. 3. Evolution of model coefficient in rotating isotropic turbulence. ——— $Ro_\lambda = 0$, - - - $Ro_\lambda = 1$, —·— $Ro_\lambda = 2$, ······· $Ro_\lambda = 3$, —···— $Ro_\lambda = 20$, ——— $Ro_\lambda = 200$.

It is important to stress that the reduction in the model coefficient with increasing Ro_λ observed in Fig. 3 is very difficult to capture using either traditional engineering turbulence models, (*e.g.*, K–ε), or SGS models that are not sensitive to the details of a particular flow, (for instance, the Smagorinsky model). Incorporation of the reduction of energy transfer due to system rotation into turbulence models is critical for obtaining accurate prediction of rotating flows. For example, Mansour *et al.* (1991) were able to account for this effect only after introducing two additional transport equations into a standard K–ε model.

In the simulations of rotating isotropic turbulence the numerator and denominator of (6) are averaged over the computational volume. On average, the model coefficient C is usually positive, reflecting the fact that there is an overall transfer of energy from large to small scales. The contribution to the numerator of (6) from each point

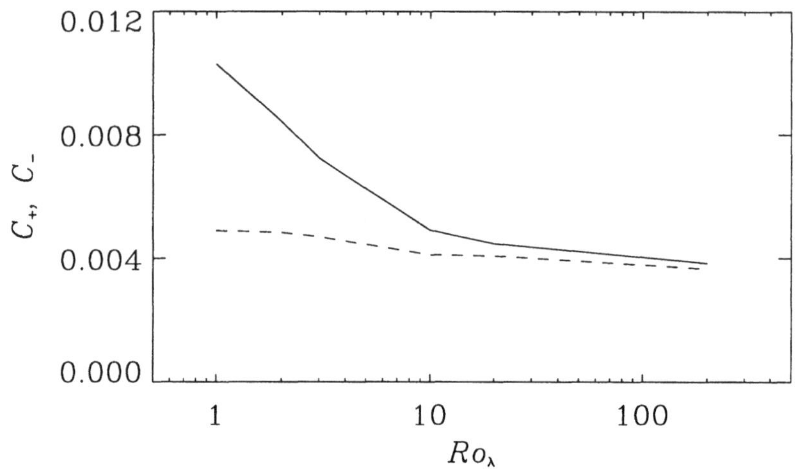

Fig. 4. Forward and backward contributions to the model coefficient in rotating isotropic turbulence at $t/\tau_0 = 4.5$. —— C_+, - - - C_-.

in the domain, however, can be positive or negative. The positive and negative contributions to C are defined as

$$C_+ = -\frac{1}{4}\frac{\langle -|\mathcal{L}_{ij}M_{ij}| + \mathcal{L}_{ij}M_{ij}\rangle}{\langle M_{kl}M_{kl}\rangle}, \qquad C_- = -\frac{1}{4}\frac{\langle |\mathcal{L}_{ij}M_{ij}| + \mathcal{L}_{ij}M_{ij}\rangle}{\langle M_{kl}M_{kl}\rangle}. \quad (8)$$

Negative values of the model coefficient correspond to regions in the flow where energy is transferred from small to large scales (backscatter). Shown in Fig. 4 are the positive and negative contributions to the model coefficient as a function of Ro_λ at the final time of the simulation. The results in Fig. 4 show that, as Ro_λ increases, the part of the model coefficient representing forward scatter, (*i.e.*, energy transfer from large to small scales) decreases, while the contribution to backscatter remains relatively unchanged by rotation. Thus, relative to forward scatter, backscatter becomes a more significant effect in the dynamic model as the rotation rate increases.

3.2 Rotating Channel Flow

As mentioned above, the interaction of rotation and a mean shear can be either destabilizing or stabilizing, depending on whether the rotation rate and mean shear have the same, or opposite sign. In a plane channel, therefore, turbulence is enhanced on one side (the lower side in this case), and damped on the opposite side. The turbulence intensities normalized by the average shear velocity (Fig. 5) show this behavior. The agreement between the LES and DNS data at $Ro_b = 0.144$ is satisfactory; the DNS of KA show that, on the unstable side, the peak rms streamwise fluctuation is maximum for $Ro_b \simeq 0.1$ and then decreases; the present calculations show the same trend. The agreement for the v and w intensities is also good.

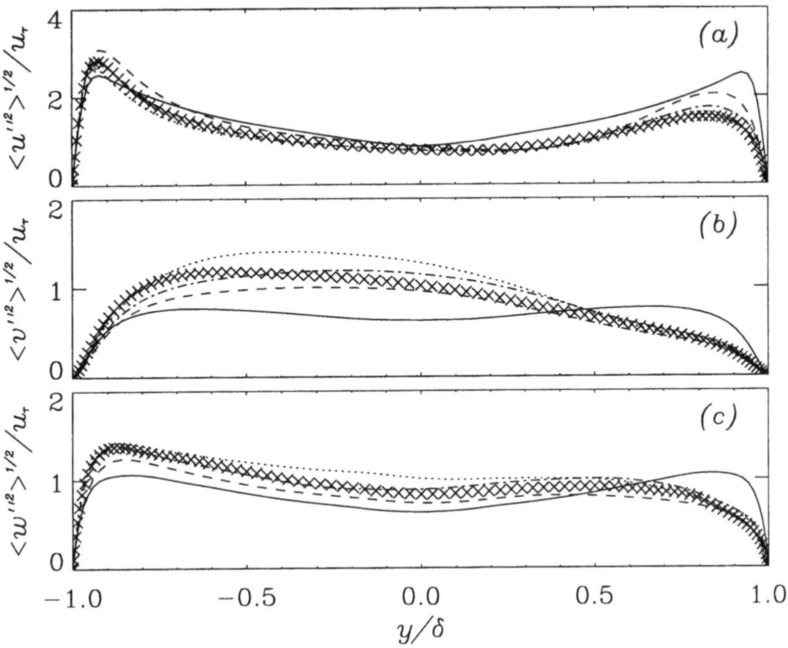

Fig. 5. Turbulence intensities in the rotating channel. DNS: × $Ro_b = 0.144$; LES: ——— $Ro_b = 0.0$, - - - $Ro_b = 0.069$, —·— $Ro_b = 0.144$, ········ $Ro_b = 0.210$. (**a**) u; (**b**) v; (**c**) w.

The mean velocity profiles follow similar trends, and the friction velocity (shown in Fig. 6, normalized by the average shear velocity) first increases with rotation rate, then reaches a saturation level; the agreement of DNS and LES data with experiments is good, especially since in the experiment the bulk velocity was obtained from the volume flow rate, which led to underestimation of the bulk velocity (and an overestimation of Ro_b) because of the presence of boundary layers on the sides of the channel (see also KA). On the stable side a more serious discrepancy between the experiments and all the numerical data is observed: in the experiment, full relaminarization was observed at high rotation rates, while both DNS and LES results show only a partial relaminarization, and a higher u_τ than measured experimentally. The reason for this difference is not clear, but one can conjecture that the relatively small aspect ratio of the experimental apparatus and the fact that the flow was not fully developed might have added a streamwise pressure gradient that could have increased the tendency of the flow towards relaminarization. The results obtained with the dynamic model are in much better agreement with the DNS results than those obtained by Tafti and Vanka (1991) with the Smagorinsky model, which tended to overdamp the fluctuations on the stable side, leading to excessively low wall stress even at the low rotation rate they examined.

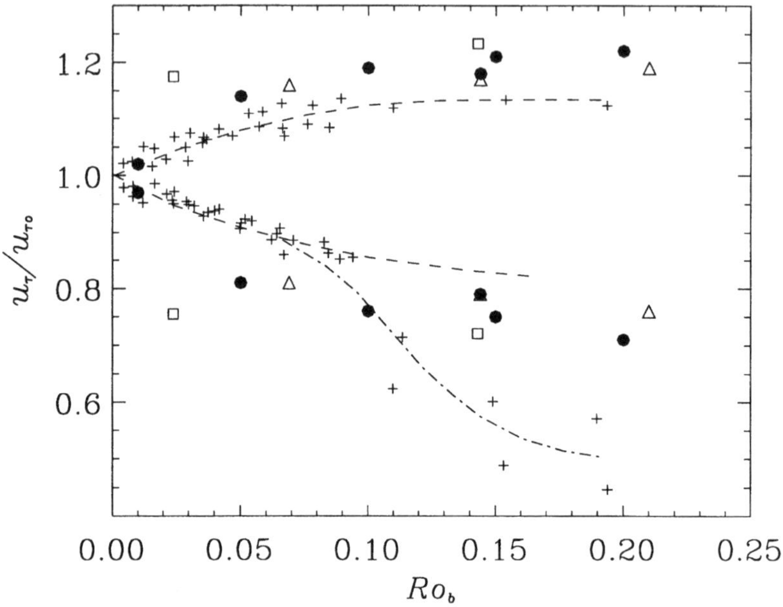

Fig. 6. Friction velocity on the two sides of the channel.
+ Experiments; • DNS (present calculation and KA); □ LES, Smagorinsky model, Tafti and Vanka (1991); △ LES, present calculation.

Compared with the rotation rates used in the isotropic case, the channel is subjected to very mild rotation; Ro_λ is varied by a factor of 3 only, and is at most $O(1)$; nonetheless, rotation affects the large-scale statistics significantly. The effect of the rotation rate on the model coefficient is, however, more subtle. In Fig. 7, for example, the model coefficient and its forward and backward parts are shown as a function of the normalized distance from the wall, $y_w = (1 - |y|)/\delta$. On the unstable side the coefficient increases with the rotation rate, reflecting the destabilization of the flow. On the stable side one can observe a significant decrease in the coefficient level, corresponding to the tendency of the flow towards relaminarization; however, no effect of the changing rotation rate can be discerned. No consistent trends in the forward and backward contributions to the coefficient are obvious.

On the unstable side, in which the mean shear is not very affected by the rotation rate, and production-dissipation mechanisms are dominated by the strain, these results indicate that the model responds appropriately to the increased turbulence fluctuations due to the shear/rotation interaction. To better understand the behavior of the model coefficient C on the stable side, in Fig. 8 it is plotted as a function of the gradient Richardson number (also known as the "Bradshaw number")

$$Ri = -\frac{2\Omega}{\mathrm{d}U/\mathrm{d}y}\left(1 - \frac{2\Omega}{\mathrm{d}U/\mathrm{d}y}\right),$$

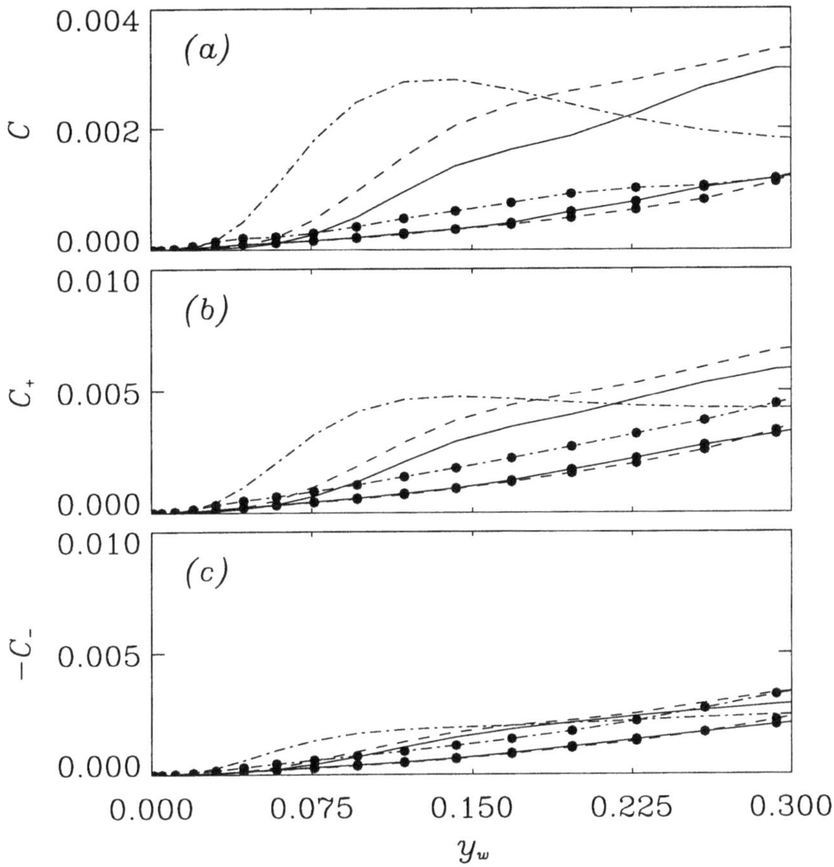

Fig. 7. Model coefficient and forward and backward contributions. ——— $Ro_b = 0.069$; - - - $Ro_b = 0.144$; —·— $Ro_b = 0.21$. Bullets denote the stable side of the channel. **(a)** C; **(b)** C_+; **(c)** C_-.

which is a measure of the relative effect of background vorticity 2Ω to mean vorticity $-\mathrm{d}U/\mathrm{d}y$. The interaction of rotation and mean shear is destabilizing for $Ri < 0$, the maximum destabilization occurring for $Ri = -1/4$. In all the cases examined here, the minimum Ri is close to $-1/4$. On the stable side, in which the background rotation plays an important role, C is clearly decreasing with rotation rate for a given Ri, a behavior consistent with the trend of the mean velocity profiles to become increasingly laminar-like.

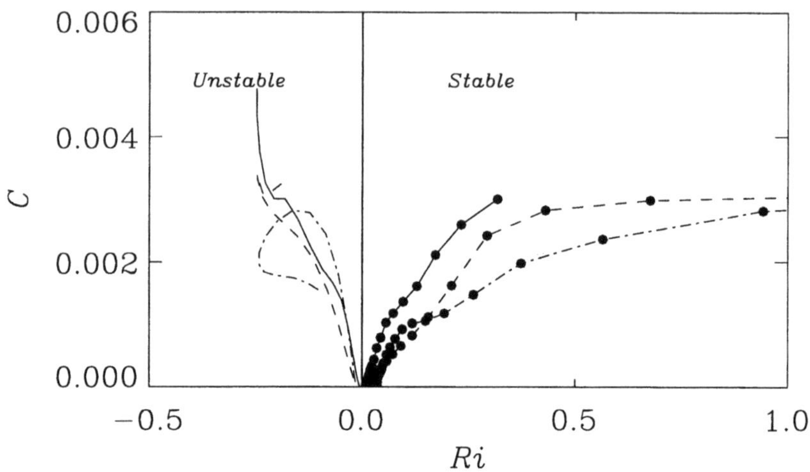

Fig. 8. Model coefficient. ——— $Ro_b = 0.069$; - - - $Ro_b = 0.144$; —·— $Ro_b = 0.21$. Bullets denote the stable side of the channel.

4. Summary

The dynamic subgrid-scale eddy viscosity model has been applied to the computation of rotating, isotropic turbulence and to rotating turbulent channel flow. A significant finding of the study is that the dynamic model is able to account correctly for the inhibited energy transfer from large to small scales in rotating isotropic turbulence, and for the stabilizing/destabilizing effect of rotation on turbulent channel flow. This is especially significant since identical formulations of the model were used in LES of two fairly different flows and the model demonstrated the correct behavior in both flows without any *ad hoc* adjustment of model parameters.

In rotating isotropic turbulence reduction of the model coefficient occurs on the rotation time scale, consistent with the time scale by which energy transfer has been observed to be reduced in the DNS study of Mansour *et al.* (1991). As the rotation rate is increased, backscatter plays an increasingly important role in determination of the model coefficient.

In rotating channel flow, SGS stresses on the stable and unstable sides of the channel were properly reduced or enhanced as the rotation rate increased. The mean-flow profiles, friction velocities and turbulence intensities were in good agreement with experimental measurements and DNS data. Reduced turbulent activity was observed on the stable side compared with the unstable one.

Acknowledgments

The authors gratefully acknowledge the financial support of the American Society for Engineering Education (KDS) and the Office of Naval Research (UP). Computer time was supplied by the Pittsburgh Supercomputing Center and the NASA Ames Research Center.

References

Bardina, J., Ferziger, J.H. & Rogallo, R.S. (1985): Effect of rotation on isotropic turbulence: computation and modeling. J. Fluid Mech. **154**, 321.

Esmaili, H. & Piomelli, U. (1993): Large-eddy simulation of relaminarizing sink flow boundary layers. In *Near-Wall Turbulent Flows*, ed. by R.M.C. So, C.G. Speziale and B.E. Launder, (Elsevier, Amsterdam), 287.

Germano, M., Piomelli, U., Moin, P. & Cabot, W.H. (1991): A dynamic subgrid-scale eddy viscosity model. Phys. Fluids A, **3**, 1760.

Greenspan, H.P. (1968): The Theory of Rotating Fluids, Cambridge University Press.

Jacquin, L, Leuchter, O., Cambon, C. & Mathieu, J. (1990): Homogeneous turbulence in the presence of rotation. J. Fluid Mech. **220**, 1.

Johnston, J.P., Halleen, R.M. & Lezius, D.K. (1972): Effects of spanwise rotation on the structure of two-dimensional fully developed turbulent channel flow. J. Fluid Mech. **56**, 533.

Kim, J. (1983): The effect of rotation on turbulence structure. in Proc. of the Fourth Int. Symp. on Turbulent Shear Flows, University of Karlsruhe, 6.14.

Kristoffersen, R. & Andersson, H.I. (1993): Direct simulation of low Reynolds number turbulent flow in a rotating channel. J. Fluid Mech. **256**, 163.

Lilly, D.K. (1992): A proposed modification of the Germano subgrid-scale closure method. Phys. Fluids A **4**, 633.

Mansour, N.N., Cambon, C. & Speziale, C.G. (1991): Single point modeling of initially isotropic turbulence under uniform rotation. in Annual Research Briefs – 1991, NASA-Stanford Center for Turbulence Research, 159.

Mansour, N.N., Cambon, C. & Speziale, C.G. (1992): Theoretical and computational study of rotating isotropic turbulence. in Studies in Turbulence, ed. by T.B. Gatski, S. Sarkar, and C.G. Speziale, (Springer-Verlag).

Moin, P., Squires, K.D., Cabot, W.H. & Lee, S. (1991): A dynamic subgrid-scale model for compressible turbulence and scalar transport. Phys. Fluids A **3**, 2746.

Piomelli, U. (1993): High Reynolds number calculations using the dynamic subgrid-scale stress model. Phys. Fluids A, **5**, 1484.

Rogallo, R.S. (1981): Numerical experiments in homogeneous turbulence. NASA TM 81315.

Smith, L.M. & Yakhot, V. (1993): Short- and long-time behavior of eddy viscosity models. Theoret. Comput. Fluid Dyn., **4**, 197.

Speziale, C.G., Raj, R. & Gatski, T.B. (1992): Modeling the dissipation rate in rotating turbulent flows. in Studies in Turbulence, ed. by T.B. Gatski, S. Sarkar, and C.G. Speziale, editors, 129.

Tafti, D.K. & Vanka, S.P. (1991): A numerical study of the effects of spanwise rotation on turbulent channel flow. Phys. Fluids A, **3**, 642.

Veeravalli, S.V. (1991): An experimental study of the effects of rapid rotation on turbulence. in Annual Research Briefs – 1990, NASA-Stanford Center for Turbulence Research, 203.

Zang, T.A. & Hussaini, M.Y. (1986): On spectral multigrid methods for the time-dependent Navier-Stokes equations. Appl. Math. Comp., **19**, 359.

Large Eddy Simulation and Second-Moment Closure Model of Particle Fluctuating Motion in Two-Phase Turbulent Shear Flows

O. Simonin, E. Deutsch, and M. Boivin

Laboratoire National d'Hydraulique, EDF, 78400 Chatou, France

Abstract

Turbulence statistics of non-settling discrete solid particles suspended in homogeneous turbulent gas shear flows generated by means of large eddy simulation (LES) are investigated for two different mean shear rates ($S = 25$ and $50\,/\mathrm{s}$) and three different particle diameters ($d = 30$, 45 and $60\,\mu\mathrm{m}$). Concurrently, a second-moment closure model of the particle fluctuating motion, based on separate transport equations for the particle kinetic stresses and the fluid–particle velocity correlations, is described and the corresponding predictions are compared with the simulation results. Large eddy simulation and closure model predictions show that the most noticeable effect of inertia is to increase the degree of anisotropy of the particle fluctuating motion with respect to the fluid one. As a matter of fact, the transverse particle turbulent velocity components are directly controlled by the dragging by the fluid turbulence and decrease with increasing particle relaxation time compared to the fluid turbulence integral time scale. In contrast, the streamwise component increases, due to the influence of both the fluid and particle mean velocity gradients, and eventually exceeds the corresponding gas turbulent component.

1. Introduction

This study is part of a more general work on the numerical prediction of turbulent dispersed two-phase flow by means of the two-fluid model approach. In this approach, separate mean transport equations are derived for both phases leading to closure problems due to the presence of turbulent correlations and interphase transfer terms. The applications of such an approach to gas–solid turbulent flows show that the predictions may be strongly dependent on the practical forms used for the particle kinetic stress tensor components (the second-order moments of the particle velocity fluctuations) and for the fluid–particle velocity correlations, through the interphase momentum transfer closure model. In the past these forms have been based on the Boussinesq or gradient hypothesis and local-equilibrium assumption, generally assuming that the particle kinetic shear stresses are proportional to the local mean shear of the dispersed phase while the normal stresses are dominated by the local dragging by the fluid turbulent motion (Tchen 1947; Hinze 1972).

Recently, Reeks (1993) and Liljegren (1993) analysed the fluctuating motion of solid particles suspended at equilibrium in a simple uniform mean shear flow in which the turbulence is homogeneous and stationary. The forms given by Reeks for the kinetic stresses are based on equilibrium solutions of a kinetic equation describing the transport of the particle phase space probability. Meanwhile, Liljegren analysed the particle fluctuating motion using Fourier transform techniques on particle equation of motion. Both approaches need to prescribe the form of the Lagrangian correlation function of the turbulent fluid velocity viewed along particle trajectories. Their analyses show that, in contradiction with predictions of previous approaches, the streamwise particle velocity variance appears to increase with increasing particle response time and may exceed the fluid velocity variance. Reeks has also pointed out the influence of a simple shear in producing particle dispersion coefficients significantly greater than those of the fluid and has noticed that the particle shear stresses depend explicitly upon the shearing of both phases.

These features are contained in a new approach proposed by Simonin (1991) for the modelling of the particle fluctuating motion and based on separate transport equations for the components of the kinetic stress tensor and for the fluid–particle velocity covariance. This approach allows one to account simultaneously for the particle dragging by the fluid turbulence, the production by the mean particle velocity gradients, the mean and turbulent transport by the particle velocity and the inter-particle collision influence, in the frame of the kinetic theory of rarefied gases. This closure model was implemented in a two-dimensional numerical code, Mélodif, and applied for validation to several practical flow configurations (Simonin 1991; Simonin and He 1992). According to existing experimental data, the predicted streamwise velocity fluctuations are found to exhibit features consistent with the above analyses. As a matter of fact, the application to a gas–solid turbulent swirling flow (Sommerfeld and Qiu 1991) shows radial and tangential predicted particle velocity variances smaller than the fluid ones and which decrease with increasing diameter. In contrast, the axial velocity variance in the highly sheared regions increases with the particle diameter up to values exceeding the fluid one. The results are qualitatively similar in a vertical gas–solid turbulent boundary layer in air (Rogers and Eaton 1990). Indeed, the predictions show that the streamwise velocity variances of 50 μm and 90 μm particles increase when approaching the wall due to the local production by the mean particle velocity shear, to values exceeding that of the air. In contrast, the transverse velocity variances are significantly smaller that the one of the air and decrease with increasing diameter. In addition, the attenuation of the transverse particle velocity fluctuations in the vicinity of the wall ($y < 2$ mm) is found to be balanced by the turbulent transport from the far-wall region which is more efficient for the larger size particles.

Such proposed approaches must, however, be regarded as heuristic to be validated by experiment or comparison with results from numerical simulation. Unfortunately, there are very few of existing experimental data to validate in detail the proposed modelling approaches, so direct numerical simulation (DNS), or large eddy simulation (LES), is providing a powerful tool to assess the basic assumptions and closure models required for practical applications. Most attention in the past has focused on the dispersion of particles by homogeneous isotropic turbulence and the question of particle velocity fluctuations in turbulent shear flow has largely been ignored. Nevertheless, Yeh and Lei (1991) have studied the particle dispersion in a turbulent shear flow generated by large eddy simulation employing a $(32)^3$ grid for two different mean shear rates. The results include a detailed study of the effect of the particle inertia and the particle's free-fall velocity in still fluid on the particle dispersion and settling velocities. In previous work, Deutsch and Simonin (1991) have applied LES to particle clouds suspended in a forced homogeneous and isotropic fluid turbulence, the application of uniform and non-uniform external force fields to the particles producing a mean drift between the phases and a mean particle velocity shear, respectively. These simulations allow one to point out the main mechanisms occurring in simple equilibrium sheared configurations and to validate a new approach for the derivation of fluid–particle velocity correlations (Simonin et al. 1993).

The purpose of this paper is to present the turbulence statistics of non-settling discrete solid particles tracked in homogeneous turbulent shear flows generated by LES and to compare with two-fluid model predictions using separate transport equations for the components of the particle kinetic stress tensor and the fluid–particle turbulent velocity covariance tensor. The approach proposed for the computation of the particle velocity correlations allows one to account for the inter-particle collisions. Nevertheless, throughout this paper, we restrict our attention to very dilute situations so that particle collisions and fluid turbulence modification can be ignored.

2. Numerical Simulation Method

2.1 Large Eddy Simulation of the Fluid Turbulence

The computation of particle trajectories in turbulent flows requires the knowledge of the instantaneous three-dimensional velocity field. Though the DNS remains the approach the most satisfactory from the theoretical point of view, LES can also be used with a good accuracy in gas–solid flows because the dominant process in heavy particle displacement is the dragging by the largest scales of the fluid turbulent motion. In the following, the fluid equations are unchanged by the presence of the particles because the solid volumetric fraction and the interphase momentum transfer are assumed to be negligible. The instantaneous turbulent velocity field u_1 is decomposed into a large-eddy field $\overline{u}_1$ (the resolved field) and a small-scale turbulence field u_1^s (the residual or subgrid field) by means of a Gaußian filter with a width Δ_f chosen to be twice the mesh size Δx. Filtering the Navier-Stokes equations, one obtains the equations for the resolved filtered incompressible flow with constant molecular viscosity ν_1, these are

$$\frac{\partial \overline{u}_{1,j}}{\partial x_j} = 0 \tag{1}$$

$$\frac{\partial \overline{u}_{1,i}}{\partial t} + \overline{\left[\overline{u}_{1,j} + u_{1,j}^s \right] \frac{\partial \overline{u}_{1,i}}{\partial x_j}} = -\frac{1}{\varrho_1}\frac{\partial \overline{p}_1}{\partial x_i} + g_i - \frac{\partial \sigma_{1,ij}^s}{\partial x_j} + \nu_1 \frac{\partial^2 \overline{u}_{1,i}}{\partial x_j \partial x_j} \tag{2}$$

where p_1 is the local instantaneous fluid pressure and g_i $(i = 1, 3)$ are the components of the external force field (gravity). The diffusive effect of the residual field appears as

$$\sigma^s_{1,ij} = \overline{u^s_{1,i}\, u^s_{1,j}} + \overline{u^s_{1,i}\, \overline{u}_{1,j}} \tag{3}$$

with the deviatoric part being modelled according to the eddy viscosity model proposed by Smagorinsky

$$\sigma^s_{1,ij} - \frac{1}{3}\,\delta_{ij}\,\sigma^s_{1,mm} = -2\,\nu_S\,\overline{S}_{1,ij} \qquad \overline{S}_{1,ij} = \frac{1}{2}\left[\frac{\partial \overline{u}_{1,i}}{\partial x_j} + \frac{\partial \overline{u}_{1,j}}{\partial x_i}\right] \tag{4}$$

where

$$\nu_S = \left\langle (C_S\, 2\,\Delta x)^2 \sqrt{2\,\overline{S}_{1,ij}\,\overline{S}_{1,ij}} \right\rangle_1 \qquad C_S = 0.12 \tag{5}$$

$\langle . \rangle_1$ is a simple arithmetic mean on the points of the fluid computational mesh and the value of the constant C_S was found by comparison of simulation results with experimental data in decaying isotropic grid-generated turbulence. In homogeneous shear flow, the mean fluid velocity shear $\partial U_{1,i} / \partial x_j$ has not to be included in the calculation of the large-scale strain rate tensor $\overline{S}_{1,ij}$.

Finally, following Bardina et al. (1983), the equation set is completed by the scale similarity closure model, representing the contribution of the subgrid scale turbulence to the convective term from the range of scales in which the resolved and residual fields overlap, then the momentum equation becomes

$$\frac{\partial \overline{u}_{1,i}}{\partial t} + \overline{\left[2\,\overline{u}_{1,j} - \overline{\overline{u}}_{1,j}\right] \frac{\partial \overline{u}_{1,i}}{\partial x_j}} = -\frac{1}{\varrho_1}\frac{\partial \overline{p}_1}{\partial x_i} + g_i - \frac{\partial \sigma^s_{1,ij}}{\partial x_j} + \nu_1 \frac{\partial^2 \overline{u}_{1,i}}{\partial x_j \partial x_j} \tag{6}$$

The numerical code used here is based on finite difference discretization and fractional step method (Laurence 1985; Khoudly 1988), leading to separate treatment of the following operators: advection solved by a weak formulation of the characteristic method, diffusion and pressure solved by Fourier transform and Gauß elimination.

2.2 Particle Trajectory Computation

If we consider a dilute dispersion of rigid heavy spheres in translation, the resulting force induced on a single particle by the surrounding nonuniform fluid flow reduces to the drag force contribution, so that the particle velocity equation may be written:

$$\varrho_2 \frac{\mathrm{d}u_{2,i}}{\mathrm{d}t} = -\,\varrho_1\,\frac{3}{4}\,\frac{C_D\,(Re_p)}{d}\,|\boldsymbol{v}_r|\,v_{r,i} + \varrho_2\,g_i \tag{7}$$

where C_D, the local drag coefficient, can be expressed in terms of the particle Reynolds number Re_p, when $Re_p \leqslant 1000$, as

$$C_D\,(Re_p) = \frac{24}{Re_p}\left[1 + 0.15\,Re_p^{0.687}\right] \qquad Re_p = \frac{|\boldsymbol{v}_r|\,d}{\nu_1} \tag{8}$$

d is the particle diameter, g_i ($i = 1, 3$) are the components of the external force field (gravity), ϱ_1 and ϱ_2 are the fluid and particle density respectively ($\varrho_2 \gg \varrho_1$), ν_1 is the fluid molecular viscosity and $\boldsymbol{v}_r$ is the local instantaneous relative velocity between the particle and the surrounding fluid.

The relative velocity is given by $\boldsymbol{v}_r = \boldsymbol{u}_2 - \widetilde{\boldsymbol{u}}_1$, where $\boldsymbol{u}_2$ is the particle translation velocity and $\widetilde{\boldsymbol{u}}_1$ is the fluid velocity, measured at the particle location, of the flow field locally undisturbed by the presence of the particle but which remains turbulent (Gatignol 1983). When the concentration of particles is small enough such that the fluid turbulence is not modified by the presence of the particles, then the undisturbed fluid velocity $\widetilde{\boldsymbol{u}}_1$ is identical with the local instantaneous turbulent velocity measured in single-phase flow. On the other hand, in this study, $\widetilde{\boldsymbol{u}}_1$ can be merged into the fluid velocity $\overline{\boldsymbol{u}}_1$ generated by LES at the particle location because the subgrid scale turbulent velocity influence remains negligible owing to the particle inertia. As a matter of fact, the particle relaxation times being of the same order as the Lagrangian integral time scale (Tables 2 and 3), the particle response function to the dragging by the fluid turbulence leads to a filtering effect on the influence of the subgrid turbulent velocity fluctuations.

Table 1. Numerical simulation conditions.

Grid points	M^3	64^3
Mesh size (mm)	Δx	3.
Simulation time step (s)	δt	6.25×10^{-4}
Tracked particles	N_p	50 000

In order to minimize numerical bias, statistical results for each case were obtained from an ensemble of about 50 000 particles and special care was taken in the treatment of the particle equation concerning the time integration and the projection of the simulated fluid velocity on the particle location. So, the particle momentum equation is time advanced using a second-order Runge-Kutta scheme with the same time step as for the fluid velocity calculation. On the other hand, analysis of the Eulerian turbulence spectrum characteristics, computed from the simulated fluid velocity field interpolated on a staggered grid, led to selection of the cubic spline interpolation method. Finally, validation of the numerical method was performed in the particular simulation case of fluid element trajectories in homogeneous and steady fluid turbulence by comparison with theoretical results in the frame of Taylor's theory of turbulent diffusion (Deutsch 1992).

2.3 Application to Homogeneous Shear Flows

Simulations of homogeneous shear flows were performed on a $(64)^3$ grid for two constant and uniform mean shear rates, $S = 25$ and $50\,/\mathrm{s}$. The mean flow direction is chosen to be the x_1-direction and the mean gradient direction is taken to be the

x_3-direction, with the center of the computational domain at $(x_1, x_2, x_3) = (0, 0, 0)$, so the mean fluid velocity is given by

$$U_{1,i}(\boldsymbol{x}) = S\, x_3\, \delta_{i1} \tag{9}$$

Periodic conditions are used for the fluctuating parts of the fluid and particle velocities. As a matter of fact, any particle crossing one of the edges of the computational domain is replaced by a new one entering through the symmetrical side with a corrected velocity according to the mean particle velocity distribution. Computations were carried out for a typical particle to fluid density ratio, $\varrho_2/\varrho_1 = 2000$ and a molecular fluid viscosity $\nu_1 = 1.45 \cdot 10^{-5}\,\mathrm{m}^2/\mathrm{s}$, with various particle diameters, $d = 30$, 45 and 60 μm, in order to modify appreciably the particle relaxation time.

The initial conditions for the gas and particle velocities are taken from numerical simulations of particle clouds suspended in a forced homogeneous and isotropic turbulence (Deutsch and Simonin 1991). The turbulent kinetic energy is kept constant artificially by adjustment of the spectrum by comparison with the theoretical von Karman-Pao form. The statistical fluid characteristics correspond to measurements performed by Wells and Stock (1983) in a grid-generated turbulence (Table 2).

Table 2. Initial conditions obtained from large eddy simulation of forced homogeneous and isotropic fluid turbulence.

Fluid rms velocity (m/s)	u'_1	0.282
Longitudinal Eulerian integral length scale (mm)	L^e_f	7.8
Lateral Eulerian integral length scale (mm)	L^e_g	3.85
Lagrangian integral time scale (s)	T_L	26.3×10^{-3}
Taylor microscale (mm)	λ	2.
Reynolds number	Re_λ	38.

The particles were allowed to disperse sufficiently before being subjected to the mean shear rate, at $t = 0$, in order to be statistically uniformly distributed in the computational domain and in equilibrium with the fluid turbulence (Table 3). In this study, the crossing-trajectory effect induced by the mean drift between the two phases is avoided by removing the external body forces from the particle motion equation, $g_i = 0$ ($i = 1, 3$) in (7), and by taking as initial particle velocity at $t = 0$ the local instantaneous particle velocity $\widehat{\boldsymbol{u}}_2$ from the forced homogeneous and isotropic simulations in addition to the linear dependent mean velocity component computed in terms of the initial particle location:

$$u_{2,i}(\boldsymbol{x}) = \widehat{u}_{2,i}(\boldsymbol{x}) + S\, x_3\, \delta_{i1} \qquad t = 0 \tag{10}$$

In this way, the mean particle velocity gradient is time independent and complies with

$$U_{2,i}(\boldsymbol{x}) = S_2\, x_3\, \delta_{i1} \qquad S_2 = S \tag{11}$$

Table 3. Initial conditions obtained from non-settling particle tracking in forced homogeneous and isotropic fluid turbulence.

Particle diameter (μm)	d	30	45	60
Particle rms velocity (m/s)	u'_2	0.25	0.22	0.20
Stokes characteristic time (s)	τ_p	7.0×10^{-3}	$16. \times 10^{-3}$	$28. \times 10^{-3}$
Mean particle relaxation time (s)	τ_{12}^F	6.5×10^{-3}	$14. \times 10^{-3}$	$24. \times 10^{-3}$
Particle Reynolds number	Re_p	0.38	0.79	1.25

3. Two-Fluid Model Approach

3.1 Statistical Average

In the two-fluid model formulation, the set of equations for each phase can be derived from the local instant single-phase conservation equations by density weighted averaging with in addition average balances of mass, momentum and energy at the interfaces (Drew 1983). Let us introduce $\langle . \rangle_k$ the averaging operator associated to phase k, for the continuous (k = 1) and dispersed phases (k = 2), respectively, and α_k the corresponding volumetric fraction defined by

$$\alpha_k \langle G \rangle_k = \langle \chi_k G \rangle \qquad \alpha_k = \langle \chi_k \rangle \tag{12}$$

where G is any physical quantity which can be defined in the phase k (velocity, density, etc ...) and $\chi_k(x_i, t)$ is the phase k characteristic function: $\chi_k = 1$ if the point (x_i, t) belongs to phase k and $\chi_k = 0$ if not. According to the random character of turbulent two-phase flows, the averaging operator $\langle . \rangle$ can be defined in any flow configuration as an ensemble average over N identical independent realizations with N tending towards infinity. But in homogeneous or steady flows $\langle . \rangle$ can also be obtained by integration over space or time domains, respectively. With regard to the following analysis of the simulation results, the gas averaging operator $\langle . \rangle_1$ is calculated as a simple arithmetic mean on the $(64)^3$ points of the computational mesh, while that of the particles $\langle . \rangle_2$ is calculated as an arithmetic mean on the ensemble of the N_p = 50 000 particles tracked in the computational domain.

Restricting our attention to dilute dispersed two-phase flows without mass transfer and assuming that the molecular viscous stresses in the continuous phase and the granular stresses due to the inter-particle collisions are negligible in comparison with the kinetic stresses (the so-called Reynolds stresses in single-phase turbulence) the average governing equations take the following simple forms,

mass balance:

$$\frac{\partial}{\partial t} \alpha_k \varrho_k + \frac{\partial}{\partial x_j} \alpha_k \varrho_k U_{k,j} = 0 \tag{13}$$

momentum balance:

$$\alpha_k \varrho_k \frac{\partial}{\partial t} U_{k,i} + \alpha_k \varrho_k U_{k,j} \frac{\partial}{\partial x_j} U_{k,i} = -\alpha_k \frac{\partial}{\partial x_i} P_1 + \alpha_k \varrho_k g_i + I_{k,i}$$

$$-\frac{\partial}{\partial x_j} \alpha_k \varrho_k \left\langle u'_{k,i} u'_{k,j} \right\rangle_k \tag{14}$$

where P_1 is the mean pressure of the continuous phase and $\boldsymbol{I}_k$ is the part of the mean interphase momentum transfer rate which remains after subtraction of the mean pressure contribution. $U_{k,i}$ the mean velocity i-component and $u'_{k,i}$ the corresponding fluctuation of the local instantaneous velocity, are given by

$$\alpha_k U_{k,i} = \alpha_k \langle u_{k,i} \rangle_k \qquad u'_{k,i} = u_{k,i} - U_{k,i} \qquad \left\langle u'_{k,i} \right\rangle_k = 0 \tag{15}$$

The fluid velocity fluctuation $\boldsymbol{u}'_1$ should take account simultaneously of the local disturbance due to the possible proximity of a particle and of the turbulent component which should remain after subtraction of this particle. But when the particle size is much smaller than the average size of the energy-containing eddies, which are of the same order as the integral length scales, the statistical velocity moments of the disturbance flow fields locally set up by the particles are negligible. So, the velocity fluctuation $\boldsymbol{u}'_1$ may be assumed to account only for the locally undisturbed fluid turbulence and written as

$$u'_{1,i} = \widetilde{u}_{1,i} - U_{1,i} \qquad \langle \widetilde{u}_{1,i} - U_{1,i} \rangle_1 \approx 0 \tag{16}$$

On the other hand, the mean and turbulent interphase transfer rates appearing in the equations governing the mean velocities and second-order moments, respectively, can be derived from the single particle dynamic equation in terms of the fluid and fluid–particle velocity correlations calculated along the particle paths (see sects. 3.2 and 3.3 below). And so we propose, for convenience, to mark with a double apostrophe the fluid and particle velocity fluctuations computed by using the dispersed phase average, so that

$$u''_{1,i} = \widetilde{u}_{1,i} - \langle \widetilde{u}_{1,i} \rangle_2 \qquad u''_{2,i} = u'_{2,i} \qquad \langle u''_{k,i} \rangle_2 = 0 \quad k = 1,2 \tag{17}$$

3.2 Mean Interphase Transfer Rate

In dilute dispersed two-phase flows, the mean interphase transfer rate can be derived by averaging from the local instantaneous expression of the force acting on a single particle. So, the interphase transfer term in the mean momentum equations for gas–solid mixture is derived from (7) as

$$I_{2,i} = -I_{1,i} = \alpha_2 \varrho_2 \left\langle \frac{\mathrm{d}u_{2,i}}{\mathrm{d}t} - g_i \right\rangle_2 = -\alpha_2 \left\langle \varrho_1 \frac{3}{4} \frac{C_D}{d} |\boldsymbol{v}_r| \, v_{r,i} \right\rangle_2 \tag{18}$$

and, in the Stokes regime ($Re_p \ll 1$), the above equation leads to the simple form

$$I_{2,i} = -I_{1,i} = -\alpha_2 \varrho_2 \frac{1}{\tau_p} V_{r,i} \qquad \tau_p = \frac{\varrho_2}{\varrho_1} \frac{1}{18} \frac{d^2}{\nu_1} \tag{19}$$

where τ_p is the so-called Stokes characteristic time and $\boldsymbol{V}_r = \langle \boldsymbol{v}_r \rangle_2$ is the mean value of the local relative velocity between each particle and the surrounding fluid. By definition, the mean relative velocity $\boldsymbol{V}_r$ is equal to the effective mean drift between the phases $\Delta \boldsymbol{U} = \boldsymbol{U}_2 - \boldsymbol{U}_1$ minus a fluid–particle turbulent drift velocity $\boldsymbol{V}_d$ due to the correlation between the instantaneous distribution of the discrete particles and the turbulent fluid velocity field:

$$V_{r,i} = [U_{2,i} - U_{1,i}] - V_{d,i} \qquad V_{d,i} = \langle [\tilde{u}_{1,i} - U_{1,i}] \rangle_2 = \left\langle u'_{1,i} \right\rangle_2 \quad (20)$$

The velocity $\boldsymbol{V}_d$ accounts for the dispersion mechanism due to the particle transport by the fluid turbulent motion (Deutsch and Simonin 1991; Simonin et al. 1993) and reduces to the simple turbulent correlation between the solid concentration and the fluid velocity in the scalar limit case, when the particle response time τ_p is very small with regard to the smallest characteristic time scales of the fluid turbulence. Because of the symmetries of the problem, one can notice that the velocity $\boldsymbol{V}_d$ must be equal to zero in the present study, involving a statistically uniform distribution of non-settling particles in homogeneous turbulence. However, the instantaneous particle distribution may be correlated with the fluid turbulent flow field and recent studies (Maxey 1987; Squires and Eaton 1991) have shown that the inertia can cause particles, with a response time comparable to the flow Kolmogorov time scale, to accumulate in local regions of low flow vorticity or high strain rate. On the other hand, following Maxey, the preferential accumulation of settling particles in regions with a negative, vertical, local-instantaneous fluid velocity component should lead to a non-zero fluid–particle turbulent drift velocity $\boldsymbol{V}_d$ and cause the mean particle settling velocity $\boldsymbol{U}_2$ to be greater than in a still fluid.

The previous expression (19) for the mean interphase transfer rate can be broadened to larger values of the particle Reynolds number Re_p using a simple model for the non-linear dependence of the drag coefficient

$$I_{2,i} = -I_{1,i} = -\alpha_2 \varrho_2 \frac{1}{\tau_{12}^F} V_{r,i} \qquad \tau_{12}^F = \frac{\varrho_2}{\varrho_1} \frac{4}{3} \frac{d}{\langle C_D \rangle} \frac{1}{\langle |\boldsymbol{v}_r| \rangle_2} \quad (21)$$

where τ_{12}^F is the mean particle relaxation time,

$$\langle C_D \rangle = C_D\left(\langle Re_p \rangle\right) \qquad \langle Re_p \rangle = \frac{\langle |\boldsymbol{v}_r| \rangle_2 \, d}{\nu_1} \quad (22)$$

$$\langle |\boldsymbol{v}_r| \rangle_2 \approx \sqrt{V_{r,i} V_{r,i} + \langle v''_{r,i} v''_{r,i} \rangle_2} \quad (23)$$

3.3 Particle Kinetic Stress Equation

The closure of the average field equations requires the computation of the second-order velocity correlations $\left\langle u'_{k,i} u'_{k,j} \right\rangle_k$ and, in practice, the particle kinetic stresses $\left\langle u'_{2,i} u'_{2,j} \right\rangle_k$ are generally derived using the Boussinesq or gradient hypothesis and local-equilibrium assumption (Tchen 1947; Hinze 1972). The particle kinetic shear stresses are given as proportional to the local mean shear of the dispersed phase flow while the normal stresses are assumed to be dominated by the local dragging by the fluid turbulent motion. However, this kind of approach requires one to neglect the production by the mean particle velocity gradient and the mean and turbulent transport mechanisms which are found to play a dominant part in many turbulent flow configurations, such as the round jet, swirling flow or vertical boundary layer (Simonin 1991; Simonin and He 1992). On the other hand, the detailed analysis of the experimental and numerical results show that an accurate separate treatment of the particle kinetic stresses is required owing to the fact that the anisotropy of the dispersed phase fluctuating motion influences the particle dispersion, appreciably. The particle kinetic stress equation can be derived by averaging (7) for the single particle velocity component $u_{2,i}$ after multiplication by $u''_{2,j}$ and then adding to it the same velocity-weighted averaged equation but with the subscripts i and j interchanged, the result is

$$\left[\frac{\partial}{\partial t} + U_{2,m}\frac{\partial}{\partial x_m}\right]\langle u''_{2,i}\, u''_{2,j}\rangle_2 = D_{2,ij} + P_{2,ij} + \Pi_{2,ij} \tag{24}$$

where the first term on the right-hand side of the equation represents the transport of the kinetic stress by the particle velocity fluctuations and vanishes in homogeneous turbulence:

$$D_{2,ij} = -\frac{1}{\alpha_2}\frac{\partial}{\partial x_m}\alpha_2\langle u''_{2,i}\, u''_{2,j}\, u''_{2,m}\rangle_2 \tag{25}$$

the second term represents the production by the mean particle velocity gradients and does not need to be modelled in second-moment closure model:

$$P_{2,ij} = -\langle u''_{2,i}\, u''_{2,m}\rangle_2\frac{\partial U_{2,j}}{\partial x_m} - \langle u''_{2,j}\, u''_{2,m}\rangle_2\frac{\partial U_{2,i}}{\partial x_m} \tag{26}$$

the third term represents the turbulent momentum transfer rate from the fluid turbulent motion:

$$\Pi_{2,ij} = -\left\langle\frac{\varrho_1}{\varrho_2}\frac{3}{4}\frac{C_D}{d}\,|\boldsymbol{v}_r|\,\left[v_{r,i}\, u''_{2,j} + v_{r,j}\, u''_{2,i}\right]\right\rangle_2 \tag{27}$$

and can be approximated by introducing the mean particle relaxation time τ^F_{12} from (21)

$$\Pi_{2,ij} = -\frac{1}{\tau^F_{12}}\left[2\langle u''_{2,i}\, u''_{2,j}\rangle_2 - \langle u''_{1,i}\, u''_{2,j}\rangle_2 - \langle u''_{1,j}\, u''_{2,i}\rangle_2\right]. \tag{28}$$

So, the particle dragging by the fluid turbulence leads to creation or destruction of the particle velocity fluctuations according to the local values of the fluid–particle velocity correlations $\langle u''_{1,i}\, u''_{2,j}\rangle_2$.

Therefore, the key point of the proposed approach is the development of a satisfactory closure model for the fluid–particle velocity covariance tensor calculated at the particle locations. In the following, we attempt to derive the corresponding transport equation from an approximate form of the fluid turbulent velocity equation written along the particle trajectories.

3.4 Lagrangian Fluid Turbulence Equation

In the conventional approach, the exact transport equations of the fluid turbulent Reynolds stresses are derived by taking the velocity-weighted average of the local instantaneous Eulerian equations of the velocity components (or Navier-Stokes equations), leading to a closure problem related to the computation of the turbulent transport, pressure correlations and viscous dissipation (Launder et al. 1975). In a similar way, one could derive the transport equations of the fluid–particle velocity correlations but the exact forms obtained by using the dispersed phase average are much more complicated because of the non-zero local relative velocity and the non-uniform instantaneous distribution of the particles. On the other hand, according to DNS and LES results in homogeneous and isotropic turbulence (Squires and Eaton 1991; Deutsch and Simonin 1991) the modifications of the fluid Lagrangian turbulent characteristics viewed by the particles appear to amount mainly to the dependence of the fluid Lagrangian integral time scale on the mean drift between the phases (the so-called crossing-trajectory effect). Therefore we propose to use a Lagrangian viewpoint to derive the fluid–particle covariance equation, assuming that a reasonable model for the fluid velocity increment encountered along the particle path can be obtained as a simple extension of the one given along the fluid element path. So, in order to insure the consistency with single-phase second-order closure models, we consider the generalized Langevin equation proposed by Haworth and Pope (1986):

$$\begin{aligned} u_{1,i}(\boldsymbol{x}+\boldsymbol{u}_1\delta t, t+\delta t) &= u_{1,i}(\boldsymbol{x},t) - \frac{1}{\varrho_1}\frac{\partial P_1^*}{\partial x_i}\delta t + \frac{\partial}{\partial x_j}\left[\nu_1 \frac{\partial U_{1,i}}{\partial x_j}\right]\delta t \\ &+ G_{1,ij}\left[u_{1,j} - U_{1,j}\right]\delta t + \left[C_0\,\varepsilon_1\right]^{1/2}\delta W_{1,i} \end{aligned} \tag{29}$$

where P_1^* is the mean fluid pressure after subtraction of the hydrostatic contribution, $G_{1,ij}$ is a second-order tensor modelled in terms of local mean quantities (Reynolds stresses, mean velocity gradients and turbulent dissipation rate), C_0 is a universal (Kolmogorov) constant, ε_1 is the dissipation rate of the fluid turbulent kinetic energy q_1^2, and $\boldsymbol{W}_1$ is an isotropic Weiner process. This is a Markovian stochastic process whose increment $\delta\boldsymbol{W}_1$ has a joint-normal distribution with zero mean and isotropic covariance matrix,

$$\langle \delta W_{1,i}\rangle = 0 \qquad \langle \delta W_{1,i}\,\delta W_{1,j}\rangle = \delta_{ij}\,\delta t\,. \tag{30}$$

By comparing (29) and the local instantaneous Navier-Stokes equations, it may be seen that the effects of viscosity and of the fluctuating pressure gradient are modelled by the terms containing $G_{1,ij}$ and $\delta W_{1,i}$.

Haworth and Pope (1986) proposed a general functional form for $G_{1,ij}$ that is linear in the mean velocity gradients and in the Reynolds stresses, but here we prefer to evaluate two simplified forms consistent with the basic pressure–strain correlation closure models. The model 1 (or simplified model) corresponds to the basic Langevin equation

$$G_{1,ij} = -\frac{1}{\tau_1^t}\,\delta_{ij} \qquad \tau_1^t = \frac{1}{\beta_1}\frac{q_1^2}{\varepsilon_1} \tag{31}$$

while the model 2 (or intermediate model) retains an additional term linear in the mean velocity gradient in order to allow a more satisfactory treatment of the effects of the fluctuating pressure gradient, including the role of the rapid pressure term,

$$G_{1,ij} = -\frac{1}{\tau_1^t}\delta_{ij} + \beta_2 \frac{\partial U_{1,i}}{\partial x_j} \tag{32}$$

with $\beta_2 = 0.6$ according to the rapid distortion constraint (Haworth and Pope 1986).

Following Haworth and Pope, a relationship among the model coefficients may be deduced by comparing the modelled transport equation for the turbulent kinetic energy q_1^2 derived from (29) to its exact counterpart obtained from the Navier-Stokes equations in homogeneous flows. It is found that both the Langevin models studied there yield the proper evolution of q_1^2 provided that β_1 satisfies

$$\beta_1 = \left[\frac{1}{2} + \frac{3}{4}C_0 - \frac{\beta_2}{4}\frac{P_{1,ii}}{\varepsilon_1}\right] \qquad P_{1,ii} = -2\left\langle u'_{1,i}\, u'_{1,m}\right\rangle_1 \frac{\partial U_{1,i}}{\partial x_m} \tag{33}$$

with $\beta_2 = 0$ for the simplified Langevin model.

Finally, the modelled transport equations for the Reynolds stresses can be derived from (29) as

$$\left[\frac{\partial}{\partial t} + U_{1,m}\frac{\partial}{\partial x_m}\right]\left\langle u'_{1,i}\, u'_{1,j}\right\rangle_1 = D_{1,ij} + P_{1,ij} + \Phi_{1,ij} - \frac{2}{3}\varepsilon_1\,\delta_{ij} \tag{34}$$

where the first term on the right-hand side of the equation represents the turbulent transport

$$D_{1,ij} = -\frac{\partial}{\partial x_m}\left\langle u'_{1,i}\, u'_{1,j}\, u'_{1,m}\right\rangle_1 \tag{35}$$

the second term represents the production by the mean velocity gradients

$$P_{1,ij} = -\left\langle u'_{1,i}\, u'_{1,m}\right\rangle_1 \frac{\partial U_{1,j}}{\partial x_m} - \left\langle u'_{1,j}\, u'_{1,m}\right\rangle_1 \frac{\partial U_{1,i}}{\partial x_m} \tag{36}$$

and the third term models the pressure–strain correlations

$$\Phi_{1,ij} = -\frac{2}{\tau_1^t}\left[\left\langle u'_{1,i}\, u'_{1,j}\right\rangle_1 - \frac{2}{3}q_1^2\,\delta_{ij}\right] - \beta_2\left[P_{1,ij} - \frac{1}{3}P_{1,mm}\,\delta_{ij}\right] . \tag{37}$$

These results show that the simplified form (31) for $G_{1,ij}$ accounts for the so-called slow pressure term in the pressure–strain correlations and is consistent with the Rotta's linear return-to-isotropy model (with a coefficient $C_1 = 2\,\beta_1$) while the intermediate form (32) accounts likewise for the mean-strain effects in the pressure–strain correlations (the so-called rapid pressure term) by means of a simple isotropization of production (IP) model.

Finally, both models contain only one degree of freedom corresponding to the determination of the universal (Kolmogorov) constant C_0 which is taken to be 2.1, following Haworth and Pope (1986). This value leads to a Rotta constant $C_1 = 4.15$ for the simplified Langevin model which is somewhat higher than the values usually recommended, $C_1 \approx 2$, but is close to the value of 4.5 which is found to yield good results in free shear flows in the absence of a model for the rapid pressure term, as Haworth and Pope point out. Figures 1*(a)-(b)* show the good agreement observed in the present work between the simplified model predictions of the separate Reynolds stresses and the corresponding LES results. The problem of determining the dissipation rate ε_1 is not considered here and ε_1 is taken directly as input of the models from the LES results using the homogeneous turbulent kinetic energy equation.

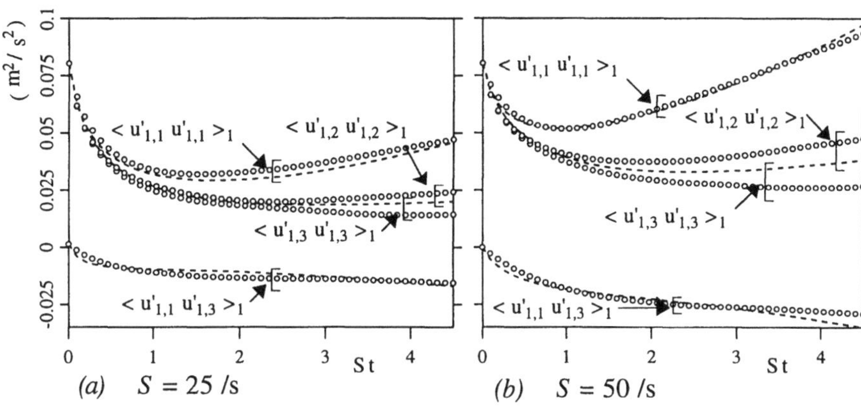

Fig. 1. Time development of the fluid turbulent Reynolds stresses: comparison between LES results (○) and basic Langevin model predictions (- - -).

Following Csanady (1963) one may assume that the local relative motion influence on the Lagrangian fluid velocity correlation function measured along the particle paths reduces to the so-called crossing-trajectory effect caused by the mean relative velocity. When the relative velocity is large with respect to the instantaneous fluid velocity Csanady recognized that the Lagrangian fluid velocity correlation function experienced by the particles amounts to a spatial Eulerian correlation function in terms of their mean relative displacement. On the other hand, for non-settling particles, the Lagrangian fluid correlation function measured along the particle paths is assumed to be nearly identical to the standard ones calculated along the fluid element trajectories. These are crude assumptions, but in fairly good agreement with DNS and LES results in homogeneous and isotropic turbulent flows (Squires and Eaton 1991; Deutsch and Simonin 1991), which allow us to model the Lagrangian fluid velocity increment along the particle path using a simple extension of the above single-phase Langevin equation (29)

$$\begin{aligned}
\widetilde{u}_{1,i}(\boldsymbol{x}+\boldsymbol{u}_2\delta t, t+\delta t) &= \widetilde{u}_{1,i}(\boldsymbol{x},t) - \frac{1}{\varrho_1}\frac{\partial P_1^*}{\partial x_i}\delta t + \frac{\partial}{\partial x_j}\left[\nu_1 \frac{\partial U_{1,i}}{\partial x_j}\right]\delta t \\
&+ \left[u_{2,j} - \widetilde{u}_{1,j}\right]\frac{\partial U_{1,i}}{\partial x_j}\delta t \\
&+ G_{12,ij}\left[\widetilde{u}_{1,j} - U_{1,j}\right]\delta t + \left[C_0\,\varepsilon_1\right]^{1/2}\delta W_{12,i} \qquad (38)
\end{aligned}$$

where $G_{12,ij}$ is a second-order tensor modelled in terms of local mean quantities, and $\delta \boldsymbol{W}_{12}$ is a random vector with a zero mean. One can notice that the covariance matrix $\langle \delta W_{12,i}\, \delta W_{12,j} \rangle$ does not need to be prescribed because the corresponding terms do not appear in the derivation of the fluid–particle velocity covariance equation. The fourth term on the right hand side of the above equation represents exactly the fluid velocity increment due to the relative particle displacement, with regard to the fluid element path, in a non-uniform mean velocity field. On the other hand, the relative motion influence on the turbulent fluid velocity viewed by the particle is assumed to be modelled through the terms containing $G_{12,ij}$ and $\delta W_{12,i}$, in addition to the effects of fluid viscosity and of the fluctuating pressure gradient. Therefore, we propose the following closure model chosen for compatibility with the single-phase approach

$$G_{12,ij} = -\frac{1}{\tau_{12}^t}\delta_{ij} + \beta_2 \left[\frac{\partial U_{1,i}}{\partial x_j} + \frac{\partial V_{d,i}}{\partial x_j} \right] \tag{39}$$

where τ_{12}^t should be affected by the crossing-trajectory effect (Simonin et al. 1993). However, restricting our attention to non-settling particles, τ_{12}^t may be identified with the corresponding fluid turbulent time scale

$$\tau_{12}^t = \tau_1^t = \frac{1}{\beta_1}\frac{q_1^2}{\varepsilon_1} \tag{40}$$

with β_1 satisfying (33).

3.5 Fluid–Particle Velocity Covariance Equation

A transport equation for the fluid–particle turbulent drift velocity $\boldsymbol{V}_d$ can be derived from the proposed Langevin equation (38) by applying directly the dispersed phase average and is consistent with the governing equation for the single-phase turbulent scalar flux (Simonin et al. 1993). On the other hand, the fluid–particle velocity covariance transport equation can be derived by averaging the sum of the fluid Langevin equation (38) and the particle dynamic equation (7) after cross-multiplication by the particle and fluid turbulent velocities, respectively,

$$\left[\frac{\partial}{\partial t} + U_{2,m}\frac{\partial}{\partial x_m} \right] \langle u''_{1,i}\, u''_{2,j} \rangle_2 = D_{12,ij} + P_{12,ij} + P_{21,ij} + \Pi_{12,ij} + \Phi_{12,ij} - \frac{1}{3}\varepsilon_{12}\,\delta_{ij} \tag{41}$$

where the first term on the right-hand side of the equation represents the transport of the velocity covariance by the particle velocity fluctuations

$$D_{12,ij} = -\frac{1}{\alpha_2}\frac{\partial}{\partial x_m}\alpha_2 \langle u''_{1,i}\, u''_{2,j}\, u''_{2,m} \rangle_2 \tag{42}$$

the second and third terms represent the separate productions by the mean particle and fluid velocity gradients, respectively,

$$P_{12,ij} = -\langle u"_{1,i}\, u"_{2,m}\rangle_2 \frac{\partial U_{2,j}}{\partial x_m} \quad (43)$$

$$P_{21,ij} = -\langle u"_{2,j}\, u"_{1,m}\rangle_2 \left[\frac{\partial U_{1,i}}{\partial x_m} + \frac{\partial V_{d,i}}{\partial x_m}\right] \quad (44)$$

the fourth term represents the production rate due to the particle dragging by the fluid turbulent motion

$$\Pi_{12,ij} = -\left\langle \frac{\varrho_1}{\varrho_2} \frac{3}{4} \frac{C_D}{d} |\boldsymbol{v}_r| \left[v_{r,j}\, u"_{1,i}\right]\right\rangle_2 \quad (45)$$

and can be approximated by introducing the mean particle relaxation time τ_{12}^F from (21)

$$\Pi_{12,ij} = -\frac{1}{\tau_{12}^F}\left[\langle u"_{1,i}\, u"_{2,j}\rangle_2 - \langle u"_{1,i}\, u"_{1,j}\rangle_2\right]. \quad (46)$$

The particle dragging by the fluid turbulence leads generally to creation of the fluid–particle velocity correlations according to the local values of the fluid velocity correlations viewed by the particles $\langle u"_{1,i}\, u"_{1,j}\rangle_2$. It is important to note that the above interphase momentum transfer rate $\Pi_{12,ij}$ appears with the opposite sign in the transport equations governing the fluid turbulent Reynolds stresses when the particle influence is taken into account in the derivation.

Finally, the two final terms derive from the term containing $G_{12,ij}$ in (38) and account for pressure–strain correlations, viscous dissipation and crossing-trajectory effects,

$$\Phi_{12,ij} - \frac{1}{3}\varepsilon_{12}\,\delta_{ij} = G_{12,im}\langle u"_{1,m}\, u"_{2,j}\rangle_2 . \quad (47)$$

By definition, $\Phi_{12,ij}$ is a return-to-isotropy term ($\Phi_{12,ii} = 0$) and ε_{12} is the dissipation rate of the fluid–particle velocity covariance $q_{12} = \langle u"_{1,i}\, u"_{2,i}\rangle_2$. According to the closure model for $G_{12,ij}$, we obtain the following forms

$$\Phi_{12,ij} = -\frac{1}{\tau_{12}^t}\left[\langle u"_{1,i}\, u"_{2,j}\rangle_1 - \frac{1}{3} q_{12}\,\delta_{ij}\right] - \beta_2\left[P_{21,ij} - \frac{1}{3} P_{21,mm}\,\delta_{ij}\right] \quad (48)$$

$$\varepsilon_{12} = \frac{1}{\tau_{12}^t} q_{12} + \beta_2 P_{21,mm} \quad (49)$$

One can notice a non-symmetrical dependence on the mean phase velocity gradients. As a matter of fact, the terms containing β_2 lead to a reduction of the production rate by the mean fluid velocity gradient given by (44). On the other hand, (49) shows that the fluid–particle turbulent time scale q_{12}/ε_{12} ($\approx \tau_{12}^t$) is about two times smaller than the corresponding fluid turbulent time scale q_1/ε_1 ($= \beta_1\, \tau_1^t$).

Finally, the closure model is completed by assuming no statistical bias in the computation of the fluid turbulent Reynolds stresses seen by the particle,

$$\langle u"_{1,i}\, u"_{1,j}\rangle_2 = \left\langle u'_{1,i}\, u'_{1,j}\right\rangle_1 . \quad (50)$$

4. Results and Discussion

4.1 Fluid Turbulence Statistics

As mentioned previously, the simulation of homogeneous shear flows was performed for two constant and uniform mean shear rates S = 25 and 50/s, the mean flow direction chosen to be the x_1-direction and the mean gradient direction taken to be the x_3-direction, with the center of the computational domain at $(x_1, x_2, x_3) = (0, 0, 0)$. The initial conditions are taken from LES of a forced homogeneous and isotropic turbulence.

Figure 2*(a)* shows the time development of the fluid turbulent kinetic energy q_1^2 normalized by its initial value and the mean shear production to dissipation rates ratio $\mathrm{P}_1/\varepsilon_1$ for the two different mean shear rates. The dissipation rate ε_1 is evaluated directly from the simulated turbulent kinetic energy time dependence,

$$\varepsilon_1 = \mathrm{P}_1 - \frac{\mathrm{d}q_1^2}{\mathrm{d}t} \qquad \mathrm{P}_1 = -\left\langle u'_{1,i}\, u'_{1,m} \right\rangle_1 \frac{\partial U_{1,i}}{\partial x_m} \tag{51}$$

As expected, the simulated turbulent intensity decreases initially and, given sufficient development time ($S\,t > 2$), increases monotonically. On the other hand, Fig. 2*(b)* shows the time development of the anisotropy of the fluid turbulence, characterized by the dimensionless tensor components

$$b_{1,ij} = \left[\left\langle u'_{1,i}\, u'_{1,j} \right\rangle_1 / 2\, q_1^2 - \frac{1}{3}\delta_{ij} \right] \tag{52}$$

which approach constant values for both cases, S = 25 and 50/s, consistent with previous simulations such as those by Yeh and Lei (1991).

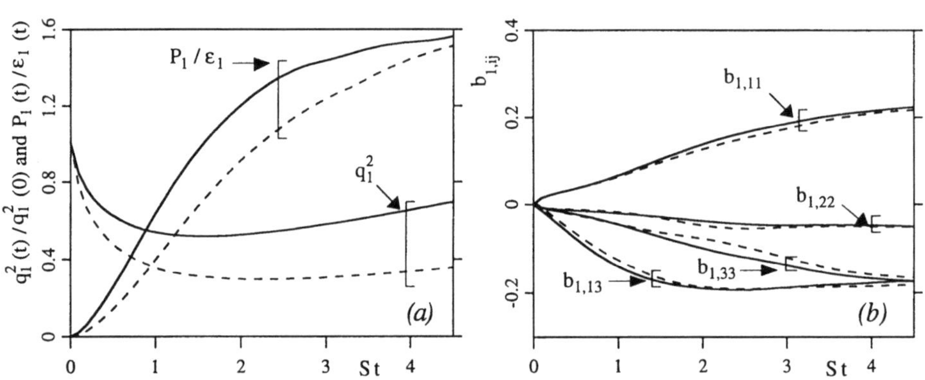

Fig. 2. Time development of the simulated fluid characteristics of homogeneous shear flows, --- S = 25/s, —— S = 50/s: **(a)** turbulent kinetic energy and ratio of production over dissipation rate, **(b)** normalized anisotropy tensor components.

Figure 3 compares the time scale of the fluid turbulence viewed by the particles $\tau_{12}^t\ (= \tau_1^t)$ from (40), for $\beta_2 = 0$ and 0.6, with the particle relaxation time τ_{12}^F from (21) for the different diameters used in the present work: d=30, 45 and 60 μm. One can notice that the time development of the characteristic time-scale ratio $\eta_r = \tau^t_{12}/\tau^F_{12}$, seen in Fig. 3, should lead to an increase of the efficiency of the particle dragging by the fluid turbulence. Finally, it is important to emphasize that, satisfying a practical closure approximation (50), no appreciable statistical bias is measured for the fluid turbulent Reynolds stresses viewed by the particles. The modifications reduce to the filtering effect induced when interpolating the fluid velocity at the particle locations using the cubic spline interpolation method.

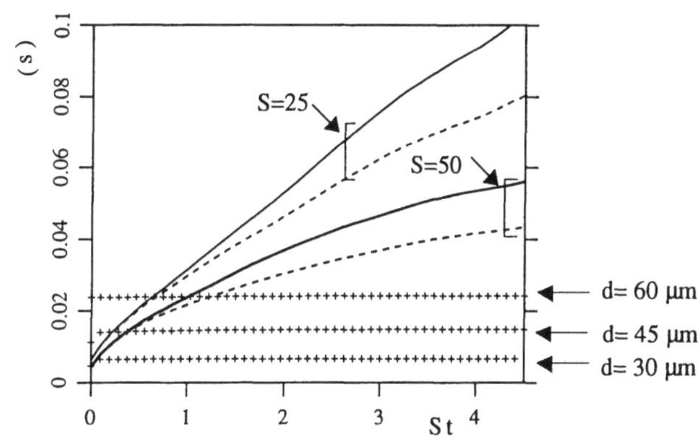

Fig. 3. Time development of the characteristic times of homogeneous shear flows: fluid turbulence time scale τ^t_{12} from closure models, --- model 1 and —— model 2; mean particle relaxation time τ^F_{12}, +.

4.2 Particle Kinetic Stresses

Figures 4*(a)*-*(e)* show the detailed simulation results for the case with the largest particles, d = 60 μm, suspended in the strongest mean shear rate, S = 50/s. This case corresponds to the most significant influence of inertia on the particle fluctuating motion measured in the present work, according to the relatively low value of the fluid turbulence to particle relaxation time scales ratio, $\eta_r = \tau^t_{12}/\tau^F_{12}$, illustrated in Fig. 3. In homogeneous and isotropic turbulent flows, the combined effects of particle inertia and dragging by the fluid turbulence lead to a decrease in the particle kinetic stresses below that of the gas and to their adjustment to the corresponding fluid–particle velocity correlations (see Tchen 1947; Deutsch and Simonin 1991). However, Fig. 4*(a)* makes it clear that the streamwise particle velocity variance $\langle u''_{2,1}\, u''_{2,1}\rangle_2$ is strongly affected by the existence of mean velocity gradients, as Liljegren (1993) and Reeks (1993) point out. Indeed, the streamwise particle velocity variance is found to be considerably higher than the level of the corresponding fluid–particle velocity correlation $\langle u''_{1,1}\, u''_{2,1}\rangle_2$, and to exceed the value of the fluid turbulent Reynolds stress $\langle u''_{1,1}\, u''_{1,1}\rangle_2$. Concurrently, Fig. 4*(d)* shows that the particle kinetic shear stress $\langle u''_{2,1}\, u''_{2,3}\rangle_2$ absolute value is always larger than those of the fluid and should lead surprisingly to a larger effective particle viscosity using the traditional form of the constitutive relations based on the Boussinesq approximation (see Hinze 1972). On the other hand, Fig. 4*(b)* and Fig. 4*(c)* show that the particle velocity variances in the spanwise and normal directions $\langle u''_{2,2}\, u''_{2,2} \rangle_2$ and $\langle u''_{2,3}\, u''_{2,3} \rangle_2$, respectively, are only indirectly affected by the existence of velocity gradients. They remain below the corresponding gas turbulent stresses and, given sufficient development time ($St > 2$), they become nearly equal to the fluid–particle velocity correlations.

The same but less pronounced features are measured for the other simulated cases. The decrease of the particle diameter, $d = 45$ and $30\ \mu$m, or of the mean shear rate, $S = 25\,/\mathrm{s}$, leads to an increase of the time scale ratio $\eta_r = \tau_{12}^t / \tau_{12}^F$ as illustrated in Fig. 3. Then the differences between the fluid, particle and fluid–particle velocity correlations are diminished as seen on Figs. 5*(a)-(e)* and Figs. 6*(a)-(e)* for $d = 30\ \mu$m ($S = 50\,/\mathrm{s}$) and $d = 60\ \mu$m ($S = 25\,/\mathrm{s}$), respectively. It is interesting to note from the comparison between Fig. 4*(a)* and Fig. 5*(a)* that the streamwise particle velocity variance shows a slight tendency to increase with the particle relaxation time τ_{12}^F. This dependence on particle response time is opposite to that measured in homogeneous and steady flow where the particle kinetic stresses all tend to zero with increasing relaxation time. On the other hand, the comparison of Figs. 4*(b)-(c)* and Figs. 5*(b)-(c)* shows that the particle velocity variances in the spanwise and normal directions decrease when the particle relaxation time is increasing. According to Fig. 4*(d)* and Fig. 5*(d)*, one cannot see a clear dependence of the particle kinetic shear stress on the particle relaxation time. The initial time development of the shear stress is found to be more effective for the larger particles. But later on, $St > 2$, the development rate is strongly reduced and the final value is about the same for both particle diameters, $d = 30$ and $d = 60\ \mu$m.

The comparison of Figs. 4*(a)-(e)* and Figs. 6*(a)-(e)* does not allow us to separate the direct influence of the strength of the mean shear on the particle kinetic stresses because of the simultaneous variation of the fluid turbulent integral time scales. Indeed, as illustrated in Fig. 3, the decrease of the mean velocity gradients leads to an appreciable increase of the corresponding fluid turbulent characteristic time scale and so reduces the influence of inertia on the particle fluctuating motion.

Finally, Figs. 10*(a)-(f)* and Figs. 11*(a)-(f)* show the time development of the dimensionless anisotropy tensor components of the particle fluctuating motion

$$b_{2,ij} = \left[\langle u''_{2,i}\, u''_{2,j} \rangle_2 \,/\, 2\, q_2^2 \; - \; \frac{1}{3}\, \delta_{ij} \right] \tag{53}$$

for the different simulated cases, which are found to increase strongly with the particle diameter and the mean shear rate, consistent with the analysis above.

In order to understand the measured changes in the trend of the particle statistics, one has to examine the dynamic behaviour of the kinetic stresses and their interactions with the fluid turbulence using the corresponding balance equations. The exact particle kinetic stress equations for homogeneous and sheared turbulent flows read

$$\frac{\partial}{\partial t}\, \langle u''_{2,1}\, u''_{2,1} \rangle_2 = \Pi_{2,11} \; - 2 \langle u''_{2,1}\, u''_{2,3} \rangle_2 \,\frac{\partial U_{2,1}}{\partial x_3} \tag{54}$$

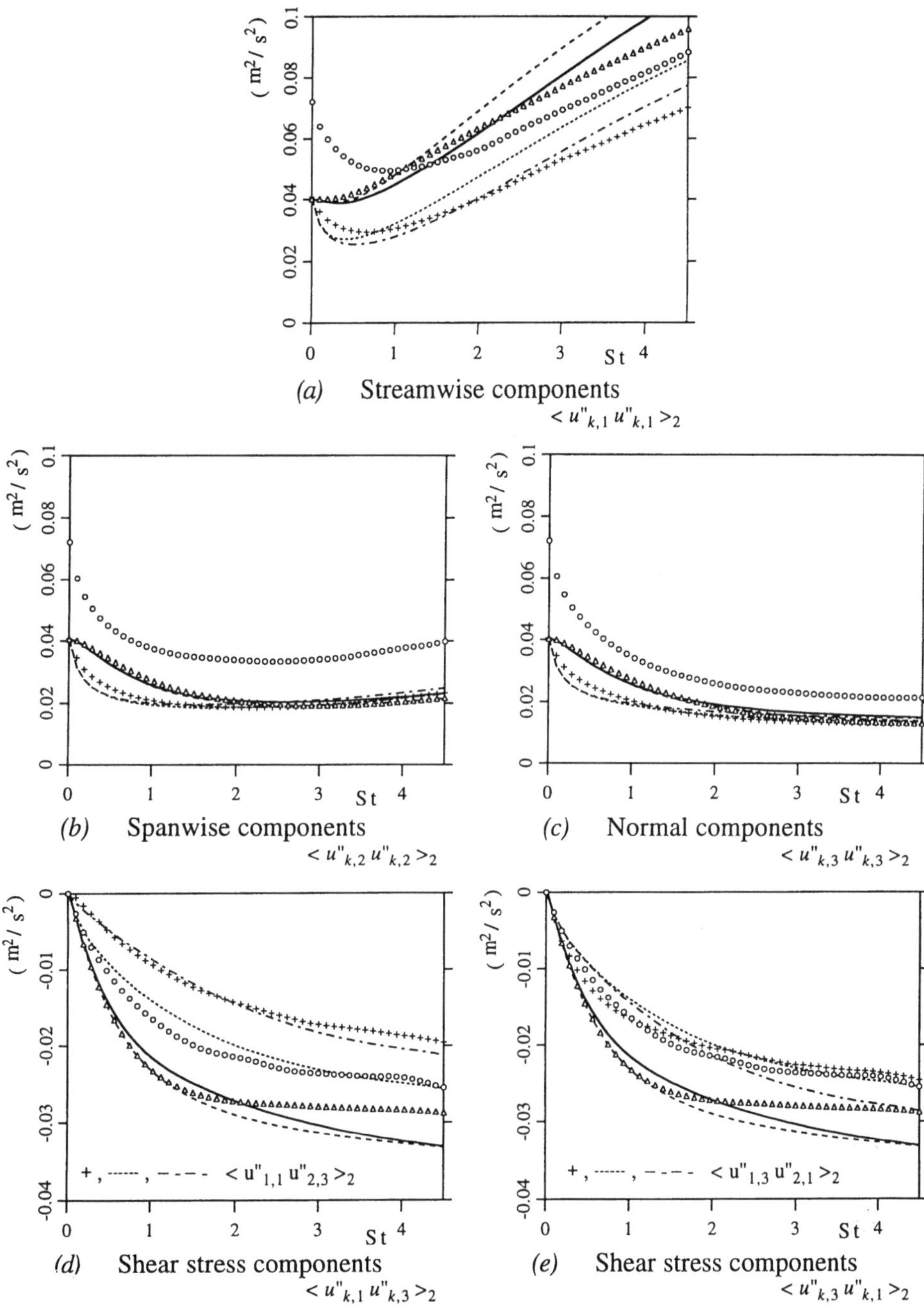

Fig. 4. Time development of the turbulent correlations (particle diameter, d = 60 μm, mean shear rate $S = 50\,/\mathrm{s}$) comparison between LES results (symbol) and two-fluid model predictions (lines) from the simulated fluid turbulent Reynolds stresses: ∘, fluid turbulent Reynolds stresses; △, --- model 1 and —— model 2, particle kinetic stresses; +, · · · · · model 1 and · — · — model 2, fluid–particle velocity correlations.

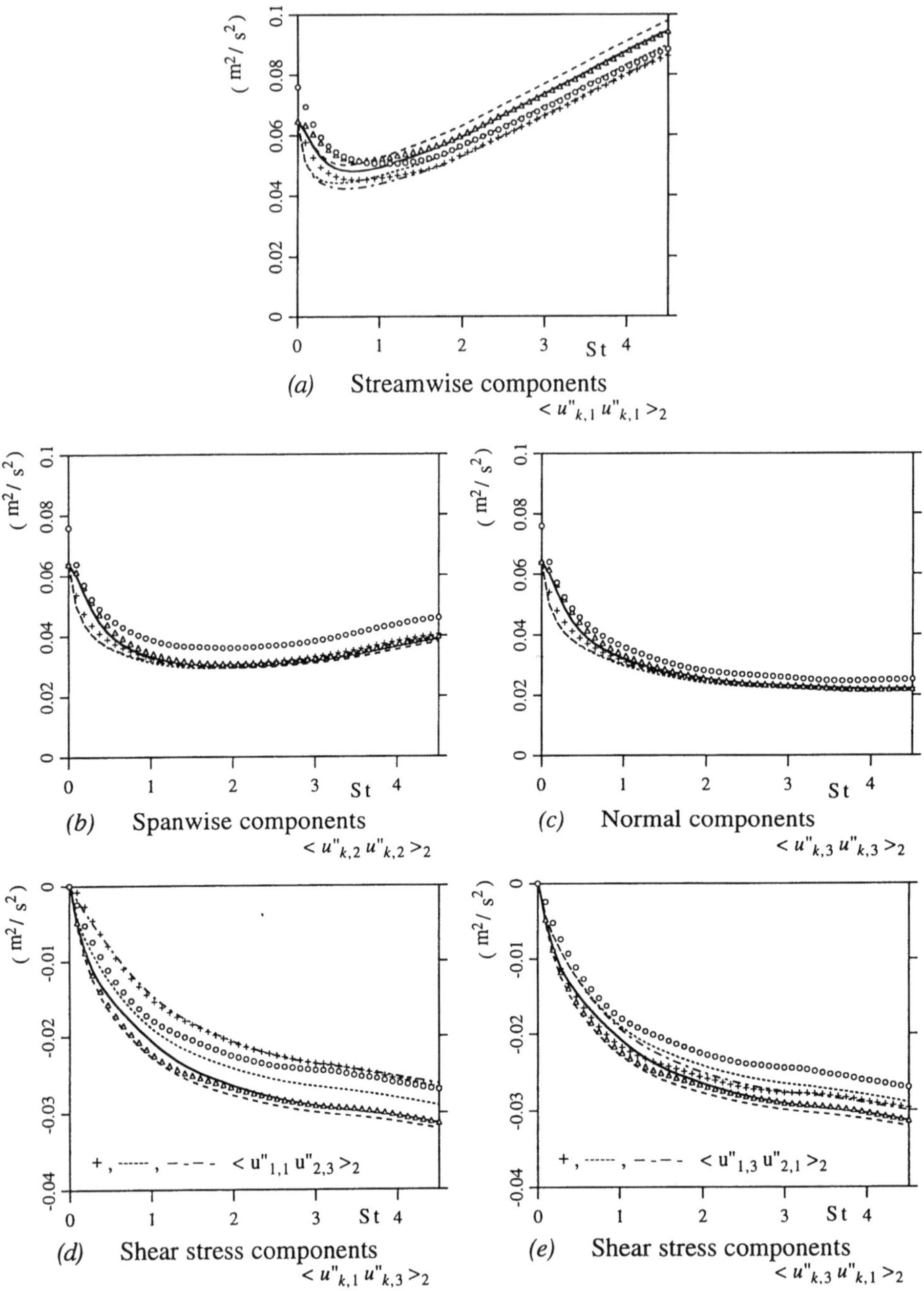

Fig. 5. Time development of the turbulent correlations (particle diameter, d = 30 μm, mean shear rate $S = 50$ /s) comparison between LES results (symbol) and two-fluid model predictions (lines) from the simulated fluid turbulent Reynolds stresses: ο, fluid turbulent Reynolds stresses; △, --- model 1 and —— model 2, particle kinetic stresses; +, · · · · · model 1 and · — · · — model 2, fluid–particle velocity correlations.

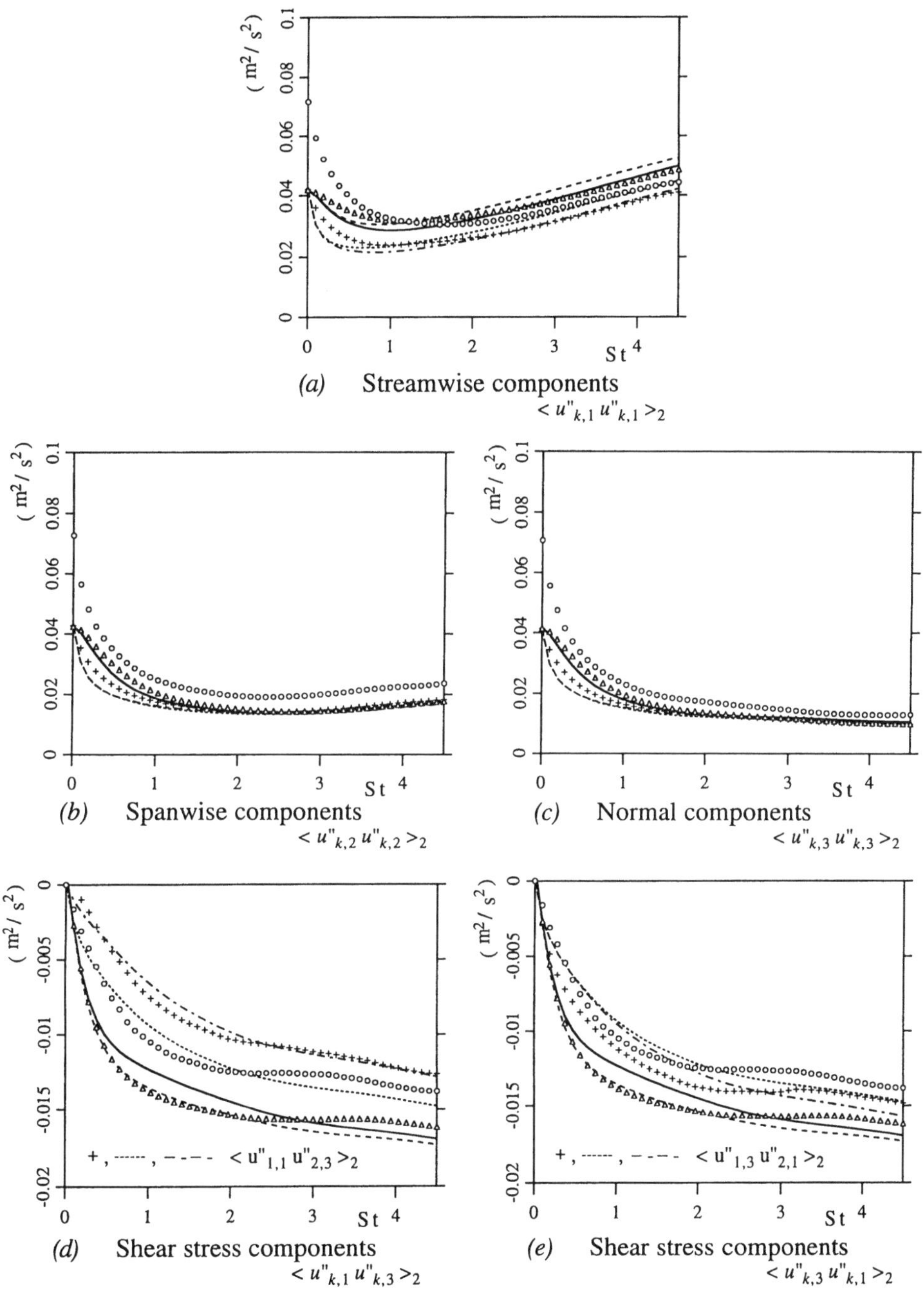

Fig. 6. Time development of the turbulent correlations (particle diameter, d = 60 μm, mean shear rate $S = 25$ /s) comparison between LES results (symbol) and two-fluid model predictions (lines) from the simulated fluid turbulent Reynolds stresses: ∘, fluid turbulent Reynolds stresses; △, --- model 1 and —— model 2, particle kinetic stresses; +, · · · · · model 1 and · — · — model 2, fluid–particle velocity correlations.

$$\frac{\partial}{\partial t} \langle u''_{2,2}\, u''_{2,2} \rangle_2 = \Pi_{2,22} \tag{55}$$

$$\frac{\partial}{\partial t} \langle u''_{2,3}\, u''_{2,3} \rangle_2 = \Pi_{2,33} \tag{56}$$

$$\frac{\partial}{\partial t} \langle u''_{2,1}\, u''_{2,3} \rangle_2 = \Pi_{2,13} \quad - \langle u''_{2,3}\, u''_{2,3} \rangle_2 \frac{\partial U_{2,1}}{\partial x_3} \,. \tag{57}$$

Equations (54) to (57) display the terms of the total change, the changes due to particle dragging by the fluid and shear production from left to right, respectively.

Before analysing the behaviour of the equations above, one can examine the accuracy of the approximate form (28) of the turbulent interphase momentum transfer rate $\Pi_{2,ij}$ adopted for the second-order model presented in this paper. This closure approximation allows one to roughly account for the non-linear dependence of the drag coefficient on the instantaneous turbulent relative velocity and can be split into two terms $\Pi^d_{2,ij}$ and $\Pi^P_{2,ij}$ linear in the particle kinetic stresses and the fluid–particle velocity correlations, respectively,

$$\Pi^d_{2,ij} = -\frac{2}{\tau^F_{12}} \langle u''_{2,i}\, u''_{2,j} \rangle_2 \tag{58}$$

$$\Pi^P_{2,ij} = \frac{1}{\tau^F_{12}} \left[\langle u''_{1,i}\, u''_{2,j} \rangle_2 + \langle u''_{1,j}\, u''_{2,i} \rangle_2 \right] \tag{59}$$

where $\Pi^d_{2,ij}$ corresponds to the particle kinetic stress dissipation rate due to the drag force which should occur in a laminar fluid flow. $\Pi^P_{2,ij}$ is a production term which brings the components of the particle kinetic stress tensor towards the ones of the symmetrical fluid–particle velocity covariance tensor, in a way consistent with the solution at equilibrium in homogeneous turbulence.

The left hand side of Fig. 7 shows a very good agreement between the approximate form (28) of $\Pi_{2,ij}$ and the exact one (27) taken from the LES results. The error induced if neglecting totally the non-linear form of the drag coefficient should be about 15%, according to the measured discrepancy between the Stokes characteristic time τ_p and the mean particle relaxation time τ^F_{12} (see Table 3). One can notice from Fig. 7 that the turbulent momentum transfer rates are found to be dominated by the destruction part, except for the spanwise component for sufficient development time. This relates initially to the fact that, due to the inertia influence, the particle turbulent stresses decrease more slowly than the fluid–particle correlations. And later, this corresponds to larger increases of the streamwise particle velocity variance and of the kinetic shear stress due to the mean velocity gradients than those of the corresponding fluid–particle velocity correlations.

The budgets of the particle kinetic stresses are shown in Figs. 8*(a)-(d)* for the largest particle diameter (d = 60 μm) and the strongest mean shear rate (S = 50 /s). We can notice a quasi-equilibrium situation for sufficient development time, the terms $\partial \langle u''_{2,i}\, u''_{2,j} \rangle_2 / \partial t$ becoming much smaller than the terms on the right-hand side of the equations. According to (55) and (56), the spanwise and normal particle velocity

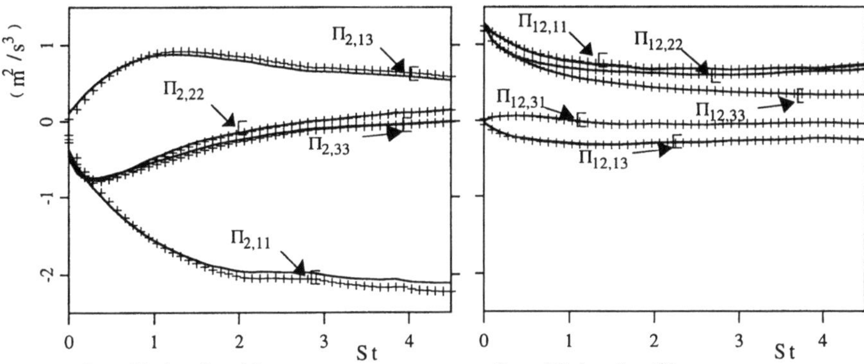

$S = 50$ /s, $d = 60$ μm $S = 50$ /s, $d = 60$ μm

Fig. 7. Turbulent interphase momentum transfer rates versus time for the particle kinetic stresses (on the left) and the fluid–particle velocity correlations (on the right). Comparison between LES results (+) and closure model predictions (——) using (28) and (46) with the simulated velocity correlations.

variances are controlled by the balance between $\Pi^d_{2,ij}$ and $\Pi^P_{2,ij}$, and tend towards the fluid–particle velocity correlations. On the other hand, Fig. 8*(a)* and Fig. 8*(d)* show the additional contributions due to the mean particle velocity gradient which act as source terms for the streamwise velocity variance and the kinetic shear stress. These contributions appear to reach constant values when the flow becomes developed.

Figures 4 to 6 compare the LES results with the predictions using the second-order model for the particle and fluid–particle velocity correlations, from (24) and (41). The problem of modelling the fluid turbulence characteristics is not considered here and the Reynolds stresses $\langle u"_{1,i}\, u"_{1,j} \rangle_2$ and dissipation rate ε_1 are taken from the LES results. Predictions are performed for the different closure approximations proposed previously for the tensor $G_{12,ij}$: model 1 and model 2, corresponding to $\beta_2 = 0$ and 0.6 in (39), respectively. One can notice that both models account for the main features of the simulated cases. In particular they predict correctly the reduction of the spanwise and normal particle velocity variances with respect to the particle inertia effect and the corresponding enlargement of the streamwise velocity variance. However, the model 1 shows a tendency to overestimate the streamwise velocity variance and this effect is reduced with the model 2. In all cases, the modelled evolution of the spanwise and normal velocity variances are practically insensitive to change in the model constant β_2. We conclude that the model yields good agreement with LES in homogeneous flows subject to a simple shear. The satisfactory quantitative agreement measured for the modelled turbulent momentum transfer rate confirms the fact that the key point of the presented second-order model is the prediction of the fluid–particle velocity correlations.

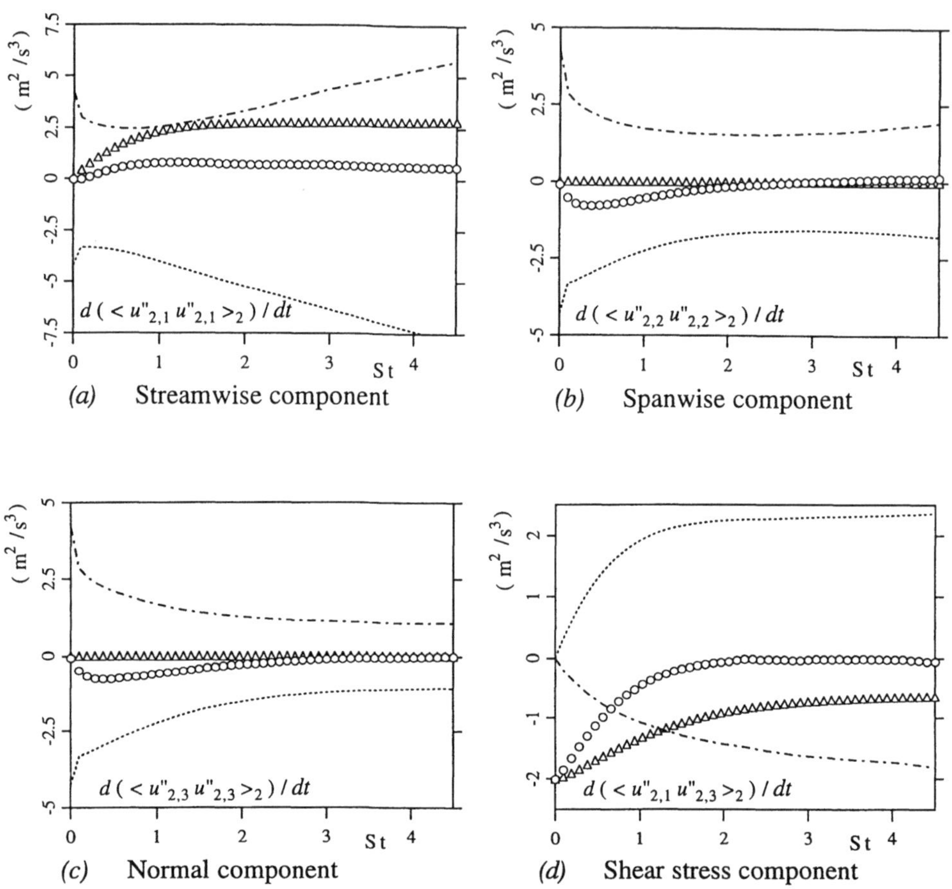

Fig. 8. Budgets of the particle kinetic stresses versus time from LES (particle diameter, d = 60 μm, mean shear rate $S = 50\,/\mathrm{s}$): ∘, derivative with respect to time; △, particle velocity shear production; · · · · ·, drag dissipation in laminar fluid flow and · — · —, production by dragging along fluid turbulence: closure model predictions using (58) and (59), respectively, with the simulated velocity correlations.

4.3 Fluid–Particle Velocity Correlations

As shown before, the fluid–particle correlations play a large part in the equations of the particle kinetic stresses through the turbulent interphase momentum transfer rates and they have to be modelled in a second-order model. Figs. 4*(a)-(c)* show that the three simulated diagonal components of the fluid–particle velocity tensor: $\langle u''_{1,1}\, u''_{2,1}\rangle_2$, $\langle u''_{1,2}\, u''_{2,2}\rangle_2$, and $\langle u''_{1,3}\, u''_{2,3}\rangle_2$, remain below the corresponding Reynolds stresses. But the gap between the fluid and fluid–particle correlations is slowly decreasing with time, probably due to the increase of the efficiency of the particle dragging by the fluid turbulence with the characteristic time scales ratio $\eta_r = \tau^t_{12}/\tau^F_{12}$. Concurrently, Figs. 4*(d)-(e)* show that the fluid–particle velocity covariance tensor is not symmetrical. As a matter of fact, the extra-diagonal component $\langle u''_{1,3}\, u''_{2,1}\rangle_2$ is nearly identical to the fluid Reynolds shear stress while the $\langle u''_{1,1}\, u''_{2,3}\rangle_2$ time development is less efficient. The same features are measured for the other simulated cases (see Figs. 5 and 6) though the differences between the fluid–particle correlations and the fluid Reynolds stresses are appreciably diminished, consistent with an increase of η_r.

The behaviour of the fluid–particle velocity correlations will now be studied by examining the corresponding budgets. The modelled fluid–particle correlation equations for homogeneous and sheared turbulent flows read

$$\begin{aligned}\frac{\partial}{\partial t}\langle u''_{1,1}\, u''_{2,1}\rangle_2 &= \Pi_{12,11} - \frac{1}{\tau^t_{12}}\langle u''_{1,1}\, u''_{2,1}\rangle_2 \\ &- \langle u''_{1,1}\, u''_{2,3}\rangle_2 \frac{\partial U_{2,1}}{\partial x_3} - [1-\beta_2]\langle u''_{1,3}\, u''_{2,1}\rangle_2 \frac{\partial U_{1,1}}{\partial x_3}\end{aligned} \tag{60}$$

$$\frac{\partial}{\partial t}\langle u''_{1,2}\, u''_{2,2}\rangle_2 = \Pi_{12,22} - \frac{1}{\tau^t_{12}}\langle u''_{1,2}\, u''_{2,2}\rangle_2 \tag{61}$$

$$\frac{\partial}{\partial t}\langle u''_{1,3}\, u''_{2,3}\rangle_2 = \Pi_{12,33} - \frac{1}{\tau^t_{12}}\langle u''_{1,3}\, u''_{2,3}\rangle_2 \tag{62}$$

$$\begin{aligned}\frac{\partial}{\partial t}\langle u''_{1,1}\, u''_{2,3}\rangle_2 &= \Pi_{12,13} - \frac{1}{\tau^t_{12}}\langle u''_{1,1}\, u''_{2,3}\rangle_2 \\ &- [1-\beta_2]\langle u''_{1,3}\, u''_{2,3}\rangle_2 \frac{\partial U_{1,1}}{\partial x_3}\end{aligned} \tag{63}$$

$$\begin{aligned}\frac{\partial}{\partial t}\langle u''_{1,3}\, u''_{2,1}\rangle_2 &= \Pi_{12,31} - \frac{1}{\tau^t_{12}}\langle u''_{1,3}\, u''_{2,1}\rangle_2 \\ &- \langle u''_{1,3}\, u''_{2,3}\rangle_2 \frac{\partial U_{2,1}}{\partial x_3}\,.\end{aligned} \tag{64}$$

Equations (60) to (64) display the terms of the total change, the changes due to particle dragging by the fluid, pressure correlations and viscous dissipation, and shear production from left to right, respectively. The modelled term $(\Phi_{12,ij} - 1/3\, \varepsilon_{12}\, \delta_{ij})$ due to pressure correlations and viscous dissipation is developed to make clear the influence of the mean fluid velocity gradient according to the given value of β_2.

From Fig. 7 it may be observed that the approximate form (46) of the turbulent interphase momentum transfer rate $\Pi_{12,ij}$ is in very good agreement with the values taken from LES results. It is interesting to note that, according to (46), $\Pi_{12,ij}$ can be split into two terms $\Pi^d_{12,ij}$ and $\Pi^P_{12,ij}$ linear in the fluid–particle velocity correlations and the fluid turbulent Reynolds stresses, respectively,

$$\Pi^d_{12,ij} = -\frac{1}{\tau^F_{12}} \langle u''_{1,i}\, u''_{2,j} \rangle_2 \qquad \Pi^P_{12,ij} = \frac{1}{\tau^F_{12}} \langle u''_{1,i}\, u''_{1,j} \rangle_2 \tag{65}$$

where $\Pi^d_{12,ij}$ corresponds to a fluid–particle dissipation rate due to the drag force and $\Pi^P_{2,ij}$ is a production term which brings the components of the fluid–particle velocity covariance tensor towards those of the fluid turbulent Reynolds stress tensor.

Shown in Figs. 9*(a)-(e)* are the budgets of the fluid–particle velocity correlations and it is apparent from the figures that a quasi-equilibrium situation occurs for sufficient development time. According to (61) and (62), the spanwise and normal particle velocity variances are controlled by the balance between $\Pi^d_{12,ij}$, $\Pi^P_{12,ij}$ and the modelled term corresponding to pressure correlations and viscous dissipation. At equilibrium, these correlations are found to satisfy the same equations as in homogeneous and steady turbulent flow

$$\langle u''_{1,2}\, u''_{2,2} \rangle_2 = \frac{\eta_r}{1+\eta_r} \langle u''_{1,2}\, u''_{1,2} \rangle_2 \tag{66}$$

$$\langle u''_{1,3}\, u''_{2,3} \rangle_2 = \frac{\eta_r}{1+\eta_r} \langle u''_{1,3}\, u''_{1,3} \rangle_2 \tag{67}$$

where $\eta_r = \tau^t_{12}/\tau^F_{12}$.

On the other hand, (60), (63) and (64) make clear the separate dependence of the streamwise fluid–particle velocity correlation $\langle u''_{1,1}\, u''_{2,1} \rangle_2$ and of the extra-diagonal components $\langle u''_{1,1}\, u''_{2,3} \rangle_2$ and $\langle u''_{1,3}\, u''_{2,1} \rangle_2$ on the fluid and particle mean velocity gradients. From Fig. 9, these contributions appear to reach constant values when the flow becomes developed. Finally, according to (63) and (64), the non-symmetrical form of the simulated fluid–particle correlation tensor can be related to the closure assumption proposed for the mean fluid velocity influence on the pressure correlations, corresponding to the so-called rapid pressure term. Consequently, model 2 gives a better prediction of the streamwise particle and fluid–particle velocity correlations by reproducing the measured gap between the two extra-diagonal components $\langle u''_{1,3}\, u''_{2,1} \rangle_2$ and $\langle u''_{1,1}\, u''_{2,3} \rangle_2$ (cf. Figs. 4, 5 and 6).

5. Conclusions

Large eddy simulation has been applied to investigate the turbulence statistics of spherical solid particles suspended in homogeneous turbulent gas shear flows. It

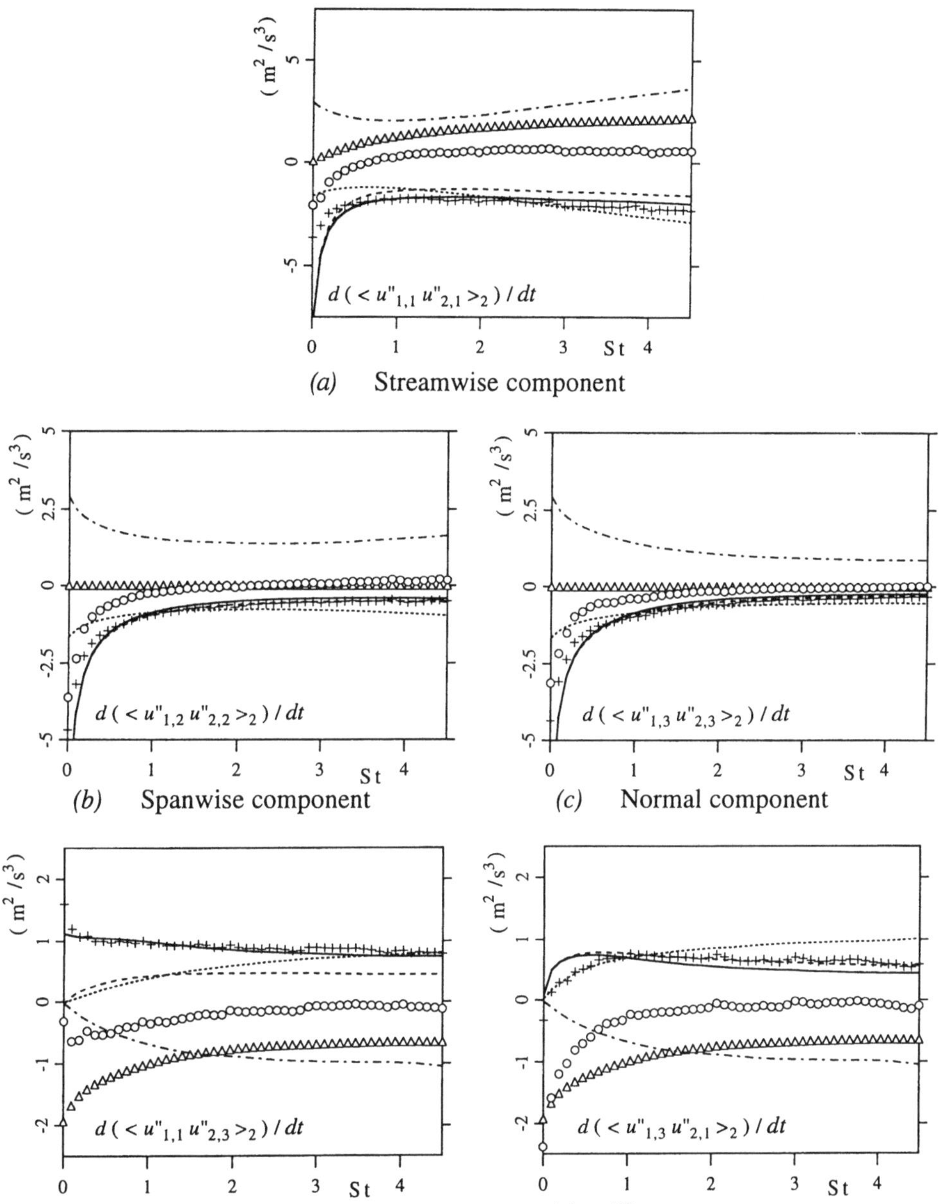

Fig. 9. Budgets of the fluid–particle velocity correlations versus time from LES (particle diameter, d = 60 μm, mean shear rate $S = 50$ /s): ∘, derivative with respect to time; △, fluid and particle velocity shear production; · · · · · and · — · —, drag dissipation in laminar fluid flow and production by dragging along fluid turbulence, respectively; +, sum of pressure–strain correlations, viscous dissipation rates and turbulent crossing-trajectory effects ($\Phi_{12,ij} - 1/3\,\varepsilon_{12}\,\delta_{ij}$), - - - and ——, corresponding closure model predictions using (48) with coefficient $\beta_2 = 0.$ and $\beta_2 = 0.6$, respectively.

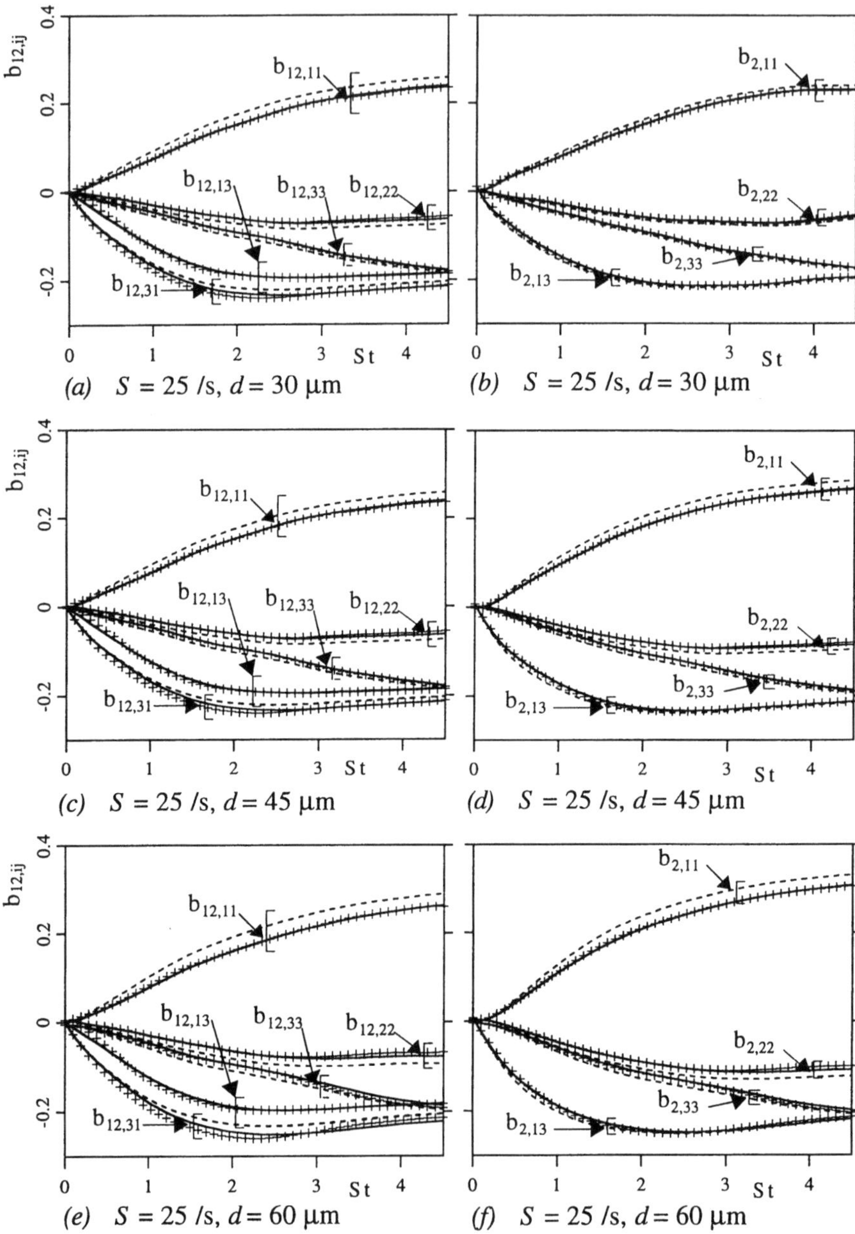

Fig. 10. Time development of the normalized anisotropy of fluid–particle velocity correlations (on the left) and particle kinetic stresses (on the right) for the mean shear rate $S = 25\,/\mathrm{s}$. Comparison between LES results (+) and two fluid model predictions (--- model 1 and —— model 2) from the simulated fluid turbulent Reynolds stresses.

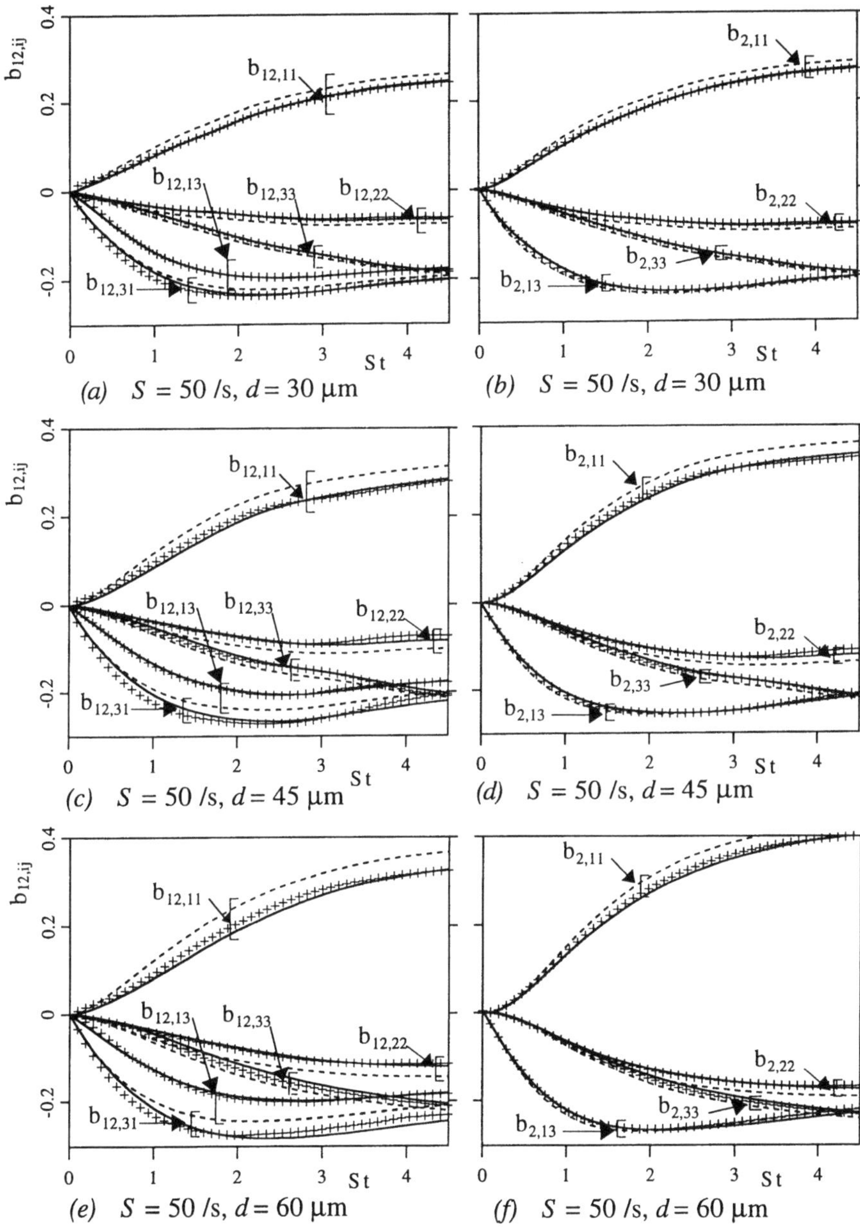

Fig. 11. Time development of the normalized anisotropy of fluid–particle velocity correlations (on the left) and particle kinetic stresses (on the right) for the mean shear rate $S = 50\,/\mathrm{s}$. Comparison between LES results (+) and two fluid model predictions (--- model 1 and —— model 2) from the simulated fluid turbulent Reynolds stresses.

was shown that the mean shear of the flow has a large influence on the particle turbulent velocities and that, in the absence of an externally imposed body force, the most noticeable effect of inertia is to increase the degree of anisotropy of the particle fluctuating motion with respect to that of the fluid. The simulation results are found to be in good agreement with the predictions of a second-order closure model based on separate transport equations for the particle kinetic stresses and fluid–particle velocity correlations. The proposed closure model comprises no extra empirical constants but those of the corresponding single-phase turbulence model.

The analysis of the equation set governing the particle kinetic stresses in very dilute gas–solid shear flows shows that the particle fluctuating motion is controlled by the dragging by the fluid turbulent motion and the production due to the mean particle velocity gradient, simultaneously. The dragging by the turbulence, which becomes dominant when the particle relaxation time to fluid turbulent time scale ratio decreases (with the particle diameter), leads to a first increase of the particle fluctuating motion anisotropy. As a matter of fact, the corresponding source term in the particle kinetic stress equation can be accurately written as a driving term towards the fluid–particle velocity correlations which show a larger degree of anisotropy than the fluid Reynolds stresses. The modelled fluid–particle velocity correlation equations show a separate dependence on the fluid and particle mean velocity gradients, while the redistribution mechanisms induced by the pressure correlations are found to be less efficient than for the fluid turbulence. On the other hand, the production term in the particle kinetic stress equation due to the particle velocity gradient induces a complementary enlargement of the streamwise particle turbulent velocity to values exceeding the corresponding fluid turbulent component.

Nevertheless, according to results of the kinetic theory of rarefied gases, we can notice that the inter-particle collision influence should lead to a return-to-isotropy contribution in the particle stress equation reducing the particle fluctuating motion anisotropy when the inter-particle collision time becomes comparable with the particle relaxation time (He and Simonin 1993). Further simulations are planned in the near future, accounting for inter-particle collisions and externally uniform, or linear dependant, body forces in the particle motion equation, in order to point out the crossing-trajectory effect and the separate dependence on the mean particle and fluid velocity gradients.

References

Bardina, J., Ferziger, J.H., Reynolds, W.C. (1983): "Improved Turbulence Models Based on Large Eddy Simulation of Homogeneous Incompressible Turbulent Flows", *Dept. Mech. Engen. Rep.*, TF 19, Standford University, California

Csanady, G.T. (1963): "Turbulent Diffusion of Heavy Particles in the Atmosphere", *J. Atm. Science*, Vol. 20, pp. 201–208

Deutsch, E., Simonin, O. (1991): "Large Eddy Simulation Applied to the Modelling of Particulate Transport Coefficients in Turbulent Two-Phase Flows", in *Proc. 8th Symp. on Turbulent Shear Flows* (Univ. of Munich), Vol. 1, pp. 1011–1016

Deutsch, E. (1992): "Dispersion de Particules dans une Turbulence Homogène Isotrope Stationnaire Calculée par Simulation Numérique Directe des Grandes Echelles", in *Collection de notes internes de la Direction des Etudes et Recherches* (Electricité de France, 92141 Clamart Cedex)

Drew, D.A. (1983): "Mathematical Modeling of Two-Phase Flow", *Ann. Rev. Fluid Mech.*, Vol. 15, pp. 261–291

Gatignol, R. (1983): "The Faxen Formulae for a Rigid Particle in an Unsteady Non-Uniform Stokes Flow", *J. de Méc. Th. et Appl.*, Vol. 1, pp. 143–160

Haworth, D.C., Pope, S.B. (1986): "A Generalized Langevin Model for Turbulent Flows", *Phys. Fluids*, Vol. 29, pp. 387–405

He, J., Simonin, O. (1993): "Non-Equilibrium Prediction of the Particle-Phase Stress Tensor in Vertical Pneumatic Conveying", in *Proc. 5th International Symposium on Gas–Solid Flows*, ASME FED, Vol. 166, pp. 253–263

Hinze, J.O. (1972): "Turbulent Fluid and Particle Interaction', *Prog. Heat and Mass Transfer*, Vol. 8, pp. 433–452

Khoudly, M. (1988): "Macrosimulation de Turbulence Homogène en Présence de Cisaillement et de Gradients Thermiques. Application aux Modèles de Fermeture en un Point", Ph.D. thesis, Ecole Centrale de Lyon

Laurence, D., (1985): "Advective Formulation of Large Eddy Simulation for Engineering Flows", *Notes on Numerical Fluid Mechanics*, Vol. 15, pp. 147–160

Launder, B.E., Reece, G.J., Rodi, W., (1975): "Progress in the Development of a Reynolds-Stress Turbulence Closure", *J. Fluid Mech.*, Vol. 68, pp. 537–566

Liljegren, L.M. (1993): "The Effect of a Mean Fluid Velocity Gradient on the Streamwise Velocity Variance of a Particle Suspended in a Turbulent Flow", *Int. J. Multiphase Flow*, Vol. 19, pp. 471–484

Maxey, M.R. (1987): "The Gravitational Settling of Aerosol Particles in Homogeneous Turbulence and Random Flow Fields", *J. Fluid Mech.*, Vol. 174, pp. 441–465

Reeks, M.W. (1993): "On the Constitutive Relations for Dispersed Particles in Nonuniform Flows 1: Dispersion in a Simple Shear Flow", *Phys. Fluids*, Vol. 5, pp. 750–761

Rogers, C.B., Eaton, J.K. (1990): "The Behavior of Solid Particles in a Vertical Turbulent Boundary Layer in Air", *Int. J. Multiphase Flow*, Vol. 16, pp. 819–834

Simonin, O. (1991): "Second-Moment Prediction of Dispersed Phase Turbulence", in *Proc. 8th Symp. on Turbulent Shear Flows* (Univ. of Munich), Vol. 1, pp. 741–746

Simonin, O., He, J. (1992): "Eulerian Prediction of the Particle Behaviour in a Turbulent Boundary Layer", in *Proc. 6th Workshop on Two-Phase Flow Predictions*, ed. by M. Sommerfeld (Univ of Erlangen), pp. 154–165

Simonin, O., Deutsch, E., Minier, J.P. (1993): "Eulerian Prediction of the Fluid/Particle Correlated Motion in Turbulent Two-Phase Flows", *Applied Scientific Research*, Vol. 51, pp. 275–283

Sommerfeld, M., Qiu, H.H. (1991): "Detailed Measurements in a Swirling Particulate Two-Phase Flow by a Phase-Doppler Anemometer", *Int. J. Heat and Fluid Flow*, Vol. 12, pp. 20–28

Squires, K.D., Eaton, J.K. (1991): "Measurements of Particle Dispersion Obtained from Direct Simulations of Isotropic Turbulence", *J. Fluid Mech.*, Vol. 226, pp. 1–35

Tchen, C.M. (1947): "Mean Value and Correlation Problems Connected with the Motion of Small Particles Suspended in a Turbulent Fluid", Ph.D. thesis, Delft

Wells, M.R., Stock, D.E. (1983): "The Effects of Crossing Trajectories on the Dispersion of Particles in a Turbulent Flow", *J. Fluid Mech.*, Vol. 136, pp. 31–62

Yeh, F., Lei, U. (1991): "On the Motion of Small Particles in a Homogeneous Turbulent Shear Flow", *Phys. Fluids*, Vol. 3, pp. 2758–2775

Part II

Free Turbulent Flows

Introductory Remarks

M. Sokolov

Department of Fluid Mechanics and Heat Transfer
Faculty of Engineering, Tel-Aviv University
Ramat-Aviv 69978, Israel

Free turbulent flows are of practical importance and have therefore received extensive investigative attention which include various configurations and viewpoints. Generally for turbulent flows, it is difficult to detach pure fundamentals from applied research. For example, discovering the important role in momentum and heat transfer of coherent structures in turbulent flows, (Brown and Roshko (1974), Winant and Browand (1974)) was a significant step toward the understanding of basic mechanism of such flows. At the same time, manipulation of such coherent structures is a powerful and practical tool for the control of the entire flow field characteristics. The investigation of coherent structures in free turbulent flows is therefore of great fundamental and practical relevance. Three of the papers in this section deal with coherent structures in wake and round jets. The other two deal with experiments in heated mixing layer and numerical simulations.

The paper by Yu, Gutmark and Schadow presents an experimental investigation of coherent large-scale structures in a compressible axisymmetric shear layer. Attention to the role of large-scale vortices in compressible mixing was drawn by complex problems like noise-generation and turbulent mixing in a combustor of scramjet engine.

In a mixing layer of two streams with velocities U_1, and U_2 and speeds of sound a_1, and a_2 respectively, the convective Mach numbers M_{c1} and M_{c2} are defined as:

$$M_{c1} = \frac{U_1 - U_c}{a_1}, \quad M_{c2} = \frac{U_c - U_2}{a_2},$$

where U_c is the convective velocity of the large scale structures. The convective Mach number characterizes the compressibility effects and was introduced by Papamoschou and Roshko (1988).

In a coordinate system moving with U_c, the existence of a saddle point between the structures implies a common stagnation pressure. For the static pressures to be equal in both streams, then:

$$\frac{P}{P_0} = \left(1 + \frac{\gamma_1 - 1}{2} M_{c1}{}^2\right)^{\frac{\gamma_1}{\gamma_1 - 1}} = \left(1 + \frac{\gamma_2 - 1}{2} M_{c2}{}^2\right)^{\frac{\gamma_2}{\gamma_2 - 1}}.$$

For specific heat ratios $\gamma_1 \approx \gamma_2$ and not too large convective Mach numbers, the convective velocity may be expressed in the form

$$U_c = \frac{a_2 U_1 + a_1 U_2}{a_1 + a_2}.$$

which is a weighted average of the speeds of sound. Published experimental works on coherent motion in compressible flows showed that for relatively small convective Mach numbers there are structures similar to the incompressible case. However coherent structures at intermediate and high convective Mach numbers were not observed.

Yu, Gutmark and Schadow have discovered a high order of organization in intermediate subsonic convective Mach number flows. The discrepancy with earlier data may be explained by the unaccounted for influence of initial disturbances due to the shape of the inner nozzle lip. The paper shows that the coupling between the compressible shear flow and the wake behind the lip is responsible for the formation of the coherent structures. Such coupling may be termed as a passive control in turbulent flows (Hussain *et al.* 1988). Similar to incompressible mixing layers where active control was achieved by applying external harmonic disturbances flow (Oster *et al.* (1978), Hussain (1978), Hussain *et al.* (1988)), the results of Yu, Gutmark and Schadow suggest that such control may be feasible also for intermediate convective Mach numbers. No evidence to the existence of organized coherent motion was found for convective Mach number above unity. Such observation is in agreement with data of McIntyre and Settles (1991). Further investigations are definitely needed in order to elucidate the conditions for existence of organized structures in flows with a supersonic convective Mach number.

Mi and Antonia study coherent large structures in the near ($x/d \lesssim 5$) and intermediate ($5 \leqslant x/d \leqslant 50$) wake of a slightly heated circular cylinder at subcritical Reynolds numbers (≈ 5500). An array of cold wires trace the thermal tags to allow characterization of the turbulent vortices. Conditional averaging (based on local temperature maximum which corresponds to the vortex center) exhibits a Gaussian shaped temperature distribution near the center of the vortex. This is similar to the vorticity distribution found by Zhou and Antonia (1993). This temperature distribution with the Oseen-like vortex model was successfully used to evaluate mean and rms velocity as well as temperature profiles with good agreement with experimental data.

Earlier experimental investigations have shown that the turbulent Kármán vortex sheet transports heat more effectively than momentum. Analysis of velocity and temperature fields, given in the paper, provides the explanation for this phenomenon. However the three-dimensional topology in the near wake of a circular cylinder as pointed by Hayakawa and Hussain (1989) is not included in such analysis. Thus, further investigation of the complicated three-dimensional heat transport in the wake is needed.

Turbulent mixing with buoyant effects is associated with vast technological applications. The problems of a vertical jet (smoke stack) discharged into stratified atmospheric cross-flow or the dispersing of thermal pollution in lakes and other water reservoirs are but few examples.

The behavior of buoyant jet depends on relative values of inertia, viscosity and buoyancy. In a buoyant jet there are three regions which could be characterized by different roles of the motive forces. In the vicinity of the jet nozzle, inertia forces are dominant and the flow may be considered as self-similar. Far from the nozzle, in the third region, the flow is governed by buoyancy and a self-similar description may also be appropriate. However, in the intermediate second region the flow is too complex to retain a self-similar description. When the jet is discharged into cross-flow, which may be also stratified, the description is obviously much more complicated.

The basic methods for modelling buoyant effects include (Schetz, 1980) 1. mean-flow models (integral analysis); 2. two-equations models ($k - \varepsilon$ models); 3. Reynolds stress models. Out of these, the first (integral models) is incapable of handling three dimensional effects which are essential in the near-field of such flow.

Development of any three dimensional computational methods which will indeed predict the complex structure of a jet discharge flow must rely on experimental data for detailed verification. The paper by Gosman, Liu and McGuirk reports on such proper experimental data. The data is used for validation of the Gibson-Launder (1978) Reynolds stresses closure procedure and a $k - \varepsilon$ model. The validation process proves that the second momentum closure performs much better than the eddy viscosity model in predicting both neutral and stable stratified cross-flow cases. Second momentum model predicts a rapid damping of the plume centerline for a jet discharged into a linearly stratified cross-flow. This result, which the integral models cannot predict, agrees well with experimental data.

Mastorakos, Taylor and Whitelaw study the mixing process in the counterflow of two turbulent jets. Such problem stems from combustion studies of stagnating flames. In their experiments, one of the jets was heated such that its temperature acted as a passive scalar. Measurements (LDA) of the velocity and temperature fields exhibit some of the new features of the scalar mixing in this flow. For example the temperature tagging elucidates the properties of mixture fractions in such mixing. It was found that the mixing layer thickness depends only on the turbulent intensity and not on the bulk velocity. Thus initial turbulence enhancement may also be used as mixing control. The probability density functions for temperature fluctuations have a bimodal character with maxima corresponding to the initial temperatures of the jets. The bimodality is governed by turbulence intensity in the jets. Such probability density distribution also means that the interface separating the two jets fluctuates randomly with fluid particles. The results by Mastorakos, Taylor and Whitelaw are supported by an eddy-viscosity model which was earlier proposed by the authors to provide detailed flow description.

The paper by Chen, Lienau and Kollman presents numerical simulation of the near field (less than fifteen diameters) turbulent flow of a round jet at $Re = 8000$. The authors consider a direct Navier-Stokes equations solution based on finite-difference method with and without the application of a large eddy simulation (LES) model. The computation commences with initial disturbances corresponding to inviscid instability at the entrance section. It is observed that the azimuthal instability of the vortex rings causes tube-like structures which collapse into counter-rotating vortex tubes which interact with the vortex pairing process.

By providing for detailed tracking of the evolution of initial inviscid instability into vortex rings and helical vortex structures, the numerical modeling is established as a powerful tool for clarifying the complexities of three-dimensional turbulent flows.

Acknowledgment

M. Sokolov wishes to express his gratitude to his colleague, Prof. A. Tumin for his very valuable help during the preparations of these introductory remarks.

References

Brown, G.L. and Roshko, A. 1974 On Density Effects and Large Structures in Turbulent Mixing Layers, J. Fluid Mech. **64**, pp. 775 - 816.

Gibson, M. M. and Launder, B. E. 1978 Ground Effects on Pressure Fluctuations in the Atmospheric Boundary Layer. J. Fluid Mech. **86**, part 3, pp. 491 - 511.

Hayakawa, M. and Hussain, A.K.M.F. 1989 Three-Dimensionality of Organized Structures in a Plane Turbulent Wake, J. Fluid Mech. **206**, pp. 375 - 404.

Hussain, A.K.M.F. 1978 Initial Condition Effect on Free Turbulent Shear Flows, *Lecture Notes in Physics*, **75**, Structural and Mechanisms of Turbulence I, Springer-Verlag, pp. 103 - 107.

Hussain, H.S., Bridges, J.E. and Hussain, F. 1988 Turbulents Managment in Free Shear Flows by Control of Coherent Structures. *Transport Phenomena in Turbulent Flows*, Editors M. Hirata and N. Kasagi, pp. 111-130

McIntyre, S.S. and Settles, G.S. 1991 Optical Experiments on Axisymmetric Compressible Turbulent Mixing Layers. AIAA-91-0623.

Oster, D., Wygnanski, I., Dziomba, B. and Fiedler, H. 1978 On The Effect of Initial Conditions on the Two-Dimensional Turbulent Mixing Layer, *Lecture Notes in Physics*, **75**, Structural and Mechanisms of Turbulence I, Springer-Verlag, pp. 48 - 64.

Papamouschou, D. and Roshko, A. 1988 The Compressible Turbulent Shear Layer: an Experimental Study, J. Fluid Mech. **197**, pp. 453 - 477.

Schetz, J. A. 1980 *Injection and Mixing in Turbulent Flow.* Progress in Astronautics and Aeronautics. **68**, AIAA.

Winant, C.D. and Browand, F.K. 1974 Vortex Pairing: the Mechanism of Turbulent Mixing Layer Growth at Moderate Reynolds Number, J. Fluid Mech. **63**, pp. 237 - 255.

Zhou, Y. and Antonia, R.A. 1993 A Study of Turbulent Vortices in the Near Wake of Cylinder, J. Fluid Mech. **253**, pp. 643 - 661.

Numerical Simulation of Low Re-Number Turbulence in Round Jets

J.H. Chen[1], J.J. Lienau[2], and W. Kollmann[2]

[1] Combustion Research Facility, Sandia National Laboratories, Livermore, CA 94551-0969,
[2] MAE Department, University of California at Davis, CA 95616

Abstract

The turbulent flow in spatially–developing round jets at low Reynolds numbers is simulated numerically in the first fifteen diameters. The simulations were started with the inviscid solution and small axisymmetric disturbances were introduced at the entrance section to accelerate the breakdown of the Kelvin-Helmholtz vortices. The development of the flow produced a rich variety of structures. The azimuthal instability of the vortex rings leads to tube-like structures which collapse into counter–rotating streamwise vortices that interact with the vortex pairing process. A passive scalar distribution also highlights the ring–braid region and reveals the severe straining of the braids near the jet axis and the formation of mushroom-like structures as a result of the counter-rotating nature of the streamwise vortices. A comparison with flow visualization experiments shows that the number of braid–like vortex structures is in agreement with the experiment.

1. Introduction

The turbulent flow in spatially–developing round jets at low Reynolds numbers is simulated numerically in the first fifteen diameters. The turbulence in round and plane jets contains a rich variety of structures. Their development is initiated by the Kelvin–Helmholtz instability of the shear layer emerging from the nozzle (Gutmark *et al.*, 1990, Liepmann, 1991, Liepmann and Gharib, 1992, Yoda *et al.*, 1991, Dahm *et al.*, 1992, Mungal *et al.*, 1992, Wicker and Eaton, 1993) and later significantly influenced by azimuthal instability modes which interact with the vortex pairing process and create longitudinal structures. These structures increase the entrainment rate significantly (Liepmann and Gharib, 1992) and are, therefore, of importance to the mixing efficiency of round jets.

The present paper reports computational results for the simulation of the turbulent flow in the initial region of a spatially–developing round jet at low Reynolds numbers. The flow is started with the inviscid solution and small disturbances are introduced at the nozzle exit to initiate the Kelvin–Helmholtz instability. The structures emerging in the development of the flow are analyzed using the local and global properties of vorticity. It is shown that the initial roll–up of the shear layer into ring vortices, the onset of circumferential instability and the pairwise interaction produce tube–like counter–rotating vortices (which are possibly braided) in agreement with experimental evidence.

The direct simulation of the flow in round jets has been carried out by several authors. Grinstein *et al.* (1987) and Grinstein *et al.* (1990) considered axisymmetric round jets and the dynamics of vortex rings in transitional subsonic jets. Melander *et al.* (1991) simulated the temporal jet, i.e. the periodic flow in the axial direction maintained by the axial motion of the inner cylindrical fluid relative to the outer fluid. This configuration allows the application of highly accurate spectral methods but is not suited for the simulation of the nonperiodic flow downstream of a nozzle. The same axially periodic configuration was investigated by Martin and Meiburg (1991) for the inviscid limit using vortex dynamics. These computations showed clearly the vortex ring instability leading to tube–like structures and the development of axial and radial vorticity. The present method for the solution of the Navier–Stokes equations is based on accurate finite difference methods (Lele, 1989, and Rai and Moin, 1991, Lienau and Kollmann, 1993) appropriate for the treatment of the spatially–developing round jets.

2. Navier–Stokes Equations and Solution Algorithm

The Navier–Stokes equations in cylindrical coordinates (r, θ, ζ) are solved using an accurate finite difference method. Simple grid stretching transformations are applied to the $r = r(\bar{r}, \beta)$ and $\zeta = \zeta(\bar{\zeta}, \beta)$ directions ($\beta > 1$ denotes the stretching parameter, Lienau and Kollmann, 1993) to concentrate the grid points in the region of interest and to move the outer boundary as far out as possible without wasting too many grid points.

The transformation of the Navier–Stokes equations (in dimensionless form, the jet pipe radius D and the bulk velocity u_0 at the entrance section are the reference scales) to the stretched coordinates is straightforward. Mass balance emerges in the form

$$\frac{1}{r}\frac{\partial \bar{r}}{\partial r}\frac{\partial}{\partial \bar{r}}(rv_r) + \frac{1}{r}\frac{\partial v_\theta}{\partial \theta} + \frac{\partial \bar{\zeta}}{\partial \zeta}\frac{\partial v_\zeta}{\partial \bar{\zeta}} = 0. \tag{1}$$

The radial momentum balance in transformed coordinates is given by

$$\frac{\partial v_r}{\partial t} + v_r\frac{\partial \bar{r}}{\partial r}\frac{\partial v_r}{\partial \bar{r}} + \frac{v_\theta}{r}\frac{\partial v_r}{\partial \theta} + v_\zeta\frac{\partial \bar{\zeta}}{\partial \zeta}\frac{\partial v_r}{\partial \bar{\zeta}} - \frac{v_\theta^2}{r} = -\frac{1}{\varrho}\frac{\partial \bar{r}}{\partial r}\frac{\partial p}{\partial \bar{r}} + \frac{1}{Re}\left\{\left(\frac{\partial \bar{r}}{\partial r}\right)^2\frac{\partial^2 v_r}{\partial \bar{r}^2}\right.$$
$$\left. + \left(\frac{1}{r}\frac{\partial \bar{r}}{\partial r} + \frac{\partial^2 \bar{r}}{\partial r^2}\right)\frac{\partial v_r}{\partial r} - \frac{v_r}{r^2} + \frac{1}{r^2}\frac{\partial^2 v_r}{\partial \theta^2} + \left(\frac{\partial \bar{\zeta}}{\partial \zeta}\right)^2\frac{\partial^2 v_r}{\partial \bar{\zeta}^2} + \frac{\partial^2 \bar{\zeta}}{\partial \zeta^2}\frac{\partial v_r}{\partial \bar{\zeta}} - \frac{2}{r^2}\frac{\partial v_\theta}{\partial \theta}\right\}. \tag{2}$$

The azimuthal momentum balance in transformed coordinates is given by

$$\frac{\partial v_\theta}{\partial t} + v_r \frac{\partial \bar{r}}{\partial r}\frac{\partial v_\theta}{\partial \bar{r}} + \frac{v_\theta}{r}\frac{\partial v_\theta}{\partial \theta} + v_\zeta \frac{\partial \bar{\zeta}}{\partial \zeta}\frac{\partial v_\theta}{\partial \bar{\zeta}} - \frac{v_\theta v_r}{r} = -\frac{1}{\varrho r}\frac{\partial p}{\partial \theta} + \frac{1}{Re}\left\{\left(\frac{\partial \bar{r}}{\partial r}\right)^2 \frac{\partial^2 v_\theta}{\partial \bar{r}^2}\right.$$
$$\left. + \left(\frac{1}{r}\frac{\partial \bar{r}}{\partial r} + \frac{\partial^2 \bar{r}}{\partial r^2}\right)\frac{\partial v_\theta}{\partial r} - \frac{v_\theta}{r^2} + \frac{1}{r^2}\frac{\partial^2 v_\theta}{\partial \theta^2} + \left(\frac{\partial \bar{\zeta}}{\partial \zeta}\right)^2 \frac{\partial^2 v_\theta}{\partial \bar{\zeta}^2} + \frac{\partial^2 \bar{\zeta}}{\partial \zeta^2}\frac{\partial v_\theta}{\partial \bar{\zeta}} + \frac{2}{r^2}\frac{\partial v_r}{\partial \theta}\right\}. \tag{3}$$

Finally, the axial momentum balance is given by

$$\frac{\partial v_\zeta}{\partial t} + v_r \frac{\partial \bar{r}}{\partial r}\frac{\partial v_\zeta}{\partial \bar{r}} + \frac{v_\theta}{r}\frac{\partial v_\zeta}{\partial \theta} + v_\zeta \frac{\partial \bar{\zeta}}{\partial \zeta}\frac{\partial v_\zeta}{\partial \bar{\zeta}} = -\frac{1}{\varrho}\frac{\partial p}{\partial \bar{\zeta}} + \frac{1}{Re}\left\{\left(\frac{\partial \bar{r}}{\partial r}\right)^2 \frac{\partial^2 v_\zeta}{\partial \bar{r}^2}\right.$$
$$\left. + \left(\frac{1}{r}\frac{\partial \bar{r}}{\partial r} + \frac{\partial^2 \bar{r}}{\partial r^2}\right)\frac{\partial v_\zeta}{\partial r} - \frac{v_\zeta}{r^2} + \frac{1}{r^2}\frac{\partial^2 v_\zeta}{\partial \theta^2} + \left(\frac{\partial \bar{\zeta}}{\partial \zeta}\right)^2 \frac{\partial^2 v_\zeta}{\partial \bar{\zeta}^2} + \frac{\partial^2 \bar{\zeta}}{\partial \zeta^2}\frac{\partial v_\zeta}{\partial \bar{\zeta}}\right\}. \tag{4}$$

The metric coefficients emerging from the stretching transformations are given explicitly in analytic form (Lienau and Kollmann, 1993); hence a reduction of the order of the discretization error is avoided. The convection-diffusion equation for a passive scalar is added to this system and solved numerically in the same fashion for the study of scalar transport.

The well known difficulty with the above equations is the absence of an explicit evolution equation for the pressure. The incompressibility constraint yields the continuity equation (1) but in practice this equation is not solved. Rather, the momentum equations are first updated and the result is then corrected in order to satisfy mass conservation. Since pressure is a phenomenon which acts at cell centers it does not generate fluid vorticity for incompressible flow. Pressure can generate vorticity for compressible flow because it causes density differences between neighboring cells which in turn can cause fluid rotation. It follows for incompressible flows that the velocity correction, $\boldsymbol{v}_c$, is irrotational and incompressible and that $\boldsymbol{v}_c$ is the gradient of a scalar

$$\boldsymbol{v}_c = -\boldsymbol{\nabla}\phi. \tag{5}$$

Enforcing mass conservation at the new time level produces a Poisson equation for ϕ

$$\boldsymbol{\nabla} \cdot \boldsymbol{v}^{n'+1} = \nabla^2 \phi \tag{6}$$

where $\boldsymbol{v}^{n'+1}$ denotes the solution of the momentum balance. Discretizing the time derivative using the three–point backward (Adams Bashforth) scheme leads to the pressure correction p_c

$$p_c = \frac{3}{2}\frac{\varrho \phi}{\Delta t}. \tag{7}$$

The pressure is corrected according to $p^{n+1} = p^n + p_c$ and velocity according to $\boldsymbol{v}^{n+1} = \boldsymbol{v}^{n'+1} - \boldsymbol{\nabla}\phi$.

The discretization of the momentum balances requires the treatment of the convective and viscous terms. The convective terms are treated in a fully windward manner using the split coefficient scalar (S.C.S.) form for the incompressible equations. This requires the use of conditional velocities defined by

$$v_\alpha^+ = \frac{v_\alpha + |v_\alpha|}{2}; \quad v_\alpha^- = \frac{v_\alpha - |v_\alpha|}{2} \tag{8}$$

where $\alpha = r, \theta$ or ζ. The convective terms are then discretized to second–order accuracy using the four–point windward biased compact (Hermitian) differencing scheme. The pressure gradient is discretized to fourth–order accuracy using the three–point central compact scheme. The viscous terms are discretized to fourth–order accuracy using five–point central differencing. The intermediate velocity $v^{n'+1}$ is obtained from the algebraic equations which result from the discretization. The divergence of the velocity field is then calculated at the $n' + 1$ time level to fourth–order accuracy using five–point central differences (details given in Lienau and Kollmann 1993) and velocity and pressure are corrected.

2.1 Poisson Solver

The novel aspect of the present finite–difference method is the treatment of the Poisson equation for the pressure correction. A hybrid method is developed for the solution of this equation. It consists of a fourth–order compact central finite difference scheme for the nonperiodic dependence on the axial and radial coordinates, that allows the application of a direct solver, and a Fourier spectral method in the periodic θ direction. The direct solver has the advantage that it provides the solution in a fixed number of computational steps per time step and it is more efficient for large problems than iterative methods.

The Poisson equation for ϕ appears in cylindrical coordinates using the stretching transformations as

$$f = \left(\frac{\partial \bar{r}}{\partial r}\right)^2 \frac{\partial^2 \phi}{\partial \bar{r}^2} + \left(\frac{1}{r}\frac{\partial \bar{r}}{\partial r} + \frac{\partial^2 \bar{r}}{\partial r^2}\right)\frac{\partial \phi}{\partial r} + \frac{1}{r^2}\frac{\partial^2 \phi}{\partial \theta^2} + \left(\frac{\partial \bar{\zeta}}{\partial \zeta}\right)^2 \frac{\partial^2 \phi}{\partial \bar{\zeta}^2} + \frac{\partial^2 \bar{\zeta}}{\partial \zeta^2}\frac{\partial \phi}{\partial \bar{\zeta}}. \tag{9}$$

The pressure correction variable, ϕ, and the divergence, $f \equiv \nabla \cdot v$, are expanded in terms of Fourier polynomials with respect to θ

$$\phi(r,\theta,\zeta,t) = \frac{1}{2}a_{\phi,0} + \sum_{k=1}^{n} \left(a_{\phi,k}\cos(k\theta) + b_{\phi,k}\sin(k\theta)\right) \tag{10}$$

$$f(r,\theta,\zeta,t) = \frac{1}{2}a_{f,0} + \sum_{k=1}^{n} \left(a_{f,k}\cos(k\theta) + b_{f,k}\sin(k\theta)\right). \tag{11}$$

The coefficients, a_f and b_f, are calculated using a standard FFT algorithm which is essentially an evaluation of the following integrals

$$a_{f,0}(r,\zeta,t) = \frac{1}{2\pi}\int_0^{2\pi} d\theta f(r,\theta,\zeta,t), \quad a_{f,k}(r,\zeta,t) = \frac{1}{\pi}\int_0^{2\pi} d\theta f(r,\theta,\zeta,t)\cos(k\theta)$$

$$b_{f,k}(r,\zeta,t) = \frac{1}{\pi}\int_0^{2\pi} d\theta f(r,\theta,\zeta,t)\sin(k\theta). \tag{12}$$

The Fourier modes, $a_{\phi,k}$ and $b_{\phi,k}$, for the pressure correction satisfy a system of $2n+1$ partial differential equations that follows at once from (9). The equation for the zeroth-order mode is given by

$$a_{f,0} = \left(\frac{\partial \bar{r}}{\partial r}\right)^2 \frac{\partial^2 a_{\phi,0}}{\partial \bar{r}^2} + \left(\frac{1}{r}\frac{\partial \bar{r}}{\partial r} + \frac{\partial^2 \bar{r}}{\partial r^2}\right)\frac{\partial a_{\phi,0}}{\partial \bar{r}} + \left(\frac{\partial \bar{\zeta}}{\partial \zeta}\right)^2 \frac{\partial^2 a_{\phi,0}}{\partial \bar{\zeta}^2} + \frac{\partial^2 \bar{\zeta}}{\partial \zeta^2}\frac{\partial a_{\phi,0}}{\partial \bar{\zeta}}, \tag{13}$$

and for the higher level modes, $a_{f,k}$ and $b_{f,k}$ modes ($k = 1, 2, ..., n$), we get

$$a_{f,k} = \left(\frac{\partial \bar{r}}{\partial r}\right)^2 \frac{\partial^2 a_{\phi,k}}{\partial \bar{r}^2} + \left(\frac{1}{r}\frac{\partial \bar{r}}{\partial r} + \frac{\partial^2 \bar{r}}{\partial r^2}\right)\frac{\partial a_{\phi,k}}{\partial \bar{r}} - \frac{k^2}{r^2}a_{\phi,k} + \left(\frac{\partial \bar{\zeta}}{\partial \zeta}\right)^2 \frac{\partial^2 a_{\phi,k}}{\partial \bar{\zeta}^2} + \frac{\partial^2 \bar{\zeta}}{\partial \zeta^2}\frac{\partial a_{\phi,k}}{\partial \bar{\zeta}} \tag{14}$$

$$b_{f,k} = \left(\frac{\partial \bar{r}}{\partial r}\right)^2 \frac{\partial^2 b_{\phi,k}}{\partial \bar{r}^2} + \left(\frac{1}{r}\frac{\partial \bar{r}}{\partial r} + \frac{\partial^2 \bar{r}}{\partial r^2}\right)\frac{\partial b_{\phi,k}}{\partial \bar{r}} - \frac{k^2}{r^2}b_{\phi,k} + \left(\frac{\partial \bar{\zeta}}{\partial \zeta}\right)^2 \frac{\partial^2 b_{\phi,k}}{\partial \bar{\zeta}^2} + \frac{\partial^2 \bar{\zeta}}{\partial \zeta^2}\frac{\partial b_{\phi,k}}{\partial \bar{\zeta}}. \tag{15}$$

The system of $2n+1$ partial differential equations for the n Fourier modes is discretized using fourth–order central compact differencing. A system of linear difference equations results for each Fourier mode. The difference equations are solved efficiently using the LAPACK linear equation direct solver. Once the solution for each spectral mode is obtained, the real pressure correction variable $\phi(r, \theta, \zeta, t)$ is calculated using an inverse FFT.

3. LES Approach

Large eddy simulation is a deterministic approach for the prediction of turbulent flow fields at high Reynolds numbers (Reynolds, 1990). In this approach the large structures in the flow are resolved and the unresolved subgrid scales are modelled. The velocity and pressure are decomposed into filtered and residual fields, $v_\alpha = \bar{v}_\alpha + v'_\alpha$ and $p = \bar{p} + p'$.

The Navier-Stokes equations are written for filtered variables

$$\bar{f}(\boldsymbol{x}, t) \equiv \int \mathrm{d}\boldsymbol{x}' G(\boldsymbol{x} - \boldsymbol{x}', t) f(\boldsymbol{x}', t) \tag{16}$$

where $G(\boldsymbol{x} - \boldsymbol{x}', t)$ denotes the filter function and $f(\boldsymbol{x}, t)$ denotes a dependent variable. The Navier-Stokes equations (in Cartesian coordinates for convenience) are filtered and emerge as

$$\frac{\partial \bar{v}_\alpha}{\partial x_\alpha} = 0 \tag{17}$$

and

$$\frac{\partial \bar{v}_\alpha}{\partial t} + \frac{\partial}{\partial x_\beta}(\bar{v}_\alpha \bar{v}_\beta) = -\frac{1}{\varrho}\frac{\partial \bar{P}}{\partial x_\alpha} + \nu \frac{\partial^2 \bar{v}_\alpha}{\partial x_\beta \partial x_\beta} - \frac{\partial}{\partial x_\beta}\tau_{\alpha\beta}. \tag{18}$$

The modified filtered pressure is defined by

$$\bar{P} \equiv \bar{p} + \frac{1}{3}\varrho\Delta_{\alpha\beta}\overline{v'_\alpha v'_\beta} \tag{19}$$

which includes the pressure and the energy of the residual turbulence. The turbulent shear stress $\tau_{\alpha\beta}$ is given by

$$\tau_{\alpha\beta} \equiv \overline{\bar{v}_\alpha v'_\beta} + \overline{v'_\alpha \bar{v}_\beta} + \overline{v'_\alpha v'_\beta}. \tag{20}$$

The present filter is the uniformly weighted average over a finite width in the x_α direction, where $\Delta x_{\alpha-}$ is the spacing to the left and $\Delta x_{\alpha+}$ is the spacing to the right of the node point. A representative filter is given by

$$G(x_\alpha - x'_\alpha; t) = \begin{cases} \dfrac{2}{\Delta x_{\alpha-} + \Delta x_{\alpha+}}, & \text{for } x_\alpha - \dfrac{\Delta x_\alpha}{2} < x'_\alpha < x_\alpha + \dfrac{\Delta x_\alpha}{2} \\ 0 & \text{otherwise} \end{cases}. \tag{21}$$

This filter is implicit in second–order accurate central differencing of the convective terms. A filter of similar form is implicit in the present differencing method.

The residual turbulence model chosen was the Smagorinsky model (1963)

$$\tau_{\alpha\beta} \cong -2\nu_T S_{\alpha\beta} \tag{22}$$

where the eddy-viscosity is defined by

$$\nu_T \equiv (c_s \Delta)^2 \left(2S_{\alpha\beta}S_{\beta\alpha}\right)^{1/2}, \quad \Delta \equiv \left(\Delta\bar{r}\frac{\partial r}{\partial \bar{r}} r\Delta\theta\Delta\bar{\zeta}\frac{\partial \zeta}{\partial \bar{\zeta}}\right)^{1/3}. \tag{23}$$

In (23), Δ denotes the filter size which is the same in all three directions and $S_{\alpha\beta}$ is the rate of strain produced by the filtered velocity field. Different values of the Smagorinsky constant have been suggested. Lilly (1967) proposed c_s=0.23 from homogeneous, isotropic turbulence with Δ being in the inertial subrange. Deardorff (1970) proposed c_s=0.1 for channel flow and found that c_s=0.23 caused too much damping. Mason and Callen (1986) proposed c_s =0.2 for channel flow with the wall approximated and Piomelli et. al. (1989) proposed c_s=0.065 for channel flow with the wall resolved. The present investigation of a free shear flow led to c_s=0.025 so that the mean centerline velocity decay is simulated properly (compared to the experiments of Call, 1992). This value of c_s=0.025 was used for the present LES.

4. Boundary and Initial Conditions

The jet issuing into stagnant surroundings operates in an infinite domain with the boundary condition that velocity is integrable. This could be simulated in the computation by mapping the infinite domain onto a finite domain and setting the values of all variables to zero at the image of infinity, or by using a finite domain with the lateral boundary sufficiently far removed from the core region of the jet. The latter possibility was chosen and boundary conditions were selected that do not affect the flow in the jet significantly. The boundary condition which was found to do this reasonably well was to set the normal derivative of velocity to zero on the boundaries of the computational region. The line $r = 0$ is not a boundary since a point on it is an interior point of the flow field. Finite difference grids can be set up such that the axis, $r = 0$, lies on the surface of a grid or such that $r = 0$ is between grid surfaces. The latter method was found to be superior.

The pressure gradient in the direction normal to the computational boundary was set to zero at the jet inlet and exit planes. On the circumferential boundary the pressure and the pressure correction were set to zero. The inlet condition for the jet was obtained using the experimental results of Liepmann and Gharib (1992). The entire flow field was initialized with the inlet profile.

Vortex roll–up occurs naturally, but the development is rather slow and would cost too much CPU–time. Hence, disturbances were introduced at the inlet to initiate and speed up the roll–up process. The disturbances $\tilde{v}_\varsigma(\theta, t)$ were chosen to be only a function of θ and time in a manner similar to Rai and Moin (1991). They are given by

$$\tilde{v}_\varsigma = \sum_{k=1}^{K} \sum_{n=1}^{N} A_{l,n} \sin(k\theta + \phi_\theta) \sin\left(\frac{2\pi n t}{\tau} + \phi_t\right) \tag{24}$$

where $A_{l,n} = c u_0 r_{\text{jet}} \delta^{k-1} \gamma^{n-1}$, $\delta = 0.25$, $\gamma = 0.5$, $c = 0.1$. Two modes in both θ and time were selected (i.e. $K = 2$, $N = 2$).

An analysis for the Adams Bashforth scheme and the compact four-point windward differencing of the convective terms revealed that the maximum stable Courant number is 0.6 (details in Lienau and Kollmann, 1993).

5. Results and Discussion

The turbulent flow in a round jet is simulated using the second/fourth–order accurate finite–difference methods discussed in the previous section for following two cases: without an LES model (case 1) and with an LES model (case 2). The nominal Reynolds number is $Re \equiv u_0 D/\nu = 8000$, where u_0 denotes the bulk exit velocity and D the jet pipe diameter. This Reynolds number is too high to allow resolution of all relevant scales; therefore, both runs must be regarded as a LES with the numerical viscosity taking on the role of the filter (Boris, 1991) in the first case. The Reynolds number is close to the experiments of Liepmann (1991), Liepmann and Gharib (1992) but larger than the direct simulations of Melander *et al.* (1991). The grid is set up in both cases with $90 \times 64 \times 190$ grid points in the radial, azimuthal and longitudinal directions and mild stretching transformations are applied to concentrate the grid in the high shear region. The computational domain covers the first fifteen diameters of the jet. The disturbances discussed in section 4.0 are introduced at the jet pipe exit to initiate the Kelvin–Helmholtz and the circumferential instabilities, which lead to the breakdown of the ring structures further downstream. The results presented in the

Fig. 1. Iso–surface of vorticity magnitude (22.5% of maximum) at $t = 66$ and a nominal Reynolds number $Re = 8000$, LES run with $c_s = 0.025$. Flow direction is from bottom to top.

figures cover the initial phase of the flow development for the dimensionless time interval $[0, 62]$. The flow is started with the inviscid solution, i.e. the entrance profile is extended through the flow field and the disturbances are introduced to initiate the flow development. A global description of the flow structures appearing in the first ten diameters of the jet (for the case $c_s = 0.0$) is presented in Fig. 1 to Fig. 3. The level surface of vorticity magnitude in Fig. 1 at 22.5% of its maximum value contains three major flow structures (flow direction is from bottom to top):

(i) vortex rings as the result of the Kelvin–Helmholtz instability;
(ii) longitudinal vortex tubes or braids;
(iii) complex interconnected structures as a result of vortex merging and reconnection.

The braids will be shown to appear in pairs producing the mushroom–shaped structures visible in the cross–sections of the flow, but only the strongest tube is visible at the chosen level value. Fig. 2 shows a close up of the third vortex ring (counting from the bottom of Fig. 1) which is beginning to deform under the influence of the circumferential instability and the shearing due to the velocity difference between the core and the outer flows. Fig. 3 presents the level surface of dissipation rate at

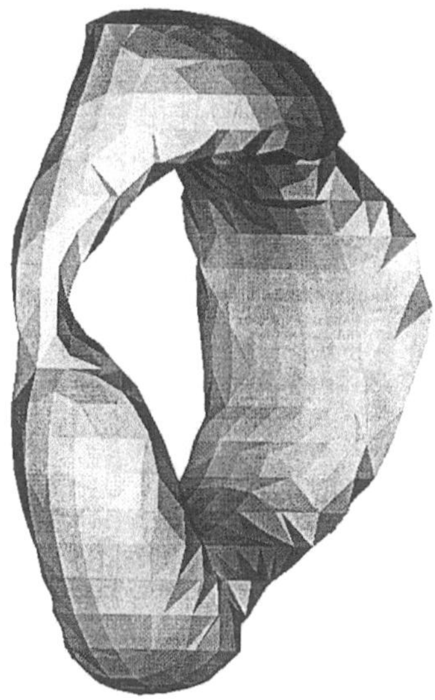

Fig. 2. Close–up of the third (counting from the bottom) vortex ring in Fig. 1.

7.5% of its maximum value. It is evident that the dissipation rate is concentrated on a smaller support than the vorticity magnitude and reveals more sheet-like structures in the downstream region of the domain.

5.1 Case 1: Without LES Model

The maximum norm of vorticity

$$\|\omega\|(t) \equiv \max_{x \in \mathcal{D}} |\omega(x,t)|$$

is known to be an indicator of the emergence of singularity in the solutions of the Navier–Stokes equations. It can be shown for the inviscid Navier–Stokes equations (see Beale *et al.*, 1984) that the onset of singularity in three dimensions is controlled by vorticity and no other singularity can appear before vorticity becomes unbounded. The development of the maximum norm is shown as a function of time in Fig. 4. During the start–up phase of the flow a sequence of small relative maxima corresponding to the roll–up of individual vortex rings is visible. After approximately five time units they become unstable and begin to interact and the maximum norm increases to more than twice its initial value. A sequence of vortex mergers can be seen as strong relative maxima. The locations of the maximum of vorticity in the axial and radial directions are presented in Fig. 5 and Fig. 6, respectively. The axial location (Fig. 5) and the radial location (Fig. 6) do not change much in the initial phase indicating that the roll–up takes place at nearly the same axial and radial location. Once the circumferential instability and the interaction of the rings have started,

Fig. 3. Iso–surface of dissipation rate (7.5% of maximum) at $t = 66$ and a nominal Reynolds number $Re = 8000$, LES run with $c_s = 0.025$. Flow direction is from bottom to top, viewing angle is approximately the same as in Fig. 1.

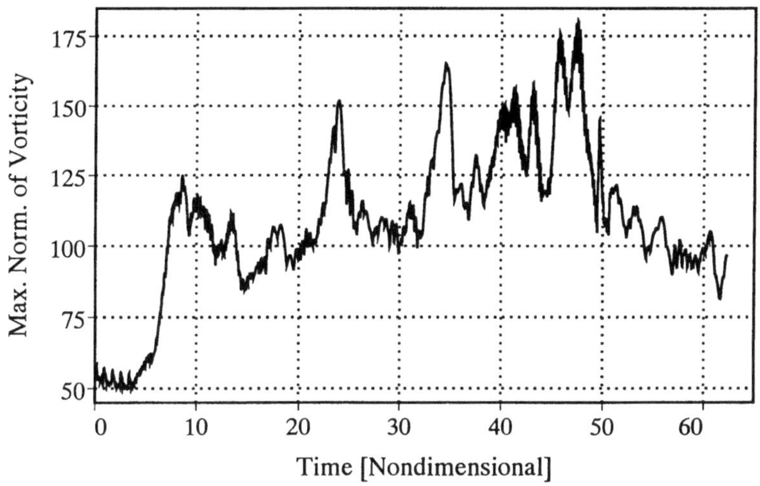

Fig. 4. Maximum norm of vorticity as a function of time for the round jet with a nominal Reynolds number $Re = 8000$ and $c_s = 0.0$.

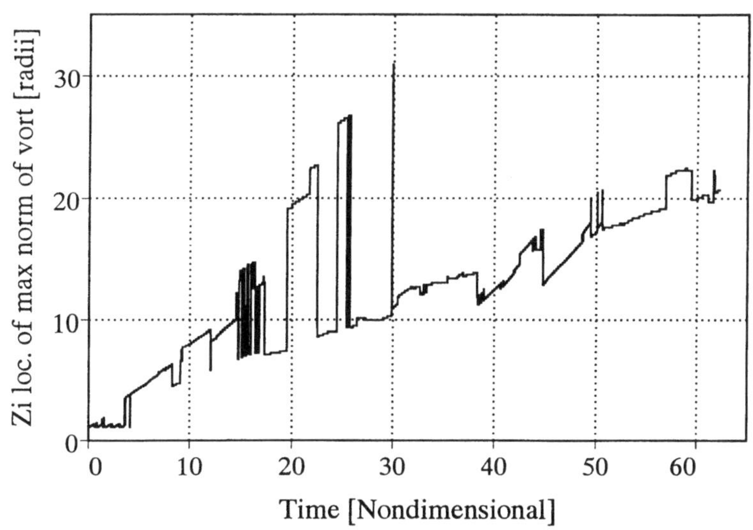

Fig. 5. Axial location of the maximum of the vorticity magnitude as a function of time (conditions of Fig. 4).

the location of the maximum moves rapidly in the axial and the radial directions. There is, however, a clear tendency of the maximum to move downstream as Fig. 5 shows and the variance of the movement in the radial direction increases in time, consistent with the spreading of the jet. The flow is not yet fully developed, even after $t = 62$ time units, since the axial and radial locations do not yet exhibit the approach of the mean values to their asymptotic limits. This is also evident in Fig. 7

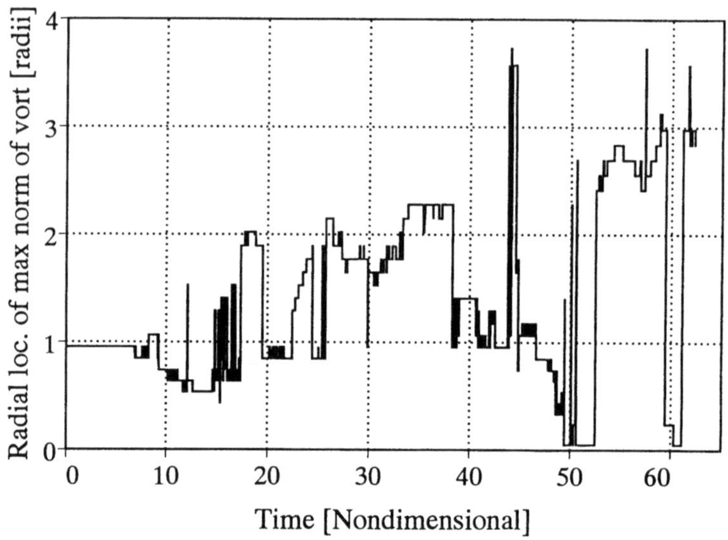

Fig. 6. Radial location of the maximum of the vorticity magnitude as a function of time (conditions of Fig. 4).

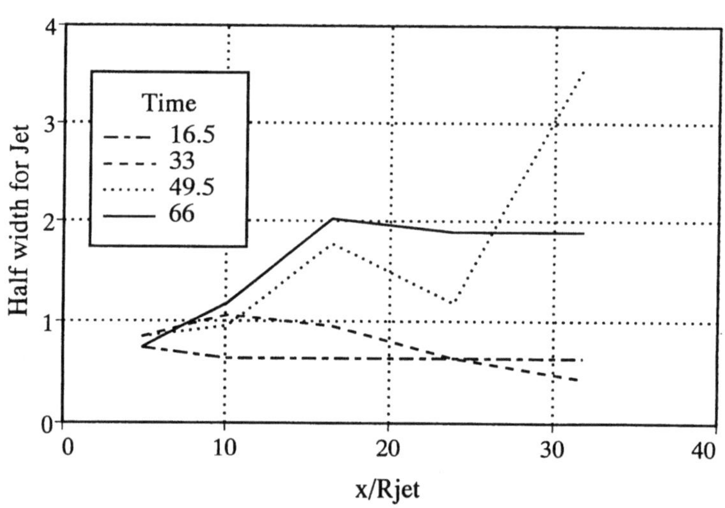

Fig. 7. Half–width of the mean velocity profile at different times (conditions of Fig. 4).

showing the half-width of the jet as a function of downstream distance for several times. There is still considerable variation in time indicating that the statistics are not yet stabilized.

The structural development of the flow is documented in Fig. 8 to Fig. 13 for the final time, t = 62, of the simulation. The distribution of vorticity magnitude in $r - \theta$ planes at several axial stations is considered first. The contour lines in Fig. 8 at x/D = 4.61 show essentially a vortex ring with weak circumferential variation. However, the outer part of the cross–section shows the remnants of the interacting vortex rings further downstream, which have been moved upstream by the velocity induced by the large downstream structure. Moving to x/D = 5.05 in Fig. 9, it becomes evident that there exist five longitudinal pairs of counter–rotating vortex tubes (as evident from Fig. 12), which are called "braids". However, strictly speaking, no braiding in the sense of algebraic topology could be ascertained in the present case, which would require winding and viscous reconnection of two or more tubes. An investigation of this aspect of the flow could be based on the properties of vortex lines and the polynomials associated with their topology. It is worth noting that Martin and Meiburg (1991) observed five longitudinal vortex structures (called braids) in their inviscid simulation of the temporally–evolving round jet. These braids collapse into counter–rotating vortex tubes as reported by Martin and Meiburg (1991) and is evident in Fig. 10 for the present viscous flow. The center of the cross–section in Fig. 9 contains vorticity concentrated in rings. Further downstream at x/D = 5.95 in Fig. 10 the five braid–like vortex structures become more prominent, whereas little vorticity is present in the center. At this station, the braid–like structures are the dominant vortex structure and this is usually called the braid region (Liepmann, 1991, Liepmann and Gharib, 1992). The number of braids is not necessarily constant as Fig. 11 (x/D = 6.88) suggests. No vortex ring survives to this downstream distance and the longitudinal vortices are no longer discernible at x/D = 8.34. A contour plot of the vorticity magnitude in Figs. 8 to Fig. 11 also suggests the existence of the longitudinal structures. Plotting the axial component of vorticity in Fig. 12 for x/D = 5.95 (same station as Fig. 10) shows that the longitudinal vortices are actually pairs of counter-rotating vortex tubes. These tubes will then produce distributions of a passive scalar in $r - \theta$ planes similar to the mushrooms observed by Liepmann and Gharib (1992). The number of counter–rotating pairs of vortices is likely to depend on the core size of the vortex rings upstream. Linear stability analysis and experiments on vortex rings (Widnall and Sullivan, 1973) show that the number of unstable waves on a vortex ring increases with decreasing core size. If such a vortex ring is exposed to the velocity field of a jet, the troughs of the disturbance waves are accelerated in the axial direction relative to the wave crests, and the formation of the braid–like structures is thus initiated. Liepmann and Gharib (1992) found that the number of braids increases with Reynolds number; the present numerical simulations agree with their lowest Reynolds number result as well as with the simulation of the temporally-developing jet by Melander *et al.* (1991). The $\zeta - r$ sections of the vorticity magnitude in Fig. 13 for the $\theta = \pi/2$-$\theta = 3\pi/2$ plane show roll-up of the vortex sheet into vortex rings and the braid-like structures in the upstream part of

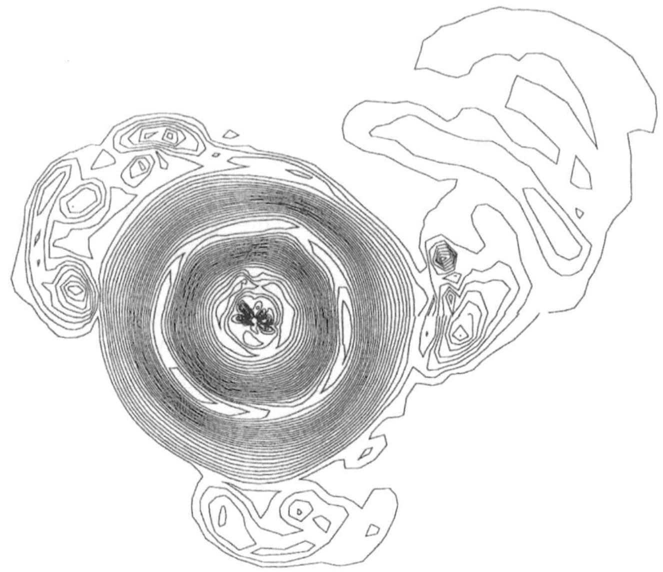

Fig. 8. Cross–section of vorticity magnitude at $t = 62$ and $\zeta / D = 4.6$ (conditions of Fig. 4).

Fig. 9. Cross–section of vorticity magnitude at $t = 62$ and $\zeta / D = 5.05$ (conditions of Fig. 4).

Fig. 10. Cross–section of vorticity magnitude at t = 62 and ζ/D = 5.95 (conditions of Fig. 4).

the flow. The interaction region further downstream exhibits more than one decade of length scales and rather complicated fluctuations.

5.2 Case 2: With LES Model

The LES model, described in section 3.0, was applied to the filtered equations to obtain a solution of the above problem. The effect of the subgrid-scale model is studied by comparing the results for case 1 and case 2. Fig. 14 is a $\zeta - r$ cross-section of the vorticity magnitude field at the $\theta = \pi/2$-$\theta = 3\pi/2$ plane. The ring–braid regions can be clearly seen with vortex pairing exhibited in some of the rings in the lower half of the figure. The cross–cuts of the braids show that they move out of the plane. The complicated vortical region to the right of the third ring shows that the breakdown process has produced a rich variety of vortical structures. A passive scalar with a Schmidt number of unity was convected and diffused in the flow field. The passive scalar distribution of Fig. 15 also highlights the ring–braid region and reveals the severe straining of the braids near the jet axis, in agreement with the passive scalar results of Liepmann and Gharib (1992). It is evident from the $r - \theta$ cross section of vorticity magnitude at x/D = 5.95 in Fig. 16 that there are four or five braids along with another vortical structure spiraling towards the axis of the jet. The counter–rotating nature of the pairs of longitudinal vortices is, once again, evident from the contour plot of the axial component of vorticity shown in Fig. 17.

Fig. 11. Cross–section of vorticity magnitude at $t = 62$ and $\zeta/D = 6.88$ (conditions of Fig. 4).

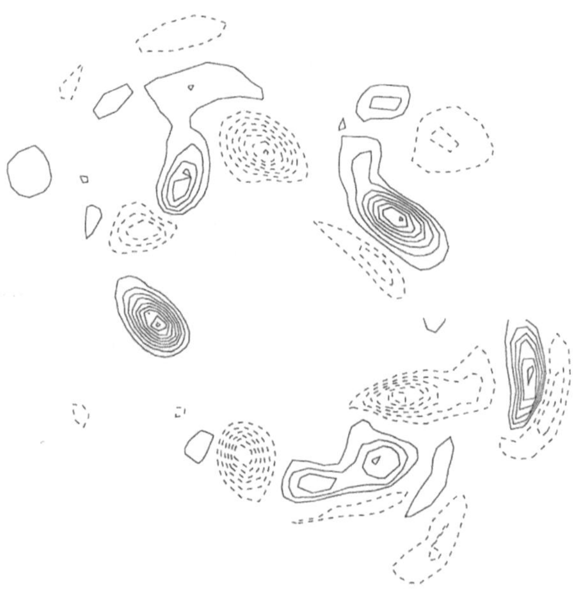

Fig. 12. Cross–section of the axial vorticity component at $t = 62$ and $\zeta/D = 5.95$ (conditions of Fig. 4).

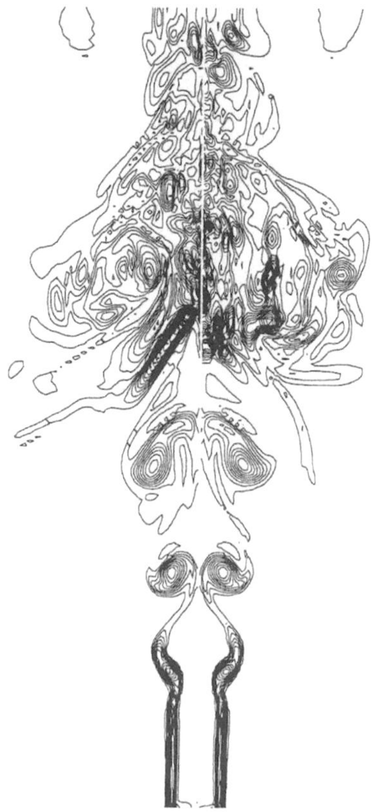

Fig. 13. Vorticity magnitude in the ζ-r plane for $\theta = 0$ at $t = 62$ (conditions of Fig. 4).

These vortices produce the mushrooms in the passive scalar distribution shown in Fig. 18. In this figure three mushrooms are clearly visible, produced by the counter–rotating pair of vortex tubes. Moving further downstream to $x/D = 7.3$ in Figs. 19 to 21 one can see that the braids and the counter-rotating pairs of vortices have lost their orderly nature, and the mushrooming process is no longer clearly discernible. The vortex tubes have been severely strained near the axis and breakdown is occurring. This is similar to the experimental findings of Liepmann and Gharib (1992).

It has been shown that the effect of the LES subgrid–scale model is to modify the dynamics of the flow. In Fig. 14, the vortical structures near the potential core for case 2 exhibit asymmetry, while the vortical structures near the potential core for case 1 (without the subgrid scale model, Fig. 12) show symmetry. The asymmetry is consistent with the appearance of an odd number of braids. A comparison of the cross-sections for the vorticity magnitude without LES model in Fig. 10 and with LES model in Fig. 16 (same location) illustrates the effect of the subgrid scale model. The vortex braids (Fig. 10) are well developed if no SGS-model is used, whereas the

Fig. 14. Vorticity magnitude in the ζ-r plane for $\theta = 0$ at $t = 66$, LES run with $c_s = 0.025$.

LES model in Fig. 16 increases the viscosity in the outer region of the jet leading to larger spacing of the tubes.

6. Conclusions

A finite difference method for the Navier–Stokes equations was developed that is second–order accurate in the convective terms and fourth–order accurate in the pressure gradient and viscous terms and second–order accurate in time. This method was then used to simulate the turbulent flow in the first fifteen diameters of a round jet. The present results were obtained for a Reynolds number of $Re = 8000$ with and without a subgrid scale model. The simulations were started with the inviscid solution and small axisymmetric disturbances were introduced at the entrance section to accelerate the breakdown of the Kelvin–Helmholtz vortices. The development of the flow produced a rich variety of structures. The azimuthal instability of the vortex rings leads to tube–like structures which collapse into counter–rotating vortex tubes

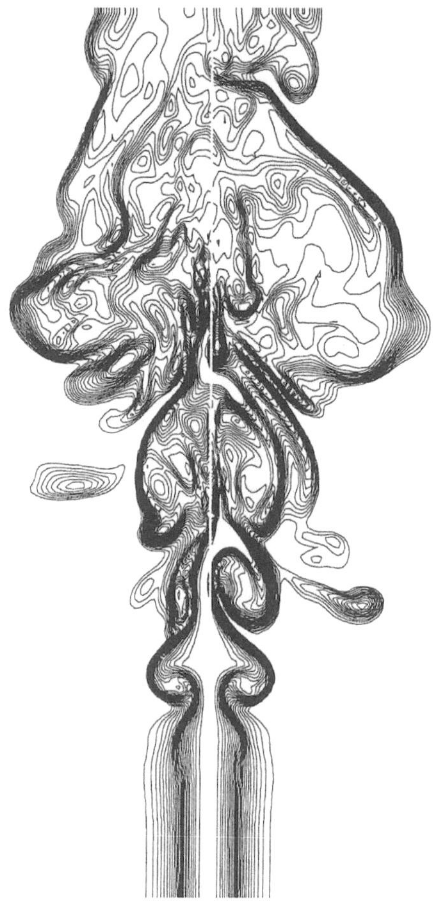

Fig. 15. Isolines for a passive scalar in the ζ-r plane for $\theta = 0$ at $t = 66$ LES run with $c_s = 0.025$.

that interact with the vortex pairing process. A comparison with flow visualization experiments shows that the number of braid–like vortex structures is in agreement with the experiment.

Acknowledgement

This research was supported by the United States Department of Energy through the Office of Basic Energy Sciences, Division of Chemical Sciences and by the Air Force Office of Scientific Research, AFOSR Grant G-F49620-92-J-0418.

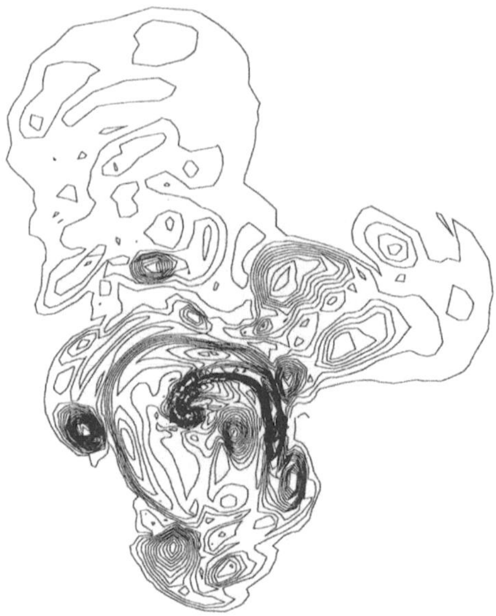

Fig. 16. Cross–section of the vorticity magnitude at t = 66 and ζ/D = 5.95, LES run with c_s = 0.025.

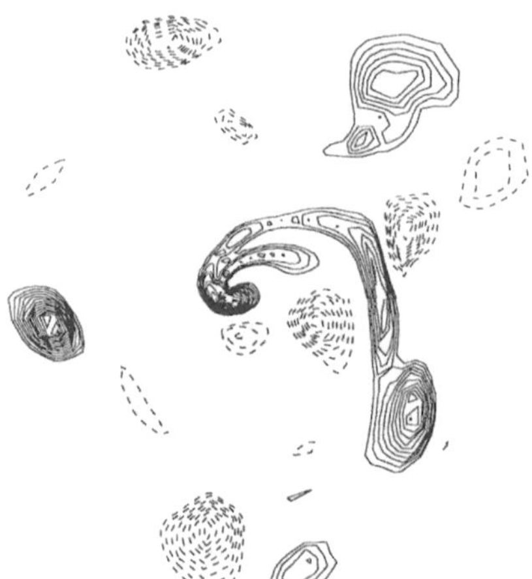

Fig. 17. Cross–section of the axial vorticity component at t = 66 and ζ/D = 5.95, LES run with c_s = 0.025.

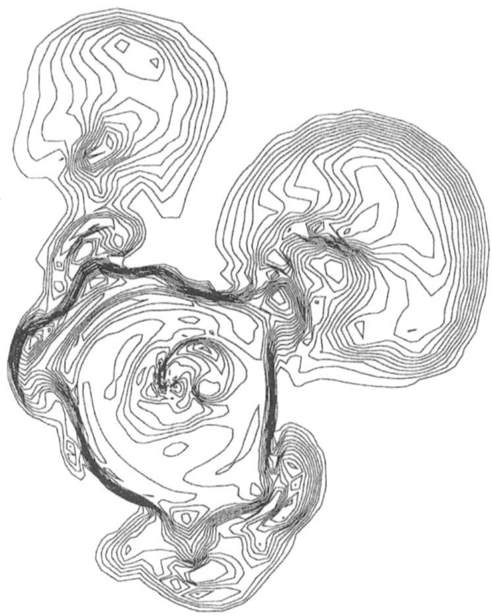

Fig. 18. Cross–section of the passive scalar at $t = 66$ and $\zeta/D = 5.95$, LES run with $c_s = 0.025$.

Fig. 19. Cross–section of the vorticity magnitude at $t = 66$ and $\zeta/D = 7.3$, LES run with $c_s = 0.025$.

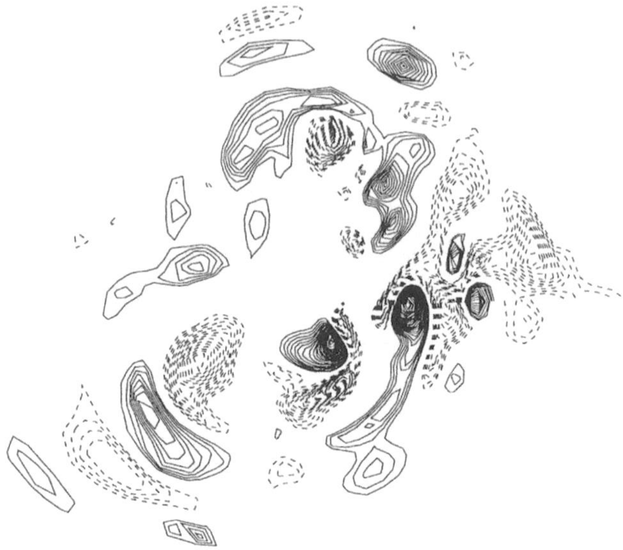

Fig. 20. Cross–section of the axial vorticity component at t = 66 and ζ/D = 7.3, LES run with c_s = 0.025.

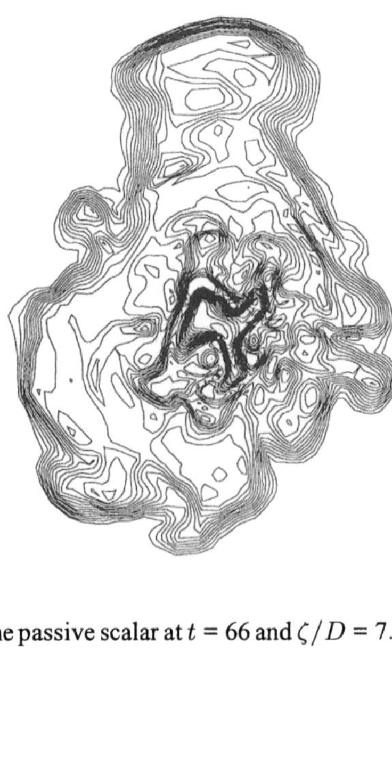

Fig. 21. Cross–section of the passive scalar at t = 66 and ζ/D = 7.3, LES run with c_s = 0.025.

References

Beale, J.T., Kato, T. and Majda, A. (1984), "Remarks on the breakdown of smooth solutions for the 3–d Euler equations", Comm. Math. Phys. **94**, 61.
Boris, J.P. (1990), "Comments on subgrid turbulence models and large eddy simulations", in *Whither Turbulence? or Turbulence at the Crossroads* (J.L. Lumley ed.), Lecture Notes in Physics vol. 357, Springer V., 344.
Call, C.J. (1992), "Particle Dispersion in Turbulent Shear Flows", PhD Dissertation, UC Davis.
Dahm, W.J.A., Frieler, C.E. and Tryggvason, G. (1992), "Vortex structure and dynamics in the near field of a coaxial jet", J. Fluid Mech. **241**, 371.
Deardorff, J.W. (1970), "A numerical study of three-dimensional turbulent channel flow at large Reynolds numbers.", J. Fluid Mech. **41**, 453-480.
Gutmark, E., Parr, T.P., Hanson-Parr, D.M. and Schadow, K.C. (1990), "Coherent and random structure in reacting jets", Experim. in Fluids **10**, 147.
Lele, S.K. (1990), "Compact finite–difference schemes with spectral-like resolution", CTR manuscript 107.
Lienau, J.J. and Kollmann, W. (1993), "Numerical simulation of the turbulent flow in round jets", AIAA Paper 93-0199.
Liepmann, D. (1991), "Streamwise vorticity and entrainment in the near field of a round jet", Phys. Fluids A **3**, 1179.
Liepmann, D. and Gharib, M. (1992), "The role of streamwise vorticity in the near-field entrainment of round jets", J. Fluid Mech. **245**, 643.
Lilly, D.K. (1967), "The representation of small scale turbulence in numerical experiments", in *Proc. IBM Scientific Computing Symposium on Environmental Sciences*, IBM, White Plains, NY.
Martin, J.E. and Meiburg, E. (1991), "Numerical investigation of three-dimensionally evolving jets subject to axisymmetric and azimuthal perturbations", J. Fluid Mech. **230**, 271.
Mason, P.J. and Callen, N.S. (1986), "On the magnitude of the subgrid–scale eddy coefficients in large eddy simulation of turbulent channel flow", J. Fluid Mech. **162**, 439-462.
Melander, M.V., Hussain, F. and Basu, A. (1991), "Breakdown of a circular jet into turbulence", Proc. 8th Symp. Turbulent Shear Flows, Munich, Germany, 15-5-1.
Mungal, M.G., Lozano, A. and Van Cruyningen, I. (1992), "Large-scale dynamics in high Reynolds number jets and jet flames", Experim. in Fluids **12**, 141.
Piomelli, U., Moin, P. and Ferziger, J.H. (1988) "Model consistency in the large eddy simulation of turbulent channel flow", Phys. Fluids **31**, 1884-1891.
Rai, M.M. and Moin, P. (1991), "Direct simulations of turbulent flow using finite-difference schemes", J. Comput. Physics **96**, 15.
Reynolds, W.C. (1990), "The potential and limitations of direct and large eddy simulations", in *Whither Turbulence? or Turbulence at the Crossroads* (J.L. Lumley ed.), Lecture Notes in Physics vol. 357, Springer V., 210.
Smagorinski, J. (1963), "General circulation experiments with the primitive equations. I. The basic experiment." Monthly Weather Review **91**, 99-164.
Wicker, R.B. and Eaton, J.K. (1993), "Flow visualization study of the structure of a co-flowing jet with and without controlled excitation", AIAA 93-0651
Widnall, S.E. and Sullivan, J.P. (1973), "On the stability of vortex rings", Proc. R. Soc. London A, **332**, 335.
Yoda, M., Hesselink, L. and Mungal, M.G. (1991), "The temporal evolution of large-scale structures in the turbulent jet", Proc. 8th Symp. Turbulent Shear Flows, Munich, Germany, 6-1-1.

Mixing in Turbulent Opposed Jet Flows

E. Mastorakos, A. M. K. P. Taylor, and J. H. Whitelaw

Thermofluids Section, Mechanical Engineering Dept.,
Imperial College of Science, Technology and Medicine, London SW7 2BX, U. K.

Abstract

This paper reports measurements of the velocity and temperature fields in the counterflow of two turbulent jets with one of the jets heated so that the temperature acted like a passive scalar, motivated by the recent use of such flows for studying non-premixed combustion. The mixing layer between the two streams had a width independent of the bulk flow strain rate but decreased with decreasing turbulence intensity, and the instantaneous structure of the scalar was characterised by sharp transitions between the temperatures of the two jets which resulted in bimodal probability density functions. These transitions were less frequent, the bimodality less pronounced, and the r.m.s. values of the normalised temperature smaller with decreasing turbulence intensities. A conceptual view of mixing in the opposed jet flow based on the thickness and movement of the thin interfaces separating the two jets explains the measured dependencies of the scalar pdfs on turbulence intensity, and the observation that the mean residence time in the present opposed jet flows is short compared to the turbulence timescale implies that special efforts are needed for modelling the scalar dissipation rate.

1. Introduction

Turbulent opposed jet flows have been used in combustion studies of the structure and extinction of stagnating flames and have been motivated by the simplifying properties that the bulk stretch rate is uniform and that the turbulence is created independently of the bulk flow by turbulence-generating grids in each of the jets. Premixed flames have been realised experimentally in this geometry by introducing a flammable mixture in one or both of the jets (Mastorakos *et al.*, 1992a, and Kostiuk *et al.*, 1993a, respectively) and diffusion flames have been established by impinging a jet of fuel against a jet of oxidant (Mastorakos *et al.*, 1992a). Mixing between the two jets is necessary for non-premixed combustion and it is important to know the mean and fluctuating properties of the conserved scalar (mixture fraction) in the opposed jet flow to assist interpretation of the structure of non-premixed flames. The mixing field in the opposed jet flow has been treated so far only theoretically (Mastorakos *et al.*, 1992b; Champion and Libby, 1993), and the present paper aims to provide experimental data on the mean and fluctuating conserved scalar in turbulent counter-flows, with the passive scalar identified with the concentration of one of the jets. An additional motivation behind our study is interest in heat transfer in stagnation-point flows (e.g. Strahle *et al.*, 1987) which has also led to extensive modelling efforts (Champion and Libby, 1991). In the next Section, the flow configuration, measurement methods, and flow conditions are described, while the results are presented and discussed in Section 3. The paper closes with a summary of the main conclusions.

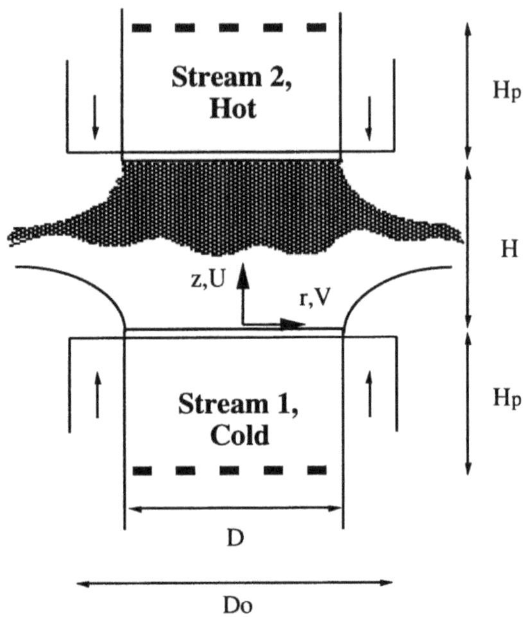

Fig. 1. The flow arrangement. An electric heater was placed 40D upstream of the upper tube to heat the upper stream at the exit plane at 80 °C. $D = 25.4$ mm; $D_0 = 51.8$ mm, $H = 20$ mm for all experiments described here, and $H_p = 55$ mm and 120 mm for conditions of high and low turbulence intensity respectively.

2. Experimental Arrangement

2.1 Flow Configuration and Experimental Methods

The flow arrangement has been described elsewhere (Mastorakos *et al.*, 1992a) and consisted of two identical vertically opposed assemblies, separated by a distance H, each comprising an inner straight pipe of diameter $D = 25.4$ mm, 40D long, concentric with a larger pipe of diameter $D_0 = 51.8$ mm and 15D long (Fig. 1). Turbulence-promoting perforated plates with solidity 45% and hole size of 4 mm were fitted inside each of the inner tubes at a distance H_p from the tube exit plane. Air, delivered by a shop compressor and filtered for particles, was metered by calibrated rotameters to within ±5% and supplied the upper and lower inner and outer flows. The outer flows were used to minimise ambient disturbances to the inner-jet stagnation flow, and the co-flow velocity was kept at 0.3 m/s. The temperature of the upper inner stream was raised up to 200 °C by an electric heater (delivering a maximum of 4.8 kW), located at the upstream end of the tube. Despite the insulation of the tube, heat losses reduced the temperature at the tube exit plane to 80 °C, and this value was used in all experiments reported here as the effective "temperature tagging" of the upper stream.

Temperatures were measured by a standard hot-wire probe operated in constant current mode with small overheat so that it acted like a resistance thermometer. The probe was made of platinum and was 5 μm thick and 1.25 mm long, and the electronic circuit was custom-made and consisted of a Burr-Brown constant- current chip providing 0.4 mA, constant to within ±0.5 μA. The voltage across the probe, of the order of one mV, was first amplified with a gain of 10, then offset, and then amplified further with a gain of 100, so that the amplified signal could be accommodated in the range 0 to 10 Volt, appropriate for the A/D board installed in an IBM-PC compatible microcomputer that was used for data acquisition. The r.m.s. electronic noise in the final signal corresponded to 0.4 °C, which was enough to resolve the large-scale features of the scalar fluctuations but not enough for the fine-scale ones. The "cold wire" responded to frequencies up to about 100 Hz (calculated following Collis and Williams, 1959) which, again, was adequate for the energy-containing, slow fluctuations, but not for the fast ones. No compensation for the finite thermal inertia of the cold wire was attempted because attention was directed to the large-scale features of the flow. The mean and fluctuating temperatures measured by the cold wire system were reasonably close to those measured previously (Mastorakos, 1993) by digitally-compensated fine-wire thermocouples, and were preferred because of the uncertainties associated with the time constant needed to compensate the thermocouple measurements.

Velocity measurements were performed by laser-Doppler velocimetry which comprised a 5 mW laser, an optical unit fitted with a rotating grating to provide frequency shift to resolve directional ambiguity, a 150 mm-focal length lens to collect the light scattered from 1 μm-diameter (nominal) alumina seeding particles, and a photodiode with a 200 μm pinhole (Mastorakos, 1993). All the optical components were mounted on a three-directional milling table. The velocity measurements were made with the upper stream unheated because the presence of seeding particles would block the narrow passages in the electric heater, but it is believed that the small overheat, and associated density difference (see below and Section 2.3), did not alter the important features of the flow field.

Temperatures are reported in terms of the normalised scalar ϑ, defined as $\vartheta = (T-T_1)/(T_2-T_1)$, with T the measured temperature and T_1 and T_2 the temperatures of the lower and the upper streams respectively; the normalised temperature is thus zero in the cold stream and unity in the upper stream. Velocities have been normalised by the bulk velocity U_0 of the cold stream. To ensure symmetric flow, the velocity of the upper stream was adjusted so that $\varrho_0 U_0^2 = \varrho_1 U_1^2$, with ϱ the density, U the bulk velocity, and the subscripts 0 and 1 denoting conditions in the cold and the hot streams respectively. The density of the upper, heated, stream was lower than the ambient by a factor of 0.83 (corresponding to 80 °C), and so the bulk velocity of the hot stream was about 10% higher than U_0. Ensemble averages of temperatures were based on about 1000 statistically independent samples giving a statistical uncertainty in the reported mean and r.m.s. values of at most 3 and 6% respectively; similar uncertainties were involved in the velocity measurements. Finally, traverses along the centreline of the flow are presented to a positional accuracy relative to the jet exit of ±0.5 mm, but to within ±0.1 mm relative to the neighbouring points along the traverse.

2.2 Initial Conditions

The temperature across the exit plane of the upper tube was found to be uniform to within ±2% up to a radius of about 7 mm with r.m.s. fluctuations in the normalised temperature of about 0.015 (i.e. 1.2 °C). The non-zero fluctuations of the temperature at the hot stream are probably due to upstream heat losses in the tube, but are more than an order of magnitude smaller than the fluctuations of the temperature in the mixing region of the two jets (presented later) so that the latter truly correspond to fluctuations due to mixing and not due to unsteady initial "source strength" of the scalar.

The resulting turbulence at the exit plane of the pipes was measured by Mastorakos *et al.* (1992a) and was found to depend on the distance of the perforated plate from the tube lip, H_p, so that at $H_p = 55$ mm, the turbulence intensity $C_0 = q'/U_0$, with $q' = \sqrt{(u'^2 + 2v'^2)/3}$, was 0.085, while at $H_p = 120$ mm, a longer time was allowed for decay from the turbulence generation to the jet exit, and C_0 was smaller and equal to 0.045. It was also found that C_0 decreased only slightly with increasing bulk velocity U_0, so that q'/U_0 was approximately independent of U_0. These measurements were made at the exit of a single jet (i.e. without an opposed flow) and it was found that in the opposed jet flow, the axial velocity profile was not uniform across the jet exit (Mastorakos, 1993), and that the turbulence intensity was larger than its value in the unopposed jet by about 20%. Despite this difference, the turbulence conditions can be adequately characterised by H_p with large values implying small values of q'/U_0. Detailed velocity measurements are presented in Section 3.1 to assist the interpretation of the mixing measurements.

2.3 Flow Conditions Examined and Scale Estimates

For all scalar measurements, the jet separation, H, was 20 mm, the upper stream temperature was 80 °C, the co-flow was 0.3 m/s, the momentum flow rates of the two jets balanced so as to keep the flow symmetric, and results are shown for the centreline only. Four cases were examined to identify the dependence of the mean and instantaneous scalar on the bulk and turbulence strain rates: same bulk velocity (3.13 m/s) and two conditions corresponding to high and low turbulence intensities

(Case A: H_p = 55 mm; Case C: H_p = 120 mm), and two conditions with the same perforated plate locations (Case B and D respectively) but at a lower bulk velocity (1.48 m/s). The velocity measurements presented in Section 3.1 correspond to Cases B and D.

Table 1. Flow conditions and estimated lengthscales and strain rates. For all cases, H = 20 mm, the integral turbulence lengthscale is taken equal to L_t = 4 mm, and q'/U_0 is taken as 0.085 for H_p = 55 mm and as 0.045 for H_p = 120 mm.

Case	H_p [mm]	U_0 [m/s]	Re_t	η_K [mm]	S_b [s^{-1}]	S_K [s^{-1}]	q'/L_t [s^{-1}]
A	55	3.13	70	0.17	313	556	67
B	55	1.48	33	0.29	148	181	31
C	120	3.13	37	0.27	313	214	35
D	120	1.48	17	0.48	148	67	17

The Reynolds number of the turbulence is estimated by $Re_\mathrm{t} = q' L_\mathrm{t}/\nu$, the bulk strain rate by $S_\mathrm{b} = 2U_0/H$, and the small-scale turbulent strain rate is taken as the inverse Kolmogorov timescale and equal to $S_\mathrm{K} = Re_\mathrm{t}^{1/2} q'/L_\mathrm{t}$ (Table 1). The integral turbulence lengthscale, L_t, is taken to be equal to the diameter of the hole in the perforated plate, L_t = 4 mm, and assumed to be independent of H_p and constant throughout the flow, and the turbulence intensities given by the q'/U_0 measured in the single jets are used. We emphasise that these estimates are based on turbulence quantities appropriate for the jet exit plane and both velocity and length scales might change at the stagnation region and so the estimates in Table 1 should be treated with caution. Finally, we do not expect buoyancy to be important because the density difference between the two jets was small and led to a Richardson number, defined as $(\Delta\varrho/\varrho)gL_\mathrm{t}/q'^2$, between 0.02 and 0.25 for the conditions of Table 1, which are smaller than the critical value of unity suggested by Breidenthal (1992) as the limit for buoyancy to affect the velocity field.

3. Results and Discussion

3.1 Velocity Field

Figure 2a shows that the mean axial velocity along the centreline decreases to zero at a point half-way between the two jets, as expected, and that the bulk velocity gradient is not far from the value estimated by the Hiemenz flow field model, i.e. $2U_0/H$, but also that a curvature in the profile exists so that the axial velocity gradient is not uniform across the flow, in agreement with the theoretical prediction of Champion and Libby (1993) (Libby, personal communication). It is also evident that the mean velocity is not affected by the turbulence because the two velocity profiles in Fig. 2a

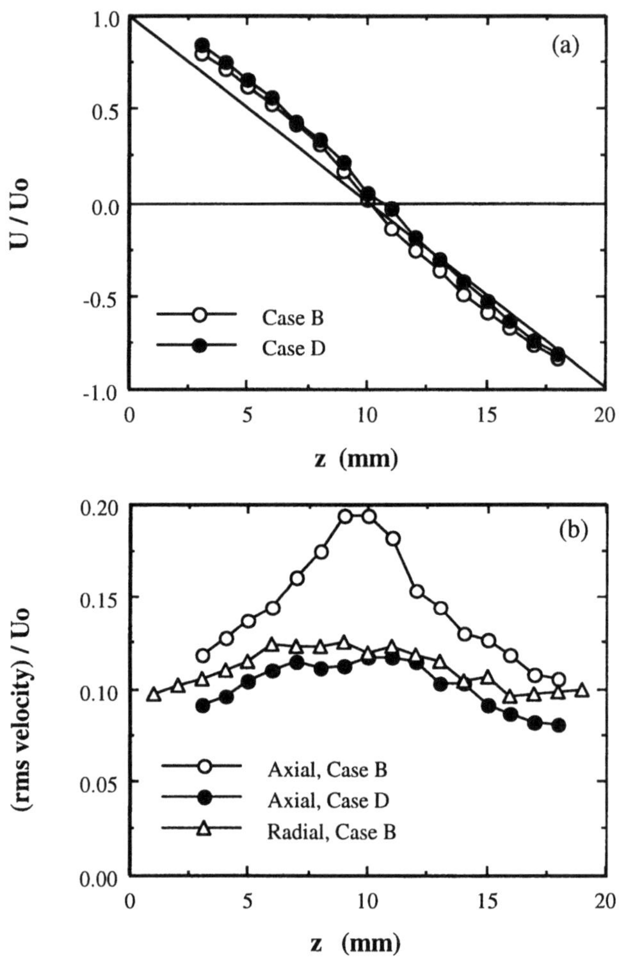

Fig. 2. Velocity measurements along the centreline for the conditions described in Table 1.

are almost identical, although they correspond to different turbulence intensities (Cases B: high, Case D: low turbulence). Figure 2b shows that the axial velocity fluctuations are higher for Case B than for Case D for all axial locations, and that they peak at the stagnation region. This increase in turbulence over its value at the jet exit is evident in the radial fluctuations (Fig. 2b, triangles) as well and is consistent with expectations from arguments of vortex stretching (for example, Tennekes and Lumley, 1972) in the straining field of the opposed jet flow. The independence of the mean flow from the turbulence suggested in Fig. 2 implies that the mean and the turbulent field can be described independently, for example by assuming a mean flow field and considering the turbulence evolving in it without influence of the latter on the former (Bray *et al.*, 1991). It is also evident from Fig. 2b that the turbulence intensities are higher than our estimates in Section 2.3 (and Table 1) and that the assumption behind a constant turbulence intensity across the opposed jet flow implemented in the theoretical model of mixing of Mastorakos *et al.* (1992b) is not valid; more discussion on this will follow in Section 3.3.

A radial profile of mean radial and axial velocities at the axial coordinate corresponding approximately to the stagnation point (z = 10 mm) is presented in Fig. 3a and shows that the flow is uniform in the radial direction, so that the mean axial velocity is a function only of the axial coordinate, and the radial velocity only a function of radial distance. Figure 3b shows that the corresponding axial and radial r.m.s. fluctuations along the stagnation plane are also uniform with radius. These results corroborate the finding of Mastorakos (1993) that the scalar fields in inert and combusting opposed jet flows are uniform with radius and confirm that, as expected, theoretical analyses can concentrate on axial variations only along the centreline. The present velocity measurements are consistent with those of Kostiuk *et al.* (1993b) and Mounaïm-Rouselle and Gökalp (1993).

3.2 Scalar Field

Mean Temperatures. The mean temperature was uniform in the radial direction for $r < 10$ mm at *all* axial locations across the opposed jet flow, as expected, and so only results along the centreline are presented here (Fig. 4a). The axial coordinate measured from the lower jet corresponding to the location of Θ = 0.5 varied between 9.5 and 10.5 mm due to the large sensitivity of the location of the stagnation point on errors in balancing the two flows; the shape of the profile, however, did not change even when differences of 20% in the momentum flow rates were deliberately introduced (Mastorakos, 1993) and so the profiles in Fig. 4 have been shifted so that the reduced coordinate z^* takes the value zero at Θ = 0.5. This way, comparison of the actual thickness of the mixing regions at different flow conditions is facilitated. Figure 4a shows that the mean temperature has a profile similar to an error function and that the mixing layer thickness depends *only* on the turbulence intensity and *not* on the bulk velocity, since the profiles for Cases A and B are almost identical and they are less steep than the profiles for Cases C and D, which correspond to H_p = 120 mm. The trend that the mixing layer is steeper with smaller turbulence intensity and the independence of the layer thickness on bulk flow velocity, both

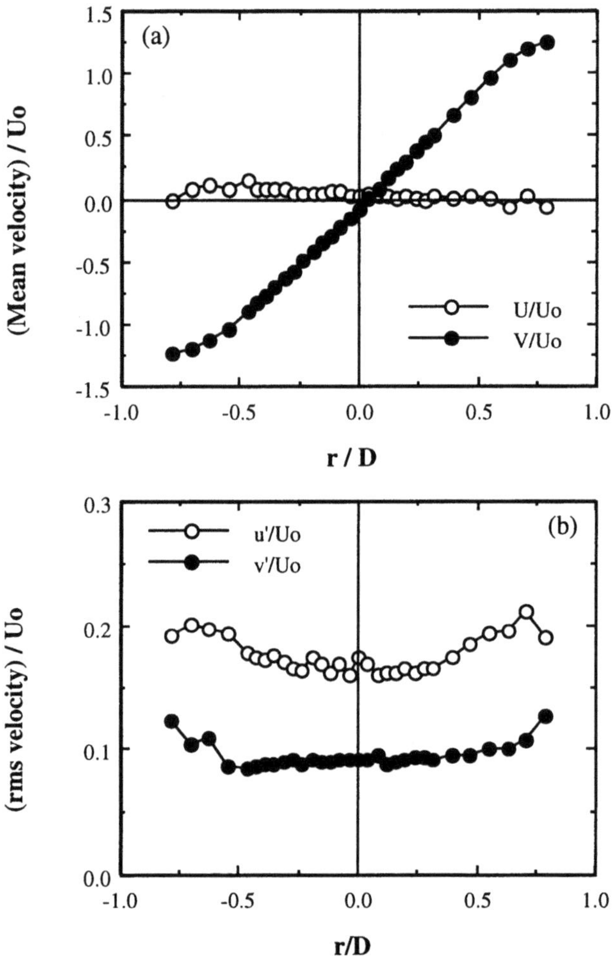

Fig. 3. Axial and radial mean (**a**) and r.m.s. (**b**) velocities at z = 10 mm for Case B as a function of radial distance.

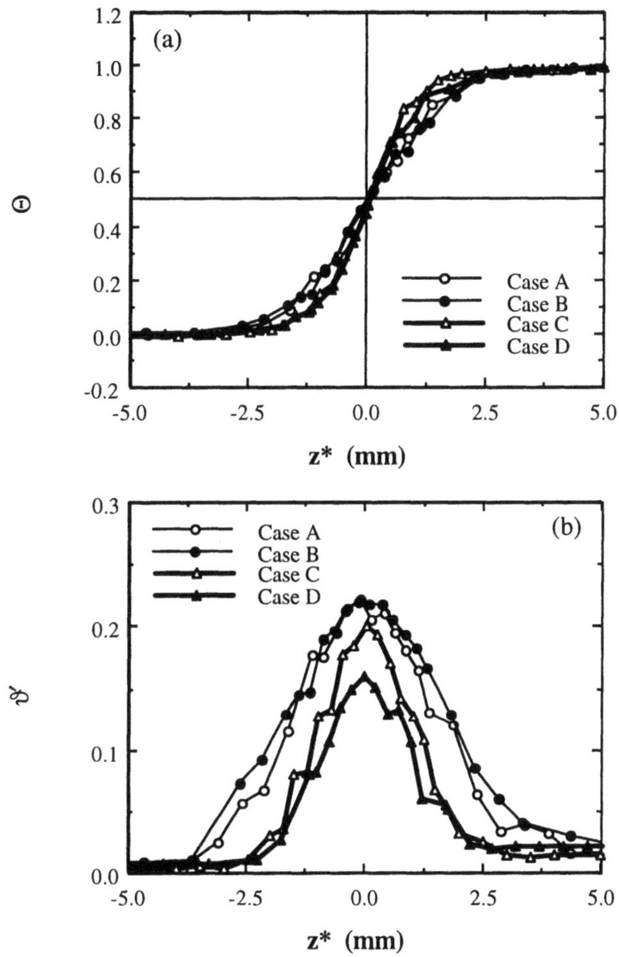

Fig. 4. Mean (**a**) and r.m.s. fluctuations (**b**) of the normalised temperature along the centreline for the four flow conditions of Table 1.

confirm the theoretical predictions of Mastorakos *et al.* (1992b) and are discussed in Section 3.3.

Fluctuating temperatures. Figure 4b shows the r.m.s. normalised temperatures, ϑ', corresponding to Cases A to D of Table 1 and the mean temperatures of Fig. 4a. The r.m.s. fluctuations peak at $z^\star = 0$, the location of steepest mean-ϑ gradient, and the smaller width of the mixing layer for Cases C and D (low q'/U_0) is evident in the width of the profile of the r.m.s. values, as well as in the mean values of Fig. 4a. The peak r.m.s. values for Cases A and B are the same, that for Case C is slightly lower, and that for Case D is significantly lower, and this difference is discussed next by examining the associated probability density functions (pdfs) corresponding to $z^\star = 0$.

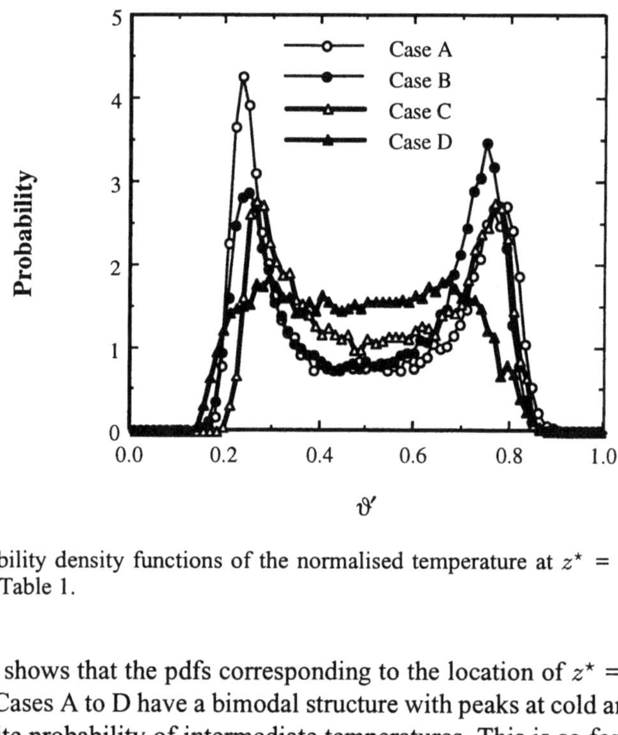

Fig. 5. Probability density functions of the normalised temperature at $z^\star = 0$ for the flow conditions of Table 1.

Figure 5 shows that the pdfs corresponding to the location of $z^\star = 0$ ($\Theta = 0.5$) for the four Cases A to D have a bimodal structure with peaks at cold and hot values and with finite probability of intermediate temperatures. This is so for Cases A, B, and C; for Case D, corresponding to low bulk velocity and low turbulence, the pdf is almost uniform without peaks. The shapes of these pdfs explain the different ϑ' of Fig. 4b with the lowest ϑ' corresponding to the uniform pdf of Fig. 5. At different axial locations, the pdfs again have a bimodal structure, but with the height of the peaks at cold and hot values changing, so that the mean depends on which peak is higher (not shown here; Mastorakos, 1993).

The bimodal structure of the pdfs is consistent with observations in the time series of the normalised temperature at the same locations that show sharp transitions

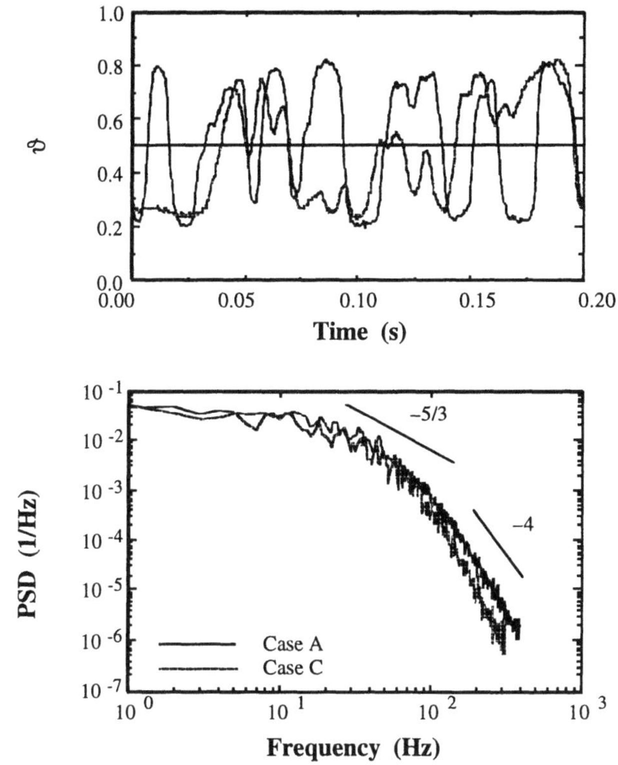

Fig. 6. (**a**) Typical normalised temperature traces for Cases A and C at $z^\star = 0$ and (**b**) the associated frequency spectra. Note the sharp transitions from cold to hot values in (**a**).

between cold and hot values (Fig. 6a), so that the fluid spends little time at intermediate values of ϑ compared to the time it spends in the cold and hot regions. It is also evident from Fig. 6a that the transitions are more frequent in the high turbulence case. The transition time from cold to hot was characterised for all four Cases corresponding to the pdfs of Fig. 5 by the Taylor time microscale, λ_ϑ, which was calculated based on the curvature of the autocorrelation curves at zero time delay, and it was found that λ_ϑ was shortest for Case A and longest for Case D (Table 2), and that it followed the velocity timescales estimated in Table 1.

Examination of the associated frequency spectra (Fig. 6b) shows that these are typical of "well-behaved" turbulent flows, without any pronounced peaks and with a region decaying with a slope of -5/3, followed by a region of steeper decay (slope: -4). Although these spectra at frequencies higher than about 100 Hz are steeper than in reality due to the finite thermal inertia of the cold wire and should not be trusted quantitatively, they are consistent with measurements of *velocity* spectra by Mounaïm-Rouselle and Gökalp (1993) in an inert opposed jet flow that had a -5/3 region followed by a steeper decay. There is a close correspondence therefore between

Table 2. Estimated Kolmogorov timescale (the inverse of S_K from Table 1) and measured Taylor timescale at $z^* = 0$.

Case	τ_K [ms]	λ_ϑ [ms]
A	1.79	4.26
B	5.52	8.76
C	4.67	8.29
D	14.93	15.92

the timescales of the scalar and the velocity field, and together with the observation that the instantaneous structure of the scalar is similar to a "telegraph signal" (Figs. 5 and 6), the present measurements suggest that the two jets mixed on the molecular level at interfaces which were randomly convected by the velocity fluctuations past the measurement point, so that the scalar fluctuations were only due to the interface displacements. In the next Section, the implications of the measurements are discussed in more detail.

3.3 Discussion

Instantaneous Scalar Structure. A conceptual model of mixing in the opposed jet flow may be based on the finding in Fig. 6a that the scalar fluctuations are characterised by well-defined transitions between cold and hot values, and that the frequency of these transitions scales with the velocity timescales. At the centreline of the opposed flow, the two jets impinge and are separated by an interface which can be viewed as a laminar diffusion layer strained in an unsteady manner by the turbulent field. The instantaneous normalised temperature profile in a direction normal to the interface can be described by an error function from zero to unity, similar to that expected from a laminar counterflow, and the thickness of this profile would be on the order of the Batchelor lengthscale (Gibson, 1968), equal to the Kolmogorov lengthscale as the Prandtl number is close to unity for the present gaseous flow. The layer would be convected in the axial and radial directions by the turbulence. Because the axial velocity fluctuations are larger than the radial ones (Fig. 2), and because an eddy turnover time is long compared with the residence time in the flow ($1/S_b < L_t/q'$; see Table 1), the interface will not have enough time to wrinkle, as would be the case if the mean flow was slow compared to the turbulence (Dimotakis, 1989). Therefore these unsteady laminar diffusion layers, which separate the two jets, retain their identity, with the instantaneous zero-fluid side being *always* towards, and at the temperature of, the cold jet, and with the instantaneous unity-fluid side of the interface *always* towards, and at the temperature of, the hot jet. This suggests that the measured scalar fluctuations are due to the axial movement of laminar diffusion layers, always separating zero- and unity-fluid, and would depend on the lengthscale of this movement and on the thicknesses of these layers.

In the measurements of Section 3.2, the Kolmogorov lengthscale corresponding to the different Cases in Table 1 varied from 0.17 mm (Case A) to 0.48 mm (Case D), and so for Case D the interface separating the two fluids is thicker than for the other Cases. Because the integral turbulent lengthscale is nominally the same for all four Cases, the axial movement of the interface is also identical for Cases A to D; hence if a thick and a thin interface move past the probe, the thick interface would result in a higher proportion of intermediate values than a thin interface, and therefore the probability of intermediates would be higher for cases with high η_K/L_t, or low Re_t. This argument offers a structural explanation why the pdf of ϑ for Case D in Fig. 5 is qualitatively different from the other cases.

The ideas behind the conceptual picture in the previous paragraphs are elaborated in the Flapping Model of scalar turbulence (Kerstein, 1991). In the context of this model, the randomness in the scalar is due to the randomness in positioning an invariant scalar profile relative to the measuring location, and the ensemble- averaged mean and r.m.s. of the scalar are given by the convolution of the profile shape with the probability of profile displacements. Kerstein (1991) used this model to predict the scalar statistics in turbulent shear flows by considering profiles of thickness of the order of the mean flow width flapping in the cross-stream direction, and Verravalli and Warhaft (1990) (see also the references in that paper) predicted many of the features of their scalar measurements of thermal dispersion from a line source in wind-tunnel turbulence using similar concepts. Here, this model is used for interpreting the large-scale scalar fluctuations based on the movement of the small-scale diffusion layers. As mentioned previously, this is a valid description for the opposed jet flow along the centreline because of the long large-eddy timescale compared to the residence time in the flow, a condition which precludes appreciable convolutions of the scalar interfaces before these are swept radially outwards by the action of the bulk strain rate. The implications of the observation that the residence time in the opposed jet flow is short compared to a turbulence timescale are discussed next through comparisons with other measurements and theoretical results.

Comparison with Other Works.

Mixing at early times. Stapountzis *et al.* (1986) measured the statistics of temperature released into decaying wind-tunnel turbulence by heating a wire, so that the experiment simulated the diffusion from a line source in homogeneous turbulence. It was found that far downstream of the source, the scalar pdfs were approximately Gaussian, but that close to the source the pdfs were bimodal with peaks at hot and cold temperatures, similar to the pdfs measured in the present experiments. Close to the line source in the experiment of Stapountzis *et al.* (1986), the time-of-flight was short compared with an eddy timescale, and so the thermal plume released from the source kept its coherence; the scalar fluctuations were mostly due to the plume "meandering", as these authors called it, in the cross- stream direction. This description of the scalar turbulence at short times from the scalar's release is very similar to the description we propose here for the scalar field in the opposed jet flow in

Section 3.3.1. Bimodal pdfs of passive scalars were also observed by Eswaran and Pope (1988) in direct numerical simulations of turbulence into which "blobs" of a passive scalar were introduced at a certain time. At times from the scalar's release shorter than about one large-eddy timescale, the pdfs of the scalar were bimodal and this bimodality became less pronounced as time passed with the pdfs becoming eventually unimodal at long times, results consistent with our data and with those of Stapountzis *et al.* (1986).

Theoretical models. Mastorakos *et al.* (1992b) formulated an eddy-viscosity model for mixing in the opposed jet flow and assumed that the mean axial velocity and scalar were radially uniform, that the axial velocity was given by $U = -S_b x$, where x is the axial coordinate measured from the stagnation point (identical to $z^\star$ for a symmetric flow), that the turbulence q' was constant, that the scalar axial fluxes were described by a constant eddy-viscosity proportional to the product $q' L_t$, and standard modelling assumptions for the fluxes and dissipation of the r.m.s. fluctuations, and then the governing equations describing the mean and variance of the conserved scalar, defined as zero in one stream and unity in the other, reduced along the centreline to

$$\frac{d^2\Theta}{d\eta^2} + 2\eta\frac{d\Theta}{d\eta} = 0 \tag{1}$$

and

$$\frac{d^2 g}{d\eta^2} + 2\eta\frac{dg}{d\eta} + \left(\frac{d\Theta}{d\eta}\right)^2 - \alpha g = 0\,. \tag{2}$$

The axial coordinate was normalised as $\eta = z^\star/Z_{ref}$, with the reference lengthscale Z_{ref} given by

$$\frac{Z_{ref}}{L_t} = C_1\alpha^{1/2}\,, \tag{3}$$

the parameter α by

$$\alpha = 2\frac{q'}{U_0}\frac{H}{L_t}\,, \tag{4}$$

and $g = \vartheta'^2$. The constant C_1 includes all the modelling constants in the eddy-viscosity formulation. Equation (1) is identical to the equation describing mixing in laminar opposed jet flows, but with a turbulent diffusion coefficient instead of a molecular one, obeys the boundary conditions $\Theta(-\infty) = 0$ and $\Theta(\infty) = 1$, and its solution is the error function:

$$\Theta = \mathrm{erf}(\eta)\,. \tag{5}$$

Equation (5) was used to calculate the mean gradient production term in Eq. (2), and based on this, Eq. (2) was solved numerically to find the dependence of ϑ' on axial location and the parameter α, which is a ratio of the large-scale turbulent to the bulk flow strain rates (Eq. 4).

The measurements in Fig. 4a show that the width of the mixing layer is independent of bulk velocity and decreases with decreasing turbulence intensity; both these trends are predicted by Eq. (3) in conjunction with the measured (and assumed) independence of q'/U_0 on U_0, and so the model performs well in predicting the qualitative dependence of the mean scalar field on the flow conditions. However, the agreement is less satisfactory for the fluctuations. Equations (2) and (5) show that ϑ' depends *only* on the parameter α and numerical solutions showed that ϑ' decreases with increasing α because the scalar dissipation rate (the last term in the l.h.s. of Eq. 2) increases with α while the other terms (production, diffusion, convection) did not. In the context of the experiments presented here, α depends on H_p (because this affects q'/U_0), on H, and on L_t. Cases A and B have $\alpha = 0.85$ and Cases C and D have $\alpha = 0.45$, so that the theoretical model predicts that the r.m.s. fluctuations for the pair A and B are identical, consistent with the data, but also for C and D, *inconsistent* with the data (Fig. 4b). Finally, the model predicts higher values of ϑ' than the experiment, and that ϑ' increases with decreasing turbulence intensity, contrary to the experiment. Some possible reasons for these discrepancies are explored below.

Figure 2 shows that q'/U_0 changes from its assumed value at the exit plane of the jet to a larger value around the stagnation region, but it is also evident that the variation of q'/U_0 around the mixing layer is not large; therefore the mixing layer could be adequately characterised by a *constant* eddy viscosity, albeit a higher one than assumed in the model. This would mean that the "effective" α in the mixing region is higher than the assumed one, but invariable, and this explains why the model works well for the mean quantities and why it predicts higher ϑ' than the experiment. However, the reason why the predicted dependence of the r.m.s. fluctuations on turbulence intensity is contrary to the measurements is not clear at this stage. Finally, we mention that the model of Champion and Libby (1993), which was based on a Reynolds-stress description of the velocity and scalar field and is thus expected to perform better than the eddy-viscosity model of Mastorakos *et al.* (1992b), predicts the same qualitative dependence of the mean mixing layer width on flow parameters as the simpler model. However, slightly different dependencies are predicted for the scalar fluctuations on flow conditions, and it would hence be interesting to compare Champion and Libby's (1993) model against the present scalar data.

The motivation for our previous modelling effort was that estimates of the scalar dissipation rate are necessary for interpreting data from non-premixed combustion, and, in particular, for explaining the occurrence of extinction of opposed jet flames. It is interesting then to discuss whether the structure of the scalar field, as revealed in the present measurements, has any implication for the scalar dissipation rate in this flow, and this is discussed next.

The scalar dissipation rate. A standard modelling assumption for the mean scalar dissipation rate, χ, is that the thermal and mechanical dissipation timescales are in a constant ratio (the last term in the l.h.s. of Eq. 2), so that

$$\chi = 2\vartheta'^2/(\text{mechanical timescale})\,, \qquad (6)$$

(Jones and Whitelaw, 1982). Warhaft and Lumley (1978) found experimentally that the timescale ratio depended on the scale at which the scalar was introduced into their homogeneous turbulent flow (i.e. the ratio of the integral scalar to velocity length-scale), so that the timescale ratio varied by up to factor of two from its commonly accepted value. Eswaran and Pope (1988) calculated the scalar dissipation rate in direct numerical simulations and found that at short times from the release of the scalar, χ was larger than the standard model would suggest by a factor of about three. At times longer than about one eddy turnover time however, χ became close to the estimate from Eq. (6).

Therefore the scalar dissipation rate depends on the time elapsed from the release of the scalar, and so questions arise about the applicability of the model of Eq. (6) for opposed jet flows where residence times are short. In conjunction with the discrepancies between the predictions and the measurements discussed previously, it is clear that models for opposed jet flows should account for the observation that the scalar turbulence is "young" compared with the velocity turbulence.

4. Conclusions

Temperature measurements in the mixing region of two counterflowing turbulent streams with one stream heated so that the temperature acted like a passive scalar were performed by resistance thermometers to show the dependence of the mean and fluctuating scalar quantities on the bulk strain rate, proportional to the bulk flow velocity, and the turbulent strain rate, with the turbulence depending mainly on the conditions of its generation inside each of the two streams.

The mean normalised temperature changed from unity in one stream to zero in the other over a characteristic distance that depended on the turbulence intensity and not on the bulk velocity for a given jet separation, and this is consistent with previous theoretical analyses of mixing in such flows.

The instantaneous temperature trace showed sharp transitions from low to high values corresponding to the initial temperatures of the two jets, and this resulted in bimodal probability density functions of the scalar and small probabilities of intermediate values, but this bimodality became less pronounced under conditions of low turbulence intensity. These results are explained by considering that the interface separating the two jets moves randomly by the velocity fluctuations, so that the scalar fluctuations were due to the interface displacements and depended on the relative magnitude of the interface thickness, of the order of η_K, to the interface displacement, of the order of L_t. This conceptual picture has counterparts in other measurements of passive scalars in turbulent flows and is valid in the opposed jet flow because the mean residence time is short compared to the time necessary for molecular mixing, which precludes large probabilities of intermediate values of the scalar and implies that special efforts are needed to model the scalar dissipation rate in this flow.

Acknowledgements

We would like to thank Prof. P. A. Libby of UCSD for helpful discussions concerning this flow and Mr. John Laker of the Thermofluids Section for constructing the cold-wire electronics. This work has been supported by CEC SCIENCE Contract № SC1-0459.

References

Bray, K. N. C., Champion, M. & Libby, P. A. (1991) Premixed flames in stagnating turbulence: Part I. The general formulation for counterflowing streams and gradient models for turbulent transport. Comb. Flame **84**, 391.

Briedenthal, R. E. (1992) Entrainment at thin stratified interfaces: the effects of Schmidt, Richardson, and Reynolds numbers. Phys. Fluids A **4**, 2141.

Champion, M. & Libby, P. A. (1991) Asymptotic analysis of stagnating turbulent flows. AIAA Journal **29**, 17.

Champion, M. & Libby, P. A. (1993) Reynolds stress description of opposed and impinging turbulent jets. Part I: Closely spaced opposed jets. Phys. Fluids A **5**, 203.

Collis, D. C. & Williams, M. J. (1959) Two dimensional convection from heated wires at low Reynolds number. J. Fluid Mech. **6**, 357.

Dimotakis, P. E. (1989) Turbulent shear layer mixing with fast chemical reactions. In *Turbulent reactive flows*, Borghi, R. & Murthy, S. N. B. (Eds.), Springer.

Eswaran, V. & Pope, S. B. (1988) Direct numerical simulations of the turbulent mixing of a passive scalar. Phys. Fluids **31**, 506.

Gibson , C. H. (1968) Fine structure of scalar fields mixed by turbulence. I. Zero- gradient points and minimal gradient surfaces. Phys. Fluids **11**, 2305.

Jones, W. P. & Whitelaw, J. H. (1982) Calculation methods for reacting turbulent flows: a review. Comb. Flame **48**, 1.

Kerstein, A. R. (1991) Flapping model of scalar mixing in turbulence. Phys. Fluids A **3**, 2838.

Kostiuk, L. W., Bray, K. N. C. & Cheng, R. K. (1993)a Experimental study of premixed turbulent combustion in opposed streams. Part II- Reacting flow field and extinction. Comb. Flame **92**, 396.

Kostiuk, L. W., Bray, K. N. C. & Cheng, R. K. (1993)b Experimental study of premixed turbulent combustion in opposed streams. Part I- Nonreacting flow field. Comb. Flame **92**, 377.

Mastorakos, E., Taylor, A. M. K. P. & Whitelaw, J. H. (1992)a Extinction and temperature characteristics of turbulent counterflow diffusion flames with partial premixing. Comb. Flame **91**, 46.

Mastorakos, E., Taylor, A. M. K. P. & Whitelaw, J. H. (1992)b Scalar dissipation rate at the extinction of turbulent counterflow non-premixed flames. Comb. Flame **91**, 55.

Mastorakos, E. (1993) Turbulent combustion in opposed jet flows. PhD Thesis, University of London.

Mounaïm-Rousselle, Ch. & Gökalp, I. (1993) Turbulent premixed combustion in counterflow geometry. The influence of a coflow. Presented at the Spring Meeting of the Western States Section of the Combustion Institute, Utah, March 1993.

Stapountzis, H., Sawford, B. L., Hunt, J. C. R. & Britter, R. E. (1986) Structure of the temperature field downwind of a line source in grid turbulence. J. Fluid Mech. **165**, 401.

Strahle, W. C., Sigman, R. K. & Meyer, W. L. (1987) Stagnating turbulent flows. AIAA Journal **25**, 1071.

Tennekes, H. & Lumley, J. L. (1972) *A first course in turbulence.* MIT Press, Cambridge, Mass..

Veeravalli, S. & Warhaft, Z. (1990) Thermal dispersion from a line source in the shearless turbulence mixing layer. J. Fluid Mech. **216**, 35.

Warhaft, Z. & Lumley, J. L. (1978) An experimental study of the decay of temperature fluctuations in grid-generated turbulence. J. Fluid Mech. **88**, 659.

Thermal Characteristics of Turbulent Vortices in the Wake of a Circular Cylinder

J. Mi and R. A. Antonia

Department of Mechanical Engineering, University of Newcastle, N.S.W., 2308, Australia

Abstract

Thermal characteristics of turbulent vortices in the near/intermediate wake of a circular cylinder have been explored using data from an array of 16 cold wires aligned in the direction of the mean shear. The data have been examined either on an instantaneous basis or conditionally via a detection method which identifies local temperature maxima. For small distances downstream of the cylinder, the conditionally averaged temperature distributions within the vortices are approximately exponential. This information has been used in conjunction with an Oseen-like vortex model to calculate mean and rms temperature profiles across the wake. There is reasonable agreement between calculation and measurement.

1. Introduction

Various aspects of the organized motion have been studied in a wide range of turbulent shear flows. In particular, the relatively well organized vortical motion in the turbulent near wake ($x/d < 4$ or 5) or intermediate wake (5 or $6 < x/d \lesssim 50$) has received significant attention (e.g. Davies, 1976; Cantwell and Coles, 1983; Kiya and Matsumura, 1985; Hayakawa and Hussain, 1989; Antonia, 1991). In a previous paper (Matsumura and Antonia, 1993), momentum and heat transport characteristics of the intermediate wake of a circular cylinder were investigated by applying a phase averaging technique to measurements of velocity (longitudinal u and lateral v) and temperature fluctuations (ϑ). It was found that the overall contribution from the organized motion to the lateral heat flux is consistently greater than that to the Reynolds shear stress, indicating that turbulent vortices transport heat more effectively than momentum. This feature has been explored in a spectral context by Antonia et al. (1993): the $u - v$ cospectrum can exhibit a change of sign near the average vortex shedding frequency but there is no corresponding sign change occurs in the $v - \vartheta$ cospectrum. Consistently, the local turbulent Prandtl Pr_t number can assume relatively small values. The previous papers have drawn attention to the significant streamwise evolution of the topologies of the velocity and temperature fields throughout the intermediate wake.

The use of temperature as a passive marker of organized structures has been exploited quite extensively in the self-preserving far-wake (e.g. Townsend, 1949, and many others since then). It has also been used in the near/intermediate wake (e.g. Wlezien and Way, 1979; Budny et al., 1979) of a circular cylinder. Wlezien and Way applied a thermal tag to one of the cylinder boundary layers prior to separation and used a y-aligned array of 6 cold wires for simultaneous temperature measurements at several spatial locations. This general approach is attractive since it allows data obtained over a region of space to be viewed either instantaneously or conditionally. Temperature fluctuations, especially in the near-wake region, should be more accurately measurable than velocity or vorticity fluctuations. In this region of the flow, relatively high turbulence intensities (15–40%) and the possibility of flow reversal make it difficult to obtain reliable data from X-wires.

In the present investigation, the combination of passive thermal tagging and an array of y-aligned cold wires is used to examine the thermal characteristics of turbulent vortices as they evolve through the near and intermediate wake regions. A preliminary investigation using a single cold wire allowed the position ($x/d \simeq 2.2$) at which vortices are completely formed to be identified. The vortex characteristics in the region $x/d > 2.2$ are inferred from the instantaneous data and/or from conventionally and conditionally averaged information. Conditional averages were obtained by detecting local temperature maxima. Assuming that the location of these maxima corresponds, albeit approximately, with the centres of the vortices, the temperature distribution through the vortex centre can serve as input for a vortex model of the near/intermediate wake. Zhou and Antonia (1993) noted that the velocity field near the vortex core could be approximated by that of an Oseen-like vortex. This model is extended here to calculate conventional mean and rms temperature profiles.

2. Experimental Details

Measurements were made in an open-return low-turbulence wind tunnel with a 2.4 m long square working section (350 mm × 350 mm). The wake was generated by a cylinder of diameter d = 12.6 mm spanning the width of the working section. The bottom wall of the working section was adjusted to obtain a zero streamwise pressure gradient. The free-stream velocity $U_1 \simeq 6.5$ m/s and the Reynolds number R_d, based on U_1 and d, was about 5500. The free-stream turbulence level was approximately 0.05%. Measurements were made in the near/intermediate wake ($x/d \leqslant 30$). The brass cylinder was heated electrically, the amount of heat being sufficiently small for temperature to be treated as a passive scalar in the near wake.

As shown in Figure 1, an array of 16 cold wires was used; the wires were at the same x and z locations and aligned in the y direction with a nominal separation Δy between wires of 2.6 mm. At x/d = 3, 5 and 20, the array covered the range $-1.5 \leqslant y/d \leqslant 1.6$ while at x/d = 10, the range was $-1.2 \leqslant y/d \leqslant 1.9$. All wires were made of 1.27 μm diameter Wollaston (Pt-10% Rh) wire, etched to a length of 0.7–0.8 mm (wire resistance ≈ 142 Ω). The cold wires were operated by in-house constant current circuits supplying 0.1 mA to each wire. The signals

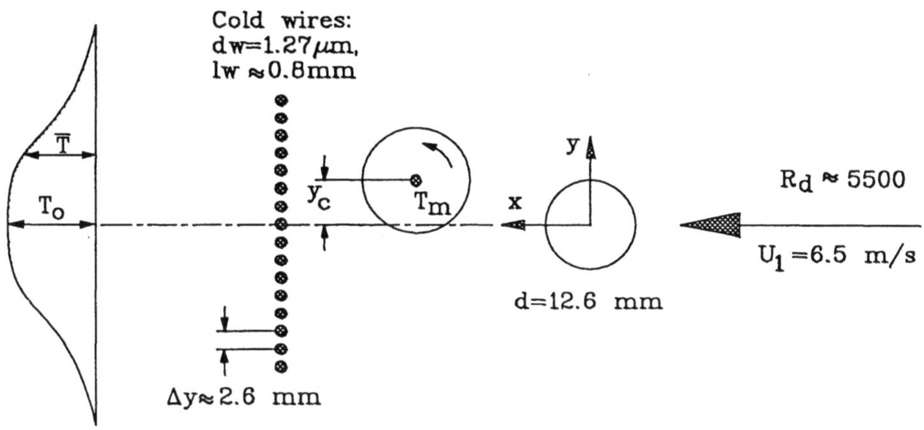

Fig. 1. Experimental configuration

from the circuits were offset, amplified and then digitized using a 16 channel, 12-bit A/D converter (RC Electronics) on a personal computer (NEC 386), at a sampling frequency f_s = 3.5 kHz per channel (filter cutoff frequency = 1.75 kHz). The record duration t_s was about 50 s. The digital signals were subsequently transferred via an Ethernet optical cable link to a VAX 8550 computer. The temperature coefficient (1.69×10^{-3} °C^{-1}) of the cold wires was estimated by calibrating a wire (etched from the same Wollaston spool used for all the 16 wires) in the exit plane of a circular jet. Mean and rms temperature profiles obtained with the array agreed, to within ±8%, with profiles made by traversing a single cold wire across the wake. Traverses with a single cold wire were used in order to establish the vortex formation region.

The value of x/d at which Kármán vortices are completely formed may be inferred from the appearance of vortex-induced peaks in the temperature signal on the wake centerline. As shown in Figure 2, these (broad) peaks are observed somewhat irregularly at $x/d = 2$ but are strongly periodic at $x/d = 2.2$ (bottom trace in Figure 2). This position is comparable to the average length of the vortex formation region obtained by Bloor (1964) from hot-wire measurements at R_d = 400 ~ 20 000 and Kourta et al. (1987) from flow visualization at R_d = 4800. The main interest here is in the region $2.2 < x/d \leqslant 30$, where the fully formed vortices evolve through the near/intermediate wake.

3. Detection Method

A wide range of methods have been used to detect the organization in the wake of a circular cylinder (e.g. Davies, 1976; Cantwell and Coles, 1983; Kiya and Matsumura, 1985; Hussain and Hayakawa, 1987; Ferré and Giralt, 1989; Antonia, 1991). The differences between these methods relate mainly to the aspects of the organized motion which are emphasized by each method. An approximation to the large-scale

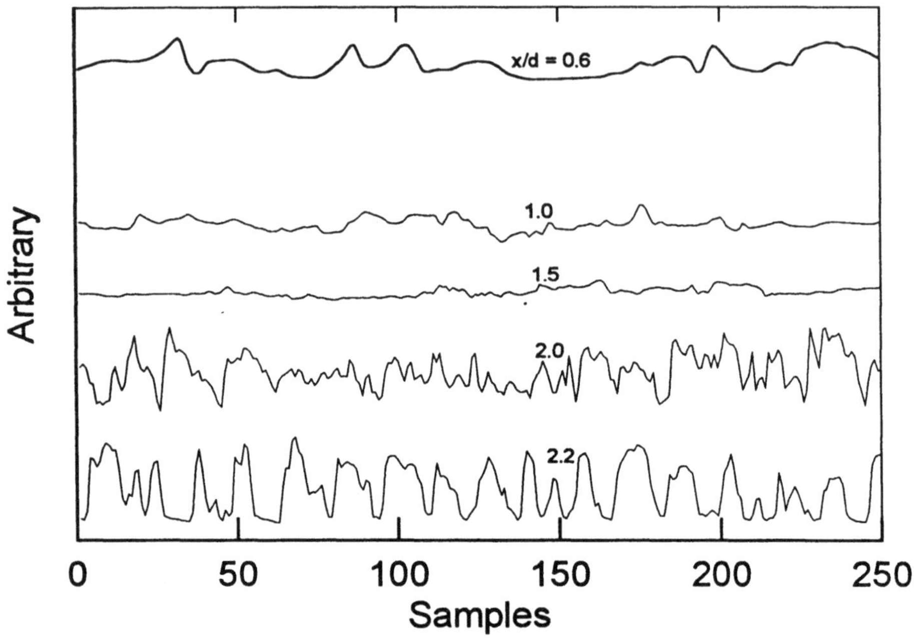

Fig. 2. Temperature signals on the wake centerline for $x/d \leqslant 2.2$

spanwise vorticity has been used in some methods (Hussain and Hayakawa, 1987; Zhou and Antonia, 1993) while others have focused on sudden changes in either velocity (Bisset et al., 1990; Matsumura and Antonia, 1993) or temperature (Browne et al., 1986) fluctuations. In general, results obtained from these methods are in reasonable agreement with each other, perhaps because of the relatively high degree of organization which characterizes this flow. In the present method, the spatial locations of local temperature maxima are identified. These locations are assumed to correspond with the centres of the vortices. Some justification for this assumption is provided by flow visualizations (e.g. Zhou and Antonia, 1992) which show that the peak concentration of dye injected near the rear stagnation point of the cylinder tends to coincide with the vortex centre, at least at small R_d. Further justification is provided by the similarity between contours of phase-averaged vorticity and those of phase-averaged temperature (Matsumura and Antonia, 1993). Although the similarity impaired as x/d increased, the locations of the phase-averaged maxima of vorticity and temperature remained in coincidence for $x/d \leqslant 40$.

Before implementing the detection criteria, some interpolation and smoothing were applied to the data. Because of the coarse separation ($\Delta y \approx 2.6$ mm) between adjacent wires in the array, an extra row of temperature data was added, using linear interpolation, between each pair of existing rows of temperature data. Some of the interpolated points were examined to ensure that the interpolation was reasonable.

Also, some smoothing of the instantaneous temperature signals was applied due to the finite values of Δy and Δx ($\equiv U_c/f_s \approx 1.6$ mm, U_c is the average vortex convection velocity). This should be acceptable since the smoothed temperature maps only facilitate the identification of the structures, but do not affect any of the conditional averages (for which only unsmoothed data were used).

The initial criterion used for detection is

$$T_m = \max\left[T(x',y')\right] \quad \text{for } |x'| \leqslant k_1 d \text{ and } |y'| \leqslant k_2 d \quad , \tag{1}$$

where T is the instantaneous temperature (relative to ambient) and k_1 and k_2 are non-dimensional thresholds. The origin of the x' ($\equiv -U_c t'$, where t' represents the time, measured in terms of samples, relative to the T_m detection; t' can be positive or negative) and y' co-ordinates is at the detection point. The average convection velocity U_c of vortices, obtained by various methods, varies only slightly across the wake and is about $0.9U_1$ in the turbulent near/intermediate wake (Zhou and Antonia, 1992) so that the influence on the detection locations is quite small. It is also important not to bias the detection with spurious values of T_m which are not associated with large-scale vortices in order only to include reliable information when conditional data are formed. To achieve this, a second criterion was added, viz.

$$0 < \frac{T_m - T_1}{T_1 - T_2} \leqslant k_3 \quad , \tag{2}$$

where T_1 and T_2 are, respectively, the averaged temperatures over all the points on the boundaries of 3×3 and 5×5 grids centred on $T_m(0,0)$. Initially, the values of the thresholds were chosen arbitrarily. They were later modified after visually comparing the detection locations on plots of instantaneous temperature contours (e.g. Figure 3). The values of k_1, k_2 and k_3 were finally chosen to be 1.2, 0.5 and 0.6, respectively. With these values, the normalized average detection frequency $(N/t_s)d/U_1$ (N being the total number of detections) on the $y > 0$ side of the centerline ranges from 0.21 ($x/d \leqslant 10$) to 0.18 ($x/d = 30$). The value for $x/d \leqslant 10$ corresponds to a Strouhal number $n_s d/U_1$ of 0.21, the value expected when $R_d \gtrsim 300$, e.g. Schlichting (1979) and Cantwell and Coles (1983) [$n_s \approx 107$ Hz was determined from the peak in the temperature spectrum]. This indicates that almost all the vortices in the region $x/d \leqslant 10$ are captured by the present detection method. Beyond $x/d = 10$, there may be some vortices which escape detection due to their centres being outside the range of the cold wire array or due to the reduced effectiveness of the thermal tagging.

4. Topology of Temperature Field

The conditional average of a quantity Q ($\equiv T$ or ϑ) is defined by

$$\langle Q\rangle(x',y') = \frac{1}{N}\sum_{i=1}^{N} Q_i(x',y') \quad ,$$

where the origin of (x',y') is at the location of the ith detection.

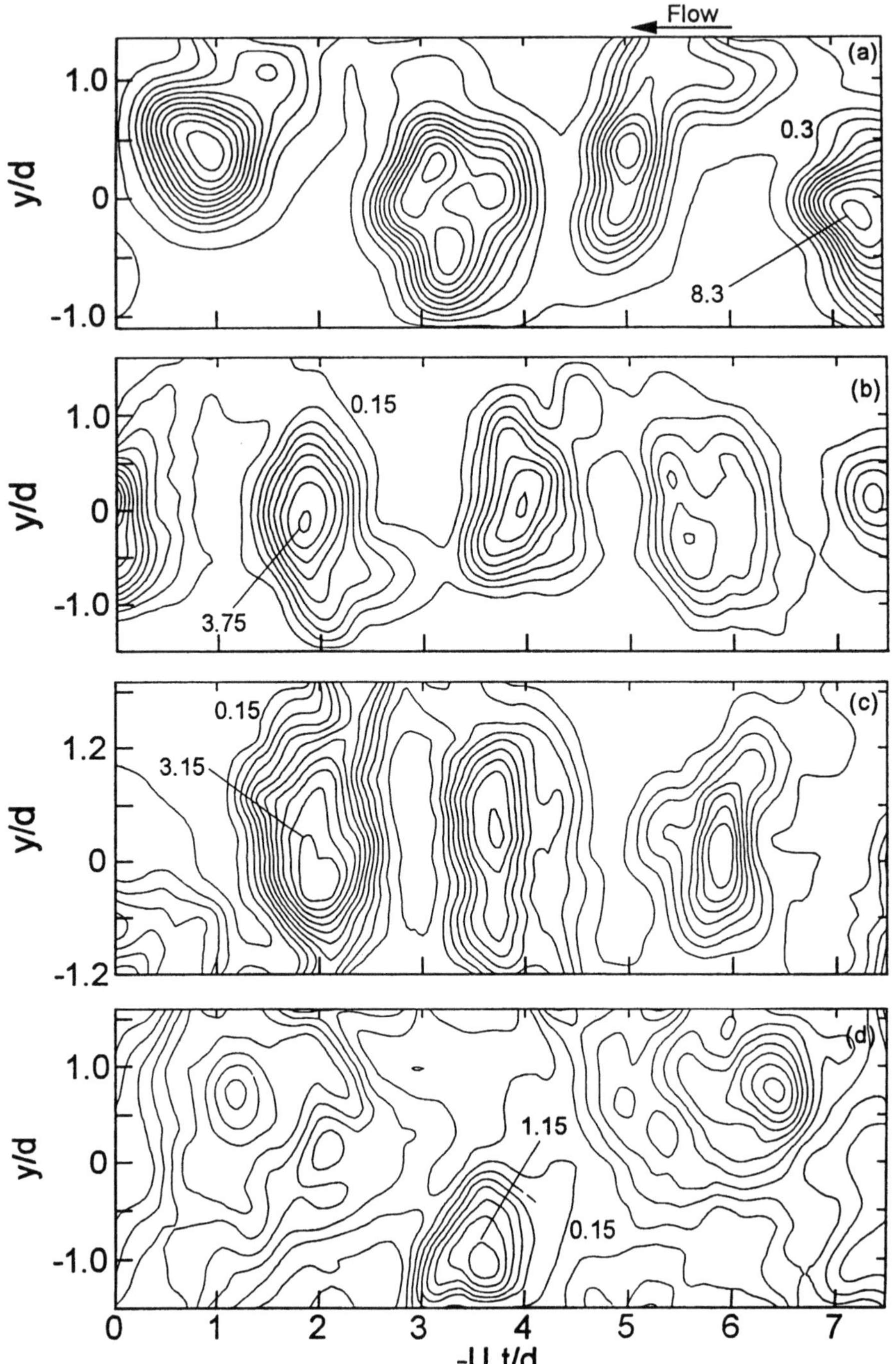

Fig. 3. Contours of constant instantaneous temperature T/T_0. **(a)** $x/d = 3$, outermost contour = 0.3 (step = 0.8); **(b)** 5, 0.15 (0.4); **(c)** 10, 0.15 (0.3); **(d)** 20, 0.15 (0.2)

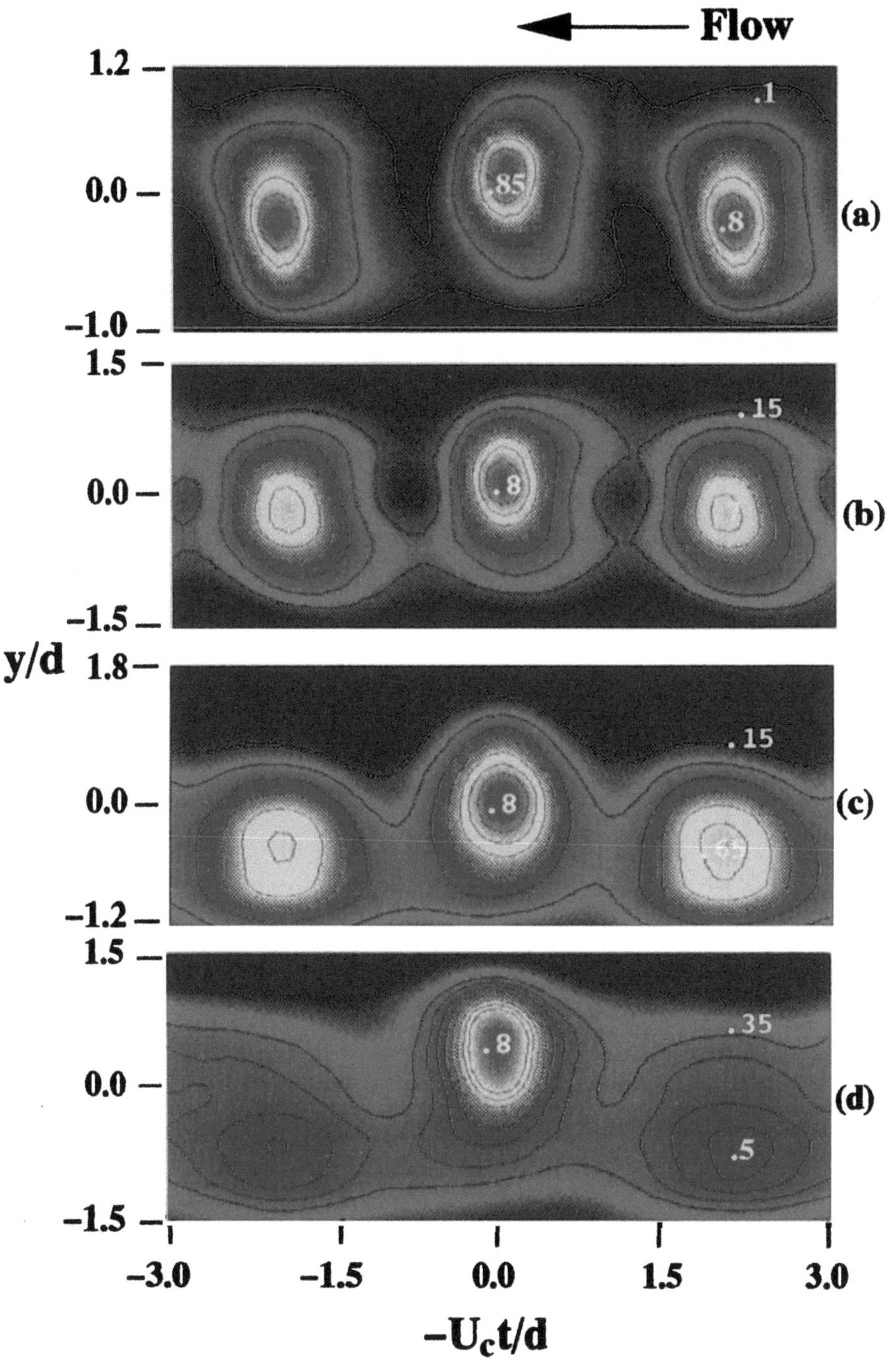

Fig. 4. Contours of constant $\langle T\rangle/\langle T_m\rangle$. (**a**) $x/d = 3$, outermost contour = 0.1 (step = 0.1); (**b**) 5, 0.15 (0.1); (**c**) 10, 0.15 (0.1); (**d**) 20, 0.35 (0.05)

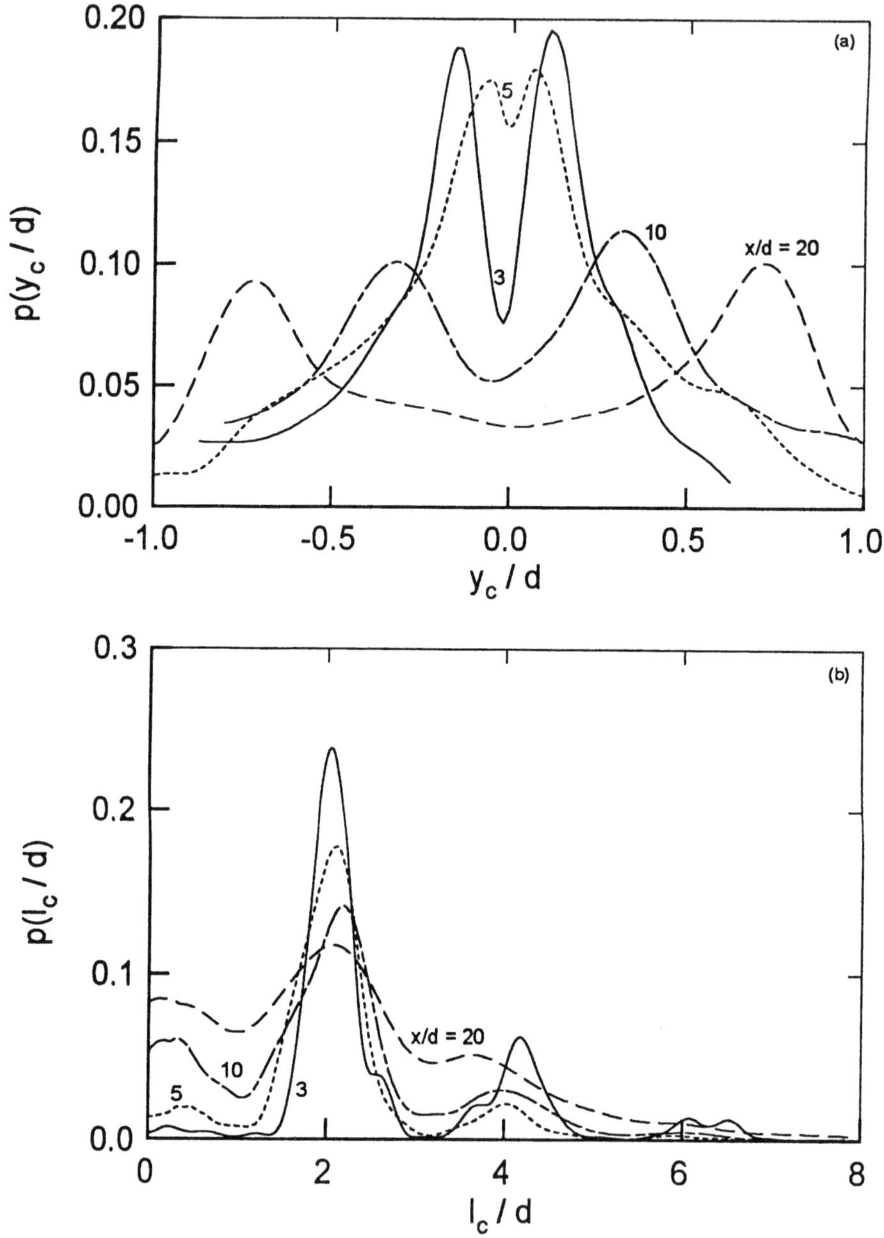

Fig. 5. Probability density functions of y_c and ℓ_c. (**a**) $p(y_c/d)$; (**b**) $p(\ell_c/d)$

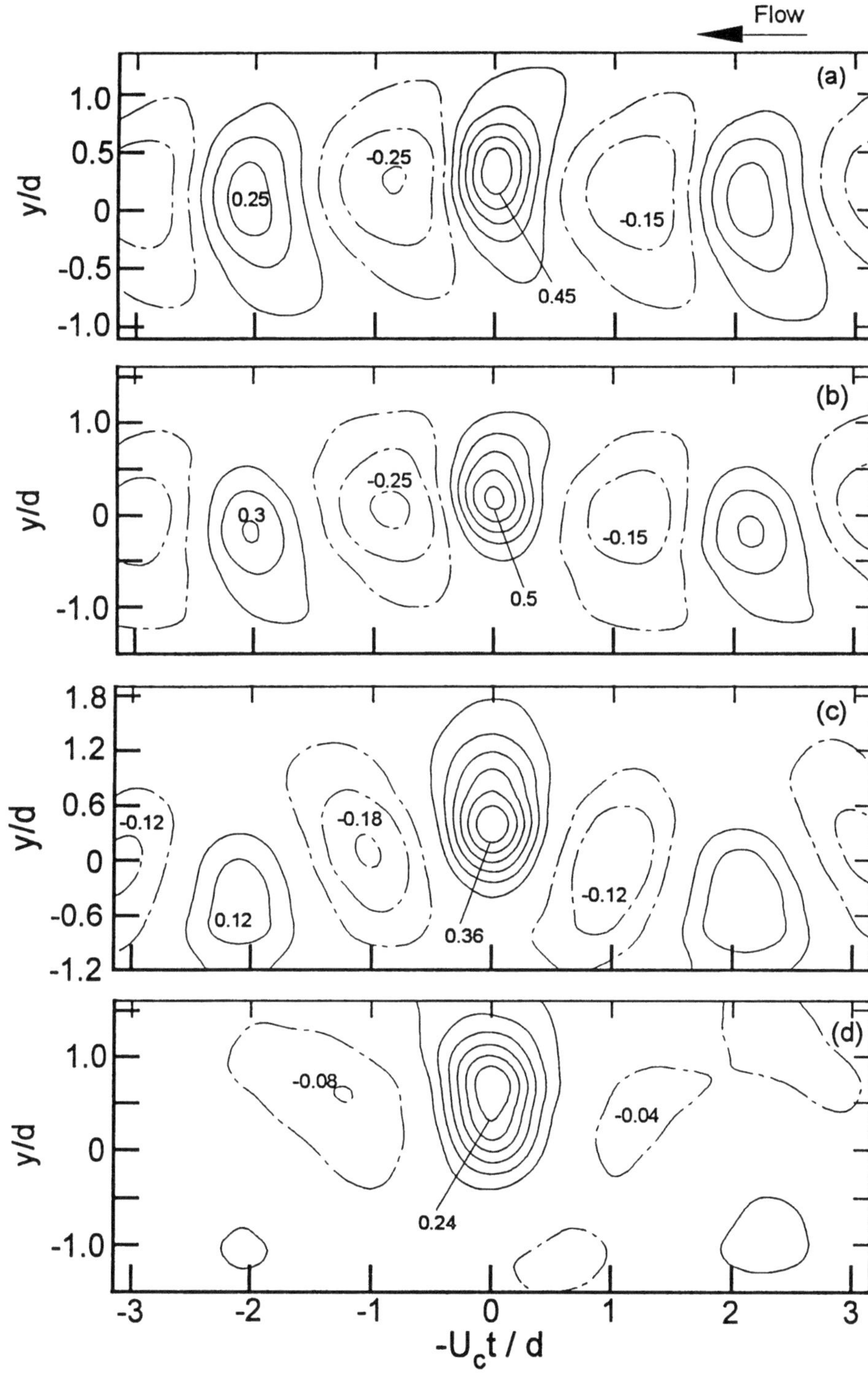

Fig. 6. Contours of constant $\langle\vartheta\rangle/\langle T_m\rangle$. Values of innermost contours are indicated on the plot. (a) $x/d = 3$, (step = 0.1); (b) 5, (0.1); (c) 10, (0.06); (d) 20, (0.04)

Colour contours of constant $\langle T\rangle/\langle T_m\rangle$, based on detections on one side ($y > 0$) of the centerline, are shown in Figure 4 for x/d = 3, 5, 10 and 20. The contours delineate the expected antisymmetrical arrangement of vortices with respect to the centerline. At x/d = 3, the peak value of $\langle T\rangle/\langle T_m\rangle$ for $y < 0$ is only marginally smaller than at the average detection location ($y > 0$), reflecting the dominance of the alternating vortex made in the initial stage of the Kármán street. Since not all vortices, after formation, move downstream with exactly the same convection velocity U_c, some of them will overtake slower opposite-signed vortices, thus resulting in a degradation of the initially antisymmetrical street. As a consequence, the difference between the peak values at $y > 0$ and $y < 0$ (see Figure 4) increases as x/d increases. This difference is also clearly reflected in $p(y_c)$ and $p(\ell_c)$, the probability density functions (pdfs) of y_c and ℓ_c (the streamwise separation between two adjacent vortices), shown in Figure 5. The peak values of $p(\ell_c)$ and $p(y_c)$ decrease as x/d increases. The value of $p(\ell_c)$ for ℓ_c = 0 increases with x/d. The latter behaviour indicates that the number of different-signed vortices which are at the same x location becomes larger as x/d increases. The magnitude of the average wavelength $2\overline{\ell}_c$ ($\approx 4d$) is emphasized by the peaks in $p(\ell_c)$ due to the quasi-periodicity of the vortex motion. Low magnitude isotherms are generally stretched across the centerline on the downstream side of each detected vortex (this is reflected by the tongues of cold fluid, indicated by royal purple in Figures 4a and b). This behaviour, which is similar to that reported by Matsumura and Antonia (1993), can also be seen in the instantaneous temperature contours for x/d = 3 and 5 (Figures 3a and b). This implies excellent marking of the vortices by temperature and a high degree of coherent vortical motion in the near-wake. As x/d increases, the thermal marking becomes less effective as can be ascertained from the instantaneous temperature contours at x/d = 20. The lack of symmetry in the conditional temperature contours is also emphasized in the contours of $\langle\vartheta\rangle/\langle T_m\rangle$: negative values of $\langle\vartheta\rangle/\langle T_m\rangle$ (Figure 6) are more pronounced downstream than upstream of the vortex centre at each measurement station. The above features of the detected vortex imply that heat is effectively transported out of the vortices from the downstream side due to the relatively strong engulfment of cold free-stream fluid; this is consistent with the model proposed by Matsumura and Antonia (1993).

The dependence on x/d of the average y position of the local temperature maxima T_m is shown in Figure 7. Immediately downstream of the cylinder, the separated boundary layer, which is primarily responsible for the transferring of heat from the cylinder to the wake, moves toward the centerline. The average position $\overline{y}_s$ which corresponds to T_m in the formation region thus decreases as x/d increases. At each y position, local temperature extrema were identified over successive periods (n_s^{-1}) in the temperature time trace and then averaged; $\overline{y}_s$ was finally identified with the position at which the maximum in the locally averaged extrema occurs. After the vortices are completely formed, they subsequently move rather sharply away from the centerline, as noted by Bloor and Gerrard (1966), perhaps because of the mutual repulsion between relatively strong vortices of opposite sign. The mean vortex location $\overline{y}_c$ increases most rapidly in the region $5 < x/d < 20$ and

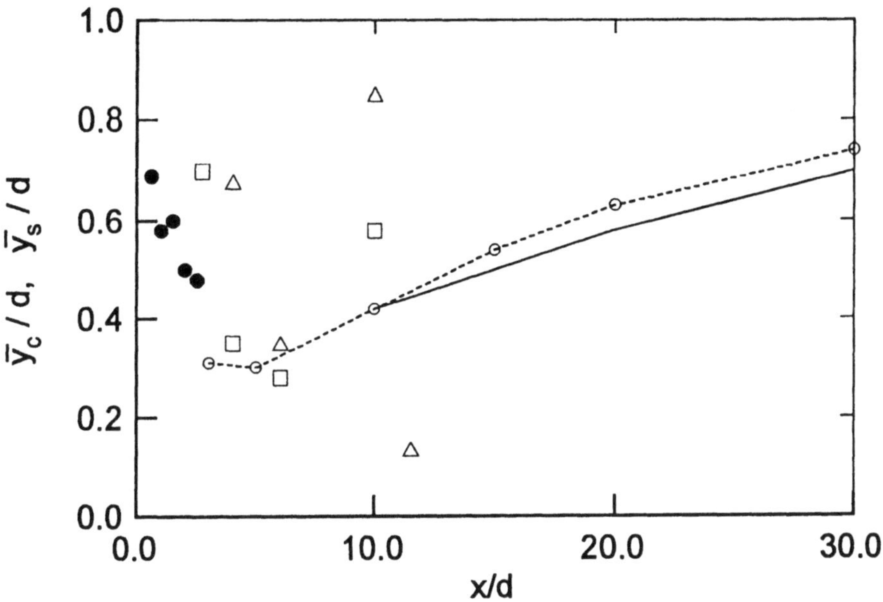

Fig. 7. Streamwise variations of $\overline{y}_c$ and $\overline{y}_s$.
Present: •, $\overline{y}_s$; o- - - -o, $\overline{y}_c$. Zhou and Antonia (1993): ———, $\overline{y}_c$. Bloor and Gerrard (1966): △, □.

is appreciably larger at x/d = 20 than for $x/d \leqslant 10$. This may be why the mean temperature $\overline{T}/T_0$ (T_0 is the value of $\overline{T}$ at $y = 0$) profile at $x/d = 20$ (Figure 8a) exhibits double peaks (albeit weak) away from the centerline. When $x/d > 20$, the $\overline{T}/T_0$ profile first becomes flat around $y = 0$ at $x/d = 30$ before once more peaking on the centerline at $x/d = 40$ (Figure 8a). This behaviour may reflect the combined effects of the streamwise increase in $\overline{y}_c$, the decrease in the magnitude of $\langle T_m \rangle$ and the increased jitter in y_c. Note that $\langle T_m(x/d = 30)\rangle/\langle T_m(x/d = 20)\rangle \approx 0.77 < T_0(x/d = 30)/T_0(x/d = 20) \approx 0.86$, indicating that the magnitude of $\langle T_m \rangle$ decays more rapidly than that of the mean temperature on the centerline. The profile of $\sqrt{\overline{\vartheta^2}}$ (Figure 8b) peaks only on the centerline at $x/d \leqslant 10$, presumably because of the small value of $|\overline{y}_c|$.

Contours of constant (2-point) temperature correlation coefficients $\varrho_\vartheta(y_1/d, y_2/d; \Delta x/d)$, with the reference point y_1/d selected to be as near as possible to the value of $\overline{y}_c/d$, are shown in Figure 9. They are very similar to the contours of $\langle\vartheta\rangle/\langle T_m\rangle$ at $x/d = 3$ and 5, reflecting the nearly perfect tagging of the vortices by temperature and the strongly intermittent nature of temperature in the near-wake, e.g. the bottom trace in Figure 2. The instantaneous temperature contours at these two stations indicate that cold fluid can be found at all y locations and the thermal interface is relatively sharp. Consistently, the value of the averaged intermittency factor $\overline{\gamma}$ obtained by Boisson et al. (1983) in the near wake of a circular cylinder (R_d = 52300) is significantly lower than the unity, even on the centerline. Not sur-

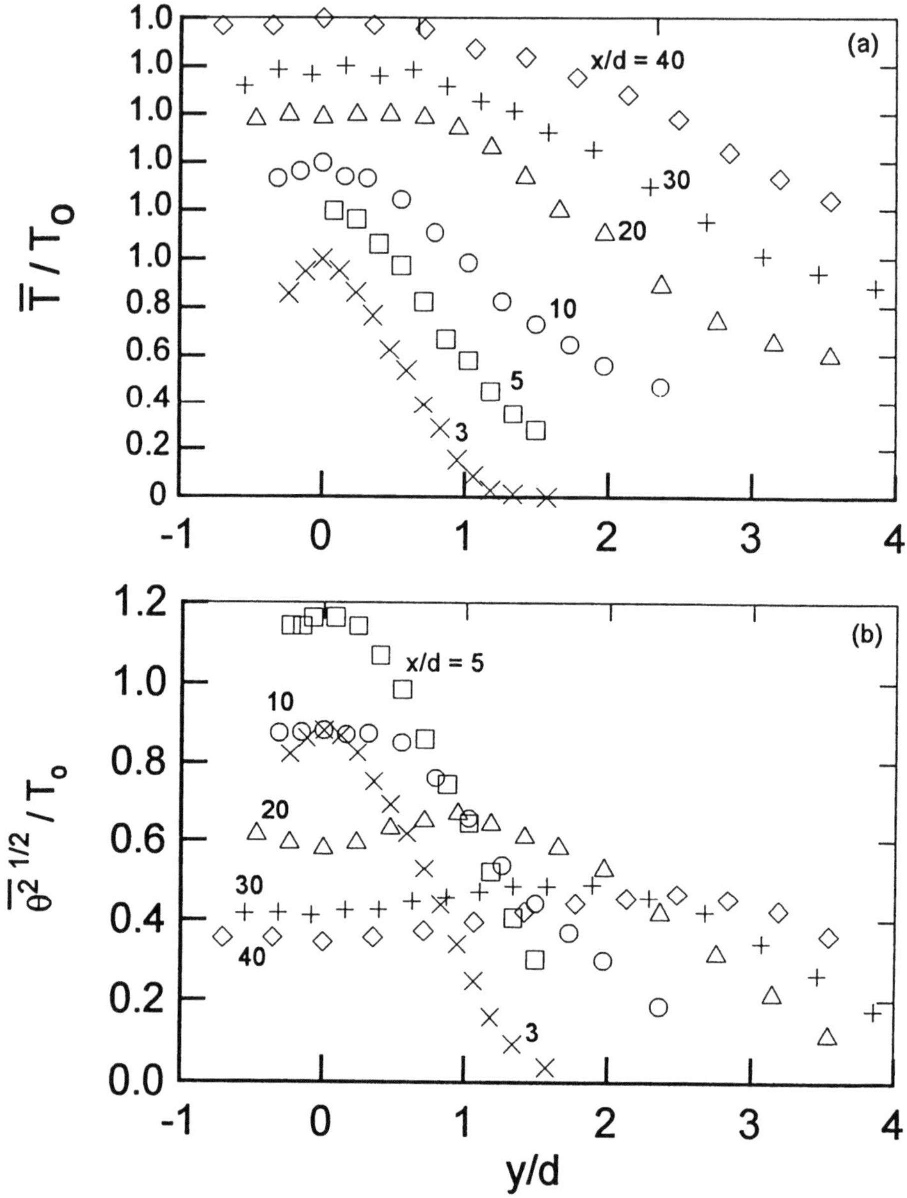

Fig. 8. Lateral distributions of mean and rms temperature.
(a) $\overline{T}/T_0$; **(b)** $\sqrt{\overline{\vartheta^2}}/T_0$. ×, $x/d = 3$; □, 5; ∘, 10; △, 20; +, 30; ◇, 40

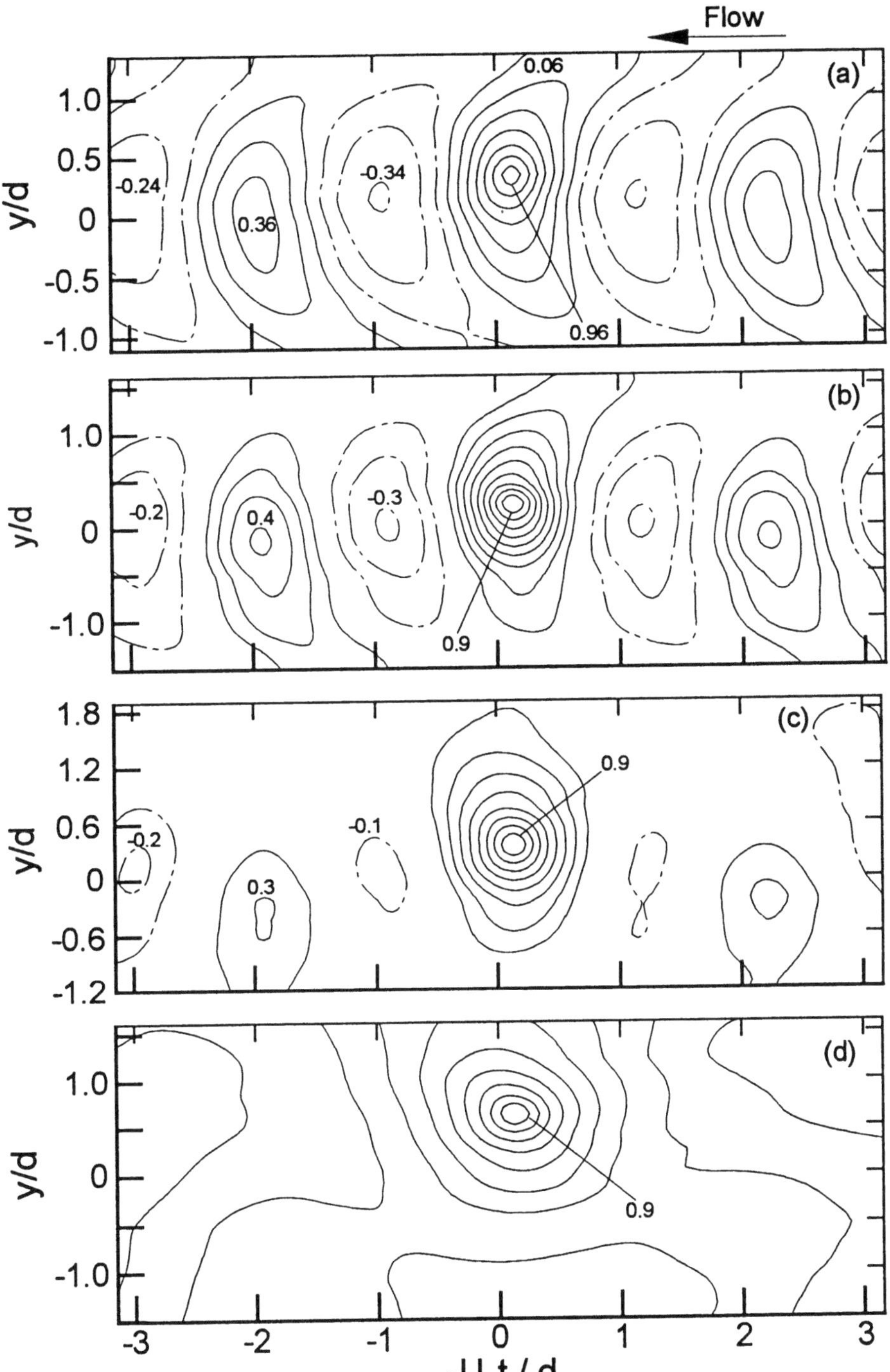

Fig. 9. Contours of constant temperature correlation coefficient $\varrho_\vartheta(y_1/d, y_2/d, \triangle x/d)$. Values of innermost contours are indicated on the plot.
(a) $x/d = 3$, step = 0.15 and $y_1/d = 0.35$; (b) 5, 0.1 and 0.16; (c) 10, 0.1 and 0.3; (d) 20, 0.1 and 0.58

prisingly, this similarity impairs as x/d increases (it has virtually disappeared at $x/d = 20$).

5. Contributions from Coherent and Incoherent Motions

The temperature fluctuation ϑ may be equated to the sum of $\langle\vartheta\rangle$, the contribution from the periodic large-scale motion and ϑ_r, the contribution from the local random (small-scale) turbulence, viz.

$$\vartheta = \langle\vartheta\rangle + \vartheta_r \quad . \tag{3}$$

Differentiating (3) with respect to x_i (where i = 1, 2 and 3; $x_1 \equiv x$, $x_2 \equiv y$ and $x_3 \equiv z$) gives

$$\frac{\partial\vartheta}{\partial x_i} = \frac{\partial\langle\vartheta\rangle}{\partial x_i} + \frac{\partial\vartheta_r}{\partial x_i} \quad . \tag{4}$$

It follows, by squaring and conditionally averaging, that

$$\langle\vartheta^2\rangle = \langle\vartheta\rangle^2 + \langle\vartheta_r^2\rangle \tag{5}$$

and

$$\left\langle\left(\frac{\partial\vartheta}{\partial x_i}\right)^2\right\rangle = \left(\frac{\partial\langle\vartheta\rangle}{\partial x_i}\right)^2 + \left\langle\left(\frac{\partial\vartheta_r}{\partial x_i}\right)^2\right\rangle \quad . \tag{6}$$

From (6), the conditional dissipation of ϑ is

$$\langle\epsilon_\vartheta\rangle = \alpha\left(\frac{\partial\langle\vartheta\rangle}{\partial x_i}\right)^2 + \alpha\left\langle\left(\frac{\partial\vartheta_r}{\partial x_i}\right)^2\right\rangle \quad ,$$

where α is the thermal diffusivity and repeated indices imply summation. The magnitudes of $(\partial\langle\vartheta\rangle/\partial x_i)^2$ and $\langle(\partial\vartheta_r/\partial x_i)^2\rangle$ are measures of the contributions from the coherent large-scale and random small-scale motions to the conditional temperature dissipation rate.

An effective way to represent the spatial distribution of these two contributions within vortices is to plot contours of both $\langle(\partial\vartheta_r/\partial x_i)^2\rangle\langle T_m\rangle^{-2}d^2$ and $(\partial\langle\vartheta\rangle/\partial x_i)^2\langle T_m\rangle^{-2}d^2$. Taylor's hypothesis $\partial/\partial x_1 = -\overline{U}^{-1}\partial/\partial t$, where the local mean velocity $\overline{U}$ is replaced by the average vortex convection velocity U_c ($= 0.9U_1$), was used to determine the streamwise derivatives $\partial\vartheta_r/\partial x_1$ and $\partial\langle\vartheta\rangle/\partial x_1$; some justification for the use of the hypothesis is provided by our unpublished data (not shown here) obtained using two parallel cold wires in the intermediate wake ($x/d \geqslant 10$). Note also that the approximation $\partial\vartheta/\partial x_2 \approx [\vartheta(y+\triangle y)-\vartheta(y)]/\triangle y$ was used but the separation $\triangle y$ ($\approx$ 2.6 mm) is about 12 Kolmogorov length scales at $x/d = 10$ and is therefore not sufficiently small to resolve all small scales; as a result, $\langle(\partial\vartheta_r/\partial x_2)^2\rangle$ is underestimated. Moreover, it was assumed that $\partial\langle\vartheta\rangle/\partial x_2 = \partial\langle\vartheta\rangle/\partial x_3$ and $\langle(\partial\vartheta_r/\partial x_2)^2\rangle = \langle(\partial\vartheta_r/\partial x_3)^2\rangle$. In this context, we emphasize that the results presented here are only qualitative. As shown in Figure 10, the contours of $\langle(\partial\vartheta_r/\partial x_i)^2\rangle$

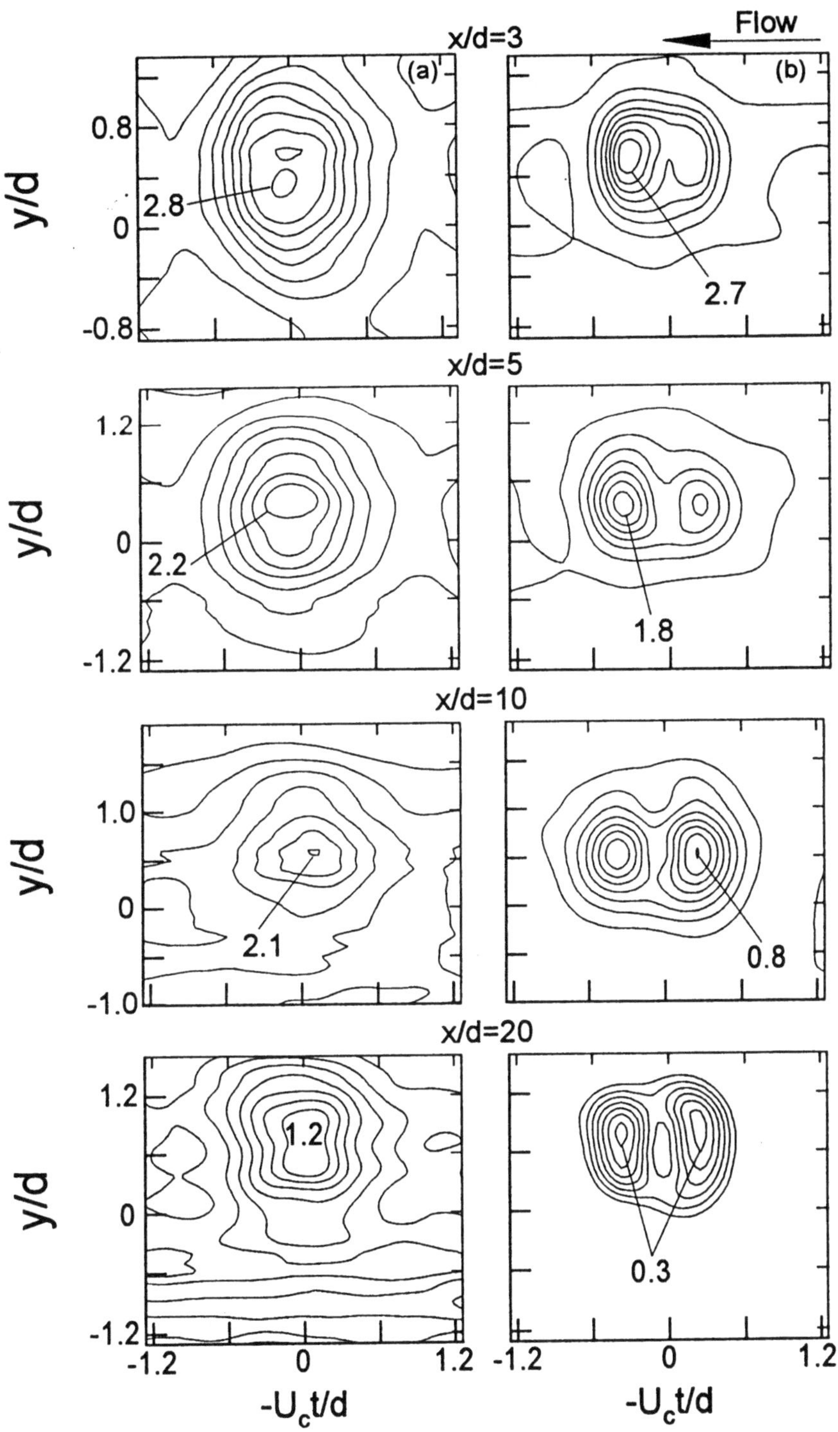

Fig. 10. Contours of constant $\left\langle(\partial\vartheta_r/\partial x_i)^2\right\rangle$ and $(\partial\langle\vartheta\rangle/\partial x_i)^2$. Values of innermost contours are indicated on the plot.
(a) $\left\langle(\partial\vartheta_r/\partial x_i)^2\right\rangle\langle T_m\rangle^{-2}d^2$; (b) $(\partial\langle\vartheta\rangle/\partial x_i)^2\langle T_m\rangle^{-2}d^2$

are approximately symmetrical with respect to the vortex centre at all the measured stations. This is not unexpected because the random small-scales should be uniformly distributed within the vortices. However, the situation is different for $(\partial\langle\vartheta\rangle/\partial x_i)^2$; this quantity is much more pronounced on the downstream side of the vortex centre at $x/d = 3$ and 5. As x/d increases, this behaviour weakens and $(\partial\langle\vartheta\rangle/\partial x_i)^2$ assumes slightly larger values on the upstream side at $x/d = 10$ and 20. Figure 10 also shows that the relative magnitudes of $\langle(\partial\vartheta_r/\partial x_i)^2\rangle$ and $(\partial\langle\vartheta\rangle/\partial x_i)^2$ depend on x.

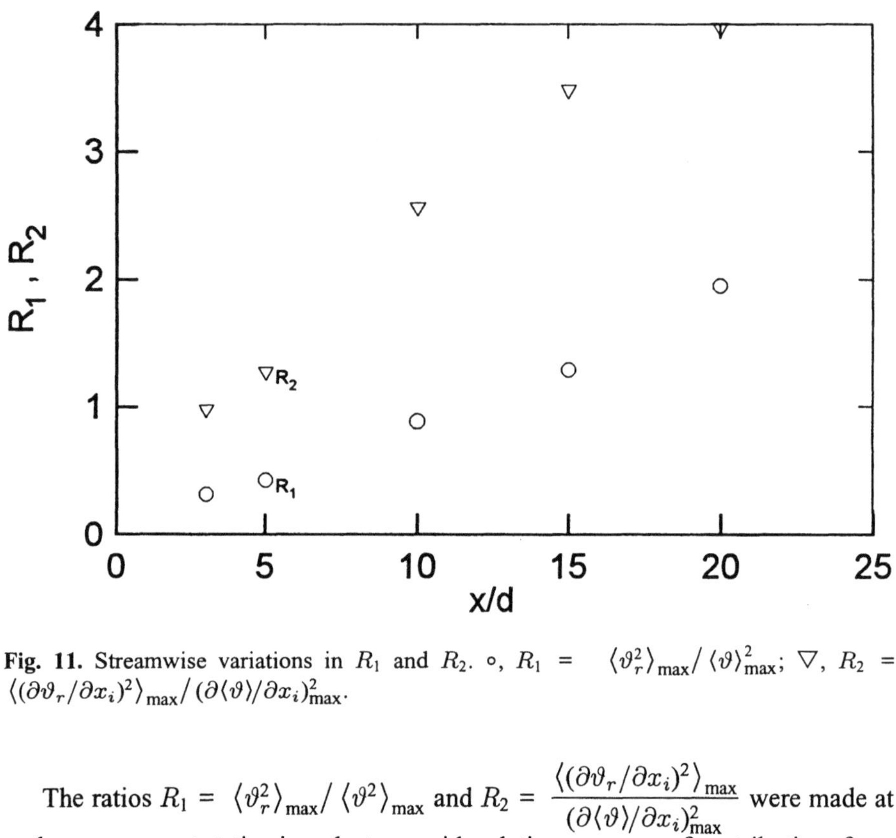

Fig. 11. Streamwise variations in R_1 and R_2. ∘, $R_1 = \langle\vartheta_r^2\rangle_{\max}/\langle\vartheta\rangle^2_{\max}$; ∇, $R_2 = \langle(\partial\vartheta_r/\partial x_i)^2\rangle_{\max}/(\partial\langle\vartheta\rangle/\partial x_i)^2_{\max}$.

The ratios $R_1 = \langle\vartheta_r^2\rangle_{\max}/\langle\vartheta^2\rangle_{\max}$ and $R_2 = \dfrac{\langle(\partial\vartheta_r/\partial x_i)^2\rangle_{\max}}{(\partial\langle\vartheta\rangle/\partial x_i)^2_{\max}}$ were made at each measurement station in order to provide relative measures of contributions from the small-scale and large-scale to $\overline{\vartheta^2}$ and $\langle(\partial\vartheta/\partial x_i)^2\rangle$. Figure 11 shows that there are significant streamwise increases in both R_1 and R_2 (for example, R_2 increases from 1 at $x/d = 3$ to 4 at $x/d = 20$). The value of R_2 is consistently greater than that of R_1, as is expected since $\partial\vartheta/\partial x_i$ is associated mainly with the small-scales whereas ϑ is essentially dominated by the large-scales. Similarly, from the near-wake data ($R_d = 14\,000$) of Cantwell and Coles (1983) who used a flying-wire combined with a phase-averaging technique, we inferred that the ratio $\langle u_r^2\rangle_{\max}/\langle u\rangle^2_{\max}$ increases rapidly as x/d increases. The trend for the above ratios generally suggests that, as the vortices move downstream, their strength decays while the superimposed small-scale motion is enhanced.

6. Temperature Distribution Within Vortices

The conditional temperature distribution $\langle T\rangle(r)$ [r is the local radius measured from the vortex centre] within vortices was estimated by applying a surface-fit, using fourth-order Chebychev polynomials, to the two-dimensional grid of temperature data centred on each vortex detection. The origin of the corresponding cylindrical co-ordinate system (r, β) is the location where $T = T_m$. To eliminate the dependence of $T(r, \beta)$ on β, we define

$$T(r) = \frac{1}{2\pi}\int_0^{2\pi} T(r,\beta)\mathrm{d}\beta \quad .$$

Distributions of $\log[\langle T\rangle(r)/\langle T_m\rangle]$ vs $(r/d)^2$ are linear for $r \leqslant r_1$ (r_1 decreases from about $0.5d$ at $x/d = 3$ to $0.3d$ at $x/d = 20$). It appears that near the centre of the vortex, $\langle T\rangle(r)/\langle T_m\rangle$ is approximately Gaußian, viz.

$$\frac{\langle T\rangle(r)}{\langle T_m\rangle} = \exp\left[-a_1\left(\frac{r}{d}\right)^2\right] \quad . \tag{7}$$

The non-dimensional parameter a_1 was estimated from a least-squares fit to $-\ln\left[\langle T\rangle(r)/\langle T_m\rangle\right]/(r/d)^2$ over the range $0 \leqslant r \leqslant r_1$. The magnitude of a_1 decreases with increasing x/d (Figure 12); this is expected because, as the vortices move downstream, the spread of $T(r)$ increases.

The present temperature distribution is similar to the conditionally averaged vorticity distribution $\langle\omega\rangle(r)/\langle\omega_m\rangle$ of Zhou and Antonia (1993). Using detections of the spanwise vorticity peaks, they noted that, in the range $10 \leqslant x/d \leqslant 60$, $\langle\omega\rangle(r)/\langle\omega_m\rangle$ is approximated reasonably well by a Gaußian distribution

$$\frac{\langle\omega\rangle(r)}{\langle\omega_m\rangle} = \exp\left[-a_2\left(\frac{r}{d}\right)^2\right] \tag{8}$$

for $r \lesssim 0.5d$, with $a_2 \approx 4.66$. For comparison, Eq. (8) is plotted in Figure 12. For $x/d = 10$ and 20, the values (≈ 0.85 and 0.51) of a_1/a_2 are reasonably close to the averaged turbulent Prandtl number values (≈ 0.9 and 0.4, estimated by averaging Pr_t across the wake) reported in Matsumura and Antonia (1993). This correspondence is somewhat similar to what has been observed in a laminar wake. Mi and Antonia (1993) showed that vorticity distributions (obtained indirectly by Okude and Matsui, 1990, at $R_d = 140$) and temperature distributions ($R_d = 94$) within laminar vortices in a circular cylinder wake are approximated quite well by the theoretical Gaußian distributions for a diffusing line vortex, i.e. $\omega/\omega_m = \exp\left[-0.693Pr^{-1}(r/R)^2\right]$ and $T/T_m = \exp\left[-0.693(r/R)^2\right]$, where Pr ($\equiv \nu/\alpha$) is the molecular Prandtl number and R is the value of r at which $T = T_m/2$. Clearly, the exponents of ω/ω_m and T/T_m differ by the factor Pr.

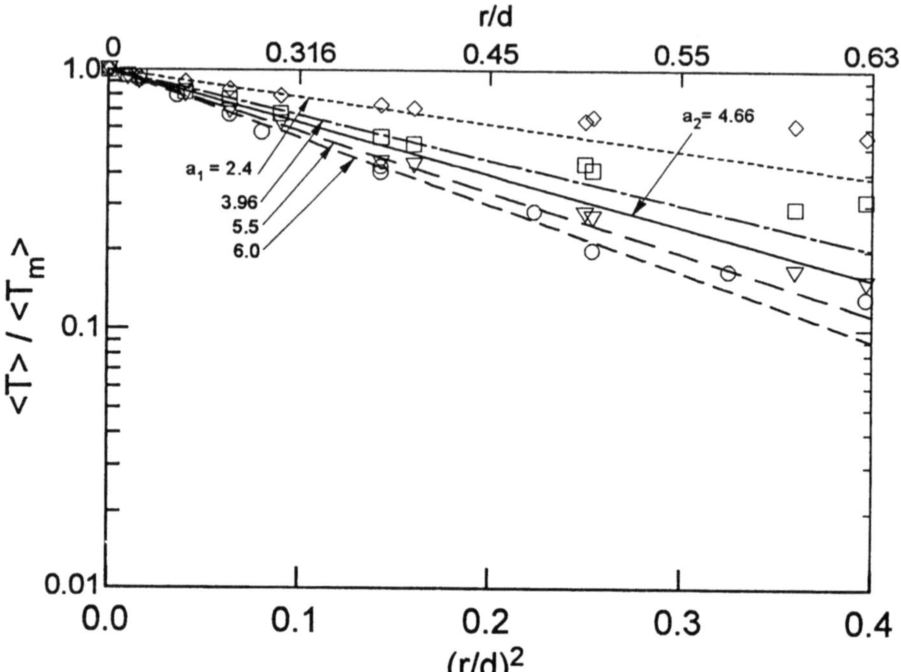

Fig. 12. Distributions of $\langle T\rangle/\langle T_m\rangle$ within vortices and comparison with the Gaußian distribution. ∘, $x/d = 3$; ▽, 5; □, 10; ◇, 20. Eq. (7): · · ·, $a_1 = 2.4$; ———, 3.96; ——, 5.5; - - - -, 6. Eq. (8): ———, $a_2 = 4.66$.

7. Heat Transfer Model

The results of the previous section and those of Zhou and Antonia (1993) provide some justification for the use of a non-isothermal Oseen-like vortex as a possible building block for a heat transfer model. The tangential velocity of the vortex which corresponds to Eq. (8) is given by

$$U_\beta = \frac{d^2\omega_m}{2a_2 r}\left\{1 - \exp\left[-a_2\left(\frac{r}{d}\right)^2\right]\right\} \quad . \tag{9}$$

The longitudinal and lateral velocities induced by the vortex located at (x_c, y_c) are then equal to $-U_\beta(r)\sin\beta$ and $U_\beta(r)\cos\beta$, where $r = \sqrt{(x-x_c)^2 + (y-y_c)^2}$ and $\beta = \tan^{-1}[(y-y_c)/(x-x_c)]$. When each vortex translates at a velocity $\boldsymbol{U} = (U_c, 0)$ with a maximum temperature (equal to T_m) at its centre, the velocity and temperature at any given point (x, y) can be calculated by superposing the contributions from all vortices, i.e.

$$U = U_c - \frac{1}{N}\sum U_\beta(r)\sin\beta\,, \qquad V = \frac{1}{N}\sum U_\beta\cos\beta \quad , \tag{10}$$

and

$$T = \frac{1}{N} T_m \sum \exp\left[-a_1 \left(\frac{r}{d}\right)^2\right] \quad , \tag{11}$$

where N (= 50) is the total number of vortices. The quantities $\overline{U}/U_1$, $\overline{u^2}/U_1^2$, $\overline{v^2}/U_1^2$, $\overline{uv}/U_1^2$, $\overline{T}/T_0$ and $\sqrt{\overline{\vartheta^2}}/T_0$ are then calculated as a function of y. For example,

$$\frac{\sqrt{\overline{\vartheta^2}}}{T_0} = \left[\frac{1}{M}\sum_{i=1}^{M}\left(\frac{T_i - \overline{T}}{T_0}\right)^2\right]^{1/2} \quad ,$$

where M (= 500) is the total number of streamwise grid points in a two-dimensional computational domain. According to Zhou and Antonia's (1993) experimental data, $a_2 \approx 4.66$ and $\omega_m \approx 1.5a_2U_1/\pi d$ at $x/d = 10$. Here, a_1 (≈ 3.98) is chosen to be the measured value at $x/d = 10$. The measured average wavelength ($\lambda \approx 4.0d$) and average vortex centre location ($\overline{y}_c \approx 0.4d$) at $x/d = 10$ are used.

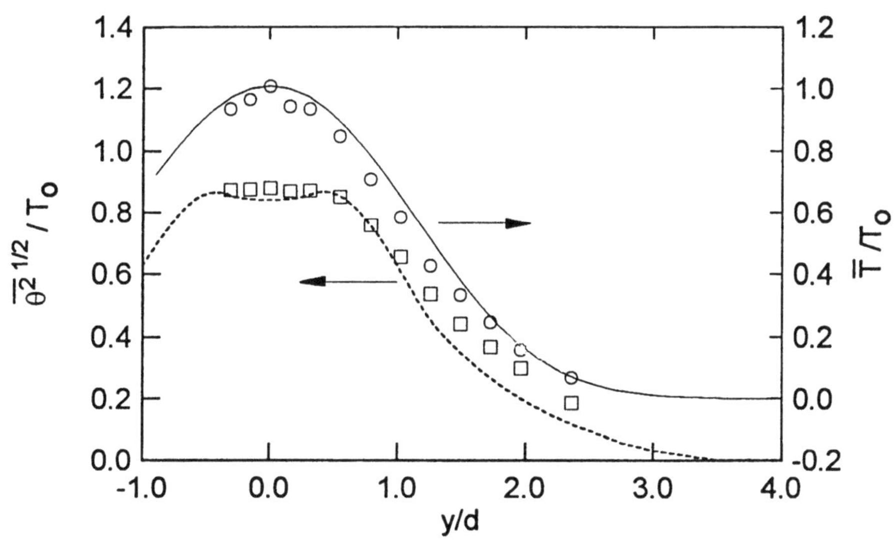

Fig. 13. Comparison between calculation and measurement for $x/d = 10$. Measurement: ○, $\overline{T}/T_0$; □, $\sqrt{\overline{\vartheta^2}}/T_0$. Calculation: ——, $\overline{T}/T_0$; - - - -, $\sqrt{\overline{\vartheta^2}}/T_0$.

Following Zhou and Antonia (1993), the lateral position y_c of the vortices was changed so as to make the variation in y_c consistent with its measured pdf (Figure 5). Also, a phase shift ψ_0 was introduced in the computation of U in Eq. (10), i.e. $U = U_c - \sum U_\beta \sin(\beta - \psi)/N$ with $\psi = \psi_0$ for $r \geqslant 0.5d$ and with $\psi = 2\psi_0(r/d)$ for $r < 0.5d$. Antonia et al. (1993) found that, for $y > 0$, v leads u by about 80° at the shedding frequency. The value of ψ_0 was therefore assigned a value of $-10°$ for $y > 0$ and $+10°$ for $y/d < 0$.

Distributions of $\overline{U}/U_1$, $\overline{u^2}/U_1^2$, $\overline{v^2}/U_1^2$ and $\overline{uv}/U_1^2$ (not shown here) are in reasonable agreement with the measurements of Matsumura and Antonia (1993) and

Antonia et al. (1993). The calculations of $\overline{T}/T_0$ and $\sqrt{\overline{\vartheta^2}}/T_0$ (Figure 13) agree reasonably well with measurement.

8. Conclusions

Thermal characteristics of vortices in the near/intermediate wake of a circular cylinder have been inferred from data obtained with a y-aligned array of cold wires. The data have been examined both on an instantaneous basis and conditionally, after identifying the spatial locations of local temperature maxima. In the region $x/d \lesssim 10$, temperature is an effective marker of the vortices and there is significant heat transfer associated with the entrainment of cold fluid downstream of each vortex. The conditional temperature distribution within the vortices is, like the conditional spanwise vorticity distribution, approximately Guassian near the vortex centre ($r \lesssim 0.4d$). Mean and rms temperatures, calculated using an Oseen-like vortex model, are in reasonable agreement with measurement at $x/d = 10$. The use of the cold wire array has also provided some insight into the contributions to the conditional temperature dissipation rate from both large-scale and small-scale motions. The small-scale contribution is significant at $x/d = 10$ and continues to increase further downstream.

Acknowledgement

The support of the Australian Research Council is gratefully acknowledged.

References

Antonia, R. A.: 1991. Organization in a Turbulent Near-Wake, *Fluid Dyn. Res.* **7**, 139-149.

Antonia, R. A., Zhou, Y. and Matsumura, M.: 1992. Spectral Characteristics of Momentum and Heat Transfer in the Turbulent Wake of a Circular Cylinder, *Exptl. Thermal Fluid Sci.* **6**, 371-375.

Bisset, D. K., Antonia, R. A. and Browne, L. W. B.: 1990. Spatial Organization of Large Structures in the Turbulent Far Wake of a Cylinder, *J. Fluid Mech.* **218**, 439-461.

Bloor, S. M.: 1964. The transition to Turbulence in the Wake of a Circular Cylinder, *J. Fluid Mech.* **19**, 290-304.

Bloor, S. M. and Gerrard, J. H.: 1966. Measurements on Turbulent Vortices in a Cylinder Wake, *Proc. Roy. Soc.* **A294**, 319-342.

Boisson, H. C., Chassaing, P. and Ha Minh, H.: 1983. Conditional Analysis of Intermittency in the Near Wake of a Circular Cylinder, *Phys. Fluids* **26**, 653-658.

Browne, L. W. B., Antonia, R. A. and Bisset, D. K.: 1986. Coherent Structures in the Far-Field of a Turbulent Wake, *Phys. Fluids* **29**, 3612-3617.

Budny, R. S., Kawall, J. G. and Keffer, J. F.: 1979. Vortex Street Evolution in the Wake of a Circular Cylinder, *Proc. Second Symp. on Turbulent Shear Flows.*, Imperial College, London.

Cantwell, B. and Coles, D.: 1983. An Experimental Study of Entrainment and Transport in the Turbulent Near Wake of a Cylinder, *J. Fluid Mech.* **190**, 321-374.
Davies, M. E.: 1976. A Comparison of the Wake Structure of a Stationary and Oscillating Bluff Body Using a Conditional Averaging Technique, *J. Fluid Mech.* **75**, 209-231.
Ferre, J. A. and Giralt, F.: 1989. Some Topological Features of the Entrainment Process in a Heated Turbulent Wake, *J. Fluid Mech.* **198**, 65-78.
Hayakawa, M. and Hussain, A. K. M. F.: 1989. Three-Dimensionality of Organized Structures in a Plane Turbulent Wake, *J. Fluid Mech.* **206**, 375-404.
Hussain, A. K. M. F. and Hayakawa, M.: 1987. Eduction of Large-Scale Organized Structures in a Turbulent Plane Wake, *J. Fluid Mech.* **180**, 193-229.
Kiya, M. and Matsumura, M.: 1985. Turbulence Structure in the Intermediate Wake of a Cylinder, *Bulletin JSME* **28**, 2617-2624.
Kourta, A., Boisson, H. C., Chassaing, P. and Ha Minh, H.: 1987. Nonlinear Interaction and the Transition to Turbulence in the Wake of a Circular Cylinder, *J. Fluid Mech.* **181**, 141-161.
Matsumura, M. and Antonia, R. A.: 1993. Momentum and Heat Transport in a Turbulent Intermediate Wake of a Circular Cylinder, *J. Fluid Mech.* **250**, 651-668.
Mi, J. and Antonia, R. A.: 1993. Temperature Distribution Within Vortices in the Wake of a Cylinder, *Int. J. Heat Mass Transfer* (to appear).
Schlichting, H.: 1979. *Boundary-Layer Theory*, New York, McGraw-Hill.
Townsend, A. A.: 1949. The Fully Developed Turbulent Wake of a Circular Cylinder, *Aust. J. Sci. Res.* **2**, 451-468.
Wlezien, R. W. and Way, J. L.: 1979. Techniques for the Experimental Investigation of the Near Wake of a Circular Cylinder, *AIAA Jnl.* **17**, 563-570.
Zhou, Y. and Antonia, R. A.: 1992. Convection Velocity Measurements in a Cylinder Wake, *Expts. in Fluids* **13**, 63-70.
Zhou, Y. and Antonia, R. A.: 1993. A Study of Turbulent Vortices in the Near-Wake of a Cylinder, *J. Fluid Mech.*, **253**, 643-661.

Second-Moment Closure Predictions of Buoyant Jets in Neutral and Stratified Cross-Flows

A. D. Gosman[1], R. Liu[2], and J. J. McGuirk[2]

[1] Dept. Mech. Eng., Imperial College, London, SW7 2BX, U.K.
[2] Dept. of Transport Technology, Loughborough University, Leicester LE11 3TU, U.K.

Abstract

The paper considers the problem of the modelling of scalar mixing in buoyant jets discharged into neutral and stratified cross-flowing streams via a second-moment closure turbulence model. Emphasis is placed on the near-field where the 3D spatial structure is most complex. Comparisons between $k - \varepsilon$ eddy viscosity predictions and calculations where the Reynolds stresses are modelled using the Gibson-Launder stress transport closure (but retaining an eddy diffusivity model for the scalar fluxes) indicate a noticeable improvement in comparison with measured data for the decay of peak mean density excess levels, plume spread and also the variance of density fluctuations. For the case of discharge into a linearly stratified crossflow, the current predictions show a rapid damping of the plume centre-line about its equilibrium level, in agreement with measurements, but in contrast to integral models for such flows, which indicate significant oscillations.

1. Introduction

Many pollution dispersion phenomena in both the marine and atmospheric environments involve a combination of four factors: (i) discharge via single, or multiple, closely-spaced round jets, (ii) a density difference between discharged and ambient fluids, (iii) ambient crossflow in the form of winds or tidal currents, and (iv) density stratification in the ambient under at least some environmental conditions. Examples are chimney and cooling tower plumes or sewage and cooling-water outfalls in rivers, estuaries and seas. Accidental releases of hot gaseous products from containments, or dense gas releases from tanker transports or pipelines can also lead to similar fluid mechanical scenarios. Safety legislation requires that design calculations for the discharge are carried out to estimate the trajectory, spread, and rate of dilution of the discharged fluid. For safety assessment it is often necessary to estimate not only the mean behaviour of these flows, but also to make predictions of the fluctuation levels (particularly of the scalar field), to allow evaluation of the likelihood of occurrence of extreme or dangerous cases (e.g. the probability of occurrence of a stoichiometric or explosive gas mixture from an open air dense gas spillage). Turbulent mixing and entrainment processes are clearly central to both the mean and fluctuating behaviour of jets and plumes, and the level of turbulence model closure needed to provide accurate predictions of these flows is an important question which forms the central theme of the current work.

It is the near-field of such discharges which forms the greatest challenge. In the near-field (defined conventionally as within the first 10 - 100 discharge length scales) the disturbance to the ambient flow is largest, the spatial structure of the plume is at its most complex, and this will therefore also be true for the turbulence field. In addition, the interactions between the density field of the discharge fluid and any ambient density stratification is strongest. As a consequence the first 90% or so of the final dilution takes place in this near-field region, and any plume entrapment or penetration through inversion layers is also determined in this zone.

Although there have been many experimental studies of jet flows, only a small number have been carried out for buoyancy-influenced jets, and very few of these have examined the effects of cross-flow or ambient density stratification. A recent review of available experimental data has been provided by Hunter (1992). The important physical characteristics of buoyant jets in crossflow are described by the three parameters: (1) jet to crossflow velocity ratio $R = V_j/U_c$, (2) jet densimetric Froude number $Fr = V_j/(\triangle \varrho g d_j/\varrho_{\mathrm{ref}})^{1/2}$, and (3) strength of ambient stratification, characterised for the linear stable density gradient commonly found in practice via the Brunt-Vaisala frequency $N = (-g\ \mathrm{d}\varrho/\mathrm{d}y/\varrho_{\mathrm{ref}})^{1/2}$. Chen and Rodi (1980) summarise the basic mean velocity, scalar and turbulence structure of undeflected jets and plumes, which have been used previously to study the performance of second-moment turbulence models (Malin and Younis (1990)). For the case of cross-flow however, measurements are restricted to information on the mean centre- line path, mean scalar profiles and rate of spread. For neutral crossflow, trajectory and dilution measurements have been made by Wright (1977). This reference also contains one of the few data sets for dilution and maximum height of rise for the case of stratified cross-flow, but insufficient measurements are reported to allow their use as a test case for thorough model evaluation; in particular, no turbulence structure data were obtained. This gap has been filled recently by Hunter ((1989) and (1992)) who has measured mean and fluctuating density statistics for a range of parameters typical of environmental release conditions; it is this data which is used in the current work to examine turbulence model performance.

As far as the prediction of buoyant jets in crossflow is concerned, although the discussion provided earlier might imply that only the most advanced computational fluid dynamics (CFD) models are likely to have true predictive value in the complex flows found in the near-field region, it is currently more common to find simple integral models used for routine calculations of plume trajectory, dilution, etc. Typical models are those of Ooms (1972) and Schatzmann (1979); Hunter (1992) has found that these models, when the entrainment functions which they contain have been suitably calibrated, work well for the case of cross flows of neutral stability. In linear stratification however, the integral models perform poorly. Experimental measurements (Ref. (5)) indicate that a buoyant jet discharged into a linearly stratified crossflow is trapped at its equilibrium height within just one cycle if an overshoot is present. Integral models however overpredict the oscillations which occur about the equilibrium rise position. The addition of an explicit (but artificial) damping term can remedy this defect, but this can have a detrimental consequence on the accuracy of predictions of plume dilution rate (see Ref. (1)).

The present paper aims to examine whether a 3D CFD model of a buoyant jet in cross-flow can improve on the near-field predictions reported above from integral models. In addition, it is of interest to examine to what extent a more complex model can predict the fluctuating scalar field, and what level of turbulence closure is required to achieve acceptable prediction accuracy. For complex 3D flows it would be advantageous if calculations could be based on the $k - \varepsilon$ model on grounds of computational economy, but the strongly buoyant nature of the present flow may call for higher level closure.

2. Mathematical Model

2.1 Governing Equations

Predictions presented below were based on numerical solution of the Reynolds-averaged Navier-Stokes (RANS) equations. For the mean properties, conservation equations for mass, momentum and the density inducing species (e.g. temperature or salt concentration differences) may be written in cartesian tensor notation as (assuming high Reynolds number, steady flow)

$$\begin{aligned}
\frac{\partial \varrho U_i}{\partial x_i} &= 0 \\
\frac{\partial}{\partial x_j}(\varrho U_i U_j) &= -\frac{\partial P}{\partial x_i} + g_i(\varrho - \varrho_{\mathrm{ref}}) \\
&\quad + \frac{\partial}{\partial x_j}\left[\mu\left(\frac{\partial U_i}{\partial x_j} + \frac{\partial U_j}{\partial x_i} - \frac{2}{3}\frac{\partial U_l}{\partial x_l}\delta_{ij}\right) - \varrho\,\overline{u_i u_j}\right] \\
\frac{\partial}{\partial x_j}(\varrho U_j \Phi) &= \frac{\partial}{\partial x_j}\left(\Gamma_\varphi \frac{\partial \Phi}{\partial x_j} - \varrho\,\overline{u_j \varphi}\right).
\end{aligned}$$

The local density was assumed to be linearly related to the mean scalar concentration Φ.

For a gradient-diffusion based closure for the Reynolds stresses ($\varrho\overline{u_i u_j}$) and scalar fluxes ($\varrho\overline{u_j\varphi}$), the $k - \varepsilon$ turbulence model was used:

$$\begin{aligned}
\varrho\,\overline{u_i u_j} &= -\mu_t\left(\frac{\partial U_i}{\partial x_j} + \frac{\partial U_j}{\partial x_i}\right) + \frac{2}{3}\mu_t\frac{\partial U_l}{\partial x_l}\delta_{ij} + \frac{2}{3}\varrho k \delta_{ij} \\
\varrho\,\overline{u_j \varphi} &= -\frac{\mu_t}{\sigma_t}\frac{\partial \Phi}{\partial x_j} \\
\mu_t &= C_\mu \varrho \frac{k^2}{\varepsilon}.
\end{aligned}$$

Values of k and ε were obtained from solution of their modelled transport equations:

$$\begin{aligned}\frac{\partial \varrho U_i k}{\partial x_i} &= \frac{\partial}{\partial x_i}\left(\frac{\mu_t}{\sigma_k}\frac{\partial k}{\partial x_i}\right) - \varrho\overline{u_i u_j}\frac{\partial U_i}{\partial x_j} - g_i\varrho\beta\,\overline{u_i\varphi} - \varrho\varepsilon \\ \frac{\partial \varrho U_i \varepsilon}{\partial x_i} &= \frac{\partial}{\partial x_i}\left(\frac{\mu_t}{\sigma_\varepsilon}\frac{\partial \varepsilon}{\partial x_i}\right) - C_{\varepsilon_1}\frac{\varepsilon}{k}\varrho\overline{u_i u_j}\frac{\partial U_i}{\partial x_j} \\ & -C_{\varepsilon_3}\frac{\varepsilon}{k}g_i\varrho\beta\,\overline{u_i\varphi} - C_{\varepsilon_2}\varrho\frac{\varepsilon^2}{k}\end{aligned}$$

where

$$\beta = -1/\varrho\;\partial\varrho/\partial\Phi\,.$$

Standard values of the model constants were used in all calculations. Note that tests were carried out with different values of the constant C_{ε_3} for the buoyancy term in the ε equation, but the predictions did not display any noticeable sensitivity to this constant. Accordingly, all calculations presented below were conducted with $C_{\varepsilon_3} = 0$.

Within the framework of this model, if it is desired to predict scalar fluctuation levels, then the following modelled equation for the scalar variance is used:

$$\begin{aligned}\frac{\partial \varrho U_i\,\overline{\varphi^2}}{\partial x_i} &= \frac{\partial}{\partial x_i}\left(\frac{\mu_t}{\sigma_t}\frac{\partial\,\overline{\varphi^2}}{\partial x_i}\right) \\ & -2\varrho\,\overline{u_i\varphi}\frac{\partial\Phi}{\partial x_i} - C_\varphi\varrho\frac{\varepsilon}{k}\,\overline{\varphi^2}\end{aligned}$$

As a higher-order closure alternative to the above model, a second-moment Reynolds stress model was used. Following earlier experiences using the Gibson and Launder (1978) model for buoyant flows reported in references (3) and (9), it was decided to adopt this basic second-moment closure model:

$$\frac{\partial}{\partial x_l}\left(\varrho U_l\,\overline{u_i u_j}\right) = D_{ij} + P_{ij} + G_{ij} + \pi_{ij} + \varepsilon_{ij}$$

where:

$$\begin{aligned}D_{ij} &= \frac{\partial}{\partial x_l}\left(C_s\frac{k}{\varepsilon}\varrho\,\overline{u_l u_m}\frac{\partial\left(\overline{u_i u_j}\right)}{\partial x_m}\right), \\ P_{ij} &= -\varrho\,\overline{u_i u_m}\frac{\partial U_j}{\partial x_m} - \varrho\,\overline{u_j u_m}\frac{\partial U_i}{\partial x_m}, \\ G_{ij} &= -g_i\varrho\beta\,\overline{u_j\varphi} - g_j\varrho\beta\,\overline{u_i\varphi}\,, \quad \varepsilon_{ij} = \frac{2}{3}\varrho\varepsilon\delta_{ij}\,, \\ \pi_{ij} &= -C_1\varrho\frac{\varepsilon}{k}\left(\overline{u_i u_j} - \frac{2}{3}\delta_{ij}k\right) - C_2\left(P_{ij} - \frac{1}{3}\delta_{ij}P_{kk}\right) \\ & -C_3\left(G_{ij} - \frac{1}{3}\delta_{ij}G_{kk}\right).\end{aligned}$$

The same values of the constants as recommended and used in Refs. (8), (9) and (3) were adopted, together with the appropriate form of the ε equation for a Reynolds stress closure (Ref. (8)).

Note that, to examine the benefits brought about by a *progressive* introduction of higher-order-closure modelling, at this stage, the transport equation model has been adopted for the Reynolds stresses in the momentum equations only; an eddy diffusivity/constant turbulent Prandtl number model has been retained for the scalar equations (where the μ_t is now defined using a k evaluated from summing the solution of the individual normal stress transport equations). This approach has been driven by the desire to assess the computational cost benefits of using a second-moment closure in a step-by-step manner.

2.2 Computational Procedure

The calculations reported were obtained with a standard, finite-volume, pressure-correction algorithm. Mesh- staggering was used to avoid pressure/velocity and stress/strain-rate uncoupling. For the prediction of the flows considered, selected from the specially conducted experiments of Ref. (1), a simple rectangular cartesian mesh was sufficient. To allow conclusions to be drawn safely on the performance of the turbulence model, care must be taken to address numerical accuracy aspects. This was studied, as reported below, using both grid refinement tests and second-order accurate quadratic upstream interpolation (QUICK) for the convection discretisation for the mean-flow variables to reduce numerical diffusion. It turned out to be particularly important to implement the QUICK scheme in the transport equation for the scalar which controlled density differences. The results discussed below may therefore be reasonably regarded as grid-independent.

3. Results and Discussion

Fig. 1 shows the flow configuration used in the experiments of Hunter (1992). It was easier to construct suitably stratified environments by filling the tank using layers of salt solution, so a towing tank facility with a downwards discharging jet was selected. For ease of understanding and relevance to the real application, in what follows the jet is, however, shown discharging vertically upwards in the positive y direction; z is the co-ordinate in the cross-flow direction, and x is the transverse co-ordinate, with the origin of co-ordinates at the jet centre. In all calculations, boundary conditions at jet exit were taken from the experiments; the cross-flow velocity was fixed at the towing tank speed, with all turbulence conditions in the cross-flow at zero values, corresponding to the stagnant ambient in the experiments. Two cases were selected for study; both have a jet/cross-flow velocity ratio $R = 6$ and a jet Froude number $Fr = 7.5$. The first test case (Case 1N) has constant density in the cross-flow (neutral stability), whereas the second (Case 1) has a linear stable density gradient, $d\varrho/dy = -100$ kg/m^4, corresponding to a Brunt/Vaisala

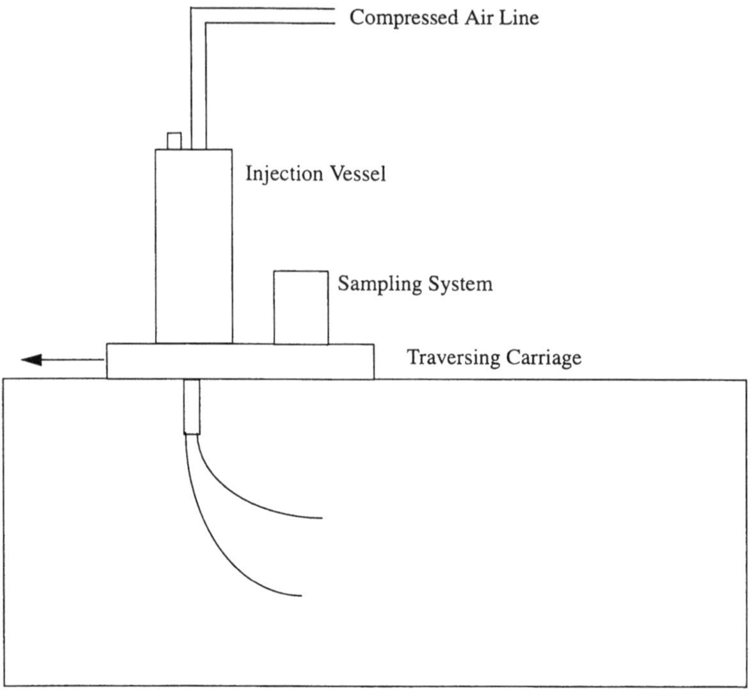

Fig. 1. Schematic diagram of experimental arrangement of Ref. (1).

frequency $N = 1.0\,s^{-1}$. Data on buoyant jet behaviour were obtained in Ref. (1) from photographic measurements, from suction probe measurements of dyed jet releases and from conductivity probe measurements. The latter two techniques provided data on mean scalar profiles along chosen vertical and horizontal traverses at selected downstream distances. The conductivity probes had sufficiently fast response that both mean and variance of the density was provided in the measurements. Further details of the experiments may be found in Refs. (1) and (5).

As an indication of the flow pattern produced in Case 1N, Fig. 2 presents particle path pictures of the predictions obtained with the Reynolds stress model. The jet penetration (almost to the floor of the tank in this case) may be clearly seen, together with the often observed feature in jets in cross-flow of very strong entrainment on the underside of the bent-over jet. The picture was very similar for the $k - \varepsilon$ predictions. The formation of the counter-rotating vortices in the jet cross-section is easily observed in the particle paths at the several downstream transverse cross-sections in Figs. 2(b) - (d); if these vortex motions are strong enough, then they can lead to a bifurcation of the jet, with the maximum scalar concentration being displaced from the jet centre-plane. Before these predictions may be compared with measured data however, we must be sure of their numerical accuracy. Comprehensive testing was carried out to study the effects of grid refinement and the switch from first-order accurate to second-order accurate discretisation of the convection terms. A noticeable change was observed when the QUICK scheme was introduced into the conservation

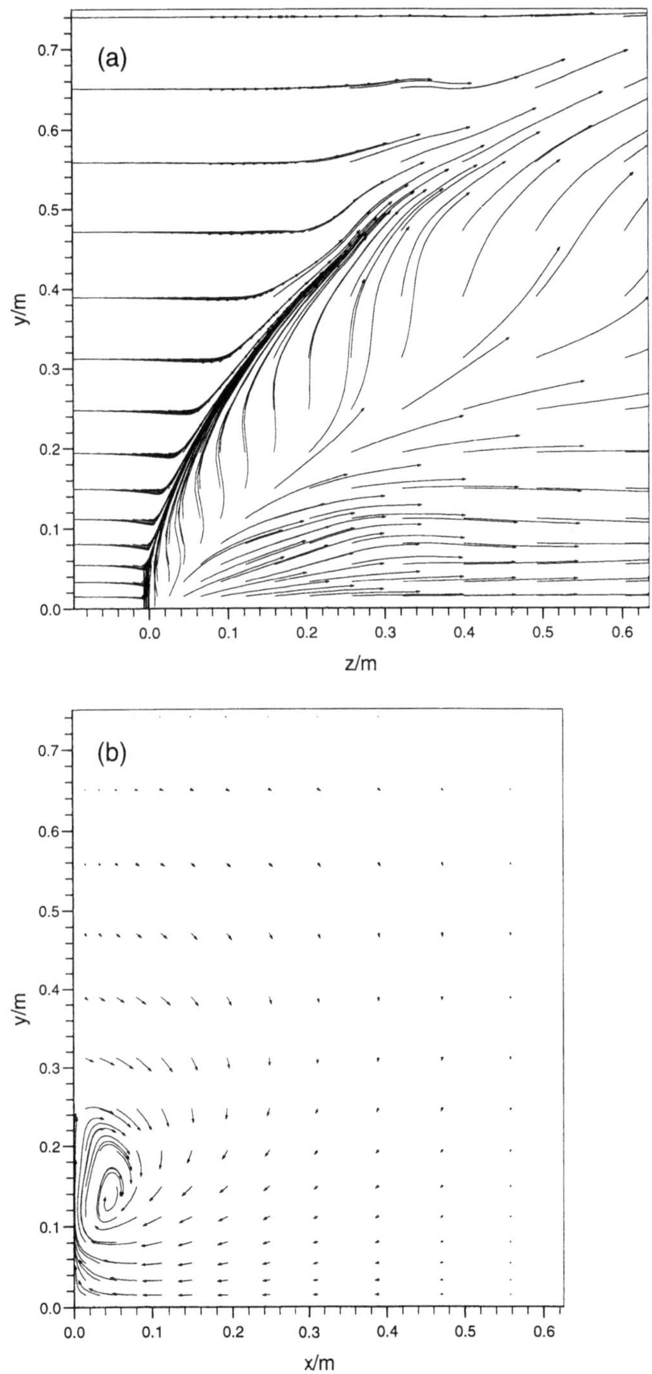

Fig. 2a, b. Case 1N: predicted flow pattern with Reynolds-stress model. (**a**) plane x = 0, (**b**) plane z = 0.053 m, (**c**) plane z = 0.093 m, (**d**) plane z = 0.253 m.

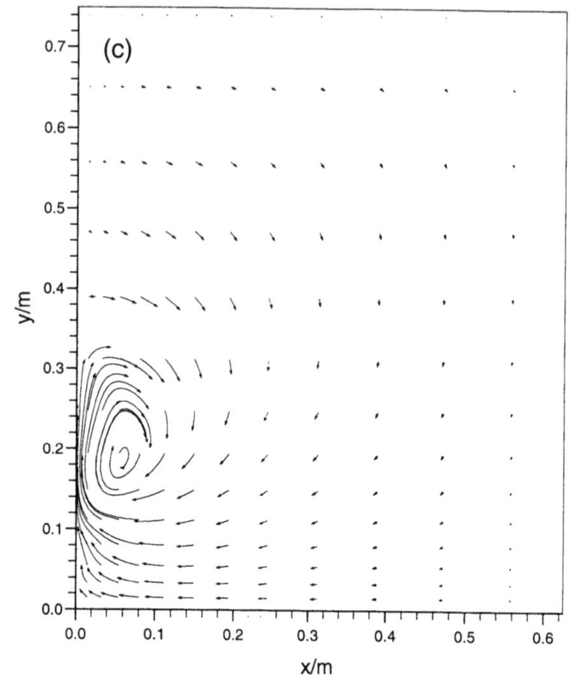

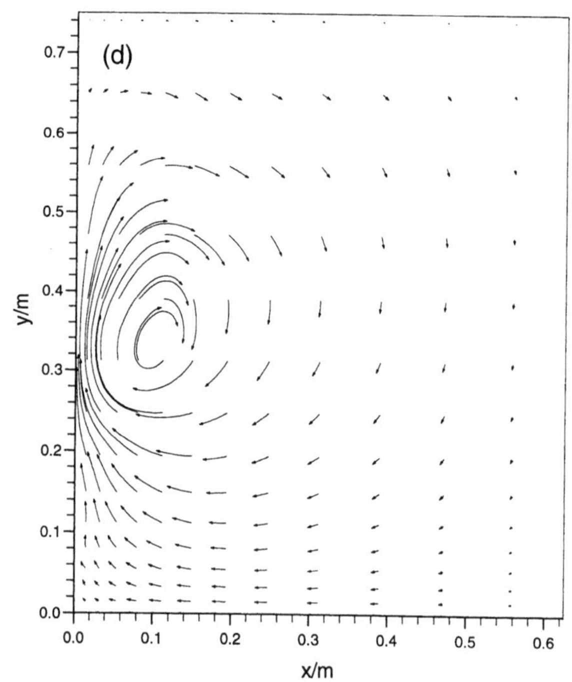

Fig. 2c, d. For caption see Fig. 2a, b.

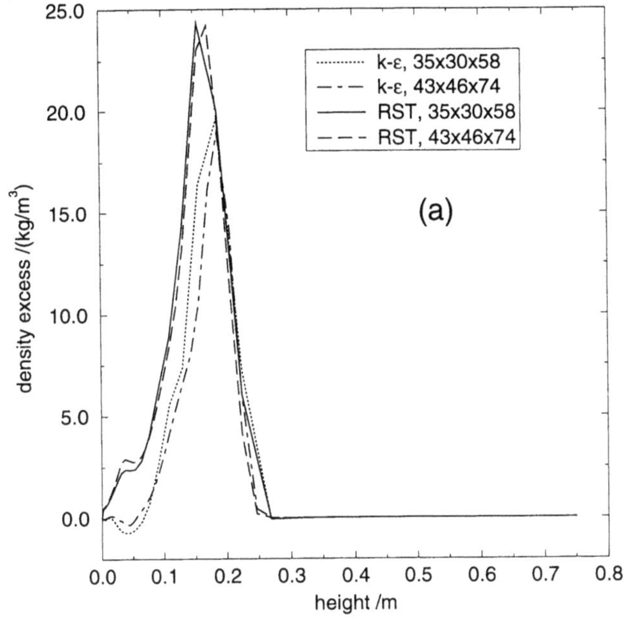

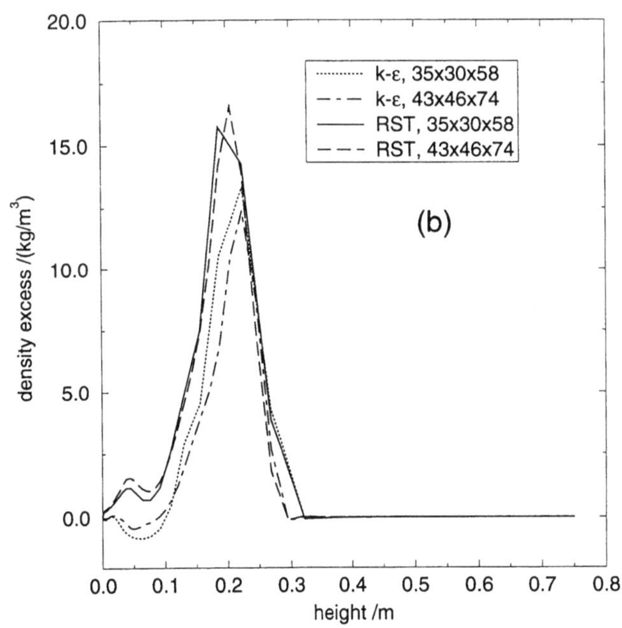

Fig. 3a, b. Case 1N: grid-refinement tests, $k - \varepsilon$ and RST models, (**a**) $z = 0.053$ m, (**b**) $z = 0.073$ m, (**c**) $z = 0.133$ m.

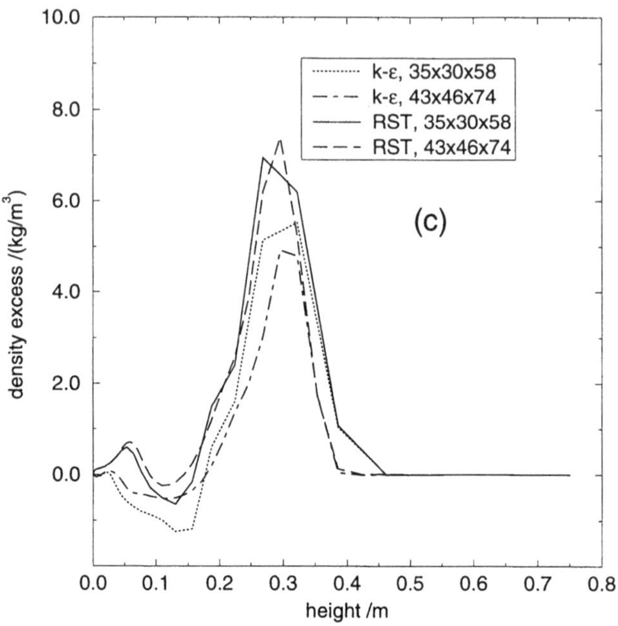

Fig. 3c. For caption see Fig. 3a, b.

equation for the mean scalar, so all grid refinement tests were conducted for this case. Fig. 3 presents vertical mean density excess profiles on the symmetry plane passing through the jet centre for three downstream stations. Results are shown for both $k-\varepsilon$ and Reynolds stress transport models with two levels of mesh density, a coarse grid of around 60 000 nodes (distributed $35 \times 30 \times 58$ in the x, y and z directions respectively) and a finer grid of around 150 000 nodes ($43 \times 46 \times 74$). These plots indicate that the grid densities used here are sufficient to produce results which show little change in mean scalar distribution; further, the differences between $k-\varepsilon$ and RST predictions are clearly sustained on grid refinement and are thus demonstrably a result of turbulence model performance. It is noticeable that the differences between the two models lie in two main areas: firstly a noticeably slower rate of decay of peak values is produced by the RST model; secondly, in terms of jet spreading, a wider plume is predicted by the second moment closure, with the main changes being brought about at the lower jet edge where the entrainment processes are strongest, and where the density stratification is stable. This implies, as has been observed before, that a buoyancy-affected eddy viscosity model overpredicts the effect of stable stratification in damping out vertical fluctuations and the associated entrainment. Examination of horizontal profiles indicates confirmation of the essentially grid-independent nature of the current predictions; the horizontal width of the plume was, however, unaffected by turbulence model choice.

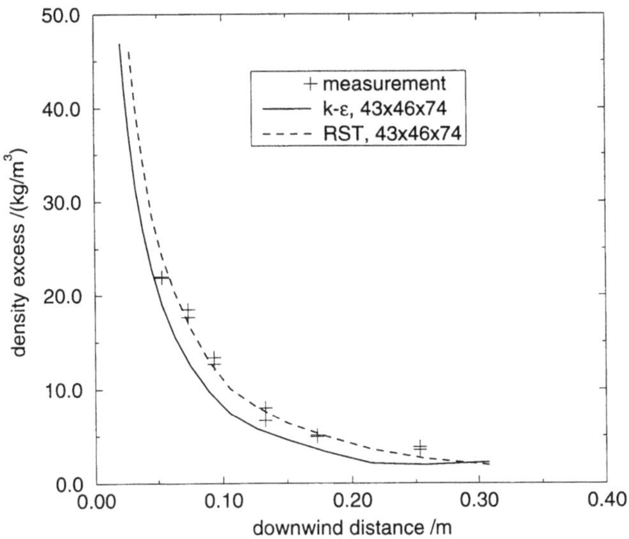

Fig. 4. Case 1N: Decay of maximum mean density excess on symmetry plane.

Once the numerical accuracy of the calculations is established, we may examine the similarity between predictions obtained with the two turbulence models and the experimental data. Fig. 4 presents a comparison on the basis of the decay of the maximum mean density excess on the symmetry plane. The larger peak values observed in the numerical tests in the second moment predictions are now confirmed as correct, indeed with this model the decay rate is very well predicted. This is further confirmed by the vertical profile comparison shown in Fig. 5, as is the extra spread on the plume lower edge returned by the RST model. Fig. 6 illustrates a comparison between predicted plume boundaries and an interpretation deduced from the photographic evidence. It should be noted that because the dye photographs represent line of sight images, the predicted plume boundaries have been deduced from laterally-averaged density values; this seems to remove most of the model differences. Nevertheless, apart from the very far downstream region, where grid density is proportionally sparser, the jet centreline and edge locations are well predicted. A complete picture of plume spread is provided by a comparison of mean density excess contours at the two downstream stations shown in Fig. 7. The agreement between measured data and the second-moment predictions is very good. Note that the plume bifurcation mentioned above and caused by the strong cross- stream vortices is represented here by the maximum scalar value being removed from the centre-line. Fig. 7 indicates that the location of this maximum, its absolute value, the vertical and horizontal widths of the plume, and the plume cross-sectional shape are adequately reproduced by the second-moment closure; the eddy viscosity prediction would be substantially worse in all these aspects as evidenced by the profiles given in Fig. 5. Finally, although measurements of turbulence qualities are not as plentiful in these

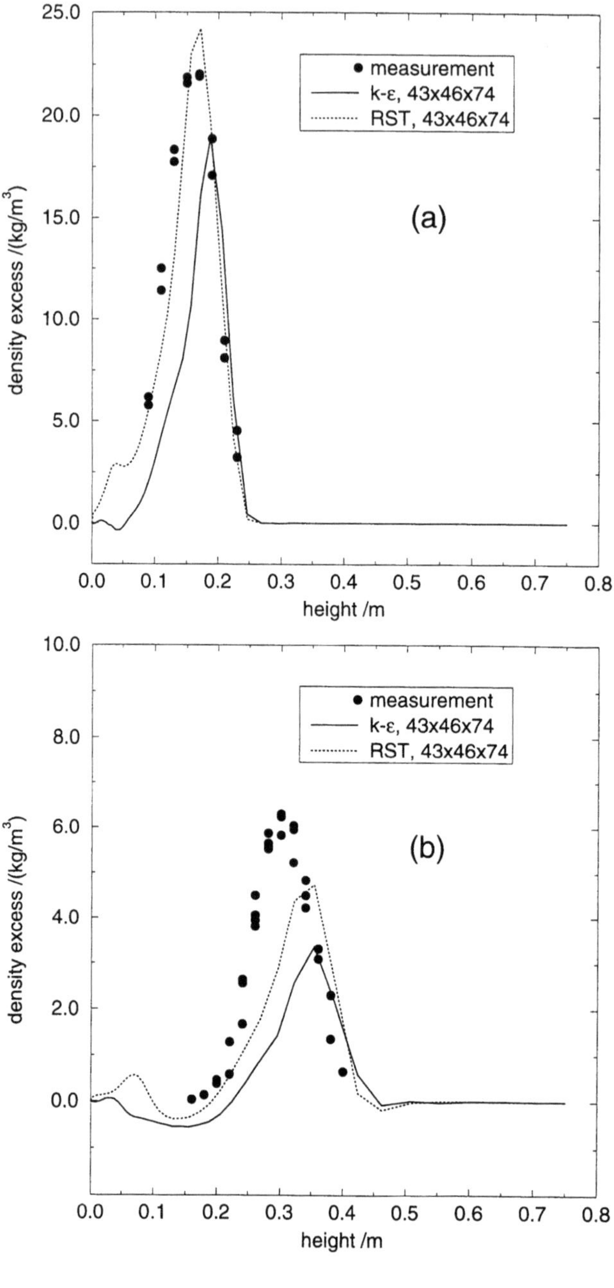

Fig. 5a, b. Case 1N: comparison of predicted and measured vertical profiles of density excess on symmetry plane (**a**) $z = 0.053$ m, (**b**) $z = 0.093$ m, (**c**) $z = 0.173$ m.

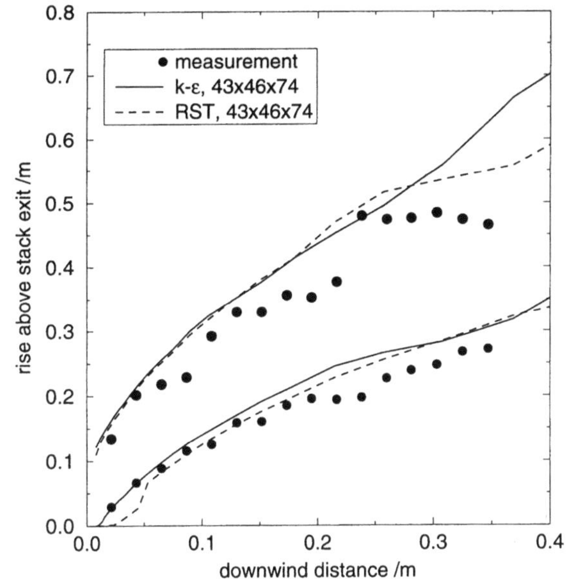

Fig. 6. Case 1N: predicted and measured plume boundaries.

experiments, Fig. 8 provides some comparison of the accuracy of prediction of the scalar variance. At the first station shown in Fig. 8a the predicted RST peak variance is about twice as high as that predicted by the $k-\varepsilon$ model. The measured data are too sparse to confirm either prediction, but the fact that both predicted peak values have been reduced by around a factor of four by the next station where the measured maxima are closer to the RST prediction does lend some support to the second moment model. However, the final profile in Fig. 8c also indicates that both of the current predictions are decaying faster than the measured data, so model performance is not as good for this parameter. This is, however, not so surprising since it should be recalled that an eddy diffusivity closure has been retained for the scalar field and Fig. 8 provides perhaps evidence that second-moment closure needs to be extended to the scalar fluxes also.

To confirm that the improved predictions noted with the second moment closure extend to other, practically more relevant, ambient conditions, a linearly stably stratified ambient was introduced (Case 1). Fig. 9 illustrates the dramatic effect this has on the flow pattern. The jet rise is interrupted as the jet fluid reaches a level where its density matches that of its surroundings. Indeed, Fig. 9 indicates that, for the parameters studied here, jet momentum leads to a slight over-penetration and a fall back to the equilibrium height; it is noticeable that the oscillation about the equilibrium rise is very heavily damped in the current predictions, confirming the suspicions raised in the introduction that a CFD model would be superior to an integral model in stratified cross-flow. The entrapment of the plume at its equilibrium height has also a dominant effect on the jet counter-rotating vortices. Fig. 9(b) shows how these continue to be formed in the initial bend-over region, but the falling back of the jet fluid to

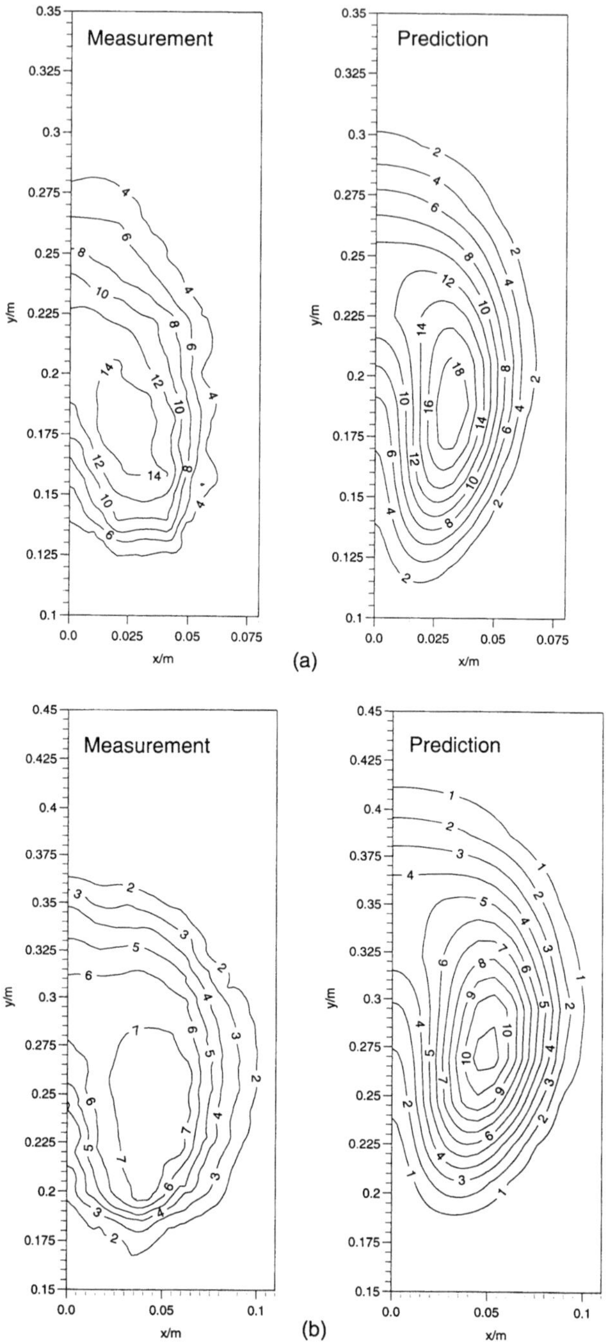

Fig. 7. Case 1N: predicted and measured mean density excess contours (**a**) $z = 0.093$ m, (**b**) $z = 0.173$ m.

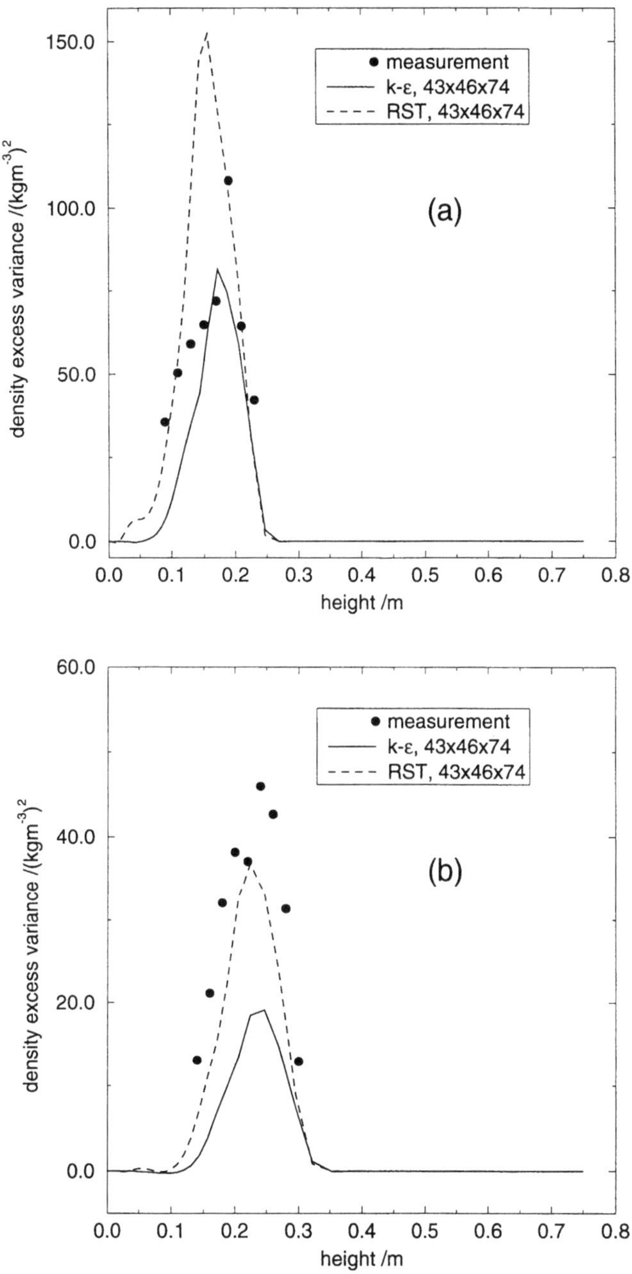

Fig. 8a, b. Case 1N: predicted and measured vertical profiles of density excess variance (**a**) $z = 0.053$ m, (**b**) $z = 0.093$ m, (**c**) $z = 0.133$ m.

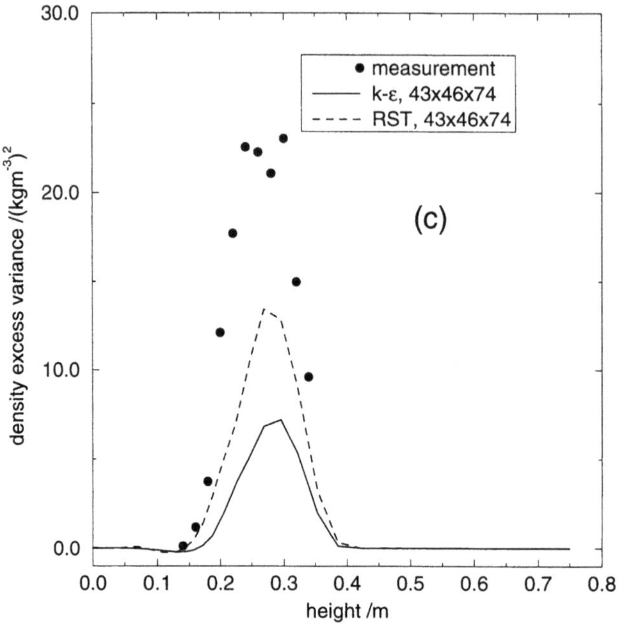

Fig. 8c. For caption see Fig. 8a, b.

a lower height after over-penetration leads to a vortex of the opposite sign being created which swamps the jet vortex - Fig. 9c. The predicted and measured plume boundaries for this case are given in Fig. 10. Once again both models reproduce the observations well in terms of plume spread, and prediction of the correct height of rise. The vertical profiles in this case, Fig. 11, confirm absolutely the superiority of the second moment closure, indeed, at $z = 0.093$ (Fig. 11b) the $k - \varepsilon$ model predicts a density excess which is the wrong sign over practically the whole width of the flow, whereas the second moment closure compares very favourably with the measured data.

4. Conclusions

The following are the main conclusions which have emerged from this computational study of buoyant jets in cross-flow:

(1) By adopting a second-order accurate discretisation of the convection terms, particularly in the mean scalar conservation equations, it was possible to obtain sensibly grid-independent solutions of the near-field of 3D jets in cross-flow using meshes of order 150 000 nodes using both $k - \varepsilon$ and Reynolds stress transport turbulence closures.

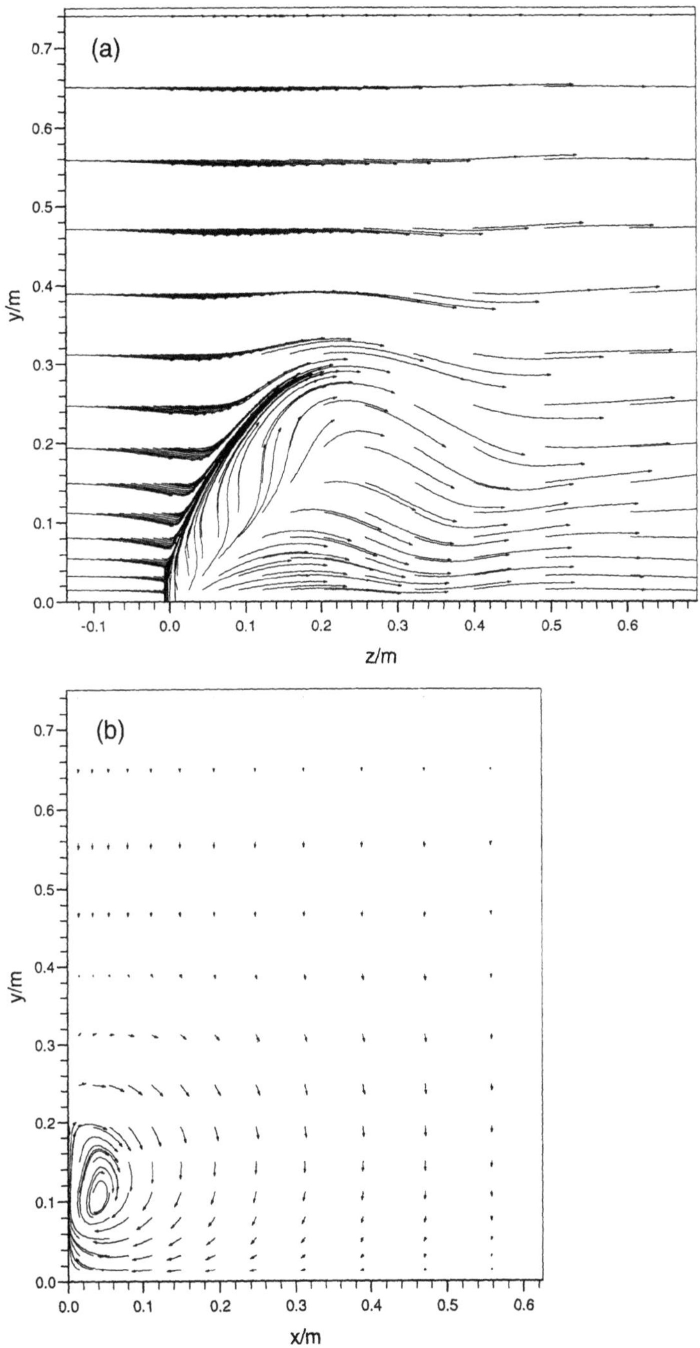

Fig. 9a, b. Case 1: predicted flow pattern with Reynolds stress model (**a**) plane $x = 0$, (**b**) plane $z = 0.053$ m, (**c**) plane $z = 0.3$ m.

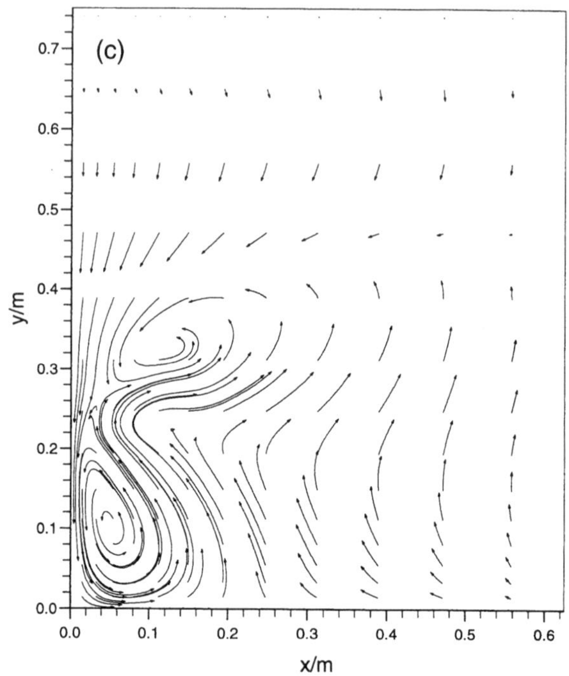

Fig. 9c. For caption see Fig. 9a, b.

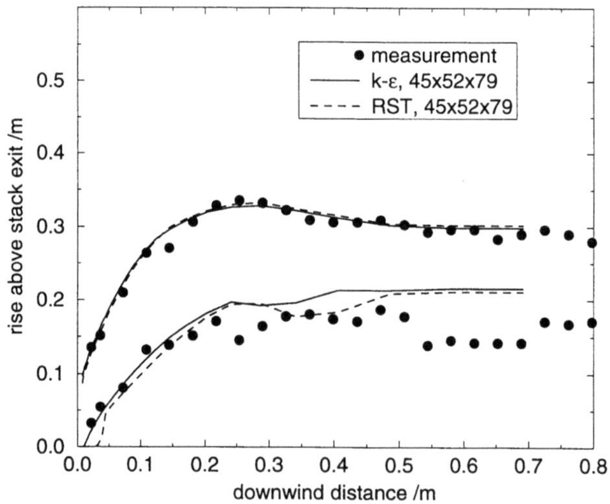

Fig. 10. Case 1: predicted and measured plume boundaries.

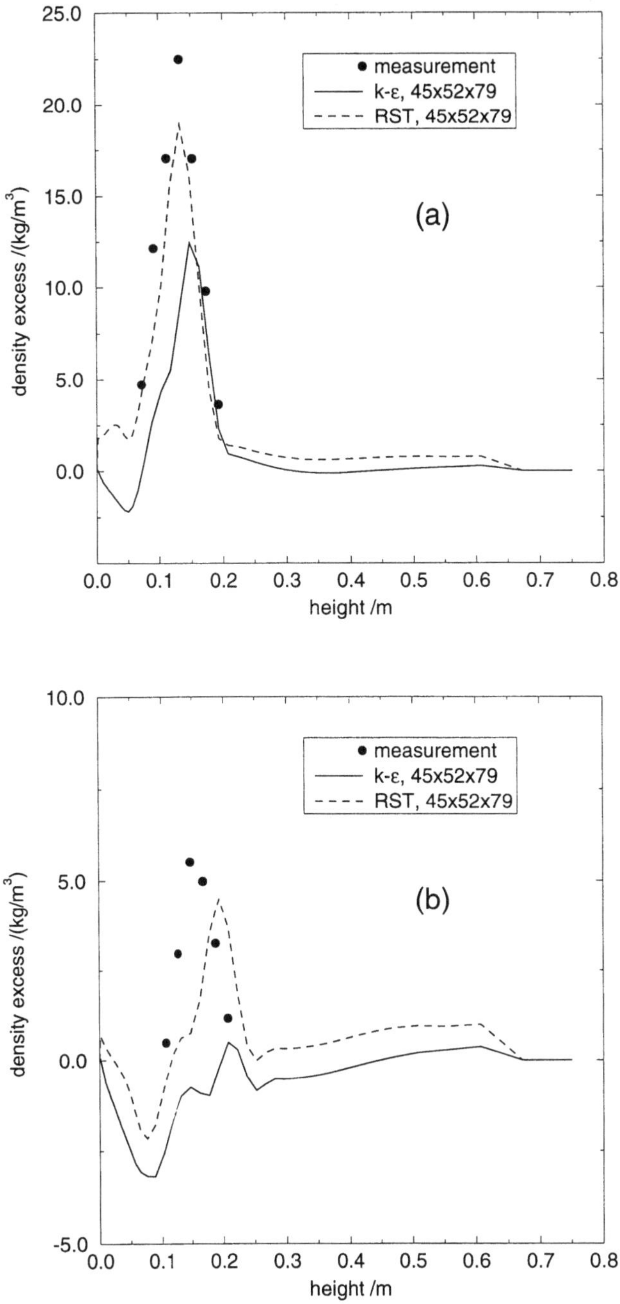

Fig. 11. Case 1: comparison of predicted and measured vertical profiles of density excess on symmetry plane (**a**) $z = 0.053$ m, (**b**) $z = 0.093$ m.

(2) The second-moment closure performed significantly better than the eddy viscosity model in both neutral and stably stratified cross-flow in predicting the mean scalar structure of the plume. Improvements were observed in terms of accuracy of prediction of peak density excess decay, plume edge locations and overall shape of density contours, in comparison with the available experimental data.
(3) Some indications were obtained that the second-moment closure model, as used here on the velocity field only (for minimum increase in computational cost), also improved prediction of fluctuating statistics such as the density excess variance. However, there was also some evidence that these plume properties would require the use of scalar flux transport equations, and this should be investigated further.
(4) The case of a linearly stratified cross-flow demonstrated a further improvement of the current CFD model approach in comparison with the more usual integral models, namely a close simulation of the heavily damped process by which the plume is captured and achieves its equilibrium rise height.

Acknowledgements

This research has been supported by SERC/ERCOS Grant GR/E 82316. Thanks are also due to Dr. A. Robins and Dr. G. Hunter of National Power Technology and Environmental Centre (Leatherhead) for many helpful discussions on the experimental study, and the provision of extra experimental data.

References

1. Hunter, G.C. 1992. The Behaviour of Plumes from Point Sources in Stratified Flows. Ph.D. Thesis, U.K. Council for National Academic Awards.
2. Chen, C.J. & Rodi, W. 1980. Vertical Turbulent Buoyant Jets: A Review of Experimental Data. Pergamon Press, Oxford.
3. Malin, M.R. & Younis, B.A. 1990. Calculation of Turbulent Buoyant Plumes with a Reynolds Stress and Heat Flux Transport Closure. Int. J. Heat & Mass Transfer **33**, pp. 2247-2264.
4. Wright, S.J. 1977. Mean Behaviour of Buoyant Jets in a Crossflow. Proc. ASCE, J. Hydraul. Div. **103**, pp. 499-513.
5. Hunter, G.C. 1989. Experimental Investigation of a Buoyant Jet in a Stratified Cross-wind. Proc. of 3rd IMA Conference on Stably Stratified Flows, Leeds.
6. Ooms, G. 1972. A New Method for the Calculation of the Plume Path of Gases Emitted by a Stack. Atmospheric Environment **6**, pp. 899-900.
7. Schatzmann, M. 1979. An Integral Model of Plume Rise. Atmospheric Environment **13**, pp. 721-731.
8. Gibson, M.M. & Launder, B.E. 1978. Ground Effects on Pressure Fluctuations in the Atmospheric Boundary Layer. J. Fluid Mech. **86**, pp. 491- 511.
9. McGuirk, J.J. & Papadimitriou, C. 1988. Stably Stratified Free Surface Shear Layers with Internal Hydraulic Jumps. In: *Stably Stratified Flow and Dense Gas Dispersion*, Ed. J.S. Puttock, Clarendon Press, Oxford.

On Coherent Vortex Formation in Axisymmetric Compressible Shear Layers

K. Yu, E. Gutmark, and K. C. Schadow

Research Department, Naval Air Warfare Center, China Lake, CA, USA

Abstract

The development of organized large-scale structures in axisymmetric compressible shear layers was studied using a planar Mie-scattering flow visualization technique. Three different combinations of fully-expanded supersonic coaxial jets were used in creating the shear layers, with convective Mach number of 0.23, 0.47, and 1.1. The disturbance characteristic of the initial shear layer was changed by varying the inner nozzle lip thickness. Under certain lip-thickness conditions, large-scale vortical structures showed uncharacteristically high degree of coherence for compressible shear layers. Further tests revealed that a coupling between the compressible shear flow and lip-created wake flow was responsible for the formation of highly organized structures. The present results suggest that this mechanism can be used to modulate turbulent compressible mixing.

1. Introduction

The industrial motivation and the fundamental interest for research into the physics of turbulent compressible mixing are well established. Over the past few years, considerable experimental effort was directed toward understanding the turbulent mixing processes in the compressible shear flows. Because the dynamics of the large-scale vortical structures play a central role in turbulent incompressible mixing (Brown & Roshko 1974, Winant & Browand 1974), these structures were also the objects of many recent focused studies involving compressible shear flows. Large-scale coherent structures are now a very familiar feature in turbulent incompressible shear layers. However, whether similar structures exist in compressible shear flows has remained an open issue due to lack of clear experimental evidence. Several recent experiments using planar flow visualization have shed some light into this issue. Clemens and Mungal (1992) used a Mie- scattering flow visualization technique to show that large-scale structures in planar compressible shear layers are similar to those in incompressible case when the convective Mach number (M_c) is low. Combining visualization with image correlation approaches, Messersmith, Dutton & Krier (1991) and Elliot, Samimy & Arnette (1993) also obtained similar results showing more rigorously the existence of organized structures in planar compressible shear

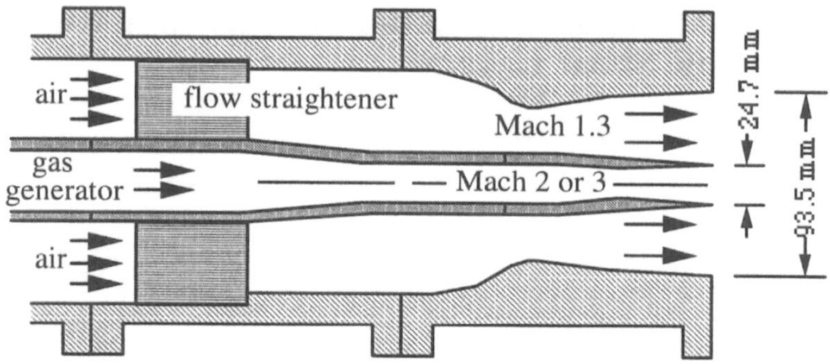

Fig. 1. Supersonic coaxial jet stand.

layers at low to intermediate M_c. Lack of coherent structure visualization at higher M_c may be due to the increased three-dimensionality of the flow. A linear stability analysis by Sandham and Reynolds (1989) showed that strongly three-dimensional structures dominate a planar mixing layer at convective Mach numbers above 0.6. The present experiment was initiated to identify large-scale structures in axisymmetric compressible shear layers of supersonic coaxial jets using a planar Mie scattering flow visualization technique. The present shear layer is different from those previously studied in that the compressible shear flow is initiated from a finite- thickness nozzle lip rather than a sharp knife-edge splitter plate. The lip effect is important in this case because a coaxial configuration is used. Preliminary visualization experiments revealed well-defined large-scale vortices which are much better organized than those seen in planar compressible shear layers. In the following, we attempt to explain this finding, further characterize the vortices, and provide additional insight into coherent vortex formation in compressible shear layers.

2. Experimental System

2.1 Facility and Test Conditions

Experiments were performed using a supersonic coaxial jet stand which included fully-expanded Mach 1.3 outer jet and Mach 2.0 or 3.0 inner jet. The outer jet was discharged into an open surrounding of ambient air while the inner jet was discharged from the core of the outer jet. The tests were carried out in an open surrounding to minimize possible disturbances from room acoustics. The exit dimensions of the nozzles are shown in Fig. 1. Total pressure of the jets was measured with a Kulite XCQ-080 pressure transducer which was mounted at the tip of a pressure probe.

Using different combinations of inner and outer supersonic jets, three different convective Mach numbers were obtained. The operating conditions for the three cases are summarized in Table 1. In the first two cases, air which was supplied from a high pressure reservoir was used for both jets, while in the last case, the product from ethylene-oxygen-nitrogen reaction was used for the inner jet. The stagnation pressures and temperatures were measured with SetraSystems Model 204E pressure transducers and Chromel/Alumel junction thermocouples respectively. The Reynolds number is based on calculated velocity difference and the vorticity thickness at the center of the images.

Table 1. Summary of experimental conditions.

M_c	M_{inner}	M_{outer}	T_{inner} [K]	T_{outer} [K]	$Re_\delta = \Delta U \delta / \nu$
0.23	2.0	1.3	157	210	$0.55 \cdot 10^5$
0.47	3.0	1.3	103	210	$1.5 \cdot 10^5$
1.1	2.0	1.3	1790[a]	220	$1.7 \cdot 10^5$

[a] from calculation

2.2 Flow Visualization

The supersonic shear flows were visualized using a planar Mie scattering technique. The inner jet was seeded using either condensed ethanol droplets with final droplet size of approximately 0.2 μm (Clemens & Mungal 1991; Messersmith, Dutton & Krier 1991) or fine alumina particles with manufacturer specified size of 0.3 μm. The particle Stokes number for large-scale structure visualization was less than 0.5 for the worst case, satisfying the correct visualization condition of Samimy & Lele (1990). A copper vapor laser (Metalaser Model 3051) with wavelengths of 511 nm and 578 nm was used as the light source. The laser has 20 nsec pulses (Fu) with maximum energy of 4 mJ/pulse and operates at a fixed 8 kHz repetition rate. Using the proper optics, the laser beam was shaped into a thin sheet that illuminated across the flow region of interest. The scattered light from passing particles was then collected by a standard CCD camera (Pulnix TM-7CN) at a right angle. The signal from the camera was recorded on a super VHS tape for storage and later playback. In order to capture only one pulse per frame, the camera's high speed shutter capability was used.

3. Results and Discussion

3.1 Identification of Large-Scale Coherent Vortices

The planar images containing the cross-sectional view along the center axis of supersonic coaxial air jets are shown in Fig. 2. The images show seeded inner jet flowing from left to right in the presence of co-flowing outer stream which was unseeded. The images extend from two to five inner jet exit diameters downstream of the exit. The convective Mach numbers of the shear layers were 0.23 and 0.47. From these

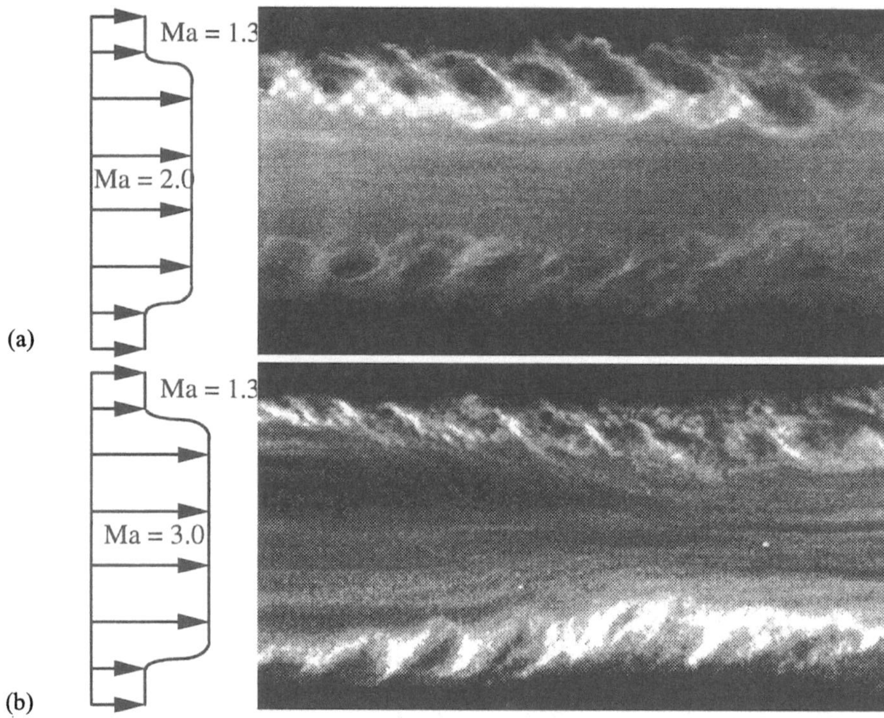

Fig. 2. Axisymmetric compressible shear layers with M_c = (**a**) 0.23 and (**b**) 0.47.

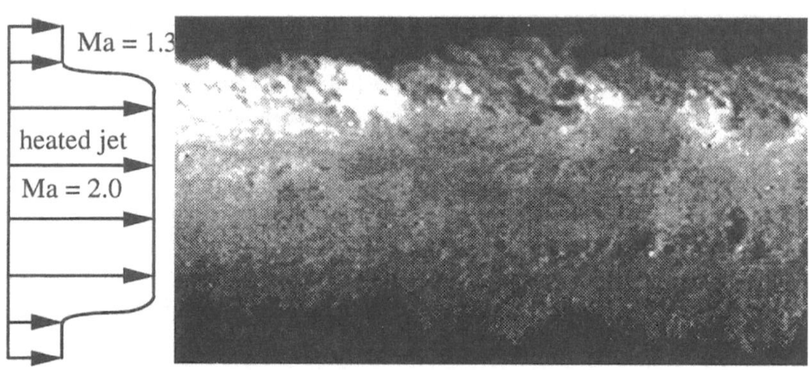

Fig. 3. Axisymmetric compressible shear layer with $M_c = 1.1$.

images, formation of large-scale coherent vortices in the shear layer can be clearly seen. The coherent vortices were repeatedly observed at these operating conditions.

Fig. 3. Axisymmetric compressible shear layer with $M_c = 1.1$. Using two-point spatial correlation of optical deflectometry data, McIntyre & Settles (1991) showed that large-scale structures in axisymmetric compressible shear layers with M_c of 0.37 and 1.5 are sporadically occurring and at best only partially organized. In contrast, the present results show remarkably high degree of organization at M_c as high as

0.47. The difference can be traced back to the use of finite thickness nozzle-lip in the present experiment. While finite lip-thickness has a strong influence on coherent structure formation at low to intermediate convective Mach numbers, the effect was much weaker at a high convective Mach number. Fig. 3 shows the flow image for the high convective Mach number case ($M_c = 1.1$). In comparison with the two previous images, the large-scale structures in this case were poorly organized. Organized structures were observed only sporadically.

4. Quantification of Vortex Characteristics

The coherent structures were characterized by processing the digitized images and quantitatively reducing the information. Two different image processing approaches were used in this experiment.

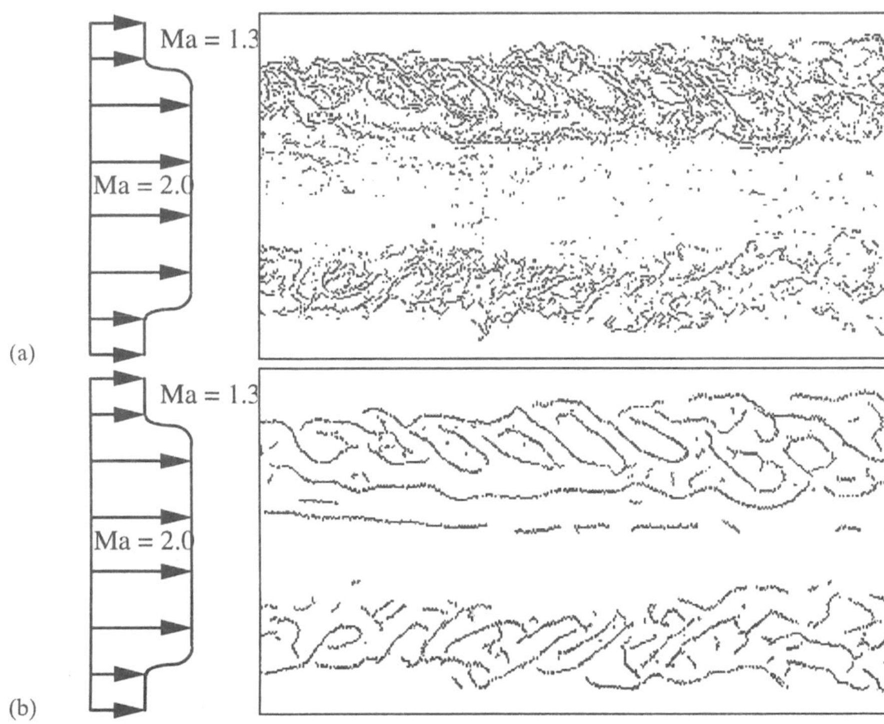

Fig. 4. Wavelet transform of the image in Fig. 2a. (**a**) scale level 1, (**b**) scale level 2.

One approach employed a two-dimensional wavelet transform technique (Gutmark et al. 1991) using an edge detector function as the analyzing wavelet. The intention was to explore the possibility of using the wavelet transform not just as a data reduction tool but as a means to identify coherent structures which may be

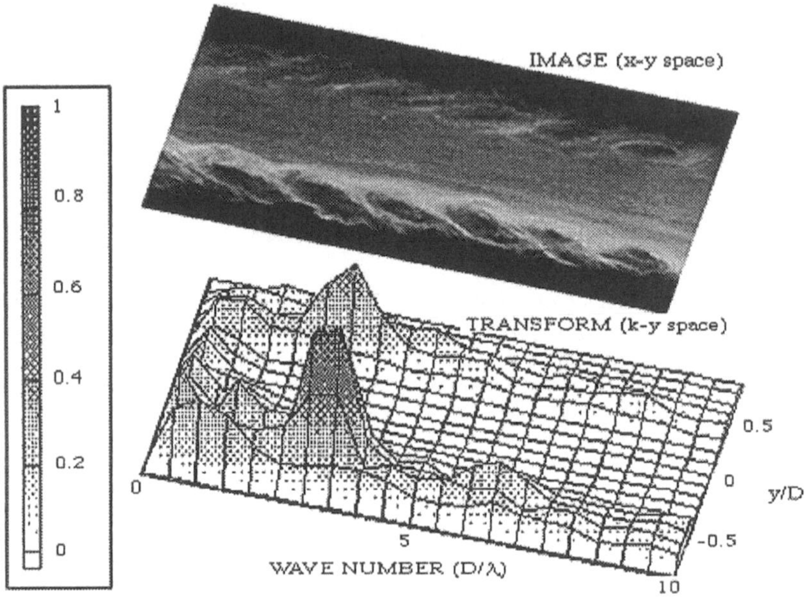

Fig. 5. Spectral intensity map obtained from FFT processing.

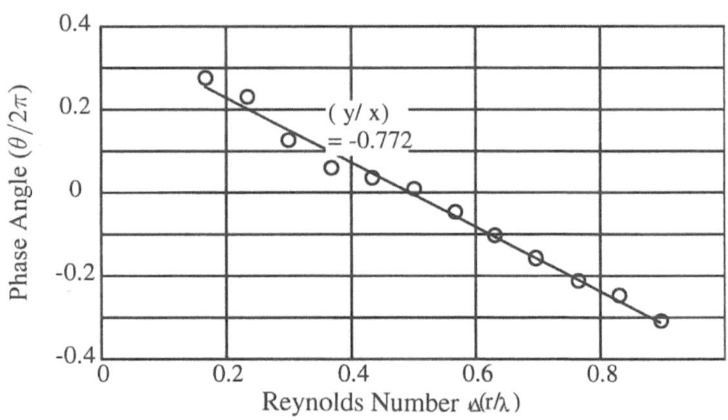

Fig. 6. Spectral phase lag at the peak wavelength for the image in Fig. 5.

obscured. The results were inconclusive. When coherent structures were apparently visible such as in Fig. 2, the wavelet transform correctly isolated the structures. Fig. 4 shows the wavelet transform of the image in Fig. 2a. At certain scale levels, the coherent structures can easily be identified. The other approach used Fast Fourier Transform (FFT) of the digitized intensity data in the axial direction. Fig. 5 shows the FFT spectral amplitude map along with the corresponding flow image. From the spectral peak wave numbers, the preferred axial wavelength of the vortices was quantified. Also, the orientation of the vortices was obtained from the phase information for the peak wave number (Fig. 6). From the slope, the rotation of the structures was measured. For this particular case with $M_c = 0.23$, the coherent structures were tilted 40 degrees toward the flow direction. The present result demonstrated the quantitative reduction of flow images using the Fourier transform approach, and the rotation information could be potentially useful in modeling efforts of the large-scale structures.

4.1 Axisymmetric Shear Layer Growth

The shear layer growth rate was measured either from the total pressure measurement or from the Mie scattering intensity measurement. For $M_c = 0.23$, the total pressure profiles were measured at several downstream locations. From the measured profiles, the shear layer thickness was defined as the radial distance between 10 and 90 percentiles of the free stream pressure difference. Fig. 7 shows the shear layer thickness as a function of downstream distance from the nozzle exit. Then, the slope of the linear fit through the data gives the spatial growth rate. For higher convective Mach number cases, the intensity measurements from the visualization data were used since the pressure and temperature of the center jet were beyond the operating range of the pressure transducer. Thirty instantaneous flow images were averaged for each flow condition, and the averaged intensity profiles were obtained. In a similar manner as before, taking 10 and 90 percentiles of the free stream intensity difference, the shear layer visual thickness was defined. The results are also shown in Fig. 7. The normalized shear-layer growth was obtained using the same incompressible growth rate as defined in Papamoschou and Roshko (1988). The reference data are for a planar configuration. Therefore, the present axisymmetric shear layer growth was compared with the published planar case data. Fig. 8 shows a very good agreement between the present axisymmetric case with the planar case.

4.2 Coupling between Wake and Parallel Shear Flow

The lowest convective Mach number case was studied in detail to better understand the physical mechanisms responsible for the uncharacteristically high degree of coherence in the present structures.

Fig. 9 shows several images of the coaxial jets when different lip thickness was used. The images clearly show the dependence of coherent structure wavelength on lip thickness. Since the center nozzle had finite lip thickness, a small wake with

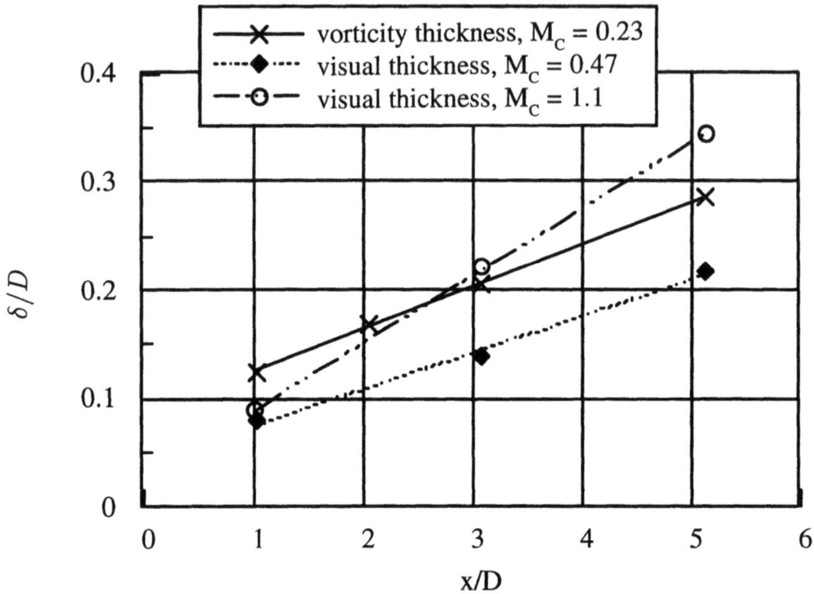

Fig. 7. Shear layer growth rate measurement.

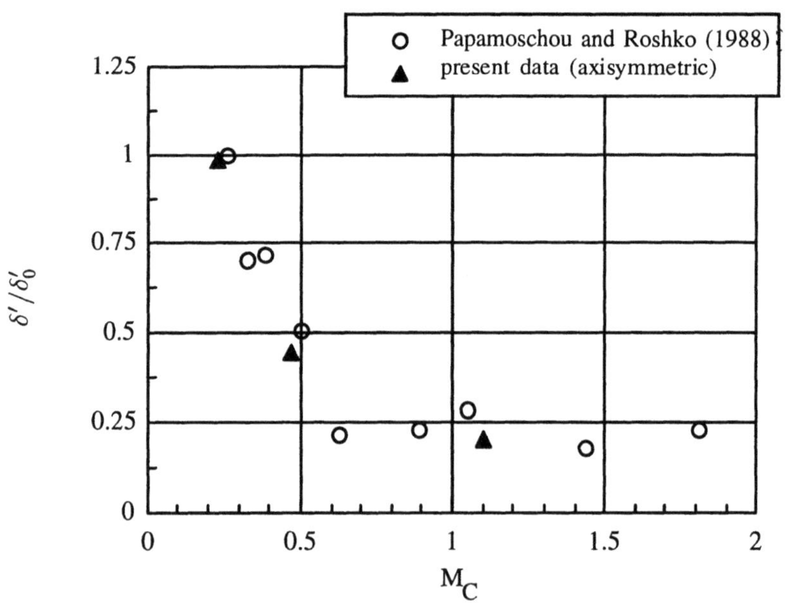

Fig. 8. Normalized shear layer growth rate vs. convective Mach number.

the size on the order of the lip thickness was created behind the lip. Therefore, the lip thickness would affect the stability characteristics of the wake. The disturbances generated by the unstable wake flow, then, would be amplified according to the stability characteristic of the initial shear layer. The coherent structure information was quantified by determining the peak wavelength from the FFT processing. The results were plotted against the lip thickness in Fig. 10. The peak wavelength increases monotonically with lip thickness. Furthermore, as evidenced by the sharpness of the spectral peaks, the case with lip thickness of 2.2 mm resulted in formation of more coherent vortices than others. This particular case also showed the least variation in the peak wavelength as a function of time.

The last result is also evident in Fig. 10 which shows the smallest scatter in the data for the lip thickness of 2.2 mm. The results suggest a possible coupling between the near-wake flow and the base shear flow. When the results were normalized by the wake characteristic parameters to obtain Strouhal number vs. Reynolds number plot (Fig. 11), the Strouhal frequency compared well with the vortex shedding frequency of a bluff body submerged in a free stream with velocity equal to the inner jet velocity.

5. Summary and Conclusions

The near-field of supersonic coaxial jets was visualized using a planar Mie scattering technique. In the compressible shear layer which formed between the two jets, large-scale coherent structures were observed. Three different combinations of supersonic jets resulted in the shear layer convective Mach numbers of 0.23, 0.47, and 1.1. At the two lower convective Mach numbers, the structures were highly coherent and at the highest convective Mach number the structures became less organized and more intermittent. The coherent structures were characterized by processing the flow images using either the wavelet transform or the Fast Fourier transform. These processing techniques reconfirmed the presence of coherent structures in the shear layer images and quantified streamwise preferred wavelength and angular orientation of the structures. Lip thickness of the inner nozzle had a strong influence on both the size and shape of the large-scale structures. The supersonic shear flow appeared to interact with the lip-created wake affecting the stability characteristics of the downstream flow. The coherent structure size increased with the lip thickness. The observation of highly coherent structures in supersonic coaxial jets was a very interesting discovery in many ways. The results provided a strong evidence that large-scale vortical structures are a dominant feature not only in incompressible mixing layers but also in compressible axisymmetric layers. The images of coherent structures, together with the characterization, provided significant insight into the physics of compressible mixing. Also, they opened up the possibility of supersonic shear flow control using vortex dynamics.

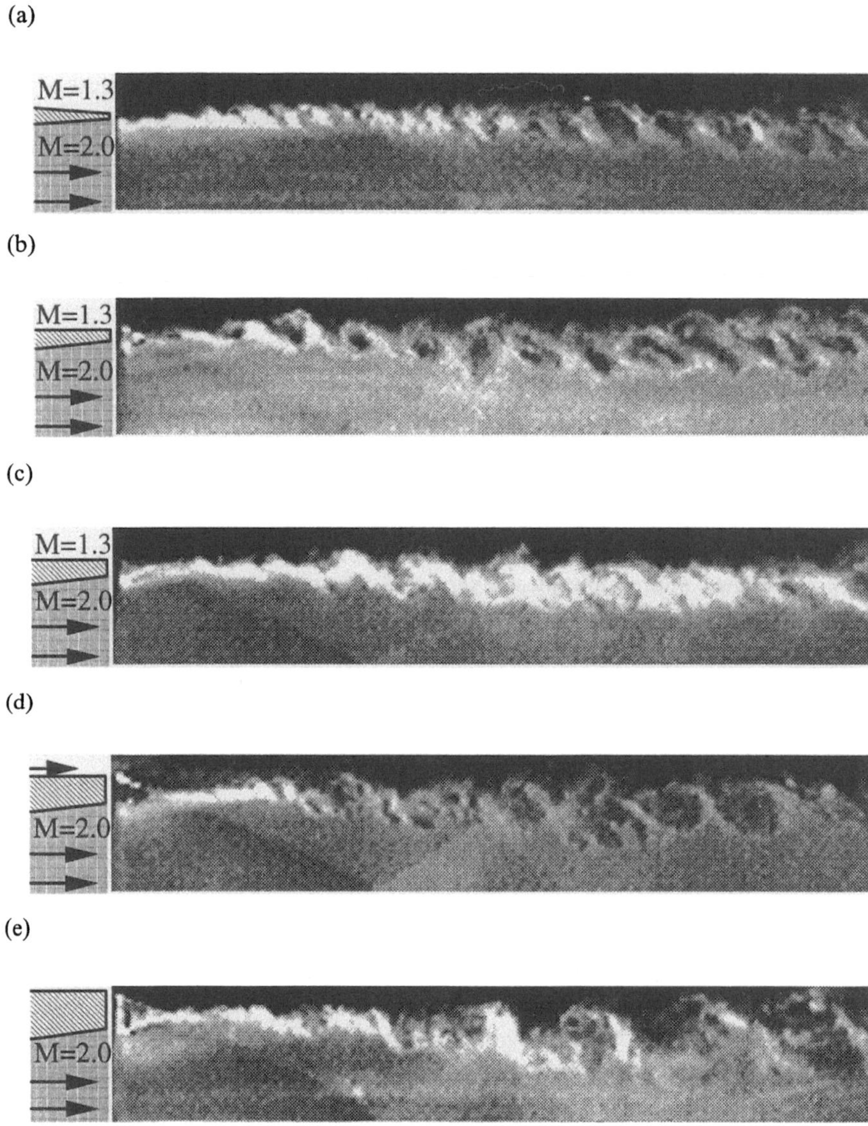

Fig. 9. Supersonic coaxial jets with $M_c = 0.23$. The lip thickness was **(a)** 1.3, **(b)** 2.0, **(c)** 2.9, (d) 4.1, and (e) 5.4 mm.

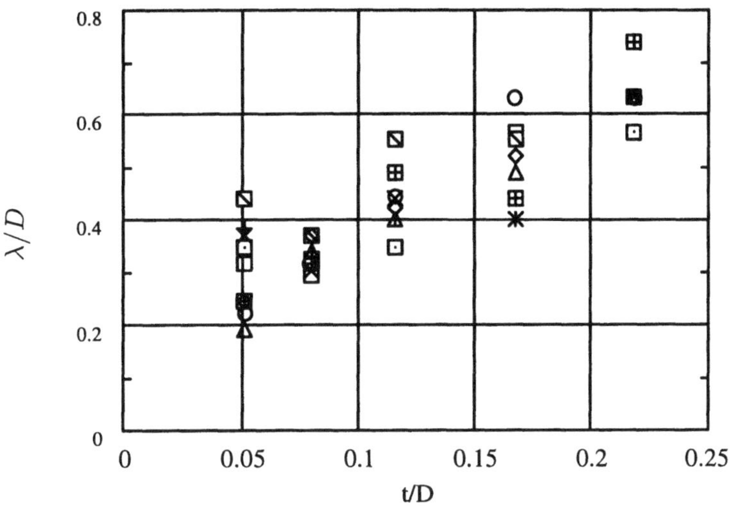

Fig. 10. Measure wavelength of coherent structures as a function of the nozzle lip thickness.

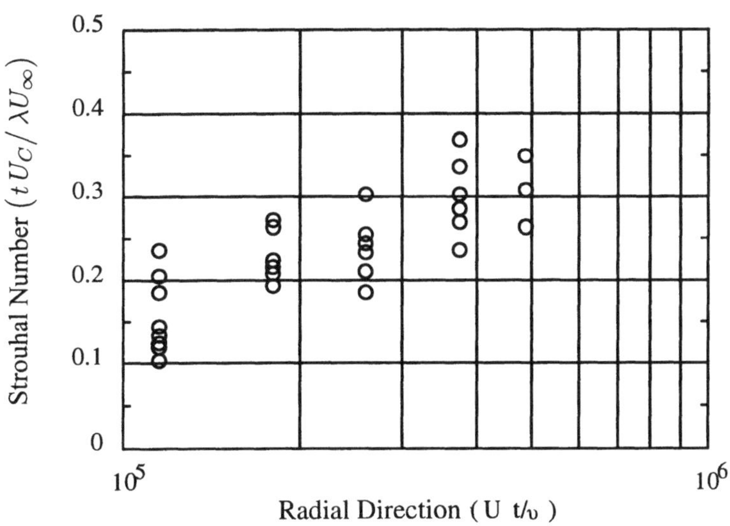

Fig. 11. Strouhal frequency of the coherent structures as a function of the wake Reynolds number.

Acknowledgment

The authors would like to acknowledge the support from the Office of Naval Research and the Scientific Officer, Dr. Gabriel D. Roy.

References

Brown, G.L. and Roshko, A. 1974 "On density effects and large structures in turbulent mixing layers", J. Fluid Mech. **64**, 775-816.

Clemens, N.T. and Mungal, M.G. 1991 "A planar Mie scattering technique for visualizing supersonic mixing flows", Exp. in Fluids **11**, 175-185.

Clemens, N.T. and Mungal, M.G. 1992 "Two- and Three-Dimensional Effects in the Supersonic Mixing Layer", AIAA J. **30**, 4, 973-981.

Elliot, G.S., Samimy, M., and Arnette, S.A. 1993 "The Evolution of Large-Scale Structures in Compressible Mixing Layers", 9th Symposium on Turbulent Shear Flows, Kyoto, Japan, Paper № 19-4.

Gutmark, E., Parr, T.P., Hanson-Parr, D.M., Schadow, K.C., and Hewer, G.A. 1991 "Wavelet Analysis of a Controlled Pulsating Flame", International Symposium on Pulsating Combustion, Monterey, CA, Paper № G-4.

McIntyre, S.S. and Settles, G.S. 1991 "Optical Experiments on Axisymmetric Compressible Turbulent Mixing Layers", AIAA-91-0623.

Messersmith, N.L., Dutton, J.C. and Krier, H. 1991 "Experimental Investigation of Large Scale Structures in Compressible Mixing Layers", AIAA-91-0244.

Papamoschou, D., and Roshko, A. 1988 "The compressible turbulent shear layer: an experimental study", J. Fluid Mech. **197**, 453-477.

Samimy, M. and Lele, S.K. 1990 "Particle-Laden Compressible Free Shear Layers", AIAA-90-1977.

Sandham, N. and Reynolds, W. 1989 "The Compressible Mixing Layer: Linear Theory and Direct Simulation", AIAA-89-0371.

Winant, C.D. and Browand, F.K. 1974 "Vortex pairing: the mechanism of turbulent mixing layer growth at moderate Reynolds number", J. Fluid Mech. **63**, 237-255.

Part III

Wall Flows

Introductory Remarks

Ron F. Blackwelder

Department of Aerospace Engineering
University of Southern California
Los Angeles, CA 90089-1191

The primary means of studying bounded turbulent shear flows is either to involve the computer in numerical simulations, use model equations or to perform experiments. The following set of papers provides excellent examples of all of these methods and includes studies ranging from the near wall region in a canonical flow to the augmentation of heat transfer in a boundary layer. In the process, the experimental and direct numerical simulation papers provide some good data that will be a challenge to many of the modelers as well as help to understand some of the fundamental problems involved.

The paper by Durst, Jovanović and Sender reports some new measurements in the near wall region of a pipe flow at a Reynolds number of 7500 using a specially designed test facility and laser doppler anemometer. These authors carefully matched the refractive index of the fluid to the index of the wall of the apparatus to avoid interference caused by the transmission of the light beam. They were able to reduce the measurement volume to less than one viscous scale in diameter and to $2.5\nu/u_\tau$ in length, thus producing a spatial resolution volume comparable to Wei and Willmarth (1989) in a channel flow. The authors primary purpose in obtaining the present data was to provide better data for comparison with the direct numerical simulations (DNS) of the wall region. Their results agree well with the DNS results except for the flatness factor of the normal velocity component, F_v. The DNS results of Kim, et al. (1987) give a sharp rise in F_v as the wall is approached whereas the experimental data shows a maximum and then a relaxation towards zero as the distance from the wall goes to zero. The authors can not explain this difference and conclude that this problem still needs further study.

Kuroda, Kasagi and Hirata address the question of how the wall influences the presence and character of the low speed streaks formed by the strong shear near the wall. They used DNS to study a turbulent channel flow in which one of the walls was moved in its own plane with velocities between zero and the velocity on the center line of the channel. This allowed them to alter the shear parameter, $S^\star = (\partial U/\partial y)q^2/\varepsilon$, as defined by Lee et al. (1990) where $q^2 = \overline{u_i' u_i'}$, while simultaneously retaining the strong boundary conditions on the normal velocity component due to the presence of the wall. As the shear at the wall was reduced, the elongated streaky structure of the u' fluctuations were reduced. When the peak value of $S^\star$ was reduced to less than 20, the streaky structure was eliminated. These results confirm the conclusions of Lee et al. that the streaky structure is due primarily to the strong shear. In addition the new results show that the presence of the wall has little to do with creating the streaks but only sets up the conditions for generating the strong shear which is responsible for the streak formation.

Large Eddy Simulation (LES) was used by Satake and Kawamura to examine the flow in concentric annuli. Even though their Reynolds numbers were relatively low, it is encouraging to see LES employed to help solve practical problems at higher Reynolds numbers. They were able to utilize Smagorinsky's sub-grid scale model to successfully calculate the flow and compare their results with experimental data obtained by Rehme (1974). Their LES data allowed them to explore the wall layers near the two concentric cylinders in much more detail than had been possible in the experiments. They found a larger degree of anisotropy near the inner cylinder than near the outer cylinder. This was due to the high speed fluid of the eddies with scales larger than the inner radius that disturbed the flow near the inner cylinder. These large eddies continued to move over the inner cylinder and formed a wake region behind it. No counterpart exists for the wall region near the outer cylinder.

A unique combination of DNS and LES is presented by Ducros, Comte and Lesieur to study transition in a boundary layer at a Mach number of 4.5. Both simulations were forced by large amplitude two-dimensional waves and small amplitude three-dimensional disturbances. The DNS calculations were utilized to study the earlier phases until the onset of the small scale turbulence. The LES calculations were able to follow the flow to higher Reynolds numbers into the transitional regime. The results suggest that transition in high Mach number boundary layers is associated with the growth of inviscid instability modes which scale with the boundary layer thickness and a simultaneous growth of a viscous mode near the wall with a spanwise wavelength of $100\nu/u_\tau$. When the scales of these two are comparable, they interact and break down into turbulence near the wall.

Unsteady flow structure in an oscillating turbulent channel flow was the subject of the study by Tardu, Feng and Binder. They concentrated on investigating the Reynolds shear stress in the wall region for a large range of imposed frequencies from the quasi- steady to frequencies higher than the ejection frequency. The mean ejection frequency itself was not affected by the unsteadiness. They conclude that the unsteady velocity effects were confined primarily to a Stokes layer near the wall of $y^+ < 2 \cdot l_s^+$ but the unsteady pressure field affects the entire flow field.

Another investigation of oscillating flows is presented by Hanjalić, Jakirlić and Hadžić. They used the two equation model for k and ε from Hanjalić and Launder (1976) and produced the only results in this section based upon model equations. They used two models differing in their coefficients and studied oscillatory flow in transitional boundary layers and pipe and channel flow. Although their agreement with the experimental and DNS data is not complete, their model appears to have captured the main characteristics of the flow fields.

The last two papers in the series concern heat transport. The transport characteristics of a developing thermal layer embedded in a fully developed turbulent pipe flow were the subject of the experimental investigation by Nagano, Sato and Tagawa.

The entrance region allowed the unheated flow to develop over 127 diameters and then heat was added over the last 45 diameters downstream. Not surprising, it takes further downstream for the fluctuation temperature field to be established than the mean field. The authors also define an intermittency signal to distinguish between the heated and unheated fluid and measure some conditioned statistics. These results will provide an interesting challenge to various models assessing non-equilibrium turbulence.

Another experiment was performed by Inaoka and Suzuki in which they studied the augmentation of the heat transfer by vortex generators attached to a LEBU plate. Although the LEBU decreased the heat transfer coefficient slightly, it did provide a platform on which to mount the vortex generator in the outer flow field where it was more useful in generating a large scale vortex. The vortex had dimensions of the boundary layer thickness and was effective in bringing lower temperature fluid from the freestream toward the wall region and hence was a useful configuration for enhancing the heat transfer.

References

Hanjalić, K. and Launder, B.E., (1976): "Contribution towards a Reynolds-stress Closure for Low Reynolds Number Turbulence", J. Fluid Mech. **74**, 593-610.

Kim, J., Moin, P., and Moser, R., (1987): "Turbulence Statistics in a Fully Developed Channel Flow at low Reynolds Number", J. Fluid Mech. **177**, 133-166.

Lee, M., Kim, J., and Moin, P. (1990): "Structure of Turbulence at High Shear Rate", J. Fluid Mech. **216**, 561-583.

Rehme, K., (1974): "Turbulent Flow in Smooth Concentric Annuli with Small Radius Ratios", J. Fluid Mech. **64**, 263-287.

Wei, T. and Willmarth, W.W., (1989): "Reynolds-number Effects on the Structure of a Turbulent Channel Flow", J. Fluid Mech. **204**, 57-95.

Detailed Measurement of the Near-Wall Region of a Turbulent Pipe Flow

F. Durst, J. Jovanović, and J. Sender

Lehrstuhl für Strömungsmechanik,
Friedrich-Alexander Universität,
Cauerstr. 4, 91058 Erlangen, Germany

Abstract

This paper presents laser-Doppler measurements of the mean velocity and statistical moments of turbulent velocity fluctuations in the near-wall region of a pipe flow at a low Reynolds number. A refractive-index matched fluid was used in a Duran-glass test section to permit optical access to the near-wall region without distortion of the laser light beams. All measurements were corrected for the influence of the finite size of the measuring control volume. The results are presented and their reliability is proven through comparison with existing experimental and numerical data available in the literature.

1 Introduction

Over the past three decades, turbulent wall shear flows have been the target of numerous experimental and lately also numerical investigations. These studies have provided information on the mean flow field and turbulence intensities for wall boundary layer, pipe, and channel flows. All of this work has resulted in general agreement regarding the flow structure away from the wall. Close to the wall, however, questions regarding the turbulence properties still remain, since the experimental data are partially inconsistent and in some details contradict the results obtained from direct numerical simulations (see, Kim *et al.* 1987; Lyous *et al.* 1991). This evidence suggests that more refined experimental work is needed in the near wall region of turbulent boundary layer, pipe and channel flows.

The objective of the paper is to present experimental data, obtained with a special LDA-system for near wall measurements, of the mean velocity and turbulence statistics. The results to be presented demonstrate that state of the art LDA-measuring techniques permit experimental data to be obtained with a much greater degree of confidence than previously obtainable with the help of hot-wire anemometry. Although measurements are provided for the entire pipe flow, emphasis in the discussion of data is given to the region of the viscous sublayer and the buffer region, since the scatter in the existing experimental data is greatest in this part of the flow.

To explore the near wall region experimentally, the application of optical measuring techniques is preferable due to the severe spatial resolution demands enforced by the special nature of the flow. Advanced laser-Doppler anemometers can be designed to satisfy the imposed requirements. However, the application of this technique requires a well designed test rig to avoid deflections of light beams near the solid surface. For the pipe flow investigations described in this paper, refractive index matched fluids were used to eliminate influences of the cylindrical pipe walls on the laser beams. An especially designed test rig was employed in the present study to meet these requirements.

To obtain reliable laser-Doppler measurements it is important to consider the influence of the finite size of the measuring control volume. The appropriate corrections can be derived analytically to account for the effects mentioned above and were applied at each measuring position in the flow.

Additional efforts were necessary in order to minimize the influence of electronic noise on the measured data and to account for the small angle misalignment of the optical system with respect to the flow. The influence of these factors turned out to be small in absolute terms, but become significant in the near wall region where the mean velocity and turbulence intensities approach zero values at the solid surface.

The authors' experimental results include distributions of the mean velocities, turbulence intensities, and higher order moments of the axial, radial and tangential, velocity components. The results obtained are also compared with the data from experimental and direct numerical simulation studies available in the literature. Some of the questions regarding the behavior of turbulence intensities in the near wall region are clarified.

2 Test Section and Instrumentation

For the near-wall measurements, a test section was designed and built. It is shown schematically in Fig. 1 and consists of a closed loop pipe flow installation driven by a screw conveyor pump operating in its suction mode. The flow was supplied from a large settling tank to the working section that was located about four meters downstream providing a development length of about 80 pipe diameters. Using the highest flow rate, a maximum Reynolds number of R_e=30 000 could be achieved based on a bulk velocity of 2.58 m/s and a pipe diameter of D=50 mm. The measurements presented here were performed mainly at a low Reynolds number, R_e=7440, corresponding to a bulk velocity of 0.64 m/s.

In order to enable laser-Doppler measurements, the test section was made of a Duran-50 glass pipe. This pipe was centrally mounted in a square viewing box as indicated in Fig. 1. To avoid any interference of the pipe with the light beams, the test fluid selected was refractive-index matched to the Duran glass, the material of the pipe wall. To achieve precise matching of refractive indices, two Diesel oils of slightly different refractive indices were selected and mixed in such a way that the refractive index of the glass pipe was achieved. In addition, the temperature of the test fluid was controlled by heating and cooling elements installed in the downstream settling chamber of the test rig. In this way the refractive indices of the test fluid and the test section wall material were matched within $\pm 5 \times 10^{-4}$.

A special laser-Doppler optical system was designed for the present measurements. The system operated with a 15 mW Helium-Neon laser and double Bragg-cell transmission optics. The arrangement was laid out to yield a measuring control volume of 70 μm in diameter and 250 μm in length. The light scattered in the forward direction was collected by the receiving optics and directed on to an appropriately sized pinhole in front of an avalanche photodiode. The signal from this diode was band-pass filtered prior to being processed by a counter processor. The band pass filter settings were chosen to be wider than ten times the root-mean-square value of the Doppler signal at each measuring location. A TSI counter model 1980 was used for all of the measurements. Its settings were controlled in such a way that no triggering signal was obtained when one of the laser beams was blocked. In this way the influence of noise was minimized on the measured data.

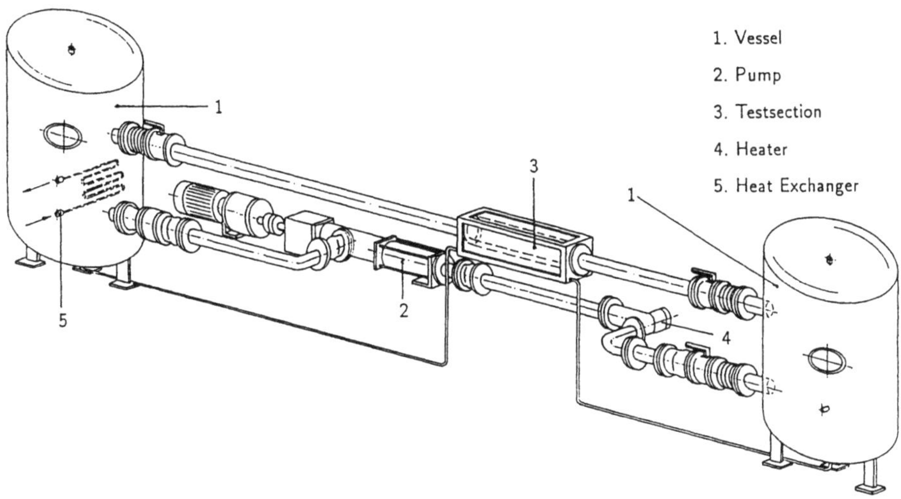

Fig. 1. Close-loop pipe flow test-section.

Sufficient particles were present in the oil to measure the turbulent fluctuations. Data rates of 100 Hz were typical for data points laying close to the wall. At other locations the average data rate changed in proportion to the local mean velocity.

The laser-Doppler optical system was placed alongside of the measuring test section. The axial velocity component could be measured in both horizontal and vertical planes. By rotating the transmitting optics by 90 degrees, the radial velocity component was measurable in the vertical plane of the pipe cross section and the tangential velocity component in the horizontal plane. Hence, the spatial resolution in the tangential direction was limited and prevented measurements closer than about 250 μm from the wall.

3 Measuring Errors and Applied Corrections

To set up the pipe test section, a precision tube was selected with a diameter variation of less than ±2 μm. The temperature variation of the refractive index matched fluid provoked positioning errors of approximately ±8 μm as maximum. The overall position error including the error of the traversing equipment was estimated to be better than ±10 μm, corresponding to ±0.02% of the pipe diameter.

The error of the LDA-system resulted mainly from the Bragg-cell driver and the counter used for the frequency evaluation. The Bragg-cell driver's stability was better than ±0.1% and the error due to the finite resolution of the counter was ±1.0%. These estimates gave an error of about ±1.0% for the velocity measurements.

In general, spatial integration of the LDA signals due to the finite size of the measuring control volume affects turbulence measurements. This effect is particularly important close to the wall, since the velocity variations across the measuring volume reaches their maximum value in this part of the flow field. In the current literature, no correction procedure exists, to take this influence into account. Recently, Durst *et al.* (1992) considered this problem analytically by representing measured mean quantities as averages in time and space:

$$\overline{U}_{i_{\text{meas.}}} = \lim_{T\to\infty} \frac{1}{T} \int_0^T \left\{ \frac{1}{V} \iiint_V f\, U_{i_{\text{vol.}}} \, \mathrm{d}V \right\} \mathrm{d}t. \tag{1}$$

In the above definition, $U_{i_{\text{vol.}}}$ corresponds to the instantaneous velocity component in the measuring volume that depends on time and space coordinates, V is the LDA measuring control volume and the weighting function f accounts for the distribution of the detected particles inside the volume of the integration. Expressing $U_{i_{\text{vol.}}}$ in the form of a truncated Taylor expansion around the center of the control volume permitted the derivation of explicit relations between measured means and true local averages.

Since the dominant effect of gradient broadening is only in the direction (y) perpendicular to the flow for the mean velocity and turbulence intensity, the following expressions were obtained for correcting the measured data:

$$\overline{U}_{i_{\text{meas.}}} = \overline{U}_{i_{\text{true}}} + \frac{d^2}{32} \left(\frac{\mathrm{d}^2 \overline{U}_{i_{\text{true}}}}{\mathrm{d}y^2} \right), \tag{2}$$

$$\overline{u_i^2}_{\text{meas.}} = \overline{u_i^2}_{\text{true}} + \frac{d^2}{16} \left(\frac{\mathrm{d}\overline{U}_{i_{\text{true}}}}{\mathrm{d}y} \right)^2, \tag{3}$$

where d corresponds to the diameter of the control volume. These corrections result from the evaluation of the following formulas:

$$\overline{U}_{i_{\text{meas.}}} = \lim_{T\to\infty} \frac{1}{T} \int_0^T \frac{1}{d} \int_{y_c-d/2}^{y_c+d/2} f(y) U_{i_{\text{vol.}}} \, \mathrm{d}y \mathrm{d}t, \tag{4}$$

$$\overline{u_i^2}_{\text{meas.}} = \lim_{T\to\infty} \frac{1}{T} \int_0^T \frac{1}{d} \int_{y_c-d/2}^{y_c+d/2} f(y) U_{i_{\text{vol.}}}^2 \, dy dt \; - \; \overline{U}_{i_{\text{meas.}}}^2 , \tag{5}$$

where y_c is the distance of the center of the measuring control volume from the wall.

The choice for an appropriate weighting function in Eq. (1) was based on the following arguments. For a given LDA-system, the probability that a particle is detected is directly proportional to the scattered light intensity. For a monodisperse particle distribution, the probability for particle detection is identical to the light intensity distribution in the measuring control volume. Therefore, the weighting function was assumed to be given by the light intensity distribution

$$f(y) = \frac{4}{\sqrt{2\pi}} \exp\left\{ -2 \frac{(y-y_c)^2}{d^2} \right\} . \tag{6}$$

The above formulas for correction show that the measured data of the mean velocity depend on the second derivative of true mean velocity profile and the correction for the turbulence intensity is proportional to the gradient of the mean velocity. It can also be shown that corrections for the higher order moments are dependent on the gradient of the mean velocity and the gradient of the lower order moment. Hence, all measured moments of the local velocity distribution need to be corrected to take the finite size measuring control volume into account.

All final measurements reported in this paper have been corrected for the influence of the finite size of the LDA measuring control volume using the methodology discussed above. The velocity gradient broadening corrections were significant only for the streamwise velocity component in the region of the viscous sublayer.

4 Experimental Results

4.1 Mean Flow

Preliminary experiments confirmed that the pipe flow was fully developed and symmetrical. To ensure the high reliability of turbulence measurements, we fixed the time interval between samples close to the integral time scale of the flow (see, Lumley 1970):

$$\Delta t \simeq L_T \simeq \frac{D}{2U_c} , \tag{7}$$

where D and U_c correspond to the pipe diameter and centerline velocity. For a sample size of $N = 40\,000$ data points the relative statistical uncertainties in the measurements were estimated (on the basis of sampling parameters according to Lumley 1970) to be 0.28% for the mean velocity, 1% for the turbulence intensity, and 2.3% for the fourth order moment.

The measured mean velocity as a function of the normalized distance from the wall is presented in Fig. 2. The normalization of measurements was performed by the inner variables:

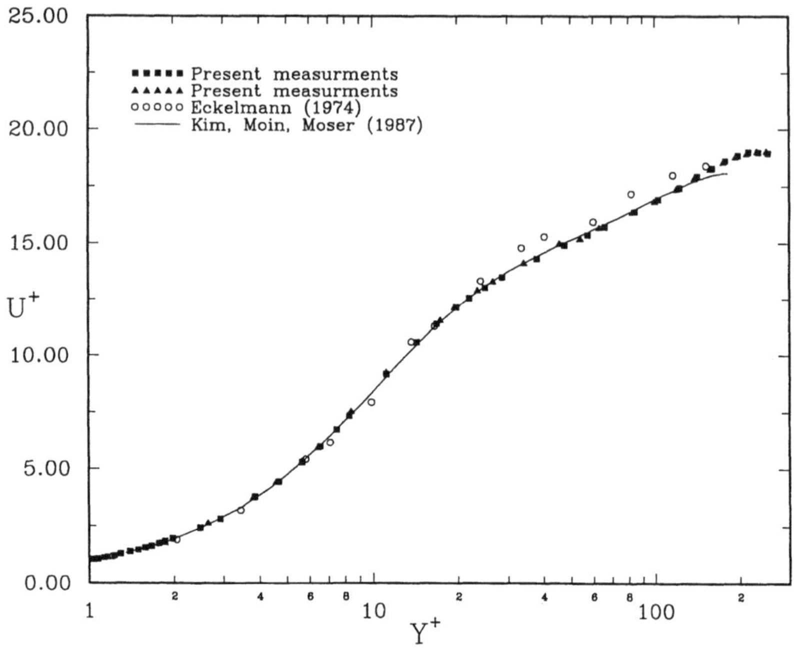

Fig. 2. Mean velocity at R_e=7440.

$$y^+ = \frac{yu_\tau}{\nu} \quad \text{and} \quad U^+ = \frac{\overline{U}}{u_\tau} \tag{8}$$

where u_τ and ν are the wall shear velocity and kinematic viscosity of the flow medium, respectively. The wall shear velocity, u_τ, was calculated directly from the slope of the velocity profile in the viscous sublayer. For these purposes, only the data points very close to the wall, i.e., $y^+ \leqslant 2.5$, were considered. This practice ensured accurate determination of wall shear stresses, since the slope of the mean velocity was slowly increasing as the wall was approached. Only in the region very close to the wall, variations of the shear stress were lower than 1%. The wall shear velocity, u_τ, obtained for R_e=7440,

$$u_\tau = 0.043 \text{ m/s},$$

was in very good agreement with the value:

$$u_\tau = 0.042 \text{ m/s},$$

computed from the Blasius formula for the friction factor.

The measured mean flow distributions shown in Fig. 2 closely follow the universal law of the wall. The linear and the logarithmic portions of the flow can be easily identified in these data. Good agreement of the presented data with the universal law of the wall indicate that the flow had reached a fully developed turbulent state. In the same diagram, we have also compared the authors measurements to the data from the direct numerical simulation of Kim *et al.* (1987). Although the simulated data

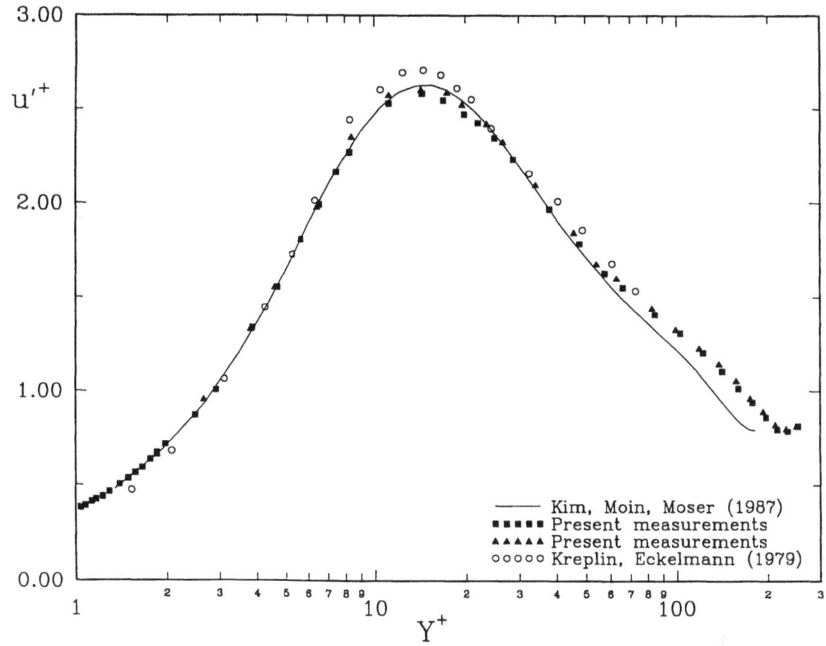

Fig. 3a. Root-mean-square of the velocity fluctuations; axial velocity component.

correspond to turbulent channel flow, the agreement is very good across the entire flow field.

4.2 Turbulence Intensity Measurements

The turbulence intensities normalized by the wall shear velocity are presented in Fig. 3. The authors' measurements agree very well with the data of Kim *et al.* (1987) across the entire wall region. The agreement is equally good for the intensities of all three fluctuating velocity components. Note that the peak in the distribution of the axial intensity is located at the inner and those in the distributions of the normal and tangential components at the outer edge of the buffer region. The data from Fig. 3 also demonstrate some deficiencies in standard hot-wire techniques for near wall applications. The experimental results of Kreplin and Eckelmann (1979) show higher intensities in radial and tangential components of velocity across the entire wall region.

The high reliability of the measured data is further supported by the data shown in Fig. 4. In this figure, the turbulence intensities normalized by the local mean are plotted in the near-wall region. These data are very sensitive, since they emphasize the measuring errors in the viscous sublayer. The present measurements are nearly identical to the simulated data of channel flow, supporting the following asymptotic values for the axial and tangential intensities:

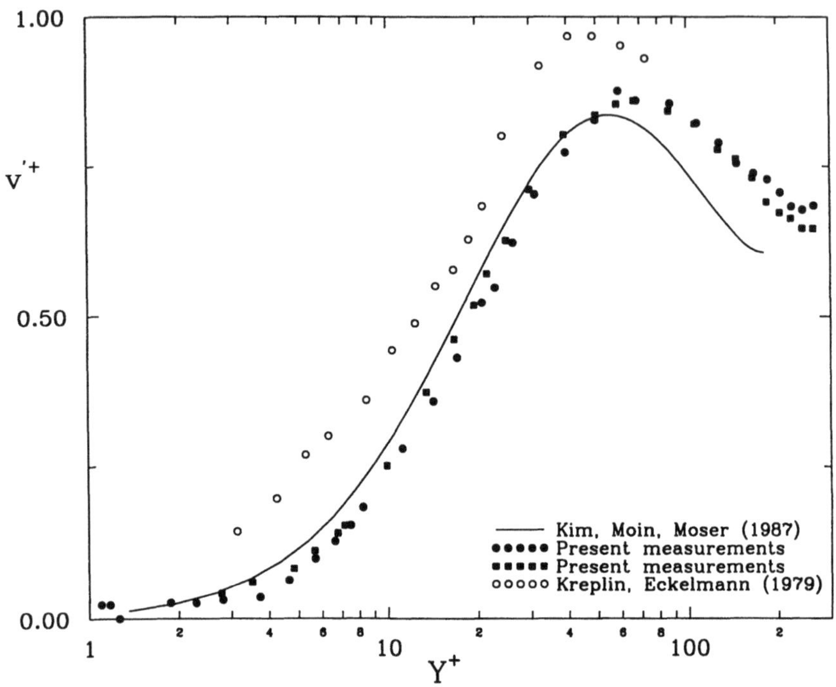

Fig. 3b. Root-mean-square of the velocity fluctuations; radial velocity component.

$$\frac{\sqrt{\overline{u^2}}}{\overline{U}} \approx 0.37 \quad \text{and} \quad \frac{\sqrt{\overline{w^2}}}{\overline{U}} \approx 0.21, \qquad y^+ \to 0,$$

as the wall is approached. The tangential intensity component is slightly lower in comparison to the simulated data close to the wall. This might be related to the limited spatial resolution of the present system for measurements in the tangential direction. The profile of the radial intensity component agrees with the data of numerical simulations down to $y^+ \approx 1$.

In the present investigations the authors have made an attempt to capture the influence of Reynolds number on the limiting behavior of turbulence intensities near the wall. The data for three different Reynolds numbers are shown in Fig. 5. From these results, we can conclude that there is no significant influence of Reynolds number on these quantities very close to the wall.

In connection to the presented results in Figs. 4 and 5, it is interesting to analyze the limiting behavior of turbulence intensities near the wall obtained in different experimental studies. The limiting behavior of turbulence intensities near the wall reproduced from various experimental studies are given in Fig. 6. These data were obtained using different measuring techniques. In general, there seems to be a fair degree of agreement among these results for the intensity of the axial velocity component. Hot-film measurements of Alfredsson *et al.* (1988) and data obtained by Karlsson and Johansson (1988) using laser-Doppler anemometry asymptotically approach slightly higher values $\left(u'/\overline{U} \approx 0.39 - 0.4\right)$ than obtained in the present

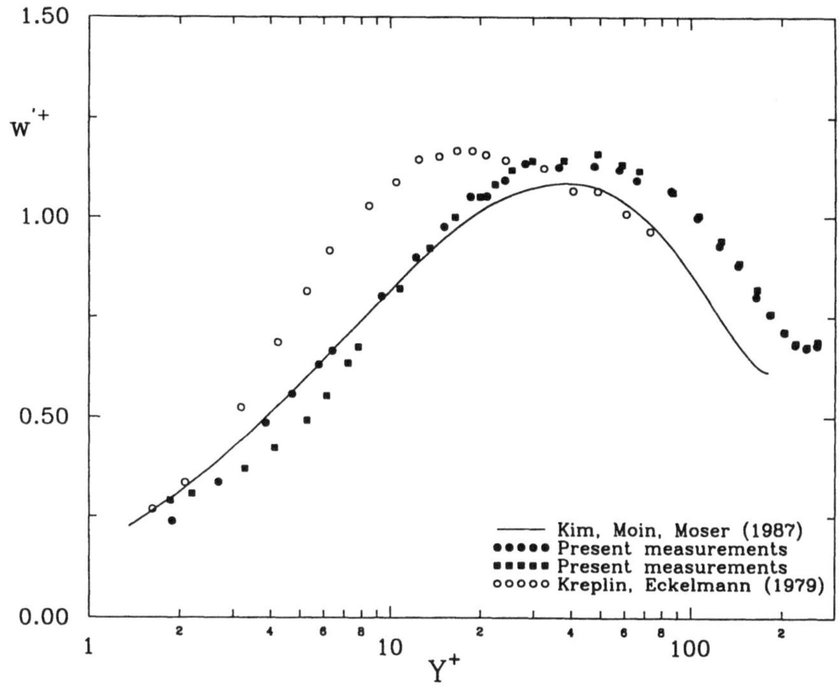

Fig. 3c. Root-mean-square of the velocity fluctuations; tangential velocity component

investigation. Good agreement exists between the data shown in Fig. 4 and the results of Niederschulte *et al.* (1990), away from the viscous sublayer.

The experimental results presented have serious implications on the further development of measuring techniques for near-wall studies. The ability of laser-Doppler anemometry to resolve the limiting behavior of turbulence intensities in the viscous sublayer opens the possibility for direct measurements of the turbulence dissipation rate at the wall. Such data would be of considerable interest, since the dissipation rate reaches a maximum value on the solid surface. In this respect, further efforts are needed in order to refine the measurements of tangential velocity component in the viscous sublayer.

4.3 Measurements of Higher-Order Moments

During the course of the present investigation, the authors have also measured higher-order moments, the skewness and flatness factors of the axial, radial, and tangential velocity components. These statistical quantities characterize the details of the probability density distributions. In wall shear flows, higher-order moments are significantly different from those corresponding to Gaussian probability distributions.

The higher-order moments of the axial velocity component are shown in Fig. 7. The non-Gaussian behavior of the turbulence fluctuations is mostly localized in the

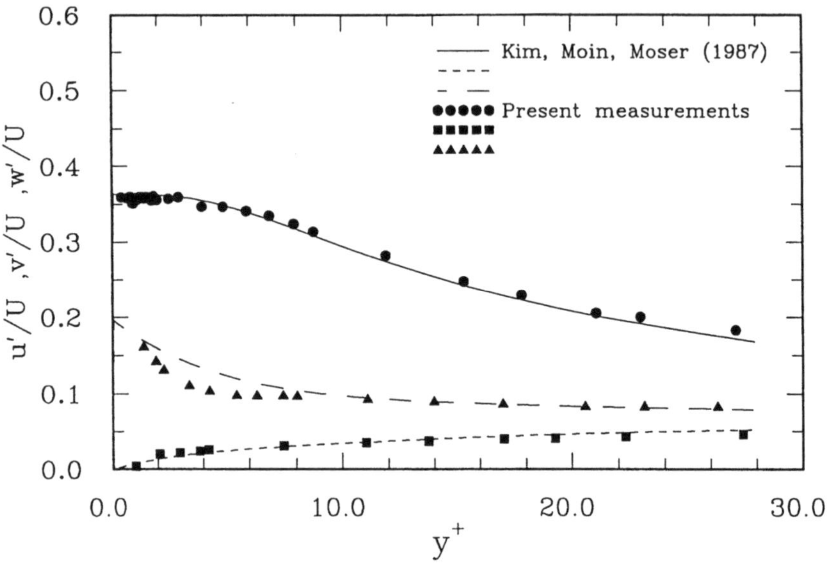

Fig. 4. Turbulence intensities near the wall relative to the local mean velocity at R_e=7440.

near-wall region. In the near-wall region, the skewness and flatness factors reach their maximum values:

$$(S_u)_w \simeq 0.85 \quad \text{and} \quad (F_u)_w \simeq 4.1.$$

The limiting behavior of the moments near the wall is in good agreement with the results of direct numerical simulations of channel flow. The present experimental results, numerical data , hot-film measurements of Kreplin and Eckelmann (1979), Alfredsson *et al.* (1988) as well as laser-Doppler measurements of Karlsson and Johansson (1988), Niederschulte *et al.* (1990) compare favorably across the entire wall region.

While there exists a fair degree of agreement on the behavior of higher-order statistics for the axial velocity component obtained from different sources, the same conclusion does not hold for the component perpendicular to the wall. This situation is illustrated in Fig. 8, where the authors have plotted the available skewness and flatness factors for the radial velocity component. The present measurements of the skewness factor of the v'- velocity component qualitatively agree with the simulated data of turbulent channel flow and data obtained from other studies of wall turbulence (see, Kreplin and Eckelmann 1979; Karlsson and Johansson 1988; Niederschulte *et al.* 1990). From the data shown in Fig. 8, it is evident that there exists noticeable discrepancies between the various results for the flatness factor of the v'- velocity component. While the simulated data show a steep rise of the flatness as the wall is approached, the authors experiments indicate that the flatness factor reaches a maximum and thereafter asymptotically decreases towards the wall. Qualitatively, this trend is also present in other experimental data obtained using hot-film

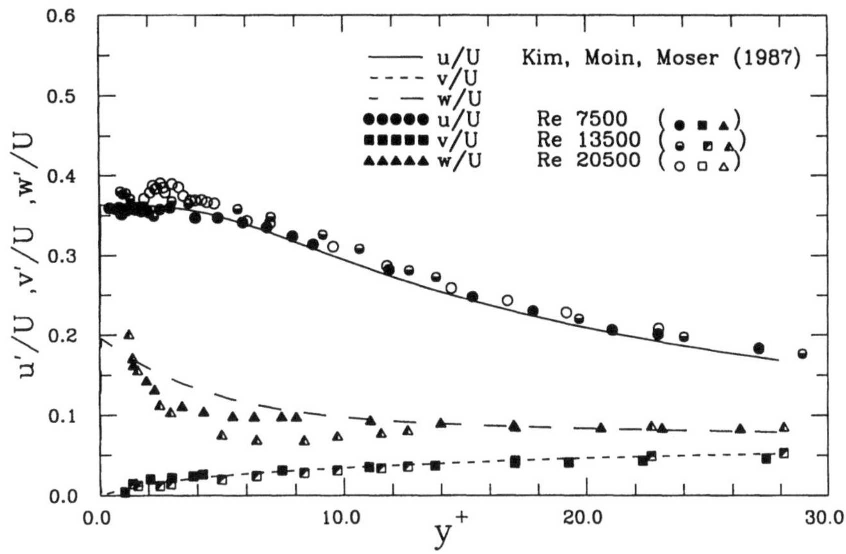

Fig. 5. Turbulence intensities near the wall relative to the local mean velocity at different Reynolds numbers.

anemometry (Kreplin and Eckelmann 1979) or laser-Doppler anemometry (Karlsson and Johansson 1988; Niederschulte *et al.* 1990).

It is not clear where the source for such big difference between the experimental and computational data lies, since repeated measurements showed high reliability in the authors' data presented in Fig. 8. A possible explanation for the strange behavior of the flatness factor near the wall is given in the recent study of Lyons *et al.* (1991). They analyzed the influence of grid resolution on various statistical quantities computed from simulated data bases of channel flow. The weak dependence on grid resolution was confirmed for all computed statistics except for the flatness of the normal velocity component in the viscous sublayer and buffer region. The flatness factor of the velocity component normal to the wall increased with grid refinement so dramatically that it was not possible to confirm the reliability of this computational result in the near wall region. Thus, we can conclude that this issue is still open for further investigations.

The distributions of the measured skewness and flatness factors of the tangential velocity components are plotted in Fig. 9. The skewness is zero across the entire flow field due to the reflection symmetry, and the measured flatness is in good agreement with the data of numerical simulations. The similar degree of agreement also exists among other available results in the literature (see, Lyons *et al.* 1991; Kreplin and Eckelmann 1979; Karlsson and Johansson 1988; Niederschulte *et al.* 1990). In the wall region, the flatness factor reaches a maximum value of:

$$(F_w)_w \simeq 6.75.$$

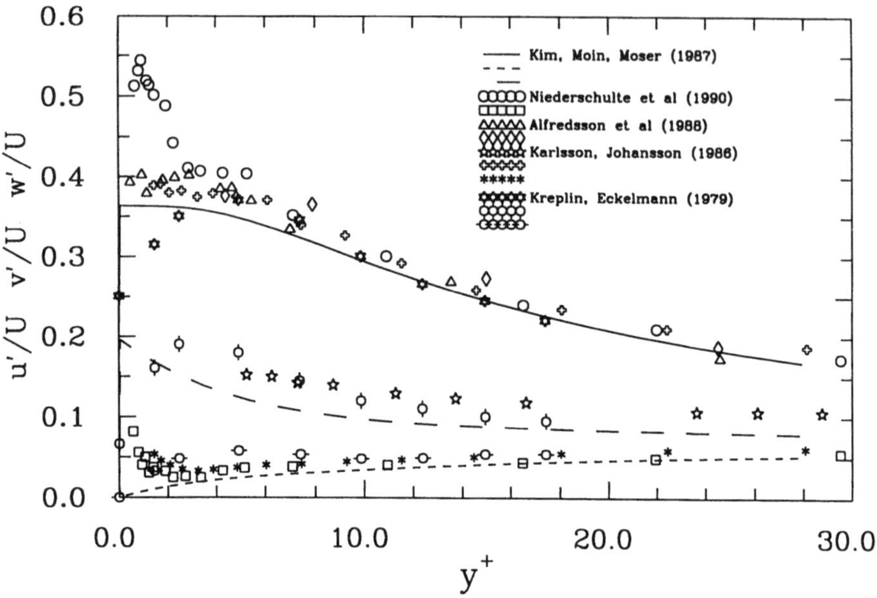

Fig. 6. The comparison of the limiting behavior of the turbulence intensities close to the wall from different experimental studies.

5 Conclusions

High resolution laser-Doppler measurements have been performed in a pipe flow in order to determine the mean velocity and turbulence statistics distributions in the near-wall region. The present experimental results for the mean velocity and turbulence intensities agree well with the results of direct numerical simulations of channel flow down to the minimum distance of $y^+ \approx 1$ from the wall. The measured and predicted profiles yield nearly identical data for the limiting behavior of the turbulence intensities as the wall is approached.

The measured higher-order statistics of the axial and tangential velocity component were found to be consistent with the data obtained from direct numerical simulations including the limiting behavior of the skewness and flatness factors toward the wall. However, significant disagreement was obtained for the higher-order statistics of the cross flow component. While these two sets of data qualitatively agree for the distribution of the skewness factor, the measured and predicted profiles of flatness completely disagree in the viscous sublayer and buffer region. It was not possible to clarify these differences on the basis of the present measurements.

Comparison of the authors' experimental results with other data, obtained using hot-wire techniques, shows good agreement for the statistics of the axial velocity component. The agreement is better away from the wall than in the region of the viscous sublayer. Large discrepancies between the data obtained using different mea-

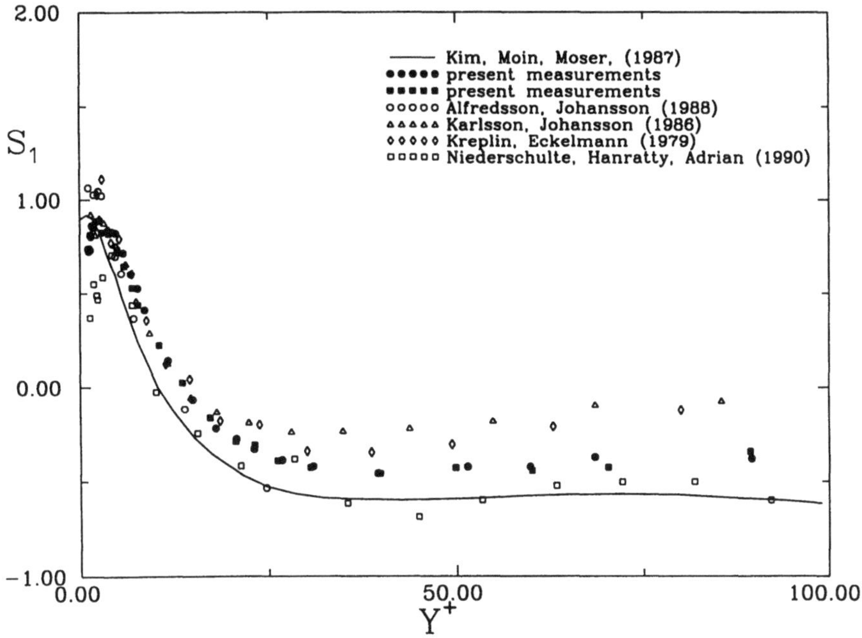

Fig. 7a. Higher-order moments of axial velocity component; skewness factor.

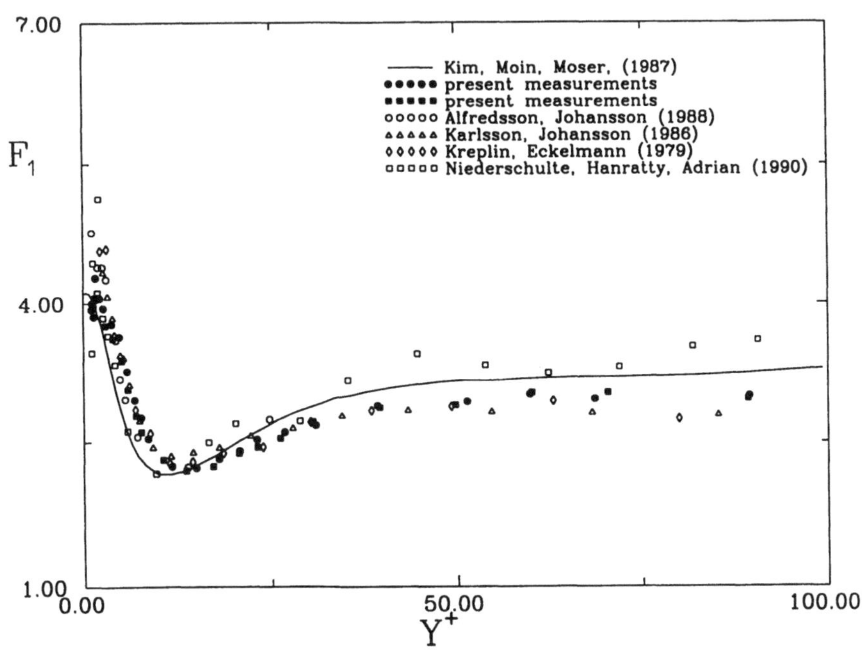

Fig. 7b. Higher-order moments of axial velocity component; flatness factor.

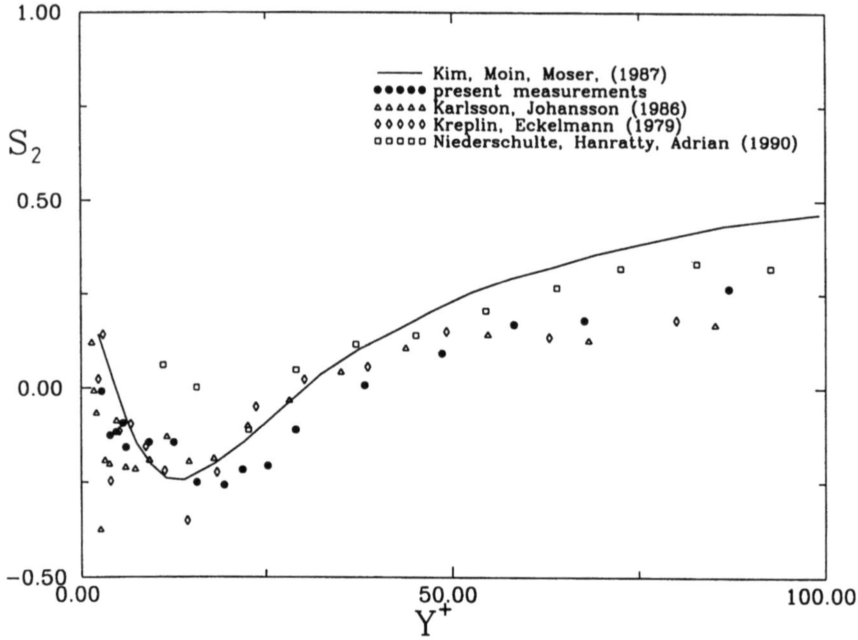

Fig. 8a. Higher-order moments of radial velocity component; skewness factor.

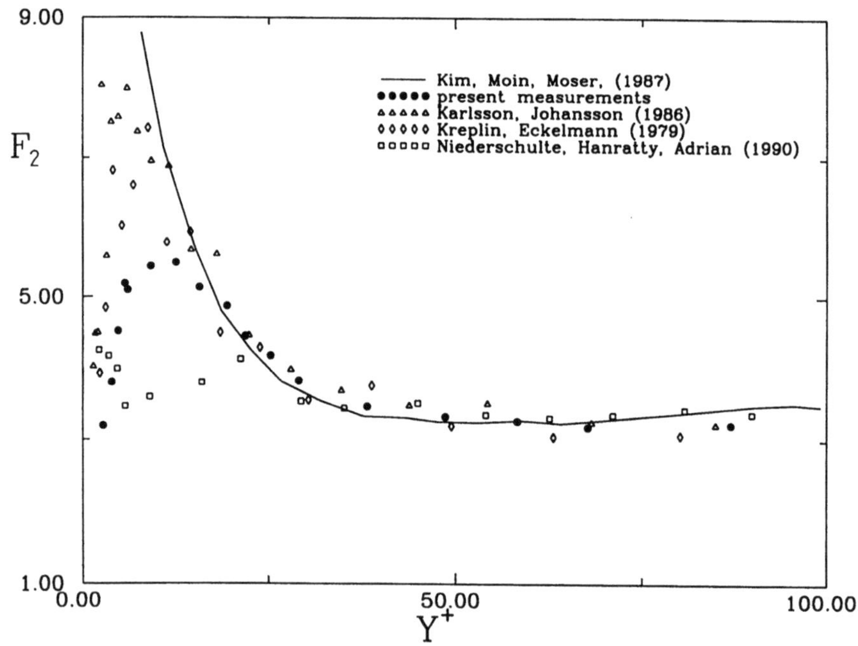

Fig. 8b. Higher-order moments of radial velocity component; flatness factor.

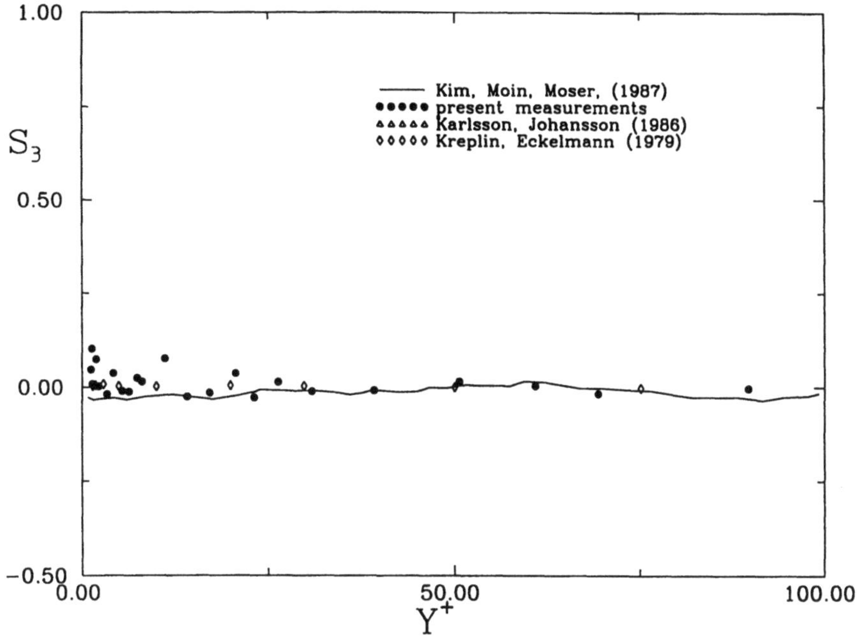

Fig. 9a. Higher-order moments of tangential velocity component; skewness factor.

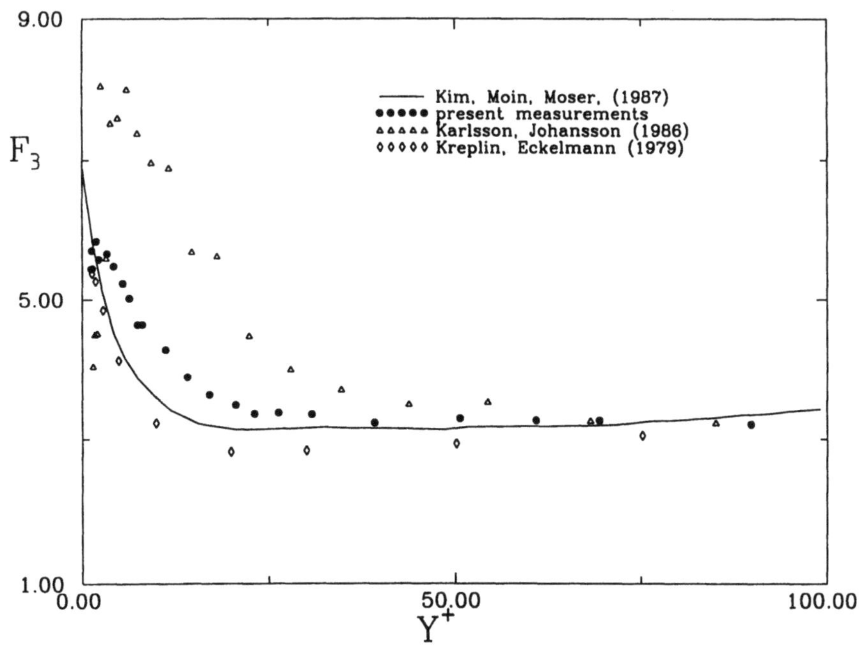

Fig. 9b. Higher-order moments of tangential velocity component; flatness factor.

suring techniques, were found for some of the measured quantities of the radial and tangential velocity components in the viscous sublayer and buffer region.

References

Kim, J., Moin, P., Moser, R. (1987): Turbulence statistics in a fully developed channel flow at low Reynolds number. Journal of Fluid Mechanics **177**, 133-166.

Lyous, S.L., Hanratty, T.J., McLaughlin, J. (1991): Large-scale computer simulation of fully developed turbulent channel flow with heat transfer. Int. Journal for Numerical Methods in Fluids **13**, 999-1928.

Durst, F., Martinuzzi, R., Sender, J., Thevenin, D. (1992): LDA measurements of mean velocity, RMS value, and higher-order moments of turbulence intensity fluctuations in flow fields with strong velocity gradients. Proceedings of Sixth International Symposium on Application of Laser Techniques to Fluid Mechanics, Lisabon, Portugal, Paper 5.1.

Lumley, J.L. (1970): *Stochastic Tools in Turbulence.* Academic Press, New York.

Kreplin, H., Eckelmann, H. (1979): Behavior of the three fluctuating velocity components in the wall region. The Physics of Fluids **22**, 1233.

Alfredson, P.H., Johansson, A.V., Haritonidis, J.H., Eckelmann, H. (1988): The fluctuating wall-shear stress and the velocity field in the viscous sublayer. The Physics of Fluids **31**, 1026-1033.

Karlsson, R.I., Johansson, T.G. (1988): LDV measurements of higher-order moments of velocity fluctuations in a turbulent boundary layer. *Laser Anemometry in Fluid Mechanics-III*, Edited by R.J. Adrian, Lisabon, Portugal, 273-289.

Niederschulte, N.A., Adrian, R.J., Hanratty, T.J. (1990): Measurements of turbulent flow in a channel at low Reynolds numbers. Experiments in Fluids **9**, 222-230.

Direct Numerical Simulation of Turbulent Plane Couette-Poiseuille Flows: Effect of Mean Shear Rate on the Near-Wall Turbulence Structures

Akiyoshi Kuroda[1], Nobuhide Kasagi[2], and Masaru Hirata[3]

[1] Department of Mechanical Engineering, Hokkaido University, Sapporo 060, Japan
[2] Department of Mechanical Engineering, The University of Tokyo, Bunkyo-ku, Tokyo 113, Japan
[3] Department of Mechanical Control Systems, Shibaura Institute of Technology, Omiya 330, Japan

Abstract

Direct numerical simulations (DNSs) of the fully developed turbulent flow between two parallel walls driven by a combination of a mean pressure gradient and the streamwise movement of one of the walls, i.e., turbulent Couette-Poiseuille flow, were carried out. The mean velocity gradient on the moving wall was reduced from that of an ordinary Poiseuille-type flow to almost zero. The turbulence statistics including the stress anisotropy and the budgets of the turbulence energy and Reynolds stresses are presented. The instantaneous flow fields visualized reveal that the structure of the near-wall turbulence under a weak mean shear does not exhibit any elongated streaks and is much different from that of the sheared wall layers.

Nomenclature

b_{ij}	stress anisotropy tensor, $\overline{u_i'u_j'} \big/ \overline{u_k'u_k'} - (1/3)\delta_{ij}$
k	turbulent kinetic energy, $\overline{u_i'u_i'}/2$
p	pressure
u, v, w	velocity components in the $x-$, $y-$ and $z-$directions
u_i	velocity component in the $i-$th direction; u_1, u_2 and u_3 denote u, v and w, respectively
u_τ	friction velocity
x, y, z	streamwise, wall-normal and spanwise coordinates
δ	channel half width
ν	kinematic viscosity
ε	dissipation rate of k
η_{w}	Kolmogorov length scale at the wall
υ_{w}	Kolmogorov velocity scale at the wall
$(\,)_{\mathrm{b}}$	normalized by bottom wall variables
$(\,)_{\mathrm{t}}$	normalized by top wall variables
$(\,)'$	fluctuating component or root mean square value
$(\,)^+$	normalized by u_τ and ν

1. Introduction

Near-wall turbulent flows arise under several dynamical effects. Recent direct numerical simulation (DNS) results of homogeneously sheared turbulence suggest that a key factor in the generation of near-wall coherent structures, e.g., streaky and vortical structures, is the high mean shear rate caused by the no-slip condition on a wall surface (Rogers and Moin 1987; Lee *et al.* 1990). Hence, it is obvious that it should be desirable to study possible dynamical effects individually, i.e., (1) viscosity (low Reynolds number effects), (2) blocking of wall-normal motion, (3) mean shear rate, in order to understand the basic mechanisms of near-wall turbulence phenomena. In the present study, we pay attention especially to effects of the mean shear rate. We carried out DNSs of turbulent plane Couette-Poiseuille flows in which the mean shear rate on one wall could be varied by changing relative wall velocities and a streamwise pressure gradient. With these DNS databases, turbulence statistics and structures under different wall shear rates were investigated. These databases should also be useful for developing and evaluating various one-point closure turbulence models.

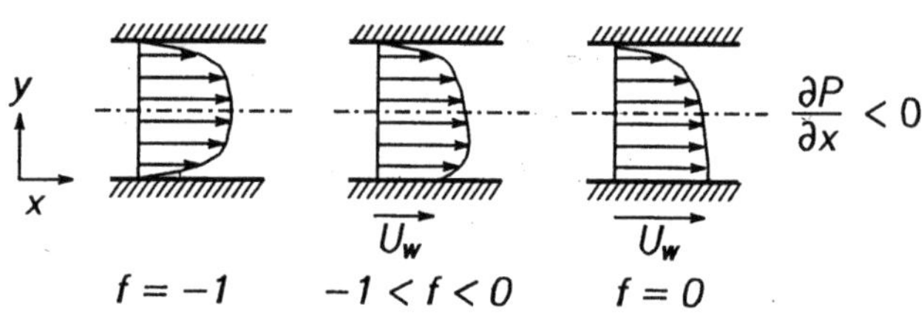

Fig. 1. Schematic of flows simulated.

2. Numerical Method

Schematic pictures of the flows simulated are shown in Fig. 1. The top wall is at rest, while the bottom wall is moving at a constant speed, U_w. The basic flow parameters are summarized in Table 1, where δ and u_τ denote the channel half width and the friction velocity, respectively. Four cases were simulated: case P is an ordinary channel flow, while cases CP's are turbulent Couette-Poiseuille flows. These parameters were chosen in order to realize different shear stresses at the bottom wall while keeping almost the same wall shear stress at the top wall (see Table 2).

Referring to the numerical method used by Kim *et al.* (1987), a forth-order partial differential equation for v, a second-order equation for the wall-normal component of vorticity and the continuity equation were used to solve the flow field. A spectral method with Fourier series in the streamwise (x) and spanwise (z) directions, and a Chebyshev polynomial expansion in the wall-normal direction were used. In all

Table 1. Basic flow conditions.

Case	$\frac{U_\mathrm{w}\,\delta}{\nu}$	$\frac{u_\tau\,\delta}{\nu}$	$-\frac{\delta}{\varrho U_\mathrm{w}^2}\frac{\partial P}{\partial x}$
P	0	150	—
CP1	1800	—	$4.34 \cdot 10^{-3}$
CP2	2640	—	$1.86 \cdot 10^{-3}$
CP3	3000	—	$1.33 \cdot 10^{-3}$

Table 2. Reynolds number based on wall friction velocity.

Case	$\frac{u_{\tau\mathrm{t}}\,\delta}{\nu}$	$\frac{u_{\tau\mathrm{b}}\,\delta}{\nu}$
P	150	150
CP1	148	79.0
CP2	152	49.0
CP3	154	17.7

cases, the computational periods were chosen to be $5\pi\delta$ and $2\pi\delta$ in the $x-$ and $z-$ directions, respectively. 128×128 Fourier modes and Chebyshev polynomials up to the order 96 were used in the wall-normal direction. The collocation grid used to compute the nonlinear terms had 1.5 times finer resolution in each direction to remove aliasing errors.

Two-point correlations and power spectra of velocity fluctuations were observed in order to check the validity of spatial period and resolution, respectively (Moser and Moin 1984). The power spectra satisfactorily decreased in the high wave number region in all cases, but the two-point correlation coefficient of u' did not decrease sufficiently and remained to be about 0.2 at half the length of the computational domain in cases CP2 and CP3. To study the effect of this unwanted feature on the numerical result, two additional computations were performed for case CP3. In one case, we adopted $64 \times 97 \times 64$ spectral modes with $5\pi\delta \times 2\pi\delta$ computational periods, and in the other case $128 \times 97 \times 128$ modes with $10\pi\delta \times 4\pi\delta$ periods. In these cases, the spacings in $x-$ and $z-$ directions were doubled. The two-point correlation did not decrease even for the latter case. Some differences were found in the calculated statistics such as mean velocity and turbulence intensities between case CP3 and the additional case. This would be mainly owing to the insufficient numerical resolution in the additional cases, because little difference was found between two additional cases.

For time integration, a second-order Adams-Bashforth scheme for the nonlinear terms and a Crank-Nicolson scheme for the viscous terms were adopted. The initial velocity field used in case P was obtained from the DNS of a lower Reynolds number channel flow (Kuroda *et al.* 1989). An instantaneous flow field in case P was modified to a Couette-type flow (the sign of fluctuations was changed in a half channel width) and used as the initial condition for case CP3. Case CP2 was developed from case CP3 and subsequently case CP1 was developed from case CP2. In each case,the statistically steady state was identified by a linear profile of the total shear stress and also by stationary behavior of the second moments of three velocity components at some distance from the wall.

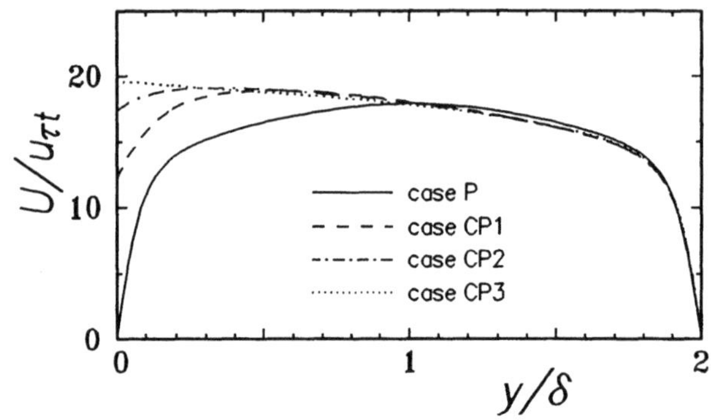

Fig. 2. Mean velocity profile.

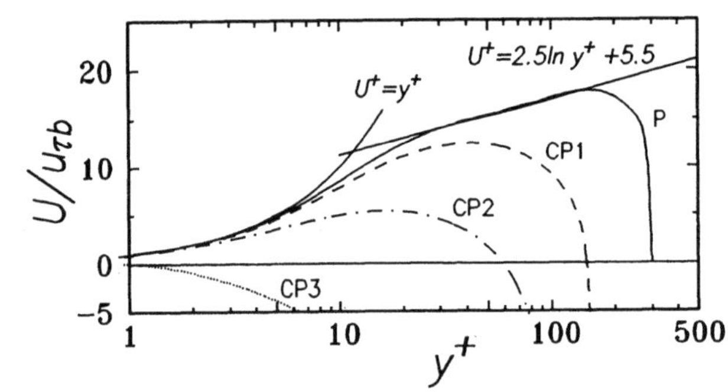

Fig. 3. Mean velocity distribution near the bottom wall.

3. Results and Discussion

3.1 Mean Velocity Field

The mean velocity distribution in each case is shown in Fig. 2, and the wall friction velocity calculated is summarized in Table 2. As mentioned previously, the friction on the top wall is kept almost the same throughout all cases, while that on the bottom

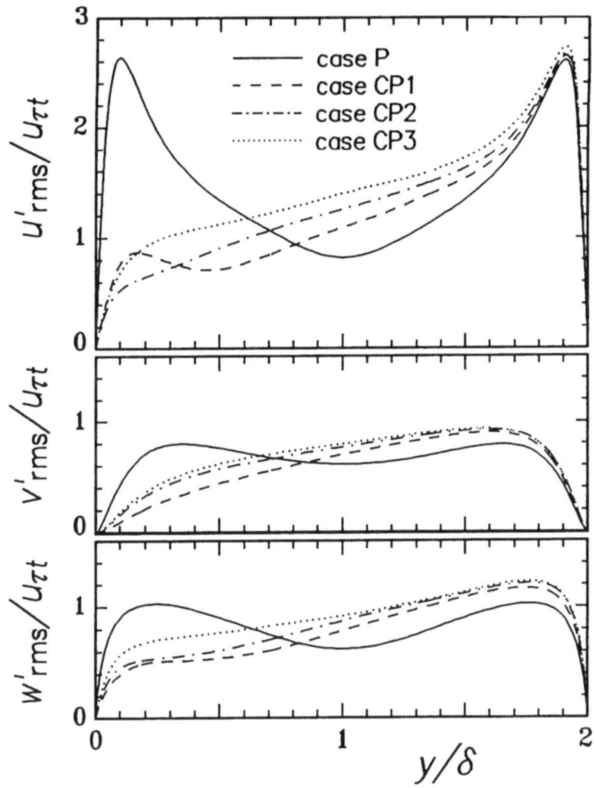

Fig. 4. R.m.s. velocity fluctuations.

wall is changed by controling the wall speed and the streamwise pressure gradient. Note that the mean shear rate becomes almost zero near the bottom wall in case CP3.

When plotted on the conventional semi-logarithmic figure (see Fig. 3), the profiles near the bottom wall show the following characteristics: (1) both the linear viscous sublayer and the logarithmic region exist in case P, (2) the linear region exists, but the logarithmic region does not in case CP1, (3) the linear region is restricted to $y^+ < 2$ in case CP2, (4) the linear region disappears in case CP3.

3.2 Reynolds Stresses

The root-mean-square values of fluctuating velocity components are shown in Fig. 4. In case CP1, the u' profile has a peak near the bottom wall but the magnitude is considerably small; $u'^+_{\text{peak}} = 1.6$ which is normalized by the bottom wall friction velocity and its location is $y^+ = 13$. In cases CP2 and CP3, the r.m.s. velocity fluctuations decrease monotonously towards the bottom wall. In these cases, the mean shear stress is too weak to dominate the local dynamical state near the bottom wall; it should be affected considerably by the turbulence generated near the top

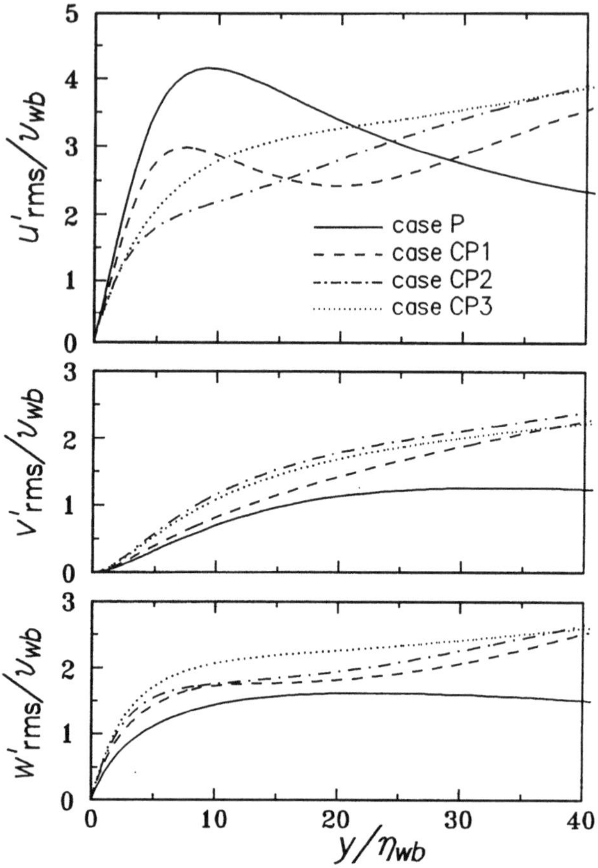

Fig. 5. R.m.s. velocity fluctuations normalized by Kolmogorov micro scale.

wall. However, this near-wall state is also affected by the blocking and viscous effects which are caused by non-permeable and no-slip conditions on the bottom wall. In this situation, the wall shear stress is no longer a useful scaling quantity. Hence, we chose the dissipation rate of the turbulent kinetic energy on the wall, which reflects the turbulence activity close to the wall, and the viscosity as a set of scaling properties, i.e., the Kolmogorov micro scale on the wall.

Figure 5 shows the new plot of the turbulence intensities with this Kolmogorov scaling. The agreement is largely improved if compared with the plot on the conventional wall coordinates, although the data do not fall into a single curve.

Figure 6 shows the anisotropy invariant map of the Reynolds stress tensor for case CP3, where $II = -(1/2)b_{ij}b_{ij}$ and $III = (1/3)b_{ij}b_{jk}b_{ki}$ (Lee and Reynolds 1985). Near the top wall where the strong mean shear exists, the component normal to the wall is suppressed and the Reynolds stresses approach the two- dimensional state as indicated by Mansour *et al.* (1988). On the other hand, the stresses approach the two-dimensional isotropic state near the bottom wall. This is caused by the

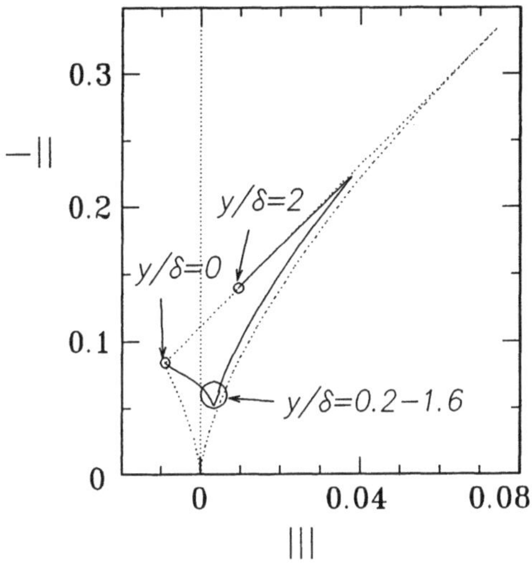

Fig. 6. Aisotropy invariant map (case CP3).

redistribution between the stress components; one is from the wall-normal component to the horizontal ones, i.e., the so-called splatting effect (Moin and Kim 1982), and the other is between the horizontal components themselves.

3.3 Budget of Turbulent Kinetic Energy

The budget of the turbulent kinetic energy near the bottom wall in each case is shown in Figs. 7(a) ~ (d). The terms are normalized by the bottom wall friction velocity and the kinematic viscosity in cases P and CP1, and by the dissipation rate at the bottom wall in cases CP2 and CP3, respectively. The profile of each term in case CP1 shows similarity to that in case P near the wall ($y^+ < 25$), though the magnitude in the latter is appreciably smaller. This suggests that the mechanism of turbulence production and transport in the sublayer to the buffer layer can exist even if the log-layer is distorted markedly as in case CP1. A similar observation was made experimentally by Maruyama and Tanaka (1987). They studied the inner-layer structure of wall turbulence in which the outer-layer is disturbed by a moving array of fences. They reported that the basic structure of the inner layer was little affected even when the tips of the disturbing fences were passing at a distance $y^+ = 45$. The result obtained in case CP1 is in qualitative accordance with their observation.

The budget terms in cases CP2 and CP3 behave in much different ways if compared with those in case P. In these cases, the diffusion terms are dominant instead of the production terms. The behavior of these terms indicate that the turbulence generated in the field far from the wall (mainly in the vicinity of the top wall) is transported to the near-wall region by the turbulence itself, and then closer to the wall by the pressure and viscous diffusions. Note that the pressure diffusion term,

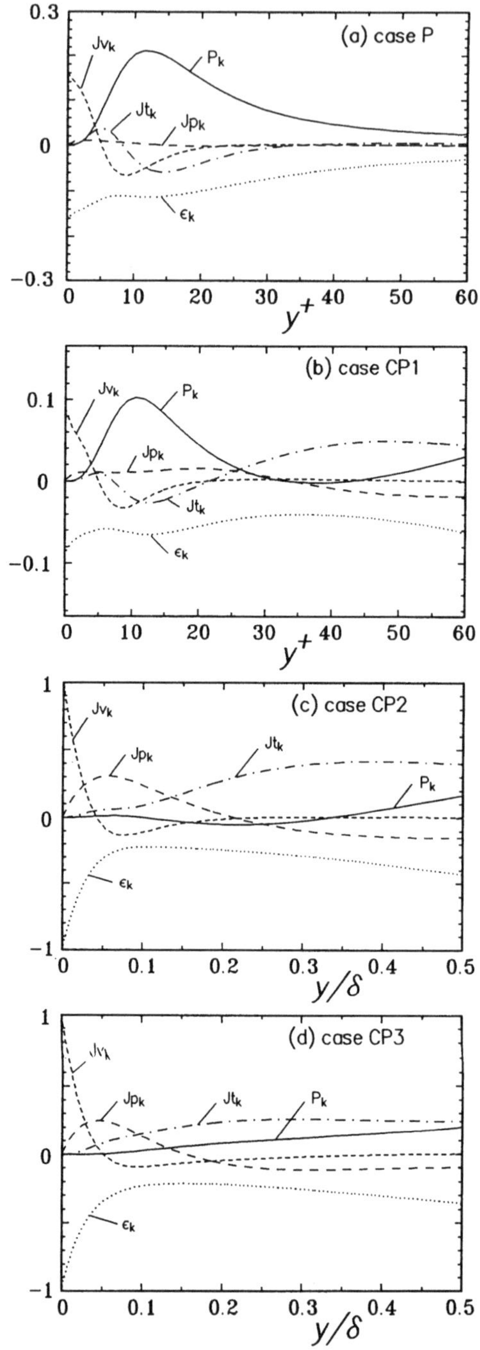

Fig. 7. Budget of the turbulent kinetic energy near the bottom wall; terms are normalized by the wall unit in cases P and CP1, and by the dissipation rate on the wall in cases CP2 and CP3. (P_k: production, ε_k: dissipation, Jv_k: viscous diffusion, Jt_k: turbulent diffusion, Jp_k: pressure diffusion).

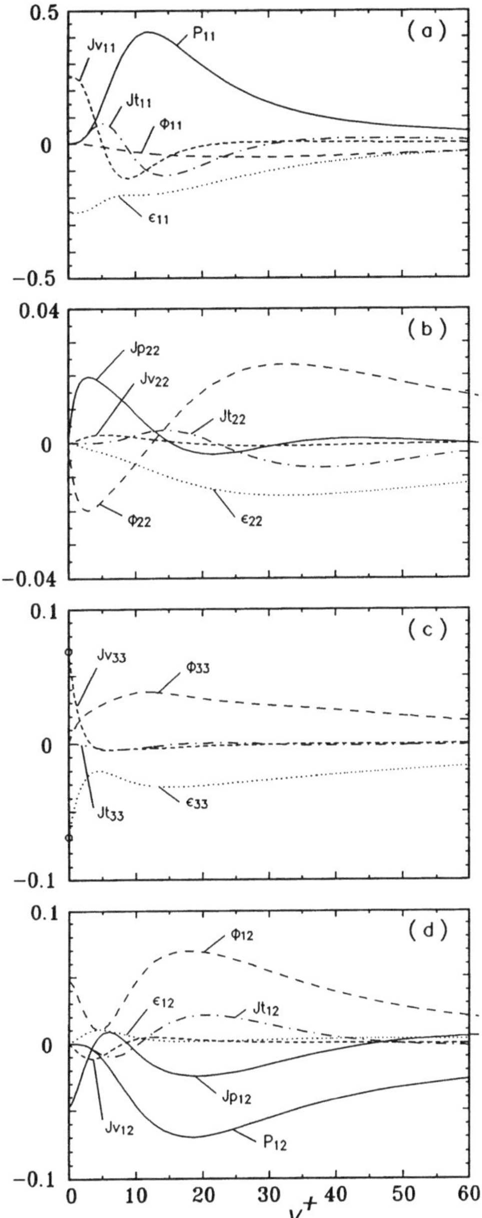

Fig. 8. Reynolds stress budget near the wall in case P; terms are normalized by u_τ and ν
(**a**) budget of $\overline{u'u'}$, (**b**) budget of $\overline{v'v'}$,
(**c**) budget of $\overline{w'w'}$, (**d**) budget of $\overline{u'v'}$
(P_{ij}: production, ε_{ij}: dissipation, Jv_{ij}: viscous diffusion, Jt_{ij}: turbulent diffusion, Jp_{ij}: pressure diffusion, ϕ_{ij}: pressure-strain correlation).

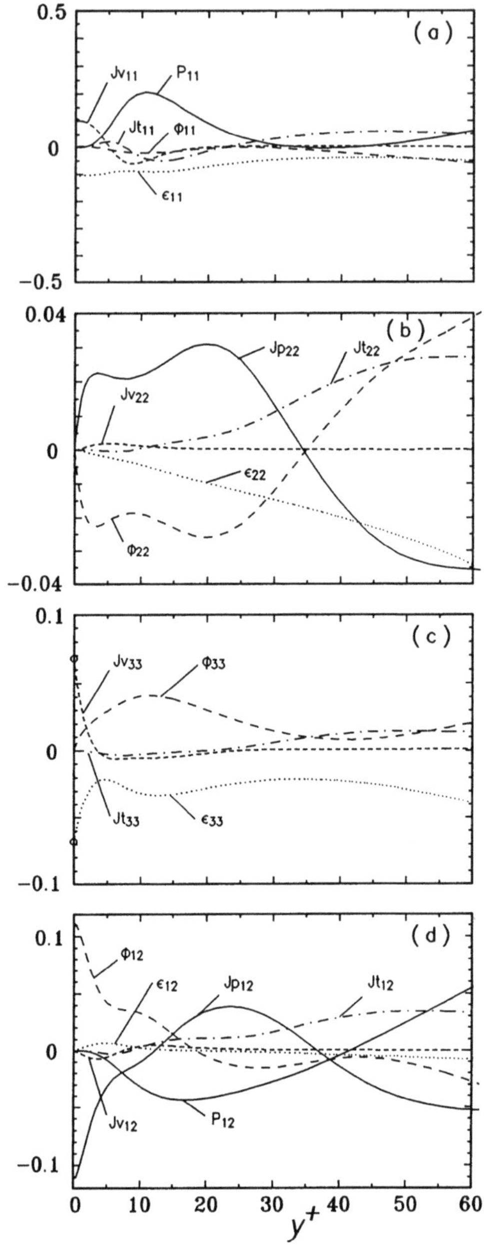

Fig. 9. Reynolds stress budget near the wall in case CP1;
terms are normalized by u_τ and ν
(**a**) budget of $\overline{u'u'}$, (**b**) budget of $\overline{v'v'}$,
(**c**) budget of $\overline{w'w'}$, (**d**) budget of $\overline{u'v'}$
(P_{ij}: production, ε_{ij}: dissipation, Jv_{ij}: viscous diffusion, Jt_{ij}: turbulent diffusion, Jp_{ij}: pressure diffusion, ϕ_{ij}: pressure-strain correlation).

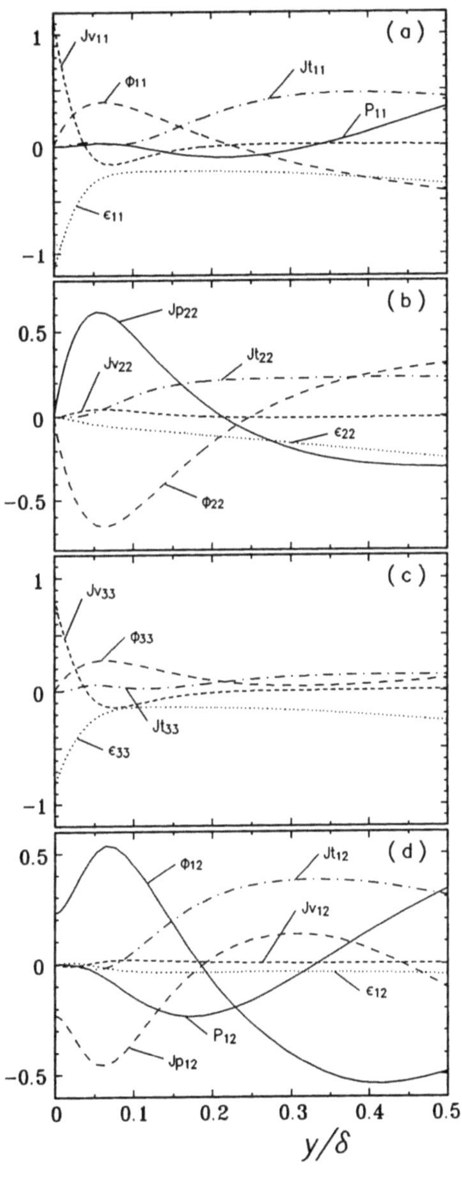

Fig. 10. Reynolds stress budget near the wall in case CP2;
terms are normalized by the dissipation rate of kinetic energy on the wall
(**a**) budget of $\overline{u'u'}$, (**b**) budget of $\overline{v'v'}$,
(**c**) budget of $\overline{w'w'}$, (**d**) budget of $\overline{u'v'}$
(P_{ij}: production, ε_{ij}: dissipation, Jv_{ij}: viscous diffusion, Jt_{ij}: turbulent diffusion, Jp_{ij}: pressure diffusion, ϕ_{ij}: pressure-strain correlation).

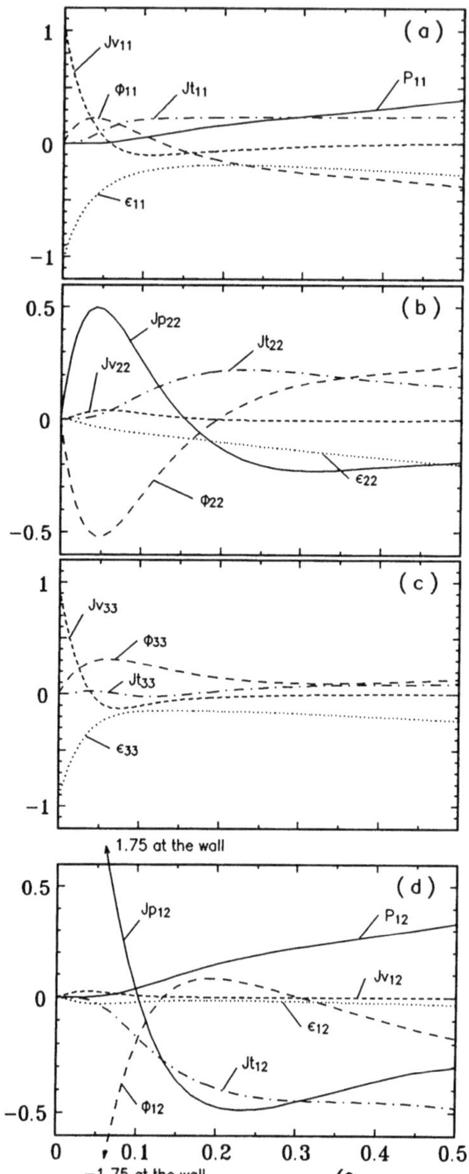

Fig. 11. Reynolds stress budget near the wall in case CP3;
terms are normalized by the dissipation rate of kinetic energy on the wall
(**a**) budget of $\overline{u'u'}$, (**b**) budget of $\overline{v'v'}$,
(**c**) budget of $\overline{w'w'}$, (**d**) budget of $\overline{u'v'}$
(P_{ij}: production, ε_{ij}: dissipation, Jv_{ij}: viscous diffusion, Jt_{ij}: turbulent diffusion, Jp_{ij}: pressure diffusion, ϕ_{ij}: pressure-strain correlation).

which is negligibly small in case P, plays an important role in these cases. The value of dissipation rate rapidly increases toward the wall. It has the local maximum value at the wall, which is about five times larger than the minimum value.

Negative production is found at $y^+ = 35 \sim 40$ in case CP1 and at $y/\delta = 0.1 \sim 0.34$ in case CP2. This is because the eddies coming from the top wall keep the positive correlation between u' and v' in these regions where the mean shear rate has a small positive value.

3.4 Budgets of Reynolds Stresses

The Reynolds stress budgets for each case are shown in Figs. 8 $\sim$ 11. It is confirmed that the result of case P agrees very well with that of Mansour *et al.* (1988).

Overall similarity is found between the terms in cases P and CP1. However, much difference in these two cases is noticed in the behavior of the pressure diffusion term and the pressure-strain correlation term. This should be because the local pressure fluctuation depends on the global state of the flow field (Kim 1989). The same trend is found in the comparison between cases CP2 and CP3. In every case, the pressure diffusion and pressure-strain correlation terms in the $\overline{v'v'}$-budget and also in the $\overline{u'v'}$-budget almost balance in the vicinity of the wall.

In the $\overline{u'u'}$-budget of case CP3 (Fig. 11(*a*)), the turbulent transport term and the pressure-strain term play important roles near the wall unlike the canonical channel flow where the production is dominant (Fig. 8(*a*)). The terms in the $\overline{w'w'}$-budget (Fig. 11(*c*)) have almost the same magnitude as those in the $\overline{u'u'}$-budget, but some differences are found between these two budgets, i.e., (1) the positive production term appears only in the $\overline{u'u'}$-budget, (2) the turbulent transport term in the $\overline{u'u'}$-budget is larger than that in the $\overline{w'w'}$-budget, and (3) the pressure-strain correlation term becomes negative at $y/\delta > 0.12$ in the $\overline{u'u'}$-budget while it is positive everywhere in the $\overline{w'w'}$-budget. These differences are owing to the anisotropic characteristics of the eddies coming from the top wall on which a high mean shear rate is exerted. In the $\overline{v'v'}$-budget (Fig. 11(*b*)), the pressure-strain correlation and pressure diffusion terms have a relatively large magnitude in the vicinity of the wall. The reversal of the sign of the pressure-strain correlation term near the wall was attributed by Moin and Kim (1982) to the splatting effect. This effect is enhanced and plays a primary role in the vicinity of the shear-free wall.

3.5 Turbulence Structures

Lee *et al.* (1990) claimed that 'shear parameter' $S^\star = (\partial U/\partial y)\overline{u_i'u_i'}/\varepsilon$ is a proper parameter which characterizes the type of turbulence structure which appears in a shear flow and showed that the streaky structures exist in a homogeneous shear flow without a wall when $S^\star > 30$. The shear parameters in the present simulations are shown in Fig. 12 and the contours of the streamwise velocity component in the $x - z$ plane near the bottom wall are shown in Figs. 13 $\sim$ 16. The elongated streaky structures are clearly observed in cases P and CP1, though they are somewhat

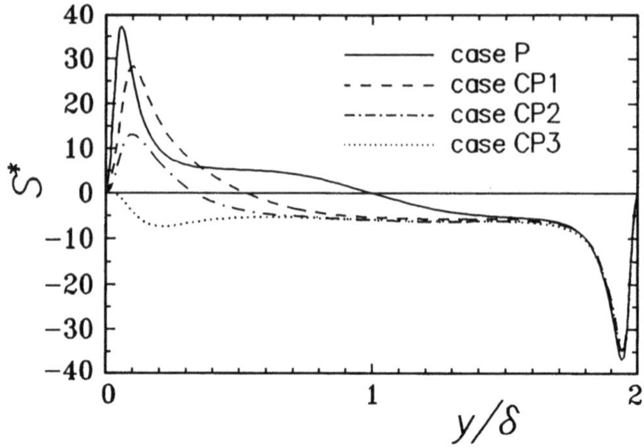

Fig. 12. Shear parameter; $S^* = (\partial U/\partial y)\overline{u_i'u_i'}/\varepsilon$.

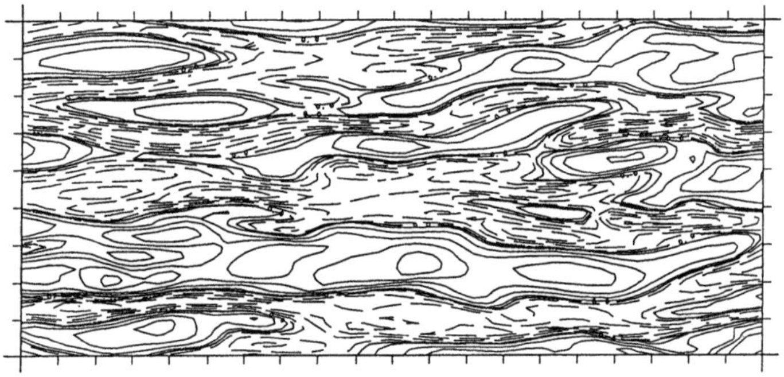

Fig. 13. Contours of streamwise velocity fluctuations in the $x-z$ plane ($1000\nu/u_\tau \times 450\nu/u_\tau$) at $y^+ = 13.4$ (case P) (solid line; positive value, broken line; negative value).

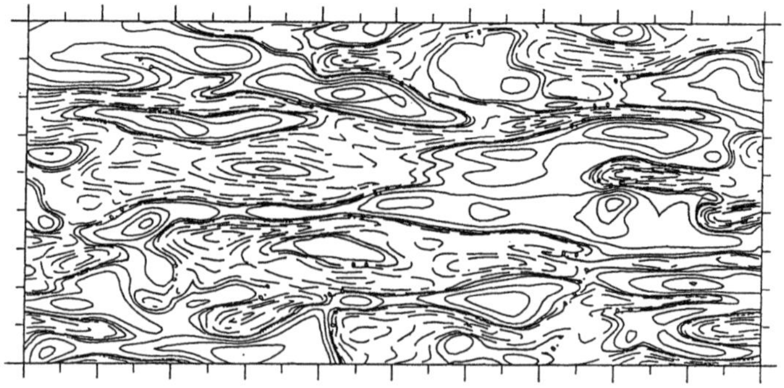

Fig. 14. Contours of streamwise velocity fluctuations in the $x-z$ plane ($1000\nu/u_\tau \times 450\nu/u_\tau$) at $y^+ = 13.3$ (case CP1).

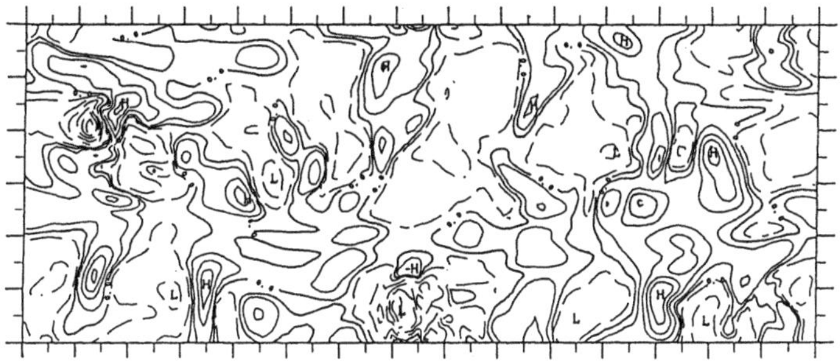

Fig. 15. Contours of streamwise velocity fluctuations in the $x - z$ plane ($15\delta \times 6\delta$) at $y/\delta =$ 0.054 (case CP2).

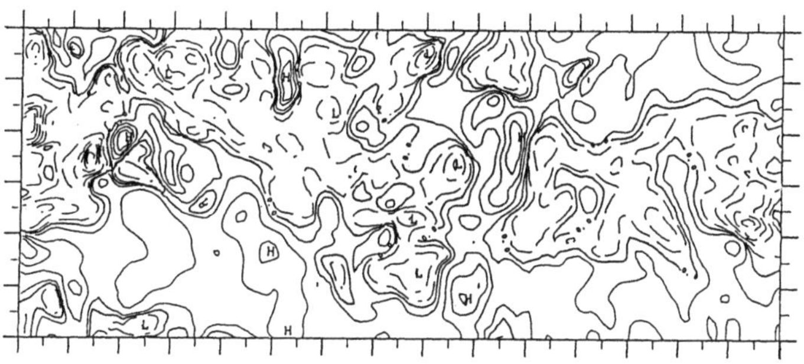

Fig. 16. Contours of streamwise velocity fluctuations in the $x - z$ plane ($15\delta \times 6\delta$) at $y/\delta =$ 0.054 (case CP3).

distorted in case CP1. The peak value of $S^\star$ near the bottom wall is about 38 in case P and 28 in case CP1, respectively. On the other hand, the streaky structures disappear in cases CP2 and CP3, where $S^\star$ is less than 20.

Figure 17 shows the velocity vectors in a near-wall horizontal plane in case CP3. A similar picture was also obtained in case CP2. Here, diverging vector patterns are found intermittently in space. It is separately confirmed that the fluid is moving toward the wall at the central part of these diverging patterns. This suggest that the redistribution of the Reynolds stresses near the wall (splatting effect) is associated with these patterns.

Figure 18 shows velocity vectors in a cross-stream plane noticed as A-B in Fig. 17. Strong inrush motions are clearly observed. In the vicinity of the wall, the wallward moving fluid changes abruptly its direction of motion and this results in a strongly strained velocity field, where large energy redistribution as well as dissipation occurs. However, it is also found that a strong outward motion is created by the strong inward motion. Thus, near the shear-free wall layer, the convective effect is considerably strong and only a fraction of the turbulent energy brought in by the eddies from the outer-layer is dissipated.

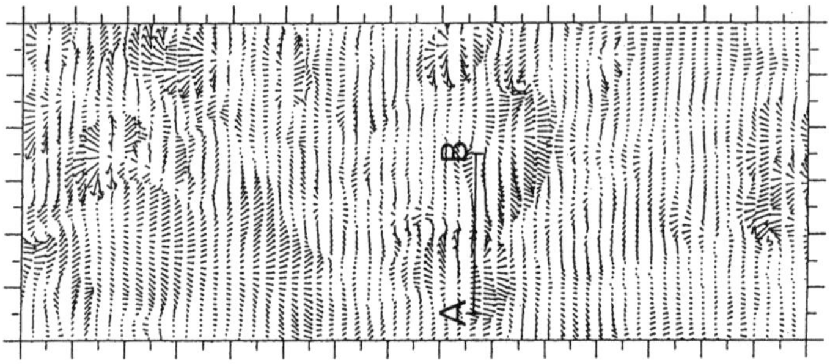

Fig. 17. Fluctuating velocity vectors in the $x - z$ plane at $y/\delta = 0.054$ (case CP3).

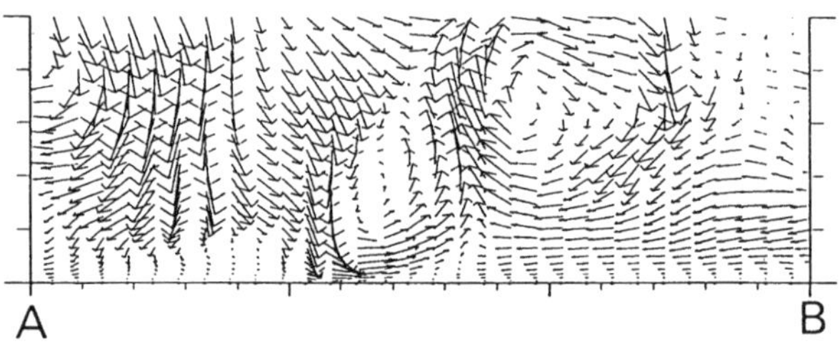

Fig. 18. Fluctuating velocity vectors in the cross-stream plane (case CP3) denoted as A-B in Fig. 17: the spanwise extent of the figure is 3δ and the wall-normal extent is δ.

4. Conclusions

The effects of the mean shear rate on the near-wall turbulent flow field were investigated by performing the direct numerical simulations of turbulent plane Couette-Poiseuille flow. As the moving speed of one of the two parallel walls was increased, the shear rate was reduced to various degrees. As a result, it was found that the structure of the shear-free wall layer is very much different from that of the sheared wall layers. The following conclusions are obtained:

(1) The near-wall streaky structures should disappear, when the mean shear rate near the wall is reduced considerably, i.e., in the present study, when the dimensionless shear parameter is less than about 20. In these cases, a typical flow pattern exists that a strong wallward fluid motion changes abruptly its direction with its impingement on the wall. This marked change in the turbulence structure leads to the emphasized 'splatting' effect in the basic turbulence mechanisms as noted below.

(2) The Reynolds stress state in the shear-free wall layer tends to be two-dimensional isotropic. This is owing to the redistribution of turbulent kinetic energy from the wall-normal to wall-parallel components, which is associated with the activated splatting motion.

(3) In the shear-free wall layer, both the pressure diffusion and turbulent diffusion terms are no longer negligible without the shear production term, and play a primary role in the turbulent energy and stress budgets.
(4) Even if the mean shear rate is reduced appreciably ($S^\star = 28$ in case CP1) so that the logarithmic law should not hold, the turbulence energy budget keeps structural similarity in its balance. However, the magnitude of each budget term becomes smaller.

References

Kim, J., Moin, P., Moser, R. (1987): Turbulence Statistics in Fully Developed Channel Flow at A Low Reynolds Number. J. Fluid Mech. **177**, 133-166
Kim, J. (1989): On the Structure of Pressure Fluctuations in Simulated Turbulent Channel Flow. J. Fluid Mech. **205**, 421-451
Kuroda, A., Kasagi, N., Hirata, M. (1989): A Direct Numerical Simulation of the Fully Developed Turbulent Channel Flow at a Very Low Reynolds Number. Numerical Methods in Fluid Dynamics **II** ed. by M. Yasuhara, H. Daiguji, K. Oshima (Japan Society of Computational Fluid Dynamics), 1012-1017
Lee, M., Kim, J., Moin, P. (1990): Structure of Turbulence at High Shear Rate. J. Fluid Mech. **216**, 561-583
Lee, M.J., Reynolds, W.C. (1985): *Numerical Experiments on the Structure of Homogeneous Turbulence.* Rep. № TF-24, Department of Mechanical Engineering, Stanford University
Mansour, N.N., Kim, J., Moin, P. (1988): Reynolds-Stress and Dissipation Rate Budgets in a Turbulent Channel Flow. J. Fluid Mech. **194**, 15-44
Maruyama, S., Tanaka, H. (1987): The Effect of Spatial Resolution on the Inner-Layer Structure of Wall Turbulence. J. Fluid Mech. **177**, 485-500
Moin, P., Kim, J. (1982): Numerical Investigation of Turbulent Channel Flow. J. Fluid Mech. **118**, 341-377
Moser, R.D., Moin, P. (1984): *Direct Numerical Simulation of Curved Turbulent Channel Flow. NASA TM-85974*
Rogers, M.M., Moin, P. (1987), The Structure of the Vorticity Field in Homogeneous Turbulent Flows. J. Fluid Mech. **176**, 33-66

Large Eddy Simulation of Turbulent Flow in Concentric Annuli with a Thin Inner Rod

Shin-ichi Satake[2] and Hiroshi Kawamura[1]

[1] Department of Mechanical Engineering, Faculty of Science and Engineering, Science University of Tokyo, Noda-shi, Chiba-ken, 278, Japan

[2] Present address: Department of Mechanical Engineering, The University of Tokyo, Hongo, Bunkyo-ku, 113, Japan

Abstract

The large eddy simulations are performed for annular flows with radius ratios of 0.1, 0.04, and 0.02. Comparison with the existing experimental data of Rehme revealed that the asymmetrical velocity field and the turbulence quantities are well reproduced by the present calculations. The production rate of turbulence energy is suppressed in the inner wall region; and thus the decrease in the peak of the turbulence energy is observed. The turbulence structures in the inner wall region are visualized by using computer graphics. The high and low speed streaks, which exist in the channel flow, are also observed over the surface of an inner rod. In the case of a thick rod, a lateral pairing of the ejection and the sweep is observed over an inner rod. In the case of a thin rod, a wake-like low pressure region is caused by a high speed streak on the opposite side of the high pressure region.

Nomenclature

A^+	coefficient in van Driest damping function
D	distance between the walls, $= R_2 - R_1$
P	modified pressure or production of Reynolds stress
p	pressure
r	distance from the axis
Re_m	bulk Reynolds number, $= u_m D/\nu$
Re_τ	friction Reynolds number, $= u_{\tau_{av}} D/\nu$
u, v, w	velocities in the directions of x, r, and θ
u_m	bulk mean velocity
$u_{\tau_{av}}$	friction velocity based on τ_{av}
u_{τ_1}	friction velocity on the inner wall
u_{τ_2}	friction velocity on the outer wall
x	axial direction
y	distance from the wall, $= r - R_1$
θ	azimuthal direction
τ	shear stress
τ_{av}	shear stress averaged over the inner and outer walls

ν	kinematic viscosity
ν_t	SGS (sub-grid scale) eddy viscosity
$(\ )_m$	position of maximum velocity
$(\ )_0$	position of zero shear stress
$(\ )_1$	inner wall
$(\ )_2$	outer wall
$(\ )^+$	nondimensionalized by ν and u_{τ_i} (i = 1, 2 or av)
$\overline{(\ )}$	GS (grid scale) quantities in Eqs. (1) to (5) or ensemble average

1. Introduction

The annular flow is often encountered in the various engineering fields such as a heat exchanger and a gas-cooled nuclear reactor. In the annular flow, the position of zero shear stress is not coincident with the position of maximum velocity. A relative decrease in the turbulent kinetic energy subjected to the strong shear on the inner wall is observed. These phenomena become more remarkable with decrease of the radius ratio.

An intensive experimental work was done by Rehme (1974) for the fully developed annular turbulent flow with a thin inner rod. Distributions of mean velocity and Reynolds stresses were measured in the annular turbulent flows with various radius ratios.

A direct numerical simulation (DNS) of axial turbulent flow over a long cylinder was performed by Neves et al. (1991). They studied the effect of transverse curvature and found boomerang-shaped structures with a large spanwise length scale. Large eddy simulations (LES) on the cylindrical coordinates were performed by Schumann (1975) for an annular tube and by Unger and Friedrich (1991) for a circular pipe. Both of the works employed the wall law as the wall boundary condition because of their relatively small radial mesh numbers.

In this paper, the LES's are performed for the annular turbulent flows with small radius ratio. The calculation procedure is based on the Schumann's numerical method but adopts the non-slip boundary condition on the wall. Particular attention is given to the flow structures in the inner wall region.

2. Numerical Procedure

The cylindrical coordinates are employed with x, r and θ in the mean flow, radial and azimuthal directions, respectively, as shown in Fig. 1. The quantities are nondimensionalized by the friction velocity averaged over the inner and outer walls $u_{\tau_{av}}$ and the distance between the walls $D = R_2 - R_1$. Then the continuity and the momentum equations in the cylindrical coordinates are

$$\frac{\partial \overline{u}}{\partial x} + \frac{1}{r}\frac{\partial (r\overline{v})}{\partial r} + \frac{1}{r}\frac{\partial \overline{w}}{\partial \theta} = 0, \tag{1}$$

$$
\begin{aligned}
\frac{\partial \overline{u}}{\partial t} + \frac{\partial \overline{u}\,\overline{u}}{\partial x} + \frac{1}{r}\frac{\partial r\overline{v}\,\overline{u}}{\partial r} + \frac{1}{r}\frac{\partial \overline{w}\,\overline{u}}{\partial \theta} = \\
- \; & \frac{\partial P}{\partial x} + \frac{\partial}{\partial x}\left[\left(\nu_t + \frac{1}{Re_\tau}\right)\left(\frac{\partial \overline{u}}{\partial x} + \frac{\partial \overline{u}}{\partial x}\right)\right] \\
+ \; & \frac{1}{r}\frac{\partial}{\partial r}\left[r\left(\nu_t + \frac{1}{Re_\tau}\right)\left(\frac{\partial \overline{v}}{\partial x} + \frac{\partial \overline{u}}{\partial r}\right)\right] \\
+ \; & \frac{1}{r}\frac{\partial}{\partial \theta}\left[\left(\nu_t + \frac{1}{Re_\tau}\right)\left(\frac{1}{r}\frac{\partial \overline{u}}{\partial \theta} + \frac{\partial \overline{w}}{\partial x}\right)\right] + 2,
\end{aligned}
\tag{2}
$$

$$
\begin{aligned}
\frac{\partial \overline{v}}{\partial t} + \frac{\partial \overline{u}\,\overline{v}}{\partial x} + \frac{1}{r}\frac{\partial r\overline{v}\,\overline{v}}{\partial r} + \frac{1}{r}\frac{\partial \overline{w}\,\overline{v}}{\partial \theta} - \frac{\overline{w}^2}{r} = \\
- \; & \frac{\partial P}{\partial r} + \frac{\partial}{\partial x}\left[\left(\nu_t + \frac{1}{Re_\tau}\right)\left(\frac{\partial \overline{v}}{\partial x} + \frac{\partial \overline{u}}{\partial r}\right)\right] \\
+ \; & \frac{1}{r}\frac{\partial}{\partial r}\left[r\left(\nu_t + \frac{1}{Re_\tau}\right)\left(\frac{\partial \overline{v}}{\partial r} + \frac{\partial \overline{v}}{\partial r}\right)\right] \\
+ \; & \frac{1}{r}\frac{\partial}{\partial \theta}\left[\left(\nu_t + \frac{1}{Re_\tau}\right)\left(\frac{1}{r}\frac{\partial \overline{v}}{\partial \theta} + r\frac{\partial}{\partial r}\frac{\overline{w}}{r}\right)\right] \\
- \; & \frac{1}{r}\left[2\left(\nu_t + \frac{1}{Re_\tau}\right)\left(\frac{1}{r}\frac{\partial \overline{w}}{\partial \theta} + \frac{\overline{v}}{r}\right)\right],
\end{aligned}
\tag{3}
$$

$$
\begin{aligned}
\frac{\partial \overline{w}}{\partial t} + \frac{\partial \overline{u}\,\overline{w}}{\partial x} + \frac{1}{r}\frac{\partial r\overline{w}\,\overline{v}}{\partial r} + \frac{1}{r}\frac{\partial \overline{w}\,\overline{w}}{\partial \theta} + \frac{\overline{v}\,\overline{w}}{r} = \\
- \; & \frac{1}{r}\frac{\partial P}{\partial \theta} + \frac{\partial}{\partial x}\left[\left(\nu_t + \frac{1}{Re_\tau}\right)\left(\frac{1}{r}\frac{\partial \overline{u}}{\partial \theta} + \frac{\partial \overline{w}}{\partial x}\right)\right] \\
+ \; & \frac{1}{r}\frac{\partial}{\partial r}\left[r\left(\nu_t + \frac{1}{Re_\tau}\right)\left(\frac{1}{r}\frac{\partial \overline{v}}{\partial \theta} + r\frac{\partial}{\partial r}\frac{\overline{w}}{r}\right)\right] \\
+ \; & \frac{1}{r}\frac{\partial}{\partial \theta}\left[2\left(\nu_t + \frac{1}{Re_\tau}\right)\left(\frac{1}{r}\frac{\partial \overline{w}}{\partial \theta} + \frac{\overline{v}}{r}\right)\right] \\
+ \; & \frac{1}{r}\left[\left(\nu_t + \frac{1}{Re_\tau}\right)\left(\frac{1}{r}\frac{\partial \overline{v}}{\partial \theta} + r\frac{\partial}{\partial r}\frac{\overline{w}}{r}\right)\right].
\end{aligned}
\tag{4}
$$

Here, ν_t is the sub-grid scale (SGS) eddy viscosity based on the Smagorinsky model (1963):

$$
\begin{aligned}
\nu_t = \; & (c_s f_s \Delta)^2 \left[2\left\{ \left(\frac{\partial \overline{u}}{\partial x}\right)^2 + \left(\frac{\partial \overline{v}}{\partial r}\right)^2 + \left(\frac{1}{r}\frac{\partial \overline{w}}{\partial \theta} + \frac{\overline{v}}{r}\right)^2 \right\} \right. \\
& \left. + \left(\frac{\partial \overline{v}}{\partial x} + \frac{\partial \overline{u}}{\partial r}\right)^2 + \left(\frac{1}{r}\frac{\partial \overline{u}}{\partial \theta} + \frac{\partial \overline{w}}{\partial x}\right)^2 + \left(\frac{1}{r}\frac{\partial \overline{v}}{\partial \theta} + r\frac{\partial}{\partial r}\frac{\overline{w}}{r}\right)^2 \right]^{1/2},
\end{aligned}
\tag{5}
$$

where the filtering width Δ can be expressed as

$$
\Delta = (\triangle x \cdot \triangle r \cdot r\triangle\theta)^{1/3}, \tag{6}
$$

The near-wall dumping function of van Driest (1956), i.e.,

$$f_s = 1 - \exp\left(-\frac{y^+}{A^+}\right), \tag{7}$$

is multiplied to the filtering width. The Smagorinsky constant c_s is 0.1 as often used for the channel flows. The common value adopted for the empirical coefficient A^+ is 25. A slight modification, however, has been found to give a better result; that is, $A^+ = 25$ for the inner surface and $A^+ = 15$ for the outer one.

3. Calculation Methods

The momentum equations (2) to (4) describe the grid scale velocity and pressure fields in a staggered grid. Non-slip boundary conditions are imposed on the velocity at the boundaries. The velocity filed is integrated in time with the second-order Adams-Bashforth scheme. The projection method by Chorin (1968) leads to a Poisson equation for the pressure, which is solved with a direct method. That is, Fast Fourier transformation in the axial and azimuthal directions reduces the 3D problem to a set of 1D problems, which are then solved by the tridiagonal matrix algorithm.

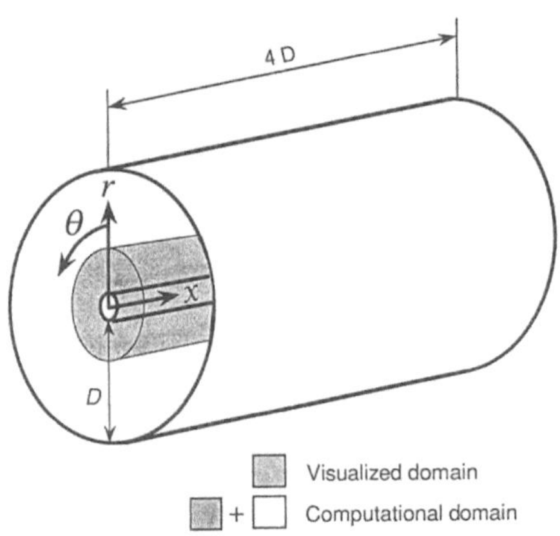

Fig. 1. The coordinates and computational domain.

4. Calculation Conditions

A calculated domain is shown in Fig. 1. Three radius ratios are studied. The computational parameters are given in Table 1. The Reynolds numbers Re_τ are selected to reproduce the experiments by Rehme (1975). Further details on the effect of the azimuthal mesh size can be found in Kawamura and Satake (1993).

Table 1. Computational parameters.

Case	R_1/R_2	Re_τ	Mesh ($x \times r \times \theta$)
Case 1	0.1	895	$64 \times 50 \times 256$
Case 2	0.04	591	$32 \times 50 \times 128$
Case 3	0.02	1080	$64 \times 50 \times 128$

The time increment Δt is 0.000 2 for all cases. About 6000 time steps were required to obtain the time averaged turbulence quantities. Numerical calculations were performed on VP-2600. Typical CPU time to calculate one case was ten to twenty hours.

5. Results and Discussions

5.1 Mean Velocity

The present results of the streamwise velocity profiles in the inner and outer regions are compared with the experimental results of Rehme (1974) in Figs. 2 to 4 for the three radius ratios investigated.

The agreement between the calculational and the experimental results is rather marginal in the outer region. This is probably because the mesh size in the outer region is inevitably large in the cylindrical coordinates. Thus, attention is mainly paid on the inner region in the rest of the paper.

In the inner region, good agreements between the calculational and experimental velocities are obtained. Among the three cases, the discrepancy is most pronounced in Case 1. In this case of experiment by Rehme (1974), spacers are employed to position the inner rod at the center of the tube. This may affect the velocity profile in the vicinity of the inner rod.

Vilemas et al. (1987) presented the mean velocity profile by use of mixing-length model considering the effects of pressure gradient and wall curvature on the velocity profile:

$$u^+ = \int_0^{y^+} \frac{2\mathrm{d}y^+}{\dfrac{R_1^+ + y^+}{R_1^+} + \sqrt{\left(\dfrac{R_1^+ + y^+}{R_1^+}\right)^2 + 4\left(\dfrac{R_1^+ + y^+}{R_1^+}\right)\kappa^2 y^{+2}\left(1 - \mathrm{e}^{-y^+/A^+}\right)^2}} \quad .(8)$$

This equation is also plotted in Figs. 2 to 4 with a solid line. The result lies lower than the ordinary universal velocity profile. This is because of the "cylindrical coordinates effect". The present results agree well with the above equation for the region close to the wall. In the region of the logarithmic law, however, the present results are fairly higher than the above equation. This is due to the so-called "laminarization effect", which will be discussed later.

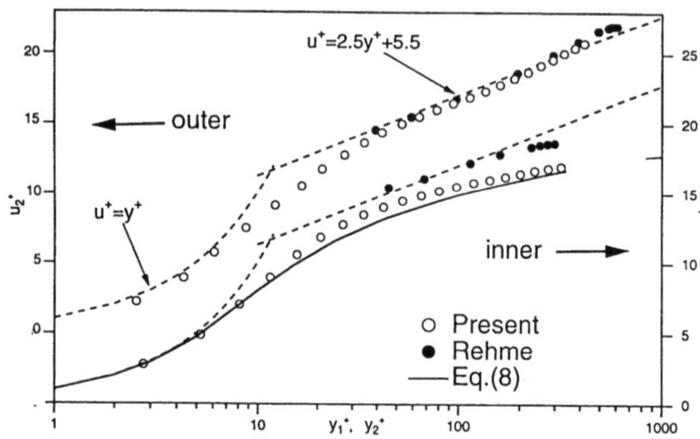

Fig. 2. Mean velocity (Case 1).

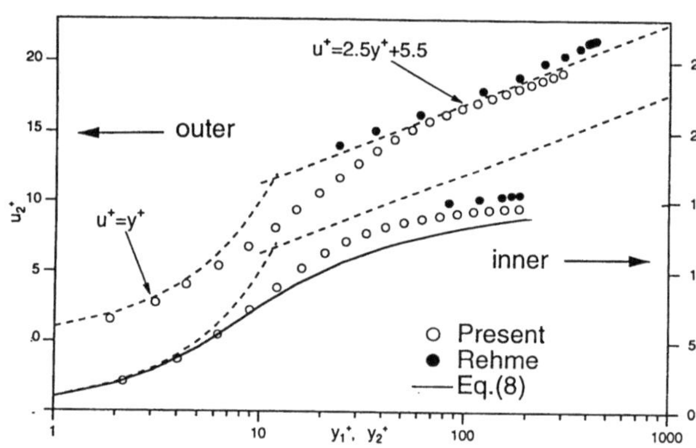

Fig. 3. Mean velocity (Case 2).

5.2 Reynolds Shear Stress and Turbulence Intensities

Figure 5 shows the total and the Reynolds shear stresses (GS+SGS) from the wall surface up to the position of zero shear stress. The stresses are normalized by the inner wall friction velocity (u_{τ_1}). The effect of radius ratio on the stress distributions is evident. The DNS of turbulent axial flow over a cylinder by Neves et al. (1991) shows a similar trend.

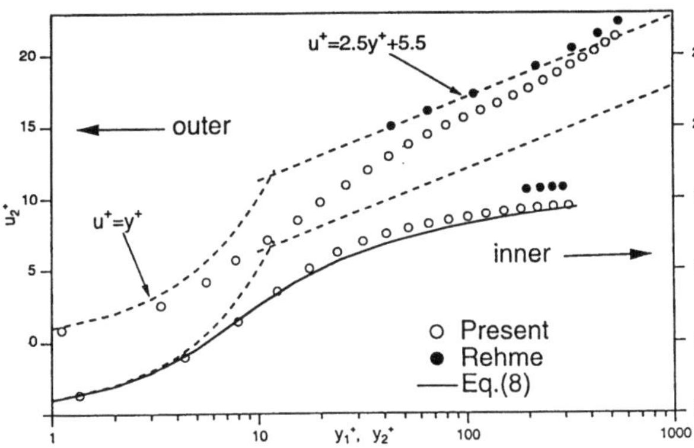

Fig. 4. Mean velocity (Case 3).

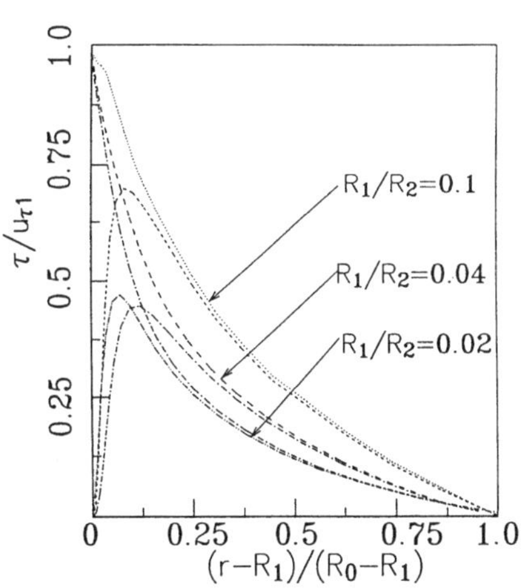

Fig. 5. Total Shear stress and Reynolds stress (GS+SGS).

Turbulent intensities normalized by $u_{\tau_{av}}$ are shown in Figs. 6 to 8. Anisotropy among the turbulence components is pronounced in the inner region. The streamwise component is the most energetic, because it is produced directly by the mean shear stress. Three components have a minimum point, which moves slightly towards the inner wall with decrease of the radius ratio. This is well reproduced by the calculation. An agreement with the experiment is relatively poor in the outer region. This is again because of the large spanwise mesh size in the outer region. The best agreement among three cases is obtained in Case 1, where the largest mesh number is employed.

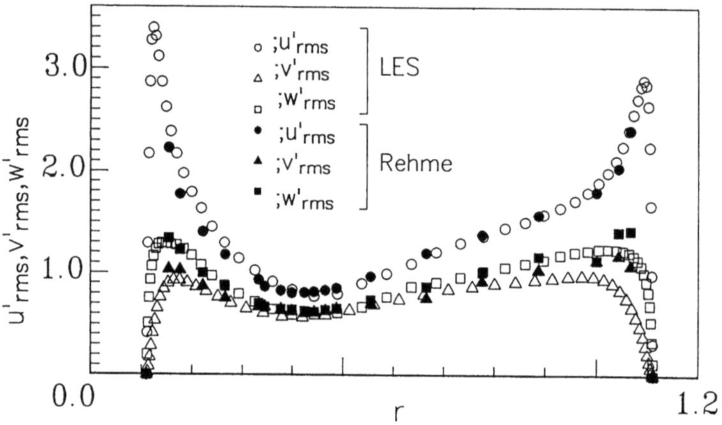

Fig. 6. Turbulent intensity (Case 1).

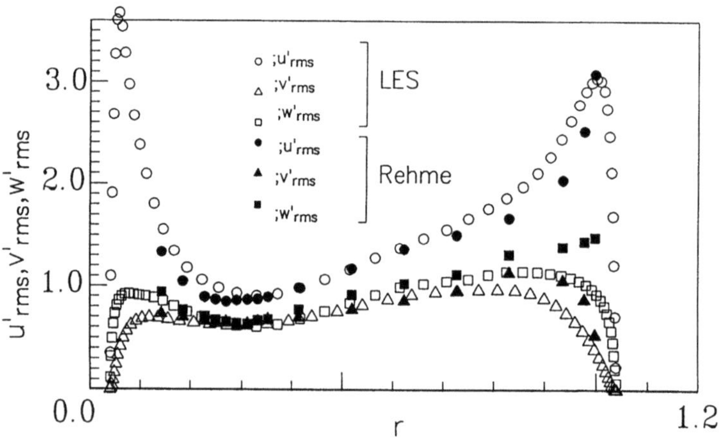

Fig. 7. Turbulent intensity (Case 2).

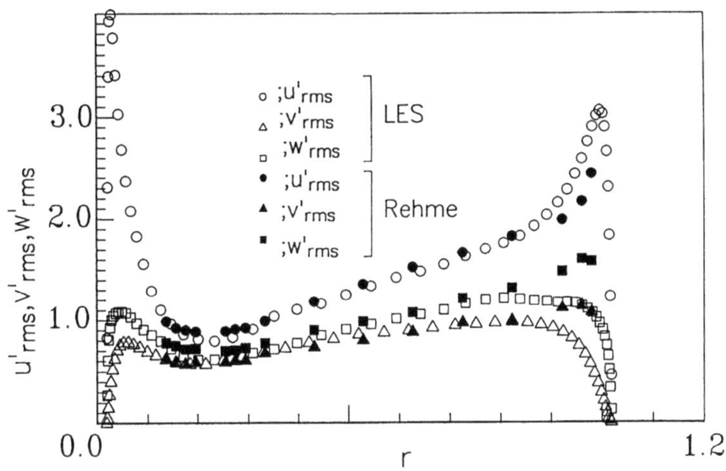

Fig. 8. Turbulent intensity (Case 3).

5.3 Positions of Maximum Velocity and Zero Shear Stress

The radial positions of the maximum velocity and the zero shear stress are in coincidence in the laminar annular flow but not in the turbulent one. Those are shown in Fig. 9. A calculation with a $k - \varepsilon$ model by Kawamura (1979) is also given in the same figure. Since the $k - \varepsilon$ model is based on the eddy viscosity concept, the two positions are forced to be coincident. The present results by LES resolve the non-coincidence between the two positions.

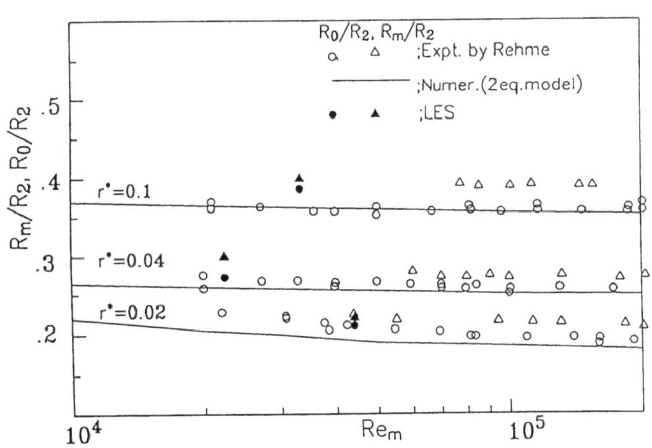

Fig. 9. Radial positions of maximum velocity and zero shear stress.

5.4 Turbulent Kinetic Energy and Production

Figures 10 and 11 show the turbulent kinetic energies and their production rates in the inner region for the three radius ratios. These figures indicate that the peaks of the turbulent kinetic energy and the production rate are strongly reduced with decrease of the radius ratio. The ones for the radius ratios of 0.04 and 0.02 indicate no appreciable difference; this is because the Reynolds number for the latter is much higher than that for the former (see Table 1). The effects of the radius ratio and the Reynolds number on the turbulent production rate are analyzed below.

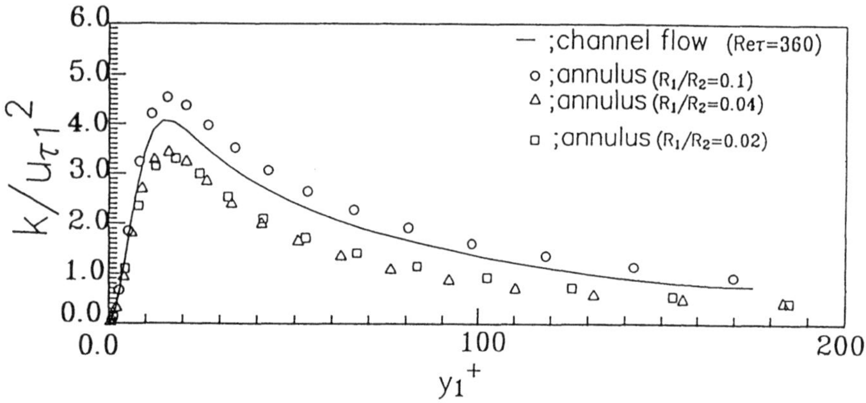

Fig. 10. Turbulent kinetic energy (inner wall region).

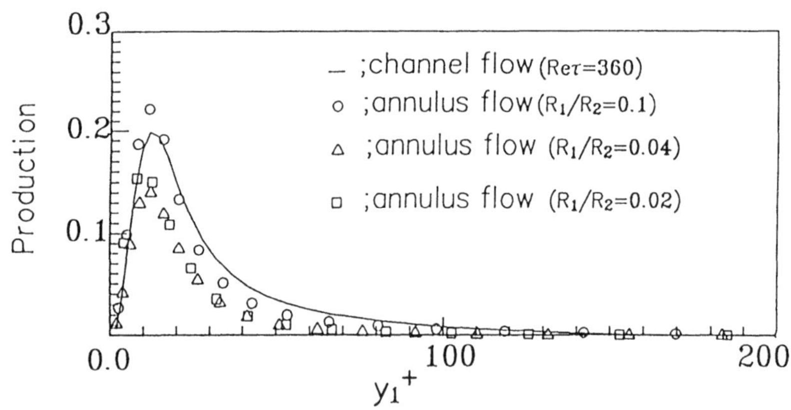

Fig. 11. Production rate of turbulent kinetic energy (inner wall region).

5.5 Laminarization Tendency

Shear Stress Distribution. The flow geometry and the coordinates system are shown in Fig. 12.

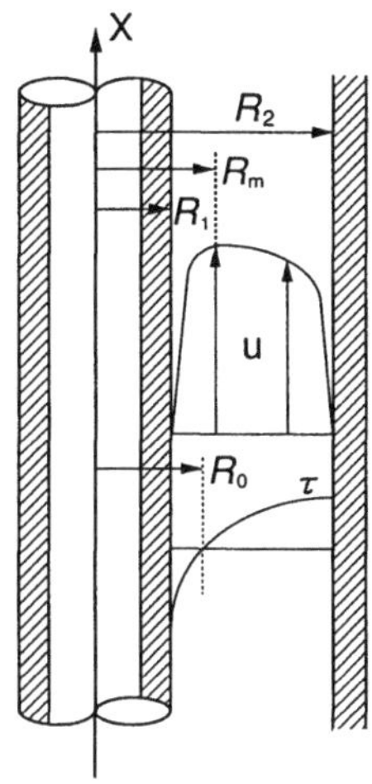

Fig. 12. Schematic of concentric annulus flow.

The relation of the shear stress distribution τ and the pressure drop for a turbulent annulus flow is expressed as

$$\frac{1}{r}\frac{\mathrm{d}}{\mathrm{d}r}\left[r\tau\right] = -\frac{\mathrm{d}p}{\mathrm{d}x}, \tag{9}$$

where the shear stress τ is given by

$$\frac{\tau}{\varrho} = \nu\frac{\mathrm{d}\overline{u}}{\mathrm{d}y} - \overline{u'v'}. \tag{10}$$

Here, the Reynolds stress is approximated as $\left(-\beta_1 y^{+3} + \beta_2 y^{+4}\right)$, and the constants β_1 and β_2 are obtained from the DNS data for the channel flow by Kim et al. (1987) as $\beta_1 = 7 \times 10^{-4}$ and $\beta_2 = 3.5 \times 10^{-5}$.

With integration of Eq. (9) from R_1 to R_2, it becomes

$$-\frac{\mathrm{d}p}{\mathrm{d}x} = \frac{2}{R_2 - R_1}\underbrace{\frac{\tau_2 R_2 - \tau_1 R_1}{R_2 + R_1}}_{\tau_{av}}. \tag{11}$$

Using $D = R_2 - R_1$ and τ_{av}, Eq. (11) is rewritten as

$$-\frac{\mathrm{d}p}{\mathrm{d}x} = 2\frac{\tau_{av}}{D}. \tag{12}$$

To obtain the shear stress distribution close to the inner wall, Eq. (9) is integrated from R_I to r ($r < R_0$):

$$\underbrace{\tau}_{\text{negative}} = \underbrace{\tau_\mathrm{I}\frac{R_\mathrm{I}}{r}}_{\text{negative}} \underbrace{-\frac{1}{2}\frac{\mathrm{d}p}{\mathrm{d}x}r}_{\text{positive}} \underbrace{+\frac{1}{2}\frac{\mathrm{d}p}{\mathrm{d}x}\frac{R_\mathrm{I}{}^2}{r}}_{\text{negative}} . \tag{13}$$

With the use of $r = R_\mathrm{I} + y$, Eq. (13) is reduced to

$$\begin{aligned} \tau &= \tau_\mathrm{I}\frac{R_\mathrm{I}}{R_\mathrm{I}+y} - \frac{\tau_\mathrm{av}}{D}(R_\mathrm{I}+y) + \frac{\tau_\mathrm{av}}{D}\frac{R_\mathrm{I}{}^2}{R_\mathrm{I}+y} \\ &= \tau_\mathrm{I}\left[\underbrace{\frac{R_\mathrm{I}}{R_\mathrm{I}+y}}_{\mathbf{A}} - \left(\frac{\tau_\mathrm{av}}{\tau_\mathrm{I}}\right)\frac{1}{D}\frac{2R_\mathrm{I}y+y^2}{R_\mathrm{I}+y}\right] . \end{aligned} \tag{14}$$

Here, the term **A** decreases with decrease of R_I for a given distance from the wall y.

Combining Eqs. (10) and (14), the following expression for the velocity gradient is obtained

$$\begin{aligned} \nu\frac{\mathrm{d}\overline{u}}{\mathrm{d}y} &= \frac{\tau}{\varrho} - \underbrace{(-\overline{u'v'})}_{\text{positive}} \\ &= \frac{\tau_\mathrm{I}}{\varrho}\left[\frac{R_\mathrm{I}}{R_\mathrm{I}+y} - \left(\frac{\tau_\mathrm{av}}{\tau_\mathrm{I}}\right)\frac{1}{D}\frac{2R_\mathrm{I}y+y^2}{R_\mathrm{I}+y}\right] - (-\overline{u'v'}) . \end{aligned} \tag{15}$$

The above equation is nondimensionarized with use of u_{τ_I} and ν as

$$\frac{\mathrm{d}\overline{u}^+}{\mathrm{d}y^+} = \left[\underbrace{\frac{R_\mathrm{I}^+}{R_\mathrm{I}^+ + y^+}}_{\mathbf{B}} - \underbrace{\frac{1}{Re_\tau{}'}\frac{2R_\mathrm{I}^+y^+ + y^{+2}}{R_\mathrm{I}^+ + y^+}}_{\mathbf{C}}\right] - \left(-\overline{u^{+\prime}v^{+\prime}}\right) , \tag{16}$$

where

$$Re_\tau{}' = \frac{Du_{\tau_\mathrm{I}}}{\nu}\left(\frac{\tau_\mathrm{I}}{\tau_\mathrm{av}}\right) = \underbrace{\frac{Du_{\tau_\mathrm{av}}}{\nu}}_{Re_\tau}\left(\frac{u_\mathrm{I}}{u_{\tau_\mathrm{av}}}\right)\left(\frac{\tau_\mathrm{I}}{\tau_\mathrm{av}}\right) = Re_\tau\left(\frac{\tau_\mathrm{I}}{\tau_\mathrm{av}}\right)^{1.5} . \tag{17}$$

Equation (16) shows that, in the vicinity of the wall, the term **B** becomes dominant with decrease of $R_\mathrm{I}{}^+$, while the term **C** with decrease of $Re_\tau{}'$.

Turbulence Production Rate. The production term of the turbulent kinetic energy k is given as

$$
\begin{aligned}
P^+ &= -\overline{u^{+\prime} v^{+\prime}} \frac{d\overline{u}^+}{dy^+} \\
&= \left(\beta_1 y^{+3} - \beta_2 y^{+4}\right) \left\{ \frac{R_1^+}{R_1^+ + y^+} - \frac{1}{Re_\tau{}'} \frac{2R_1^+ y^+ + y^{+2}}{R_1^+ + y^+} \right. \\
&\quad \left. - \beta_1 y^{+3} + \beta_2 y^{+4} \right\} ,
\end{aligned}
\tag{18}
$$

where u_{τ_1} is the friction velocity on the inner wall and $y^+ = y u_{\tau_1} / \nu$. The inner wall shear stress τ_1 can be expressed by means of a force balance as

$$
\tau_1 = \left| \frac{dp}{dx} \right| \frac{R_0{}^2 - R_1{}^2}{2R_1} . \tag{19}
$$

To obtain the inner wall shear stress, the position of zero shear stress R_0 is required. Hudina (1979) suggested that R_0 for high Reynolds numbers can be approximated as

$$
\frac{R_{0,\infty}}{R_1} = \frac{(R_2/R_1)^n + (R_2/R_1)}{(R_2/R_1)^n + 1} \quad (Re \rightarrow \infty). \tag{20}
$$

Here, the exponent n is found empirically to be 0.36. The radius R_0 for the laminar flow is obtained analytically as

$$
\frac{R_{0,L}}{R_1} = \sqrt{\frac{(R_2/R_1) - 1}{2 \ln(R_2/R_1)}} \quad \text{(Laminar)}. \tag{21}
$$

For intermediate Reynolds numbers, an empirical equation proposed by Kawamura (1972) gives a good fit of the zero-shear positions:

$$
\frac{R_0}{R_1} = \frac{R_{0,\infty}}{R_1} + \left(\frac{R_{0,L}}{R_1} - \frac{R_{0,\infty}}{R_1} \right) \left[\frac{2300}{Re_m} \right]^{0.8} \quad (Re_m > 2300). \tag{22}
$$

The maximum production rates calculated with equation (18) are shown in Fig. 13 in comparison with the LES calculations. A good agreement is obtained. This indicates that the decrease in the turbulent kinetic energy in the inner region is primarily caused by the so called "cylindrical coordinates effect". That is, in the cylindrical coordinates, the shear stress and thus the velocity gradient decrease rapidly with the distance from the wall (see Fig.5). Tanaka et al. (1982) suggested that the laminarizations in the accelerating boundary layer and in the heated gas flow are caused by the large shear stress gradient close to the wall. The present study provides that the decrease of turbulent intensities on a thin rod is caused by the same process.

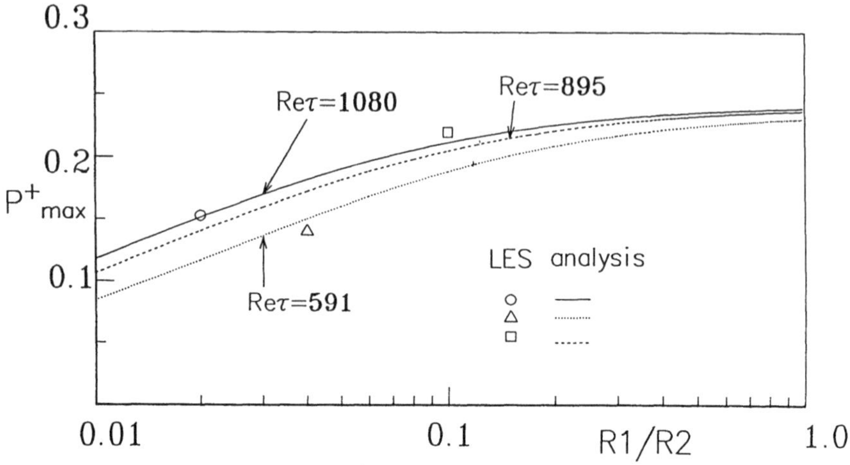

Fig. 13. Maximum turbulent production rates.

5.6 Flow Structures

The velocity and pressure fields in the inner wall region are visualized to examine the flow structure. The visualized subvolume is shown schematically in Fig. 1. The volume visualized is up to 20 grid points from the inner surface.

Fig. 14. Contour of axial velocity fluctuation in the $x - \theta$ plane over the inner rod ($-2 < u'^+ < 2$: white<gray<black)($y^+ = 11.5, R_1/R_2 = 0.1$).

Fig. 15. Contour of axial velocity fluctuation in the $x - \theta$ plane over the inner rod ($-2 < u'^+ < 2$: white<gray<black)($y^+ = 17.8, R_1/R_2 = 0.02$).

Figures 14 and 15 show the instantaneous plan view of the velocity contour over the inner surface enveloped in the $x - \theta$ plane for Cases 1 and 3, respectively.

The high-speed streaks, where the axial velocity is higher or lower than the local mean, are elongated streamwise. Lueptow and Jackson (1991) found the appearance of the streaky structure similar to the planner channel near the wall of a cylinder. They suggested a similarity in the mechanism for the generation of turbulence in the boundary layer over a flat plate and over a cylinder. Further evidence of streaky structure in the turbulent boundary layer over a cylindrical surface is given by Neves et al. (1991). In the present study, approximately six to seven streaks can be counted in the spanwise direction in Fig. 14. If the length is nondimensionalized with the friction velocity and the kinematic viscosity, the circumference of the inner rod, i.e., the spanwise width of Fig. 14, is about 800.

Thus the average spacing of the high speed streaks is about 100-150, which is in good agreement with the one found usually in the turbulence over a plane wall. In Fig. 15, the streaks are observed even over the thin rod of R_1/R_2 = 0.02. The nondimensional spanwise width is about 200 and the number of streaks observed is one to two. Thus there seems to be no great change in the spacing of the streak even over such the thin inner rod; but the structure of the streaks is found different for the thinner rod as discussed later.

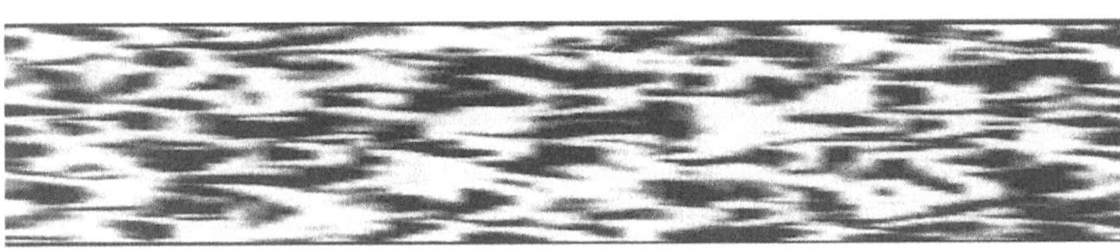

Fig. 16. Contour of azimuthal vorticity in the $x-\theta$ plane over the inner rod $(-0.1 < \omega_\theta'^+ < 0.1$: white<gray<black$)$ (y^+ = 11.5, R_1/R_2 = 0.1).

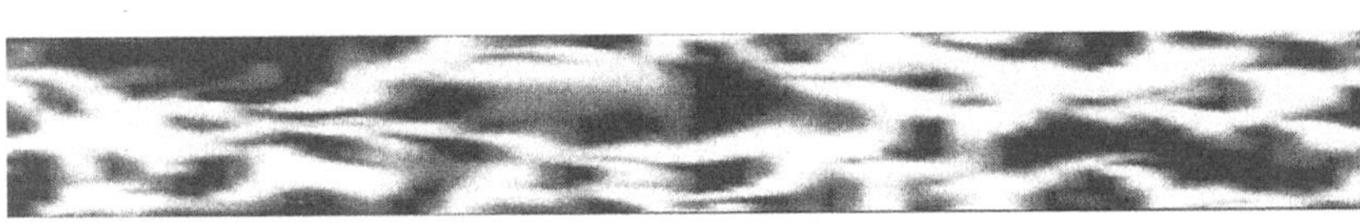

Fig. 17. Contour of azimuthal vorticity in the $x-\theta$ plane over the inner rod $(-0.1 < \omega_\theta'^+ < 0.1$: white<gray<black$)$ (y^+ = 17.8, R_1/R_2 = 0.02).

Figures 16 and 17 show the instantaneous plane view of the azimuthal vorticity contours over the inner surface enveloped in the $x-\theta$ plane for the Cases 1 and 3, respectively. The azimuthal width of Fig. 17 is enlarged five times. Even with consideration of the enlargement, only fewer boomerang-shaped structures are observed in Fig. 17 compared in Fig. 16. This supports the finding by Neves et al. (1991) on the effect of increasing the transverse curvature.

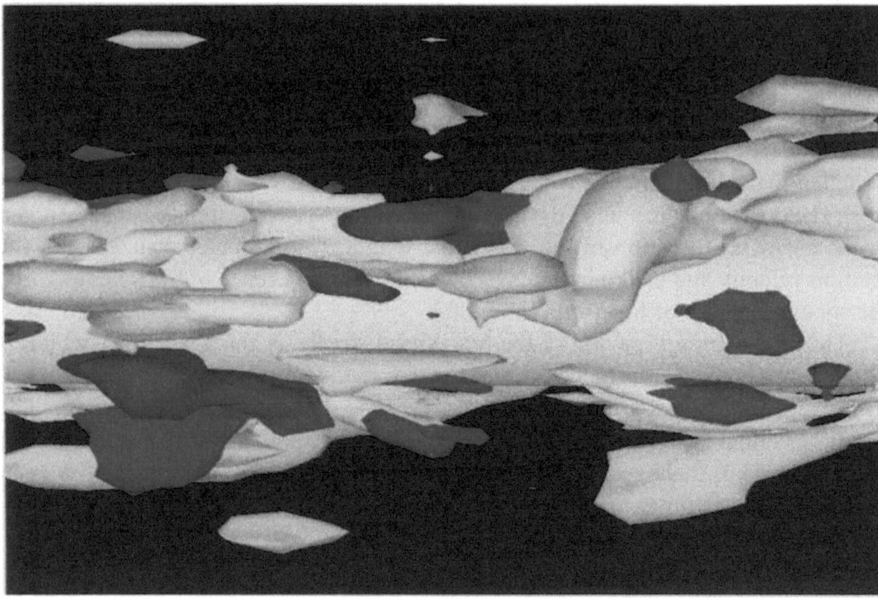

Fig. 18. 3D contour surface of shear product: $(R_1/R_2 = 0.1)$ White: $p'^{+} < -2.5$, Red(sweep): $(u'^{+}v'^{+})_4 < -2.0$, Yellow(ejection): $(u'^{+}v'^{+})_2 < -2.0$.

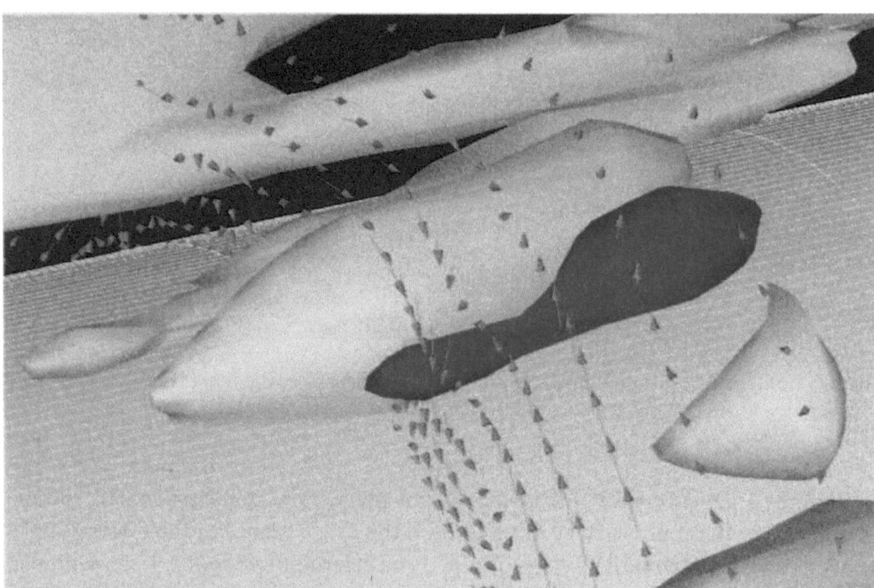

Fig. 19. 3D contour surface of shear product: $(R_1/R_2 = 0.1)$ White: $p'^{+} < -2.0$, Red(sweep): $(u'^{+}v'^{+})_4 < -1.8$, Yellow(ejection): $(u'^{+}v'^{+})_2 < -1.8$.

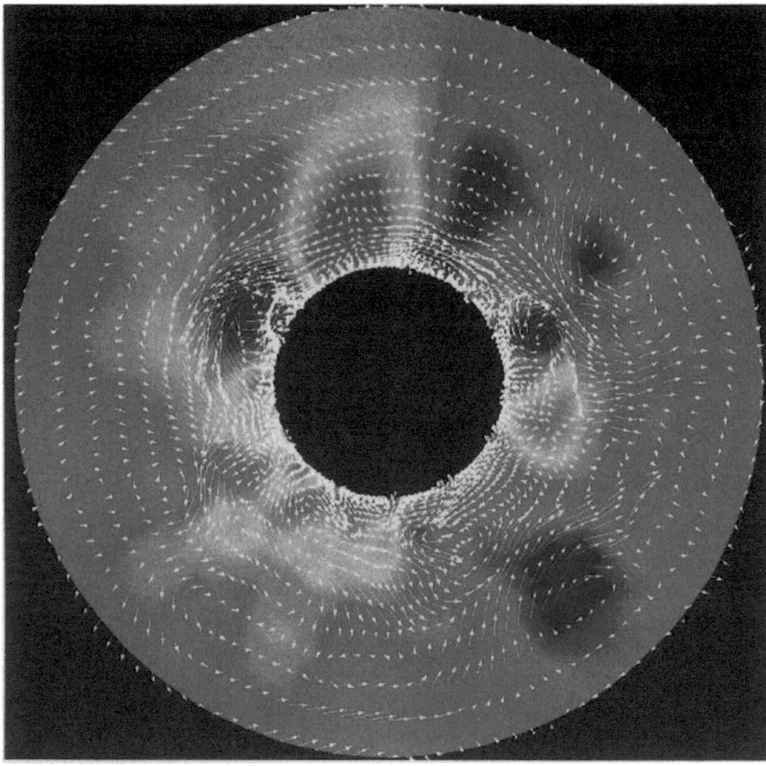

Fig. 20. v'^+, w'^+ velocity vectors and fluctuating pressure ($-2 < p'^+ < 2$: blue<green<red)(R_1/R_2 = 0.1).

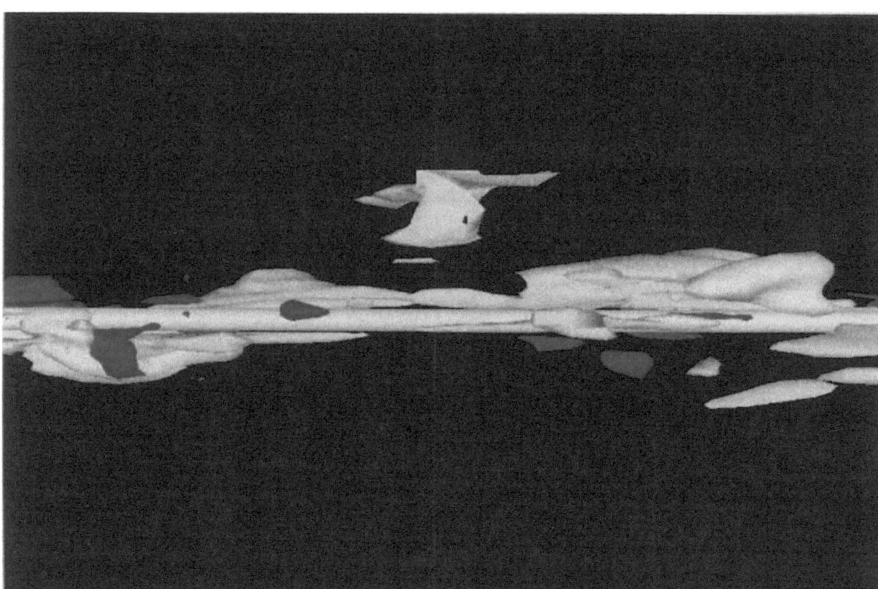

Fig. 21. 3D contour surface of shear product (R_1/R_2 = 0.02) White: $p'^+ < -1.5$, Red(sweep): $(u'^+v'^+)_4 < -2.0$, Yellow(ejection): $(u'^+v'^+)_2 < -2.0$.

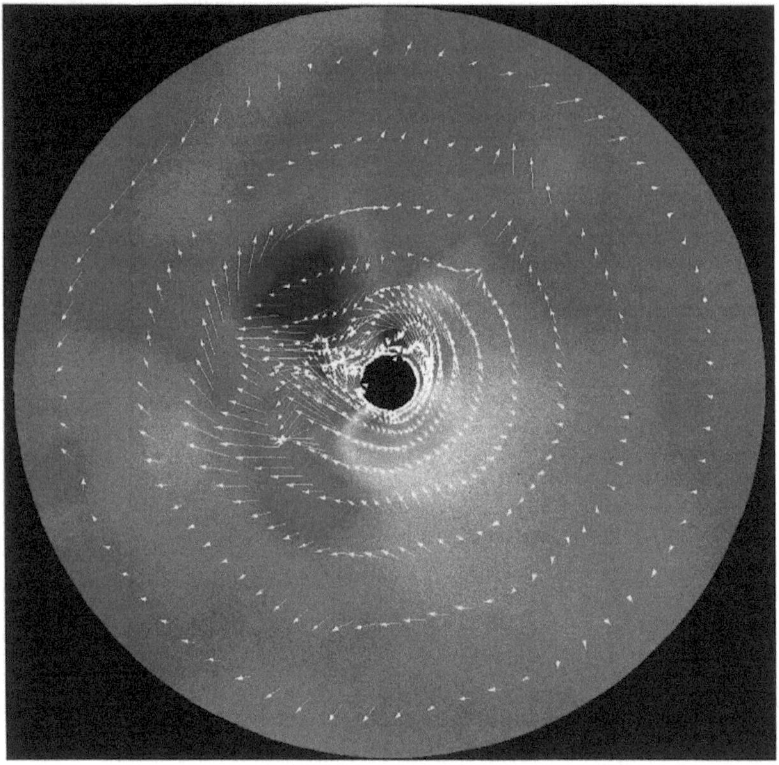

Fig. 22. Velocity vectors and fluctuating pressure ($-2 < p'^{+} < 2$: blue<green<red)($R_1/R_2 = 0.02$).

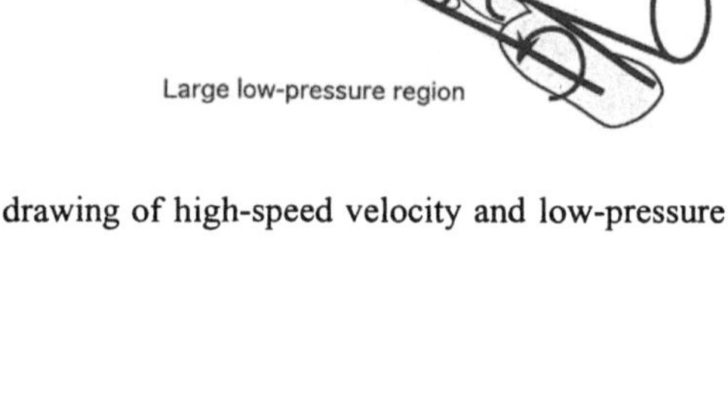

Fig. 23. Schematic drawing of high-speed velocity and low-pressure regions around a thin inner rod (Case 3).

Figure 18 shows the contour surfaces of the low pressure region, the ejection and the sweep in the inner wall region for Case 1. Robinson (1991) suggested the ejection and the sweep are often laterally associated with both sides of the low-pressure regions in turbulent boundary layer. The pairing is observed also in the DNS data base for the channel flow by Kasagi et al. (1993). The lateral pairing of the ejection and the sweep over the inner wall is clearly seen in Fig. 19. Instantaneous velocity vectors and pressure fluctuation contour in the cross section perpendicular to the axis are given in Fig. 20 for Case 1. A large high pressure region caused by the impinging flow, is observed over the top of the rod. It is accompanied by a pair of low pressure regions on its both sides. The vortex motion on the left one is counterclockwise, which strongly indicates that this vortex is caused by the impinging flow in the large high pressure region.

The structures over the thin inner rod (Case 3) are shown in Fig. 21. The low pressure region is again observed in this case; however, the structure seems to be rather different from the one over the thicker rod. A characteristic size of the low pressure region is even larger than the diameter of the inner rod. Inspection of Fig. 22 indicates that the high speed fluid impinging against the rod flows across the inner rod surface and forms a wake-like region behind the rod. Thus, the high and low pressure regions appear not in side by side but in the opposite side of the thin rod. The structure is schematically illustrated in Fig. 23. The production terms for normal u'^2 and the shear $u'v'$ stresses in the cross section perpendicular to the axis are shown in Figs. 24 and 25. Since the production for $u'v'$ is expressed as $-v'^2\partial\overline{u}/\partial r$, it is large (in its absolute value) where either in or outward motion (v'^2) is prominent. Inspection of Figs. 22 and 25 indicates that the regions of the high production rate (negative in its sign) are associated with high and low pressure regions. The high pressure region roughly overlaps with the high production area, because the high pressure is caused by the collision of the inward motion ($v' < 0$) against the inner rod. In the case of the low pressure region, the high production areas exist on the both sides of the low pressure region, because this low pressure is caused by the large vortical motion and the in and outward motions are prominent on both sides of the vortex. The pair of the high production region around the low pressure region has been found also in the planar turbulent channel flow (Robinson (1991), Kasagi et al. (1993)) and is recognized as the typical production process of the shear stress.

Comparison of Figs. 24 and 25 indicates that the high production regions for u'^2 and $u'v'$ roughly overlaps each other (although the sign is inversed). This is because the production for u'^2 is proportional to the shear stress $-u'v'$, while the shear stress itself is high where its production rate is large.

The redistribution terms of normal stresses; i.e., the pressure strain correlations, for u'^2,v'^2 and w'^2 are shown in Figs. 26, 27 and 28. In Fig. 26, the pressure-strain correlation term for u'^2 is shown. In this figure, this term is negative in almost the whole field. The negative value is large in the high- and low-pressure regions so that u' is decelerated ($\partial u'/\partial x < 0$) in the high-pressure region and is accelerated ($\partial u'/\partial x > 0$) in the low-pressure region. Comparison of Figs. 27 and 28 shows that the positive and negative regions are roughly symmetric in these two figures; that is,

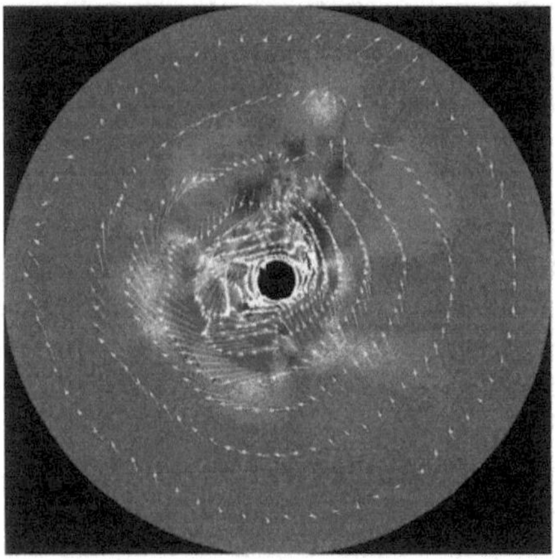

Fig. 24. Velocity vectors and production for $u'^2(-0.2 < 2u'^+v'^+(\partial\overline{u}^+/\partial y^+) < 0.2$: blue<green<red)($R_1/R_2$ = 0.02).

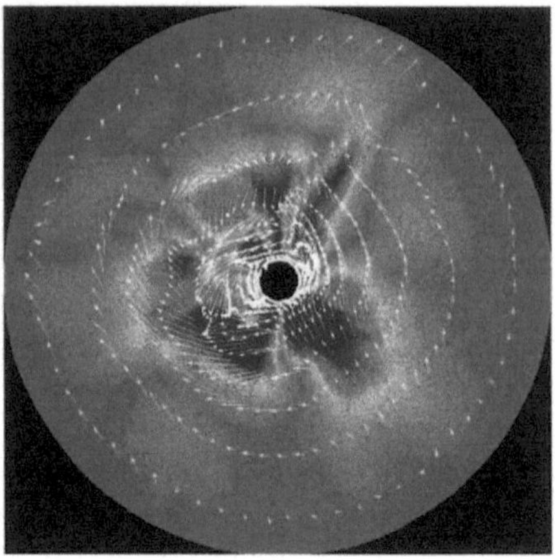

Fig. 25. Velocity vectors and production for $u'v'(-0.2 < -v'^{+2}(\partial\overline{u}^+/\partial y^+) < 0$: blue<green<red)($R_1/R_2$ = 0.02).

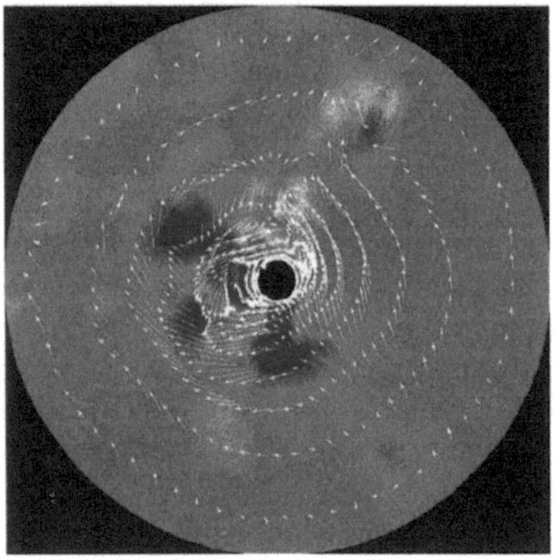

Fig. 26. Velocity vectors and pressure strain for u'^2 $(-0.02 < 2p'^+(\partial u'^+/\partial x^+)$. < 0.02: blue<green<red)(R_1/R_2 = 0.02)

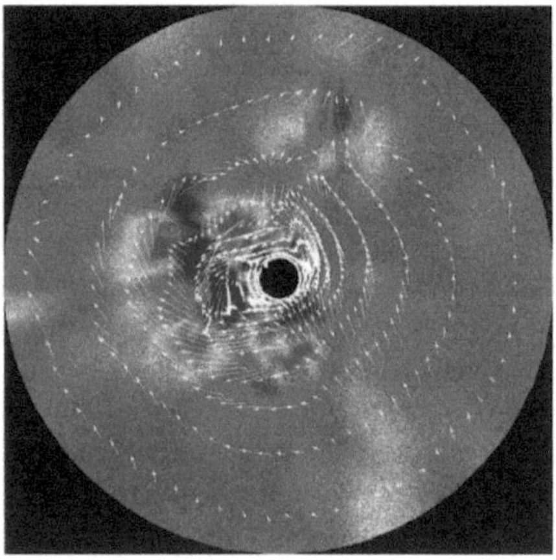

Fig. 27. Velocity vectors and pressure strain for v'^2 $(-0.02 < 2p'^+(\partial v'^+/\partial r^+)$. < 0.02: blue<green<red)(R_1/R_2 = 0.02)

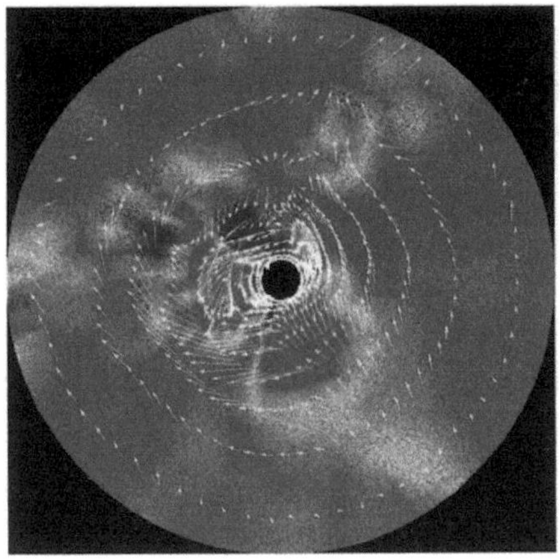

Fig. 28. Velocity vectors and pressure strain for w'^2 ($-0.02 < 2p'^+(\partial w'^+/r\partial\theta^+) < 0.02$: blue<green<red)($R_1/R_2 = 0.02$).

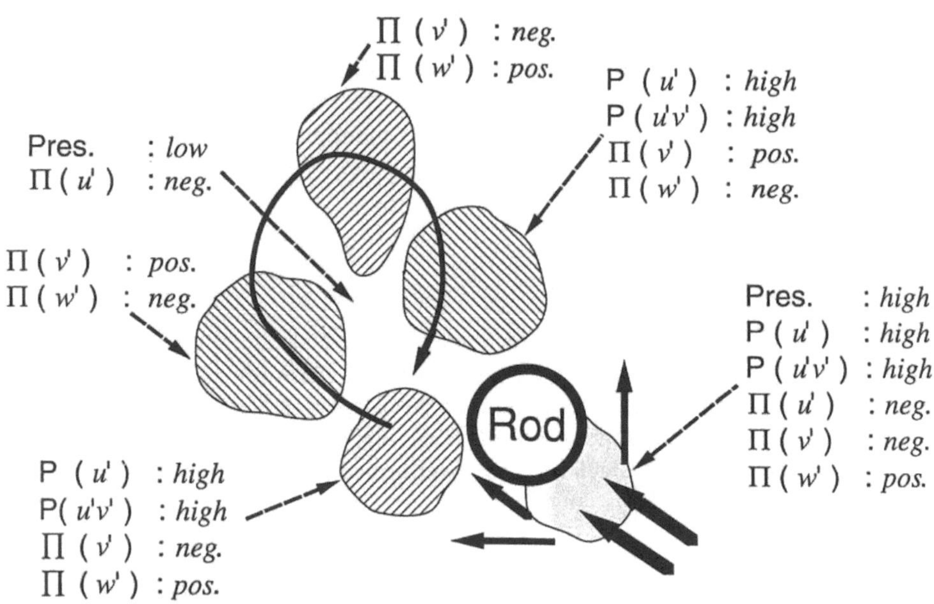

Fig. 29. Production and redistribution processes of Reynolds stresses around a thin inner rod (Case 3) (P: Production, Π: Pressure strain, Pres.: Pressure).

the positive region in one figure corresponds roughly to the negative region in the other one. Firstly, in the region where the inrushing flow impinges on the inner rod and thus the pressure is positive, the radial inward flow is decelerated ($\partial v'/\partial r < 0$) while the azimuthal flow is accelerated ($\partial w'/r\partial\theta > 0$). Consequently, the radial energy is converted to the azimuthal component; this is often called as the "splatting" effect. Secondly, in the low pressure region the large vortical motion exists and thus the signs of $\partial v'/\partial r$ and $\partial w'/r\partial\theta$ are roughly inverted. So, the positive and negative regions appear symmetrically in these two figures. This is again similar to the case of the planar wall turbulence, although the symmetry is less prominent in the present case.

The above discussions are schematically illustrated in Fig. 29. In summary, the production of u'^2 and $u'v'$ are prominent in both the impinging and the wake-like vortical regions. The redistribution of the energy from u'^2 and v'^2 to w'^2 occurs in the impinging region. In the vortical region the redistribution between v'^2 and w'^2 is roughly symmetrical while u'^2 loses its energy in almost the whole region.

6. Conclusions

The LES performed reproduced well the experimental results obtained by Rehme (1974). As the radius ratio decreases, the turbulent normal stresses become more anisotropic and the inner peak of the turbulent kinetic energy is reduced. In the case of a thick rod, a pairing vortex similar to that over the planar wall is observed; while in the case of a thin rod, a high speed streak causes a wake-like low pressure region on the opposite side of the high pressure one. The production and redistribution processes of the Reynolds stresses are, however, found similar to those for the plane wall turbulence.

Acknowledgments

The authors would express their appreciation to Mr. S. Nakamura (graduate student) for his help in preparing graphics and also to Prof. Y. Morinishi at Nagoya Institute of Technology for his helpful suggestions on the calculation methods. The structural aspect is indebted to discussions in Prof. N. Kasagi's laboratory at The University of Tokyo. The numerical calculations were performed with use of VP-2600 at Japan Atomic Energy Research Institute.

References

Chorin, A.J. (1968): "Numerical solution of the Navier-Stokes equations", Mathematics of Computations 22, pp. 745-762.

Hudina, M. (1979): "Prediction of displacement between the maximum velocity and zero shear position in asymmetric annular flows", SL-R/27.

Kasagi, N., Sumitani, Y., Suzuki, Y., Iida, O. (1993): "On the kinematics of the vortical structure in near wall turbulence ", 3rd World Conference on Experimental Heat Transfer, Fluid Mechanics and Thermodynamics, Honolulu, pp. 235-242.

Kawamura, H. (1972): "Heat transfer of fully developed in concentric annulus", Proc. 9th National Heat Transfer Symp. of Japan, C207, pp. 273-276.

Kawamura, H. (1979): "Analysis of turbulent annular flow using a phenomenological turbulence model", Proc. 16th National Heat Transfer Symp. of Japan, A110, pp. 28-30.
Kawamura, H., Satake, S. (1993): "Large eddy simulation of turbulent flow in an annular tube", Proc. Joint International Conference on Mathematical Methods and Supercomputing in Nuclear Applications, Karlsruhe, pp. 248-259.
Kim, J., Moin, P., Moser, R.D. (1987): "Turbulent statistics in fully developed channel flow at low Reynolds number", J. Fluid Mech., Vol. 177, pp. 133-166.
Lueptow, R.M., Jackson, C.P. (1991): "Near-wall streaky structure in a turbulent boundary layer on a cylinder", Phys. Fluids A, Vol. 3, pp. 2822-2824.
Neves, J.C., Moin, P., Moser, R.D. (1991): "Numerical study of axial turbulent flow over long cylinder", Proc. 8th Symp. Turbulent Shear Flows, Munich, pp. 5.2.1-5.2.6.
Rehme, K. (1974): "Turbulent flow in smooth concentric annuli with small radius ratios", J. Fluid Mech., Vol. 64, pp. 263-287.
Robinson, S.K. (1991): "The kinematics of turbulent boundary layer structure", NASA TM-103859.
Schumann, U. (1975): "Subgrid scale model for finite difference simulation of turbulent channel flows in plane channels and annuli", J. Comput. Phys., Vol. 18, pp. 376-404.
Smagorinsky, J. (1963): "General circulation experiments with the primitive equations", Mon. Weather Rev. 91, pp. 99-164.
Tanaka, H., Kawamura, H., Tateno, A., Hatamiya, S. (1982): "Effect of laminarization and re-transition on heat transfer for low Reynolds number flow through a converging to constant area duct ", J. Heat Transfer, Vol. 104, pp. 363-371.
Unger, F., Friedrich, R. (1991): "A Large eddy simulation of fully-developed turbulent pipe flow", Proc. 8th Symp. Turbulent Shear Flows, Munich, pp. 19.3.1-19.3.6.
Van Driest, E.R. (1956): "On turbulent flow near a wall", J. Aero. Sci., Vol. 23, pp. 1007-1011.
Vilemas, J., Cesna, B., Survila, V. (1987): "Heat transfer in gas-cooled annular channels", A. Zukauskas, eds., p. 58, Springer Verlag.

Direct and Large-Eddy Simulations of Transition of a Supersonic Boundary Layer

Frédéric Ducros, Pierre Comte, and Marcel Lesieur

Laboratoire des Ecoulements Géophysique et Industriels,
Institut de Mécanique de Grenoble, CNRS - INPG - UJF
B.P. 53X, F38041 Grenoble-Cedex, France

Abstract

The forced transition of a temporal boundary layer over an adiabatic flat plate is simulated by means of direct and large-eddy simulations, for an external Mach number of 4.5 and an initial Reynolds number of 10 000. The basic state is composed of a laminar flow, 2D perturbations mostly composed of Mack's second mode, plus 3D white noise. In both simulations, a first transient is observed: Kelvin-Helmholtz-like vortices develop at the height of the generalized inflection point. Due to oblique subharmonic modes, they start three-dimensionalizing as in mixing layers at high convective Mach numbers. Then, the turbulent activity abruptly shifts towards the wall, in the forms of streaks mostly. Immediately after, the LES shows these streaks breaking down into turbulence at smaller scale, with coherent hairpin vortices (DNS are no longer possible). The flow then presents strong analogies with turbulent incompressible boundary layers.

1. Nomenclature

x, y and z	streamwise, vertical (normal to the wall) and spanwise coordinates.
$U_\infty, T_\infty, \varrho_\infty$, M_∞, μ_∞	streamwise velocity, temperature, density, Mach number and dynamic viscosity at $y \to \infty$.
L_x, L_y and L_z	physical dimensions of the computational domain.
k_x, k_z	streamwise and spanwise wavenumbers.
$\varrho, u, v, w, \mathcal{T}$	density, x, y and z components of velocity field, temperature.
Δ_x and Δ_z	size of the mesh in the streamwise and spanwise directions.
$\boldsymbol{x}, \boldsymbol{r}$	points of the physical space.
ω_x, ω_y and ω_z	vorticity in the x, y and z directions.
$\bar{F}_2(\boldsymbol{x}, \Delta_x, t)$	second order structure function at point $\boldsymbol{x}$ at t.
$\nu_t(\boldsymbol{x}, t)$	turbulent cinematic viscosity at point $\boldsymbol{x}$ at t.
$\mu_t(\boldsymbol{x}, t)$	turbulent dynamic viscosity at point $\boldsymbol{x}$ at t.
$\bar{u}(\boldsymbol{x}, t)$	resolved velocity field at point $\boldsymbol{x}$ at time t.
$\hat{u}(k_x, y, k_z, t)$	Fourier Transform of $\bar{u}(\boldsymbol{x}, t)$.
$\hat{u}^*(k_x, y, k_z, t)$	complex conjugate of $\hat{u}(k_x, y, k_z, t)$.

$E(k_x, k_z, t)$	energy at the streamwise and the spanwise wavenumbers k_x and k_z at time t.
$v'_{(k_x,k_z)}(y,t)$	modulus of the vertical velocity at height y for the wavenumber k_x and k_z at time t.
$\langle\varrho\rangle\langle u'u'\rangle(y,t)$, $\langle\varrho\rangle\langle u'v'\rangle(y,t)$	streamwise and spanwise average of the Reynolds stress profiles.
$y(j)$	height of the jth grid point.
δ_i	initial displacement thickness.
δ_1	displacement thickness.
δ_2	momentum thickness.
y_S	height of the generalized inflection point.
A	amplitude of the 2D forcing.
$\widehat{M}(y,t)$	convective Mach number at height y at t.
c, a	phase speed and sound speed.
ω_i	temporal amplification growth rate.
c_f	Skin-friction coefficient.
H_{12}	shape factor.

2. Introduction

The understanding of the fundamental mechanisms of transition to turbulence of boundary layers at high Mach numbers is crucial for the design of future hypersonic spaceplanes. For Mach numbers larger than 4, the linear stability theory predicts that the most unstable mode is two-dimensional and inviscid in nature (Mack's second mode, Mack 1969). For in-flight conditions, the level of perturbations being extremely low, it is generally assumed that three-dimensionalization occurs through the secondary-instability mechanisms, for which oblique subharmonics of Mack's second mode seem to be dominant (Ng and Erlebacher 1992). In the transitional regime, laboratory (*e.g.*, Fischer and Weinstein 1972) and numerical experiments (Pruett and Zang 1992) show "rope-like structures" for a wide range of configurations (cones and flat plates) and Mach numbers (4.5 to 14). The most recent work, to our knowledge, in direct numerical simulation on this subject shows the evolution of instabilities near the critical layer up to the roll-up of Λ vortices (Adams and Kleiser 1993a and 1993b). However, information relative to the coherent structures in the turbulent regime is relatively sparse. Because of the resolution requirement which increases when small scales turbulence starts developing (see, *e.g.*, Adams and Kleiser 1993a), we believe that LES (*i.e.*, numerical simulations in which subgrid-scale turbulence is parameterized) are among the best solutions available nowadays. In this context, Normand and Lesieur (1992) performed the large-eddy simulation of a forced transition of a supersonic spatial boundary layer, but the resolution was too low to reveal the mechanisms of transition.

The purpose of the present study is to give, by means of direct and large-eddy numerical simulation, a fully nonlinear counterpart to the secondary stability analysis of Ng and Erlebacher (1992), with an insight into the turbulent regime. We thus consider a case of forced transition of a boundary layer at Mach = 4.5 over an adiabatic flat plate, assuming streamwise and spanwise periodicity. The Reynolds number based on the initial displacement thickness $Re_{\delta_i} = \varrho_\infty U_\infty \delta_i / \mu_\infty$ is 10 000, with an external temperature $T_\infty = 180$ K. The initial conditions of the simulations consist of:

- a laminar flow solution of the similarity equations,
- two-dimensional fluctuations essentially featuring the Mack's second mode. They are obtained at the end of the linear development of a two-dimensional simulation, in which the laminar flow is perturbed by 2D white noise of amplitude $10^{-4}\, U_\infty$. The amplitude of these fluctuations is then rescaled to $A = 8\%$ of U_∞,
- small amplitude 3D white noise ($10^{-4} U_\infty$).

Two calculations are presented: a direct simulation performed till the onset of small-scale turbulence and a large-eddy simulation in which the turbulent regime is reached.

3. Numerical Aspects

The complete Navier-Stokes equations for an ideal gas are solved in their conservative form by means of a code developed by Normand and Lesieur (1992) on the basis of an explicit Mc-Cormack-type scheme which is 4th order accurate in space and 2nd order accurate in time (Gottlieb and Turkel 1976). Sutherland's law (see, *e.g.*, Schlichting 1979 pp. 328-329) is prescribed for viscosity. As in Normand and Lesieur (1992), the viscous terms of the equations are modified in such a way that the laminar profile imposed initially is a steady solution of these equations. Otherwise, it would rapidly degenerate into an irrelevant error-function profile.

For the LES, the *structure-function* model of Métais and Lesieur (1992) is used as in Normand and Lesieur (1992). This model introduces an eddy-viscosity defined by:

$$\nu_t(\boldsymbol{x}, t) = 0.063\ \Delta_x \sqrt{\bar{F}_2(\boldsymbol{x}, \Delta_x, t)} \tag{1}$$

and,

$$\bar{F}_2(\boldsymbol{x}, \Delta x, t) = \left\langle \|\bar{u}(\boldsymbol{x} + \boldsymbol{r}, t) - \bar{u}(\boldsymbol{x}, t)\|^2 \right\rangle_{\|\boldsymbol{r}\| = \Delta x} \tag{2}$$

with the averaging being done over the four immediate neighbours of the point considered in a plane parallel to the wall.

The meshes are uniform in the streamwise and spanwise directions. Stretching is applied in the direction normal to the wall (y): the height of each grid point is defined by:

$$y(j) = L_y \frac{C\,\eta}{1 + C - \eta}, \quad \eta = \frac{j-1}{n_y - 1} \tag{3}$$

in which L_y denotes the domain's height and n_y the number of points in the y direction. The constant C controls the stretching.

4. Two-Dimensional Calculation

For the parameters we consider, the laminar boundary layer profile has a generalized inflection point at a height $y_S = 1.05\ \delta_i$, the initial displacement thickness. The most amplified mode predicted by the linear stability theory is a two-dimensional inviscid mode (Mack's second mode) developing about y_S. Its wavelength is $\lambda_a = 2.8\ \delta_i$. The length of our domain is $L_x = 4\ \lambda_a$. Its height is $L_y = 20\ \delta_i$, most of the points being concentrated between the wall and y_S: we have set $C = 0.3$, which puts the first point at $y = 0.024\ \delta_i$, with an aspect ratio $\Delta_x/\Delta_y = 5.8$. For the mesh around y_S, one has $\Delta_x/\Delta_y = 2.6$

Between $t = 60$ and $400\ \delta_i/U_\infty$, the velocity fluctuations grow exponentially with time at a rate $w_i = 2.4\ 10^{-2} U_\infty/\delta_i$, which is very close to the growth rate of Mack's second mode (see, in particular, Ng and Erlebacher 1992). Modal analysis (as described in the next section) shows that the mode which dominates has the phase speed about $c = 0.94\ U_\infty$ (versus $0.91 U_\infty$ in Ng and Erlebacher 1992). Figure 1 shows the vorticity map at $t = 400\ \delta_i/U_\infty$.

Fig. 1. Vorticity map for the 2D calculation at $t = 400\ \delta_i/U_\infty$, showing "rope-like structures" which look like Kelvin-Helmholtz vortices.

The extremal values are reached near the wall, with both signs alternatively. This seems to be due to reflections of acoustic waves between the wall and the sonic line, coupled to baroclinic effects. Vorticity is also concentrated in four structures located about y_S and reminiscent, on the one hand, of the Kelvin-Helmholtz vortices observed in mixing layers and on the other hand, of the "rope-like" structures obtained experimentally by Fisher and Weinstein (1972) and numerically by Pruett and Zang (1992).

5. Three-Dimensional Calculation

5.1 Time Evolution of Energy

We now clip the grid defined in the previous section to $L_x = 2\ \lambda_a = 5.5\ \delta_i$ and $L_y = 7.15\ \delta_i$, in order to minimize the computational cost. We extend it in the spanwise direction with a width $L_z = 6.28\ \delta_i$, the preferential wavelength of the subharmonic mode of secondary instability found by Ng and Erlebacher (1992); it is the most unstable mode they find, and its wavelength does not depend much on A (see their Fig. 12). Our resolution is of $40 \times 70 \times 36$.

With the initial and boundary conditions given in the introduction, we perform simultaneously a direct simulation and a large-eddy simulation. The former blows up at $t = 390\ \delta_i/U_\infty$, but the latter continues further. It is stopped at $t = 450\ \delta_i/U_\infty$, after transition is completed. The longest run takes about 70 hours on a 4-proc. Stardent/Kubota Titan 3000.

We plot the time evolution of the energy of the prominent modes for the DNS (Fig. 2a) and the LES (Fig. 2b). The energy is defined by:

$$E(k_x, k_z, t) = \frac{1}{L_y} \int_0^{L_y} \sqrt{\hat{u}(k_x, y, k_z, t)\ \hat{u}^*(k_x, y, k_z, t)}\ \mathrm{d}y \quad , \tag{4}$$

in which $\hat{u}$ denotes the streamwise and spanwise Fourier transform of the velocity field and $\hat{u}^*$ its complex conjugate. The wavenumbers are normalized in such a way that k_x or $k_z = 1$ corresponds to the box size. Mack's second mode then appears as $(2, 0)$ and the oblique subharmonic of Ng and Erlebacher (1992) as $(1, 1)$.

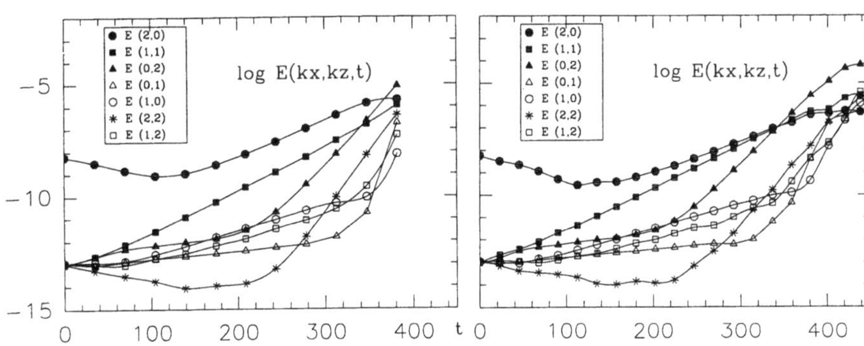

Fig. 2a. Energy of the prominent modes as a function of time for the DNS.

Fig. 2b. Energy of the prominent modes as a function of time for the LES.

For both DNS and LES, the mode $(2, 0)$ dominates all the others up to $t \approx 350\ \delta_i/U_\infty$. Its amplitude profile for v', defined as:

$$v'_{(2,0)}(y, t) = \sqrt{\hat{v}(2, y, 0, t)\ \hat{v}^*(2, y, 0, t)}\ , \tag{5}$$

is plotted at several instants in Fig. 3.

Up to $t = 350\ \delta_i/U_\infty$, $v'_{(2,0)}(y, t)$ grows in an approximately self-similar manner, with the characteristic shape found in the stability calculations by Erlebacher and Hussaini (1989), *viz.* with a large peak at $y \approx 0.4\ \delta_i$ and a secondary peak at $y \approx 1.3\ \delta_i$. Figure 4 shows phase profiles of $u'_{(2,0)}$, $v'_{(2,0)}$ and $p'_{(2,0)}$ taken at $t = 315\ \delta_i/U_\infty$, that is when the second mode is amplified and still dominant. Notice the phase jumps occurring about y_S. This figure matches very well with Fig. 1e of Erlebacher and

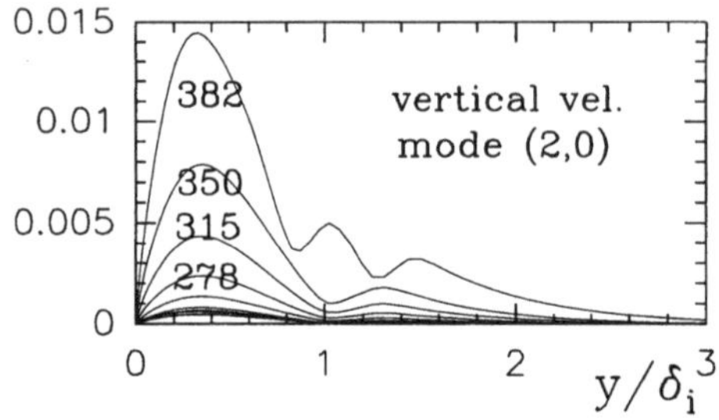

Fig. 3. Temporal growth of Mack's second mode in direct simulation: profiles of $v'_{(2,0)}/U_\infty$ at different instants, $t = 382\ \delta_i/U_\infty$ corresponding to the saturation of this mode.

Hussaini (1989), which can be seen as a validation of our numerics. Notice, however, that $E(2,0,t)$ remains slightly lower in the LES case than in the DNS case. This is due to an excess of eddy-viscosity brought about by the subgrid-scale model in the region of the generalized inflection point. An improvement of the model in this respect is in process.

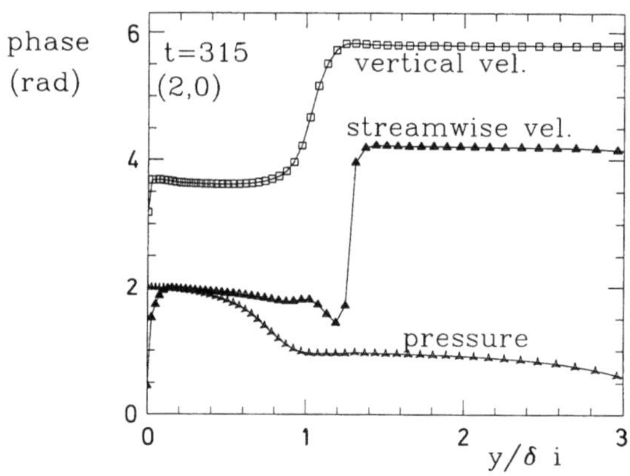

Fig. 4. Phase profiles $\phi(y) = \mathrm{Arctg}\dfrac{\Im(\hat{f})}{\Re(\hat{f})}$ at $t = 315\ \delta_i/U_\infty$, f standing for $u'_{(2,0)}$, $v'_{(2,0)}$ and $p'_{(2,0)}$ respectively.

Nevertheless, the second dominant mode $(1,1)$ does not seem to be affected at all by the model: both in the DNS and LES cases, it grows exponentially with time from $t = 0$ up to $t = 350\ \delta_i/U_\infty$ at a rate $1.7\ 10^{-2}U_\infty/\delta_i$, in acceptable agreement with Ng and Erlebacher (1992) who find $2.5\ 10^{-2}U/\delta_i$ for $A = 6\%$.

The most remarkable feature revealed by this kind of modal analysis is the behaviour of the purely spanwise mode $(0,2)$, which does not grow much between $t = 0$ and $200\ \delta_i/U_\infty$ but shoots up immediately after (see, Fig. 2). It overwhelms the other modes from $t = 350\ \delta_i/U_\infty$, and this confirms that this mode could be essential for transition, as stressed by Pruett and Zang (1992).

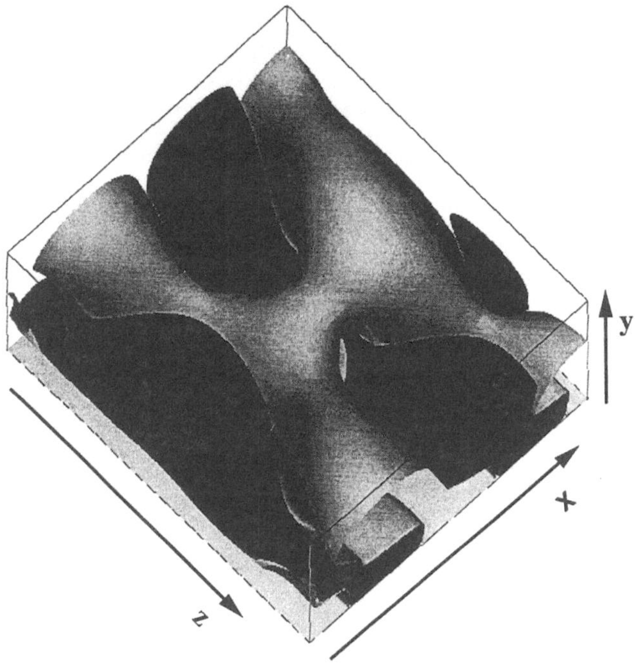

Fig. 5. Isosurfaces of vertical velocity at $t = 350\ \delta_\mathrm{i}/U_\infty$ in DNS (clear surface $v = 0.0005U_\infty$; dark surface $v = -0.0005U_\infty$) showing the modes (2,0) and (1,1) about y_S.

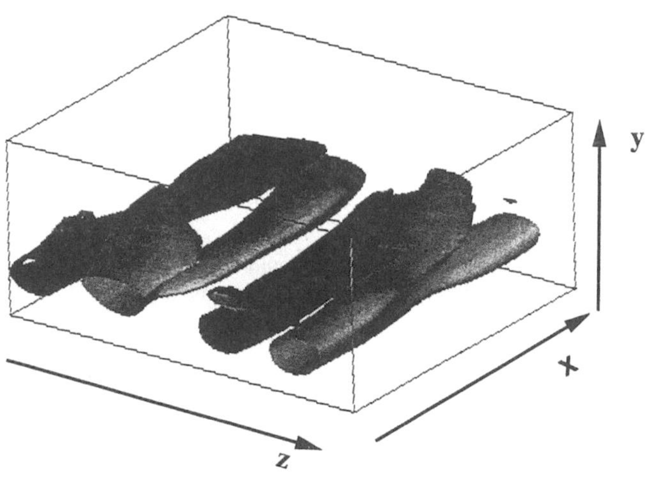

Fig. 6. Isosurfaces of vertical vorticity at $t = 350\ \delta_\mathrm{i}/U_\infty$ in direct simulation: clear surface $\omega_y = 0.03U_\infty/\delta_\mathrm{i}$; dark surface $\omega_y = -0.03U_\infty/\delta_\mathrm{i}$.

5.2 Time Evolution of the Organized Structures

Figure 5 shows isosurfaces of the vertical velocity at $t = 350\ \delta_\mathrm{i}/U_\infty$ for the DNS case (the LES counterpart is almost identical). Seen from the top, it shows clearly that the prominent mode is still the Mack's mode $(2,0)$, but that it is about to be caught up on by $(1,1)$. A side view of the same picture (not presented here) shows that these surfaces lay around y_S.

The mode $(0,2)$ is not visible in Fig. 5. It develops between the wall and y_S. It concerns mostly ω_y, the vertical component of the vorticity (Fig. 6).

Despite the streaky pattern shown, the vortical structure of the flow at this moment remains dominated by Kelvin-Helmholtz-like vortices at y_S, slowly going three-dimensional as mixing layers at $M_c \approx 1$ (Sandham and Reynolds 1991, Comte *et al.* 1992, Fouillet 1992), in the form of Λ-vortices facing each other; Fig. 7 shows the most intense vortex lines together with isosurfaces of streamwise vorticity ω_x.

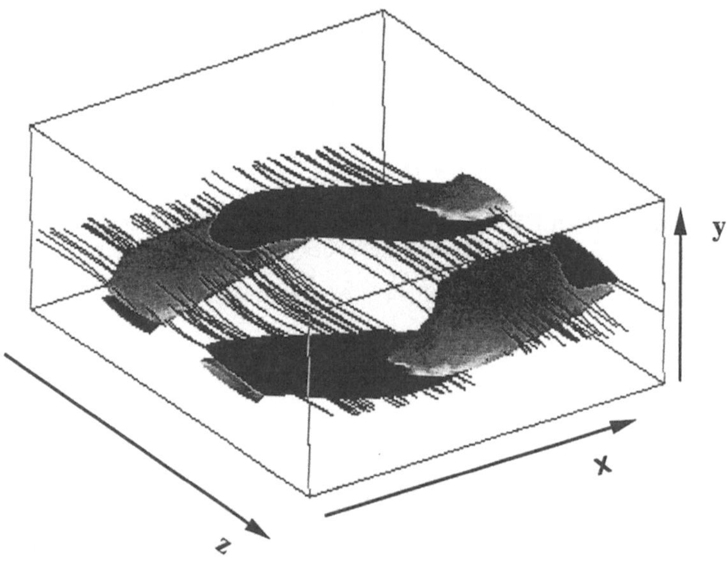

Fig. 7. Vortex lines at the same instant as Figs. 5 and 6, with isosurface $\omega_x = 0.1U_\infty/\delta_\mathrm{i}$ in clear and $\omega_x = -0.1U_\infty/\delta_\mathrm{i}$ in dark. This resembles very much mixing layers for convective Mach number about 1.

Another representation of the flow at the same instant is given in Fig. 8: from isosurfaces of spanwise vorticity (or vorticity norm), one can make out staggered Λ-vortices stretched into *Y-layers*, as proposed by Adams and Kleiser (1993a).

However, at least in our case, this is not the eventual stage of transition; from $t = 390\ \delta_\mathrm{i}/U_\infty$ onwards, most of the three-dimensionality will be concentrated near the wall. In the meantime, small-scale turbulence develops, which makes our DNS calculation blow up. Figures. 9a and 9b show the counterpart of Fig. 7 for the LES

Fig. 8. Isosurface of spanwise vorticity $\omega_z = -U_\infty/\delta_i$, still at $t = 350\ \delta_i/U_\infty$. In order to increase the resemblance with Fig. 11 of Adams and Kleiser (1993a), the domain has been instanced twice thanks to spanwise periodicity.

case at $t = 390\ \delta_i/U_\infty$ (vortex lines lying about y_S have to be shown separately from those close to the wall to avoid overcrowding).

The Λ-vortices at y_S are not visible anymore. Instead, one sees streaks developing near the wall, with the same spanwise spacing, *viz.* $s \approx L_z/2 \approx 3.14\ \delta_i$. At $t = 390\ \delta_i/U_\infty$, the wall friction is such that 1 wall length unit $\approx 1/33\ \delta_i$. Thus, one also has $s \approx 100$ wall units, as in most incompressible experiments.

Notice that each vortex line shown in Fig. 9 has a hairpin shape (with an angle of $\approx 45°$ with respect to the wall and a rounded tip kinked upwards due to self induction) but there is no tendency for these lines to bundle up into vortex tubes yet. Figures 10a and 10b are a sketch of the expected effect of such lines upon the vertical vorticity and velocity components, to be compared with Figs. 11 and 12.

This streaky pattern does not persist, as from $t = 400\ \delta_i/U_\infty$, vortex line plots become very complicated. The lines gather into distorted hairpin vortex tubes in the presence of a high level of three-dimensional turbulence (Fig. 13). From $t = 400\ \delta_i/U_\infty$ to the end of the simulation ($t = 450\ \delta_i/U_\infty$, Fig. 14), the general flow pattern does not evolve much while the spectra fill up. Notice that Figs. 13 and 14 resemble very much the vortex lines obtained at Mach = 0.5 by Normand (1990) in the beginning of the turbulent regime (see, his Figs. 13-24). This meets the experimental statement that high- and low-Mach-number (at least up to $M \approx 5$) turbulent boundary layers have many common trends (such as the velocity profile and the normalized stress profile $\tau/\tau_w(y)$, see, in particular Morkovin 1961; Maise and McDonald 1968)

5.3 Mean Quantities and Fluctuations

Figures 15 and 16 show the time evolution of the skin friction coefficient $c_f = \frac{\tau_w}{\frac{1}{2}\varrho_\infty U_\infty^2}$ and the shape factor $H_{12} = \delta_1/\delta_2$ respectively.

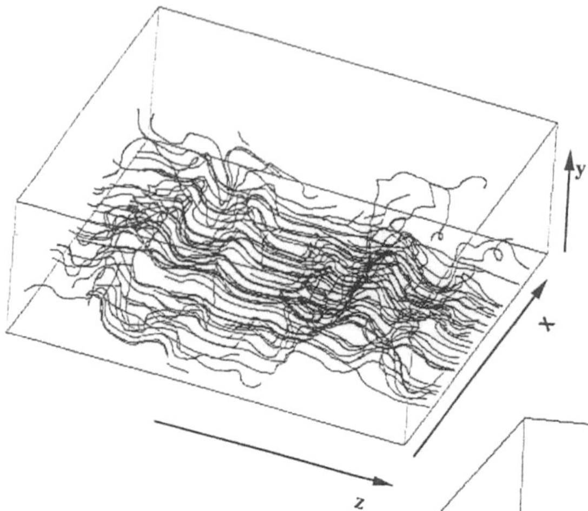

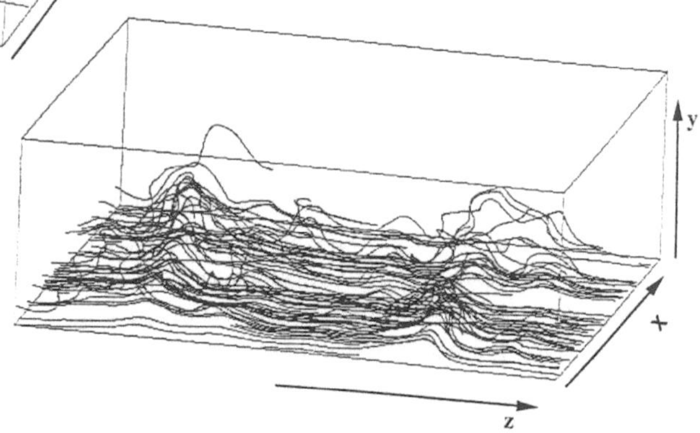

Fig. 9a. Vortex lines located just below y_S, at $t = 390\delta_i/U_\infty$.

Fig. 9b. Vortex lines near the wall, also at $t = 390\delta_i/U_\infty$.

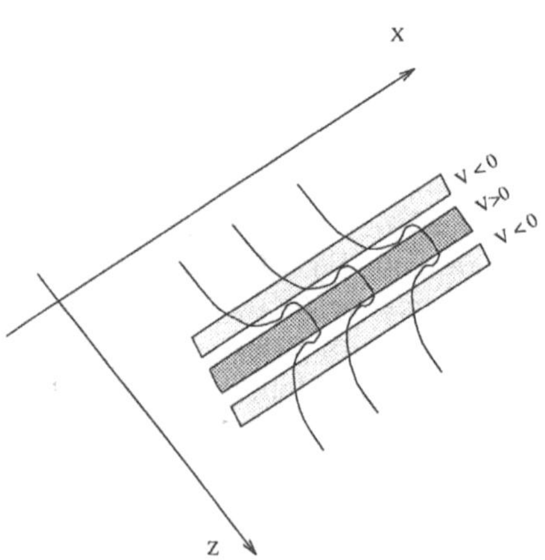

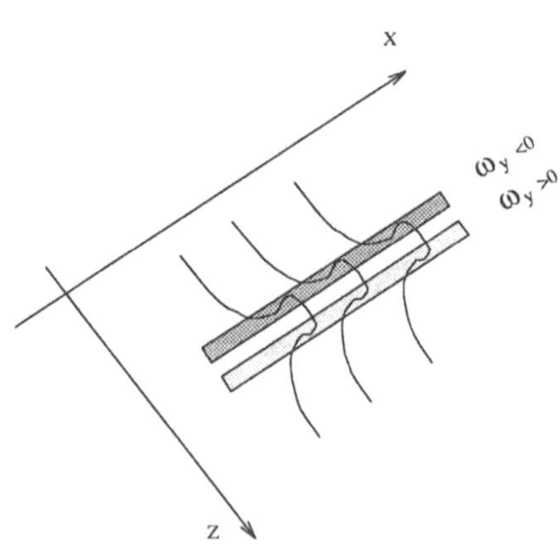

Fig. 10a. Sketch of the consequences on v of Fig. 9.

Fig. 10b. Sketch of the consequences on ω of Fig. 9.

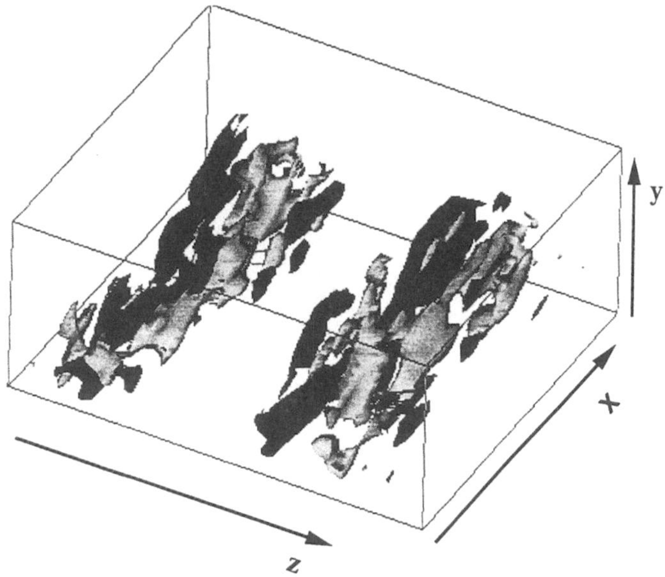

Fig. 11. Isosurfaces of $v = 0.05U_\infty$ in clear and $-0.05U_\infty$ in dark.

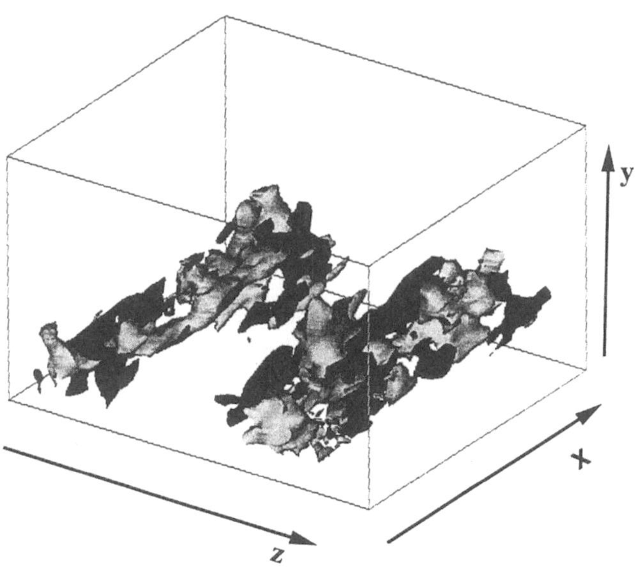

Fig. 12. Isosurfaces of $\omega_y = 0.7U_\infty/\delta_\mathrm{i}$ in clear and $-0.7U_\infty/\delta_\mathrm{i}$ in dark.

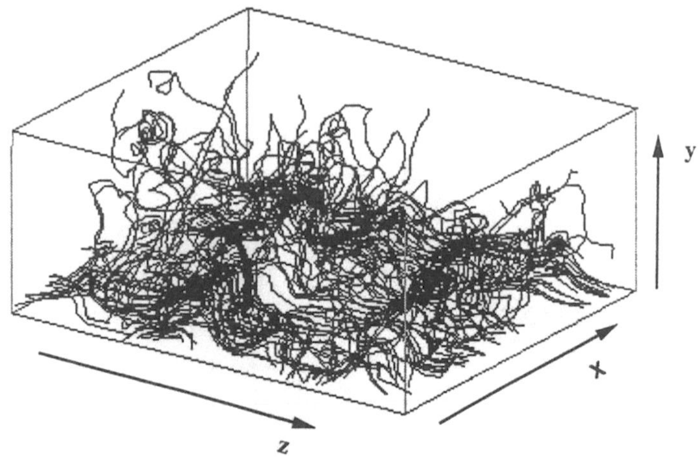

Fig. 13. Vortex lines at $t = 400\delta_i/U_\infty$.

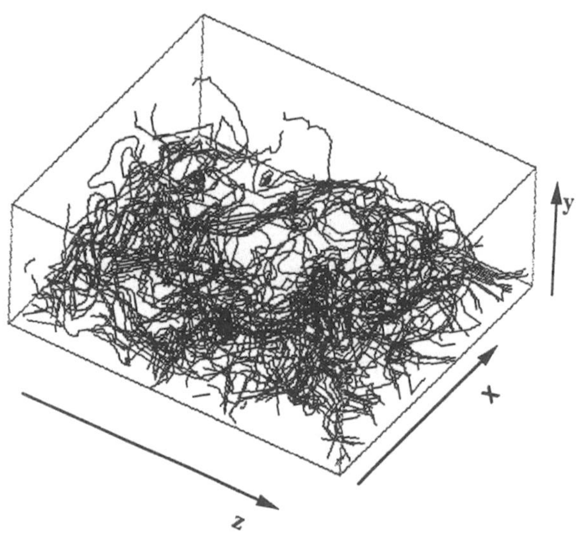

Fig. 14. Vortex lines at $t = 450\delta_i/U_\infty$. The resemblance with Fig. 13 suggests that the developed turbulent regime has been reached, without any more transient.

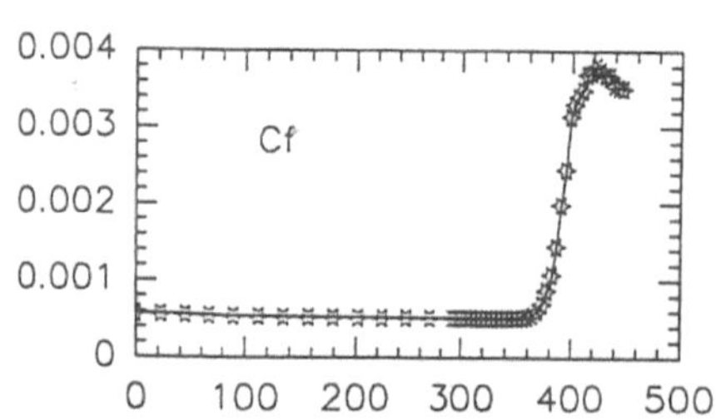

Fig. 15. Skin-friction coefficient c_f as a function of time.

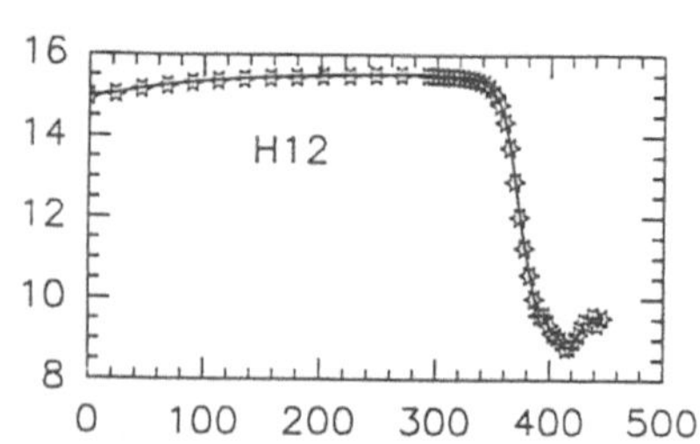

Fig. 16. Shape factor $H_{12} = \delta_1/\delta_2$ as a function of time.

Because of the modification of the viscous terms mentioned in section 2, the initial profile is a steady solution of the equations solved. Thus, c_f remains approximately constant until transition occurs, instead of decreasing in $\sqrt{t}$ as it should do. The same trend is observed for H_{12}, which gradually increases from 15 to 15.5. At about $t \approx 360\ \delta_i/U_\infty$, the skin friction coefficient starts increasing abruptly while H_{12} decreases. This moment corresponds to the formation of the streaks. At $t \approx 420\ \delta_i/U_\infty$, both curves reach extrema: a maximum of $3.8\ 10^{-3}$ for c_f and a minimum of 8.75 for H_{12}. Immediately after, c_f starts decreasing while H_{12} increases and levels off to 9.5, the value given by the empirical formula of Michel (1967):

$$H_{12} = H_{\mathrm{inc}} + 0.4\ M_\infty^2 + 1.222\ \frac{T_{\mathrm{w}} - T_{\mathrm{ad}}}{T_\infty} \quad , \tag{6}$$

where $H_{\mathrm{inc}} = 1.4$ denotes the incompressible counterpart of H_{12} and T_{ad} the adiabatic recovery temperature at the wall, equal to the wall temperature T_{w} in our case. We have no explanation for the undershoot of H_{12}, that we have not encountered so far at low Mach numbers. However, a similar drop (but even more severe: $H_{12} \approx 6$) is found in the direct simulations by Pruett and Zang (1992) with no tendency to go up before the blowup, while their c_f keeps on growing up to $3.4\ 10^{-3}$ in good agreement with ours.

Figure 17 shows the instantaneous Reynolds stress profiles $\langle\varrho\rangle\langle u'u'\rangle(y,t)$ (Fig. 17a) and $\langle\varrho\rangle\langle u'v'\rangle(y,t)$ (Fig. 17b), in which $\langle\rangle$ denotes the streamwise and spanwise average over the box.

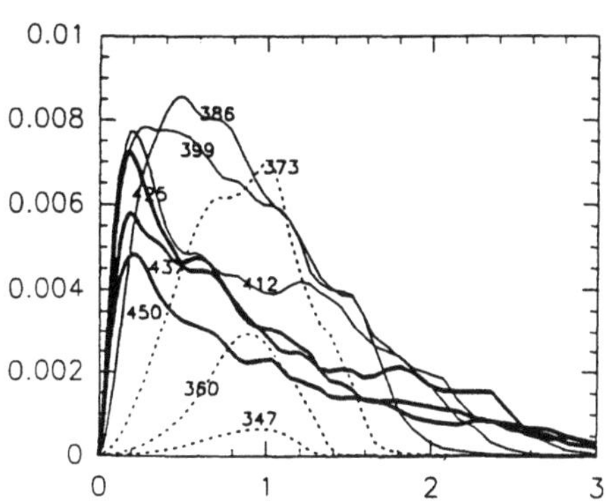

Fig. 17a. Time evolution of the Reynolds stress profiles $\langle\varrho\rangle\langle u'u'\rangle(y,t)$.

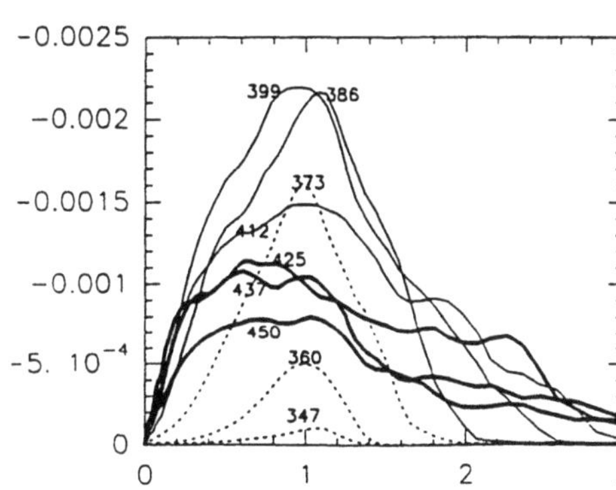

Fig. 17b. Time evolution of the Reynolds stress profiles $\langle\varrho\rangle\langle u'v'\rangle(y,t)$.

As stressed in Pruett and Zang (1992), $\langle\varrho\rangle\langle u'u'\rangle$ remains larger than $\langle\varrho\rangle\langle u'v'\rangle$, but only by a factor of 4 although they mention the stress are "nearly an order of magnitude". During the laminar stage (*i.e.*, up to $t \approx 360\ \delta_i/U_\infty$), the curves grow almost self-similarly with a peak around y_S. Between 360 and 420 δ_i/U_∞, the peak of $\langle\varrho\rangle\langle u'u'\rangle$ gets more and more acute. Meanwhile, it shifts towards the wall and settles down at about $y \approx 0.2\,\delta_i \approx 10$ wall units. In contrast, the profile of $\langle\varrho\rangle\langle u'v'\rangle$ flattens in time with no visible shift towards the wall.

To illustrate the drastic change in the shape of mean streamwise-velocity and temperature profiles $\langle u\rangle(y,t)$ and $\langle T\rangle(y,t)$ during transition, these profiles are plotted in Figs. 18 and 19 respectively, at nine time steps between $t = 0$ and $450\ \delta_i/U_\infty$.

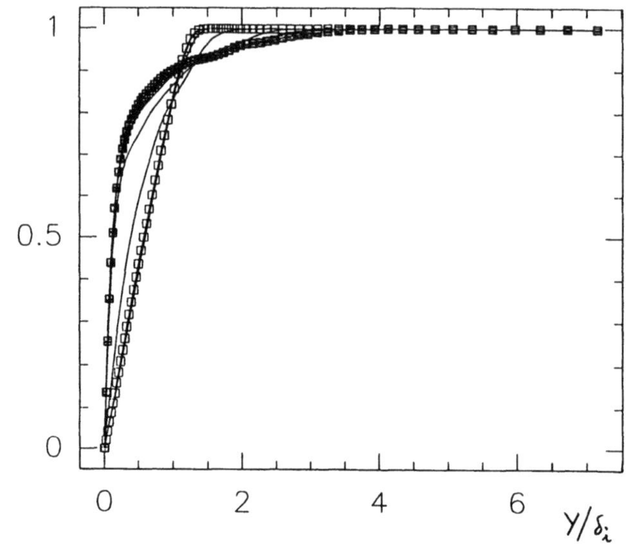

Fig. 18. Time evolution of the mean velocity profile $\langle u\rangle(y,t)$.

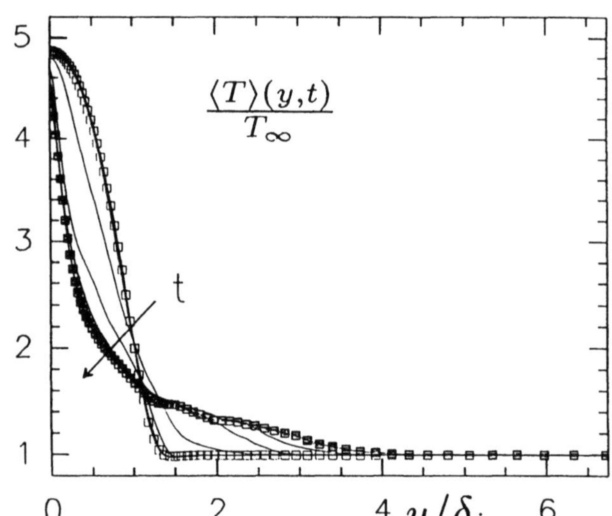

Fig. 19. Time evolution of the mean temperature profile $\langle T\rangle(y,t)$.

The eventual turbulent mean-velocity profile (onto which collapse several curves in Fig. 18) is very close to the turbulent profiles measured experimentally in the incompressible regime (see, *e.g.*, Schlichting 1979, p. 567). This is confirmed by Fig. 20, in which this profile is plotted in the wall coordinates $\langle u^+\rangle(y^+)$, together with the logarithmic law commonly accepted in the incompressible regime (see, *e.g.*, Schlichting 1979):

$$\langle u^+\rangle = \frac{\ln y^+}{0.41} + 5 \quad . \tag{7}$$

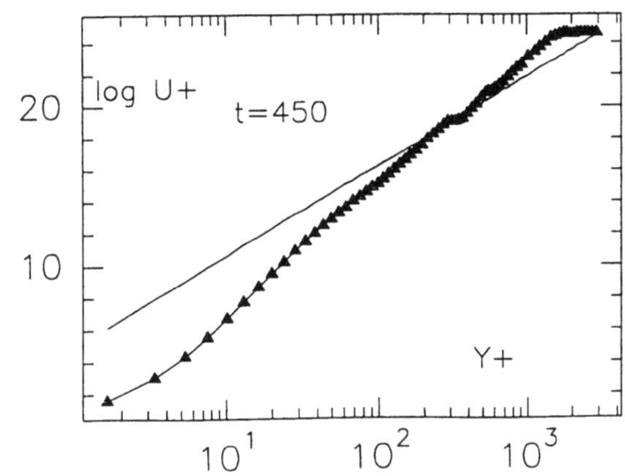

Fig. 20. $\langle u^+\rangle(y^+, t = 450\ \delta_i/U_\infty)$ (turbulent regime).

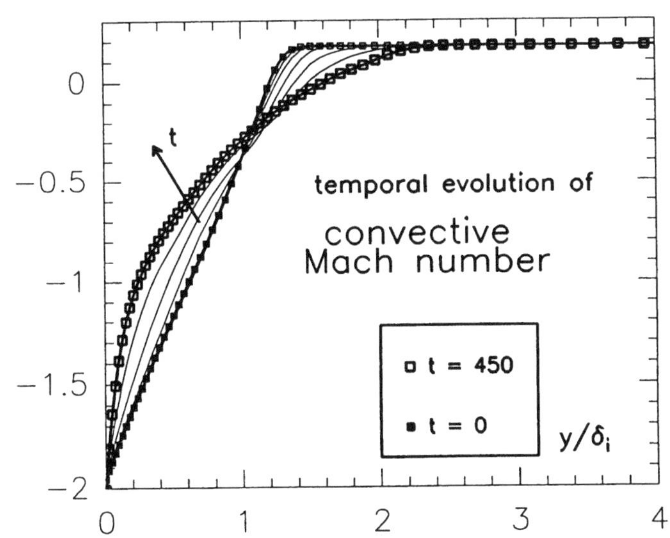

Fig. 21. Time evolution of $\widehat{M}$ for $c = 0.94\ U_\infty$.

From Fig. 19, one can see that for any given distance from the wall y smaller than y_S, $\langle T\rangle(y,t)$ decreases continuously with time throughout transition. The same trend is observed on the average speed of sound $\langle a\rangle(y,t) = \sqrt{\gamma\frac{\mathcal{R}}{\mathcal{M}}\langle T\rangle(y,t)}$. In the same vein, it is straightforward to plot:

$$\widehat{M}(y,t) = \frac{\langle u\rangle(y,t) - c}{\langle a\rangle(y,t)} \quad , \tag{8}$$

the convective-Mach-number profile relevant to a frame of reference travelling at a constant speed c with respect to the wall. This is what has been done in Fig. 21 for $c = 0.94\,U_\infty$, the phase-speed of Mack's second mode. One can see that the distance to the wall y such that $|\widehat{M}(y,t)| = 1$ also decreases with time, which means that the zone dominated by supersonic instabilities shrinks during transition, as observed by Adams and Kleiser (1993a). It would of course have been more convincing not to assume c constant and educe profiles $c(y,t)$ for all modes of importance.

6. Conclusions

The transition to turbulence in a forced compressible boundary layer at Mach 4.5 over an insulated flat plate is simulated at $R = 10\,000$ by means of direct and large-eddy calculations. The initial perturbations consist of Mack's second mode developing at the height y_S of the generalized inflection point, plus small-amplitude 3D white noise. Both calculations predict the development around y_S of a subharmonic mode of secondary instability at the rate given by the stability analysis of Ng and Erlebacher (1992). At the beginning of the transition (shown by the rise of c_f and the drop of H_{12}), all the turbulent activity shifts towards the wall, in the form of streamwise streaks (of spanwise spacing $\approx$ 100 wall units), and small-scale turbulence. This suggests a possible scenario for the transition of high-Mach number boundary layers:

- growth of 3D inviscid modes about y_S (either as a secondary instability of a 2D mode or a 3D primary instability, depending on the Mach number and the nature of the initial perturbations). Their spanwise (and streamwise) wavelength scales with the local thickness of the layer.
- simultaneous growth (but much slower, possibly linearly in time as suggested by Landahl (1980)) of viscous wall instabilities, essentially in the form of purely spanwise Squire modes. Their wavelength scales with the instantaneous wall unit ($\approx$ 100 *wall units*).
- resonance between the two (possibly the *direct resonance* of Benney and Gustavson 1981; see also Schmid and Henningson 1992, Gathmann *et al.* 1993) occurs when their respective length scales commensurate. This brings about turbulence near the wall, which causes an increase of the wall friction and a "devaluation" of the wall unit. The streaks then break down into turbulence at smaller scale.

Direct simulations are then no more possible at such a Reynolds number (10 000). The LES continues and seems to converge towards a self-similar turbulent regime (with coherent hairpin vortices) having strong analogies with its incompressible counterpart.

Acknowledgments

We wish to acknowledge the financial support of CNES and Dassault-aviation. One of us (F.Ducros) is indebted DRET for a grant.

References

Adams, N.A., Kleiser, L. 1993a: Numerical simulation of transition in a compressible flat plate boundary layer, *Transitional and Turbulent Compressible flows*, ASME 1993, *151*, pp. 101-110.

Adams, N.A., Kleiser, L. 1993b: Numerical Simulation of Fundamental Breakdown of a Laminar Boundary Layer at Mach 4.5, *AIAA paper* 93-5027.

Benney, D.J., Gustavsson, L.H. 1981: A new mechanism for linear and nonlinear hydrodynamic instability, *Stud. Appl. Math.*, **64**, 185-209.

Comte, P., Fouillet, Y., Lesieur, M. 1992: Simulation numérique des zones de mélange compressibles, *Revue Scientifique et Technique de la Défense*, **3**, 43-63.

Erlebacher, G., Hussaini, M.Y. 1989: Non-linear evolution of a second mode wave in supersonic boundary layers, *ICASE report* № 89-15.

Fisher, M.C., Weinstein, L.M. 1972: Cone transitional boundary-layer structure at *Me* = 14, *AIAA J.*, **10**, 5, 699-701.

Fouillet, Y. 1991: *Contribution à l'étude par expérimentation numérique des écoulements cisaillés libres: effets de compressibilité*, Thèse de l'Institut National Polytechnique de Grenoble.

Gathmann, R.J., Si-Ameur, M., Mathey, F. 1993: Numerical simulation of three-dimensional natural transition in the compressible confined shear layer, *Phys. Fluids* A **5** (11) 2946-2968

Gottlieb, D., Turkel, E. 1976: Dissipative two-four methods for time-dependent problems, *Math. Comp.*, **30**, (136), 703-723.

Landahl, M.T. 1980: A note on an algebraic instability of inviscid parallel shear flows, *J. Fluid Mech.*, **98**, 243-251.

Mack, L.M. 1969: *Boundary-layer stability theory*, Jet Propulsion Lab., Pasadena, Calif., rep. № 900-277.

Maise G., Mac Donald, M. 1968: *AIAA J.*, **6**, 73.

Métais, O., Lesieur, M. 1992: Spectral large-eddy simulations of isotropic and stably-stratified turbulence, *J. Fluid Mech*, **239**, 157-194.

Michel, R. 1967: *Couches limites - frottement et transfert de chaleur*. Cours E.N.S.A.E.

Morkovin, M. V. 1961: Effects of compressibility on turbulent flows. In *Mécanique de la Turbulence*, colloque CNRS № 108, ed. C.N.R.S., pp. 367-380.

Ng, L.L., Erlebacher, G. 1992: Secondary instabilities in compressible boundary layers, *Phys. Fluids A*, **4** (1992) 710.

Normand, X. 1990: *Transition à la turbulence dans les écoulements cisaillés compressibles libres et pariétaux. Thèse*, Institut National Polytechnique de Grenoble.

Normand, X., Lesieur, M. 1992: Direct and large-eddy simulations of transition in the compressible boundary Layer, *Theor. and Comp. Fluid Dyn.*, **3** p. 231.

Pruett, C.D., Zang, T.A. 1992: Direct numerical simulation of laminar breakdown in high-speed, axisymmetric boundary layers, *Theor. and Comp. Fluid Dyn.*, **3** 345-367.
Sandham, N.D., Reynolds, W.C. 1991: Three-dimensional simulations of the compressible mixing layer, *J. Fluid Mech.*, **224**, 133.
Schlichting, H. 1979: *Boundary-layer theory*. McGraw-Hill, New-york, seventh edition.
Schmidt, P.J., Henningson, D.S. 1992: A new mechanism for rapid transition involving a pair of oblique waves, *Phys. Fluids A*, **4** (1992) 1986.

Fine Structure of Reynolds Shear Stress in an Unsteady Turbulent Channel Flow

S. Tardu, M. Q. Feng, and G. Binder

Institut de Mécanique de Grenoble, Laboratoire des Ecoulements Géophysiques et Industriels,
INPG, UJF, CNRS, B.P. 53, 38041 Grenoble Cedex 9, France

Abstract

The modulation characteristics of the Reynolds shear stresses in an unsteady channel flow are presented. The imposed frequency covers a large range going from the quasi-steady limit to 2 times the ejection frequency at $y^+ = 15$. The time mean contributions of the quadrants to the Reynolds shear stress are unaffected by the imposed unsteadiness. The contributions of the second and fourth quadrants are strongly modulated near the imposed frequency in wall units $f^+ \approx 0.002$, where the modulation of the ejection frequency is weak and the phase shift with respect to the centerline velocity is maximum. In the low frequency regime the ejection frequency is modulated as the wall shear stress according to the quasi-steady behaviour, and the decline in its response in the high imposed frequency regime stretches from the near-wall to the outer layer.

Nomenclature

$A_{\tilde{q}}$	Amplitude of the first Fourier mode
$a_{\tilde{q}}$	Relative amplitude of the first Fourier mode
f	Imposed frequency
$\tilde{f}_e$	Ejection frequency modulation
$\tilde{H}_2$	Unsteady threshold in the quadrant 2 scheme
$l_s^+ = 1/\sqrt{\pi f^+}$	Imposed frequency parameter
$\tilde{c}_i$	Modulation of the contribution of quadrant i to the turbulent shear stress
$\tilde{c}_E$	Contribution per ejection to quadrant 2
$\partial\tilde{p}/\partial x$	Oscillating pressure gradient
T	Imposed period
$\tilde{U}_c$	Oscillating centerline velocity
$\tilde{u}$	Oscillating velocity
u', v'	Instantaneous streamwise and normal velocities
$\widetilde{u'u'}$, $-\widetilde{u'v'}$	Modulation of wall turbulent intensity and of Re stress
$\bar{u}_\tau$	Friction velocity

$\partial\tilde{u}/\partial y$	Oscillating shear
x, y	Longitudinal and wall normal coordinates
$\Delta t^+ = \Delta\Phi/\omega^+$	Time lag
$\Phi_{\tilde{q}}$	Phase of the first Fourier mode
ν	Kinematic viscosity
ϱ	Density
$\tilde{\tau}$	Wall shear stress modulation
τ'	Instantaneous wall shear stress
$\omega = 2\pi f$	Imposed angular frequency
$()_{qs}$	Quantity in the quasi-steady regime
$()^+$	Variable normalized by $\bar{u}_\tau$ and ν

1. Introduction

Previous investigations (Tardu *et al.* 1987; Tardu and Binder 1992) of the response of the bursting process to forced oscillations were based on the detection of ejections and bursts via the u' or τ' signals. They showed that the amplitude and phase of the modulation of the ejection and bursting frequencies depend strongly on the frequency f of the imposed oscillations. They also revealed the existence of a critical forcing frequency at which the response of the structures goes through a marked minimum together with large phase shifts. The decline of the amplitude response with the imposed frequency was so far explained by the finite relaxation time of the turbulent structures. This decline however was not noted in the response of the clusters of ejections (called as multiple ejection bursts) which responded in quite a uniform manner as the wall shear stress in a large imposed frequency range. Since the measurements of the Reynolds stress were not available in these studies, further investigations were needed in order to clarify the reaction of the near wall turbulence to such forcing. In particular it is important to know how the unsteady organization of the Reynolds shear stress takes place during the oscillation cycle. Only a limited number of studies concerning this aspect of the unsteady flows have been carried out in the past (Brereton *et al.* 1990; Brereton and Reynolds 1991) and more details and new data are needed in order to clear up some unanswered questions.

This research furthered by the acquisition of a data set of u' and v' signals for several forcing conditions with a double aim, firstly, to determine how the phase averaged shear-stress $-\langle u'v' \rangle$ is related to the oscillating shear during the cycle and, secondly, to reassess the previous results on the response of the coherent structures on the basis of $u'v'$ detections. The response of the modulation of the Reynolds shear stresses and the related quantities such as the production and the eddy viscosity is partly discussed in Feng *et al.* (1993). The results associated with the fine structure of the unsteady near wall turbulence are the subject of the present study.

2. Experimental Set-up and Data Reduction

The experiments were performed in a 100 mm wide 2600 mm long unsteady water channel with a span of 1000 mm as described by Tardu *et al.* (1994). The sinusoidal velocity oscillations which are superimposed to the fully developed turbulent channel flow are generated by a specific pulsating device. Both the mean flow, the period and the amplitude of the imposed velocity oscillations can be independently changed. The mean centerline velocity can be varied from 0 to 50 cm/s. The corresponding maximum Reynolds number based on the half height of the channel is $Re_h = 25 \cdot 10^3$. The amplitude can be varied up to 80% of the centerline mean velocity and the imposed period can be changed from 2.5 sec to infinity.

Table 1. Flow Conditions.

	$\overline{U}_c$ cm/s	$a_{\tilde{u}_c}$	l_s^+
◊	17.4	0.20	42
□	17.4	0.20	30
×	12.0	0.14	21
+	12.0	0.14	15
△	17.4	0.20	12
■	17.4	0.20	8.0
•	12.0	0.14	5.4
♦	12.0	0.14	4.3

The flow conditions corresponding to the experiments discussed here are given in Table 1. Two Reynolds numbers, which are respectively $Re_h = 8700$ and $Re_h = 6200$, together with two imposed amplitudes (20 and 14% of the centerline velocity) are considered. The imposed frequency parameter $l_s^+ = \sqrt{T\bar{u}_\tau^2/\pi\nu}$ is varied from 42, to 4.3 covering, therefore, a large range going from the quasi-steady to high imposed frequency regime ($\bar{u}_\tau$ is the friction velocity and ν is the kinematic viscosity).

Simultaneous measurements of streamwise $u(t)$ and wall normal $v(t)$ velocities were performed by means of a TSI 2148-10w x-hot film probe. The probe was calibrated vs. velocity and angle and $u(t)$ and $v(t)$ were obtained from the hot-film signals by use of the look-up table technique (Lueptow et al. 1988). Data acquisition was performed on a MASSCOMP computer with a 16 bit accuracy. The sampling frequency was $2\,\bar{u}_\tau^2/\nu$. The minimum record length was $6000\,h/\overline{U}_c$ where h is the channel half width and $\overline{U}_c$ is the mean centerline velocity.

The phase averages were determined by dividing the cycle into 30 to 40 bins. The classical triple decomposition is used. A quantity q is decomposed into a mean $\bar{q}$, oscillating $\tilde{q}$ and fluctuating q' components. The phase average of q is defined as $\langle q \rangle = \bar{q} + \tilde{q}$. In a similar fashion all quantities such as the ejection frequency, the contribution of quadrants to the Reynolds stress, the time scales and the threshold levels are decomposed into mean value plus modulation. The modulation characteristics of $\langle q \rangle$ are defined as the amplitude $A_{\tilde{q}}$ and phase $\Phi_{\tilde{q}}$ of the first Fourier mode of $\langle q \rangle$, i.e., $\langle q \rangle \approx \bar{q} + A_{\tilde{q}} e^{i(\omega t + \Phi_{\tilde{q}})}$, where $\omega = 2\pi f$ and f is the imposed frequency. The relative amplitude $a_{\tilde{q}} = A_{\tilde{q}}/\bar{q}$ is introduced for convenience. Here, $()^+$ designates variables normalized by the inner scale related to the friction velocity $\bar{u}_\tau$ and the viscosity ν, i.e the length and the frequency scales are respectively $l_\nu = \nu/\bar{u}_\tau$ and $f_\nu = \bar{u}_\tau^2/\nu$.

3. Results

3.1 Modulation of Reynolds Shear Stress

Figure 1 shows the relative amplitude of the modulation of the Reynolds shear stress versus y^+. The distribution of the relative amplitude of $\widetilde{u'u'}$ is also shown in Fig. 1a for comparison. At low and medium forcing frequencies, the shear stress modulation characteristilcs (i.e. the amplitude and the phase shift) is close to the value inferred from the quasi-steady assumption $a_{-\widetilde{u'v'}}/2\, a_{\tilde{u}} \approx 1$, in particular near the wall (Fig. 1a). There is also reasonable correspondance between the relative amplitude of $-\widetilde{u'v'}$ and $\widetilde{u'u'}$ in this frequency range. At larger frequencies $l_s{}^+ = 1/\sqrt{\pi f^+} = 8$ and 5, these modulations are quite small especially near the wall (Fig. 1b); because of its finite relaxation time, the turbulence cannot follow the imposed unsteadiness once the imposed period is close to its mean relaxation time near the wall, as shown earlier by Finnicum and Hanratty (1988) and Tardu and Binder (1993). The relaxation regime wherein the cyclic variation of the turbulence weakens, takes place when the imposed frequency is of the same order of the bursting frequency of the corresponding steady flow (Tardu and Binder, 1993).

When the imposed frequency is further increased, the reaction of the near-wall turbulence changes significantly. At the highest imposed frequency $l_s{}^+ = 4$, for instance, the amplitude ratio $a_{-\widetilde{u'v'}}/2\, a_{\tilde{u}}$ is quite large compared with $l_s{}^+ = 8$ and 5, in the entire inner layer. It is seen in Fig. 1b that this ratio is even larger than the one corresponding to the low imposed frequency regime. This result too, is in agreement with the previously published data which showed quite a different reaction of the turbulence once the imposed frequency is increased above the bursting frequency (Tardu and Binder, 1993).

The time lag of the $-\widetilde{u'v'}$ modulation increases linearly away from the wall in the small imposed frequency range (Fig. 2a), but the propogation speed is not independent of the imposed frequency once $l_s{}^+ < 8$ (Fig. 2b). Considering that the time lag is due to the finite response time of the near wall turbulence, one can conclude a substantial decrease of the relaxation time in the high imposed frequency regime.

3.2 Organization of Quadrant Contributions

The phase average of the contribution of each quadrant to the turbulent shear stress $\langle u' v' \rangle$ is denoted by $\langle c_i \rangle = \langle u' v'_i \rangle / \langle u' v' \rangle$ where i refers to the quadrant 1

a)

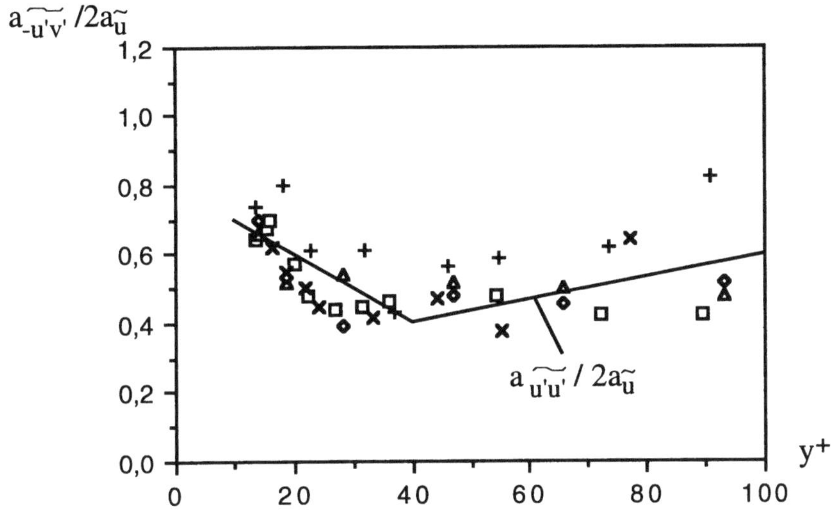

b)

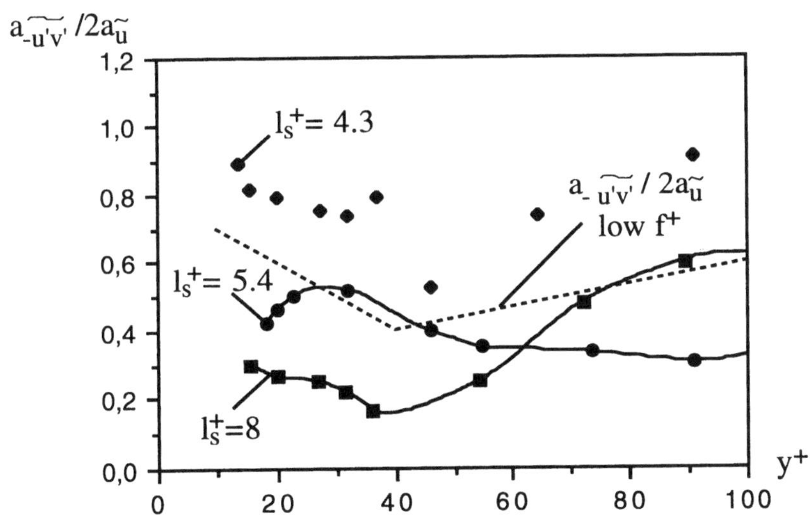

Fig. 1. Relative amplitude of the modulation of the Re shear stress **a**) low imposed frequency range; **b**) high imposed frequency range.

a)

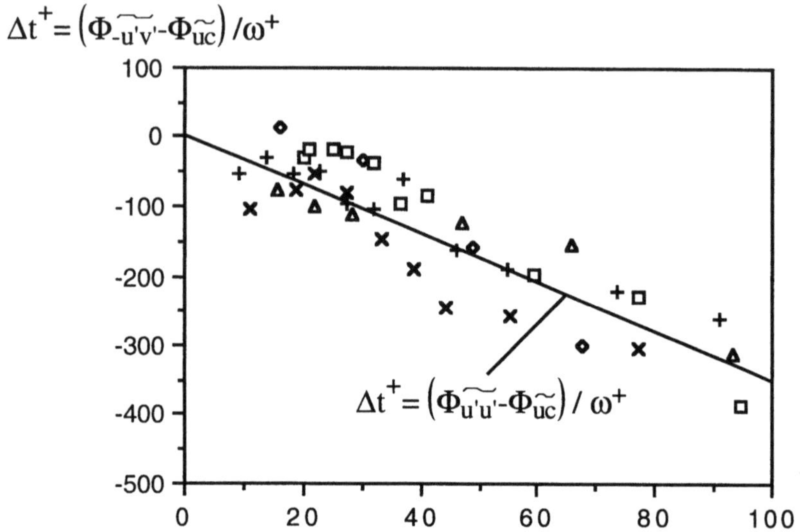

b)

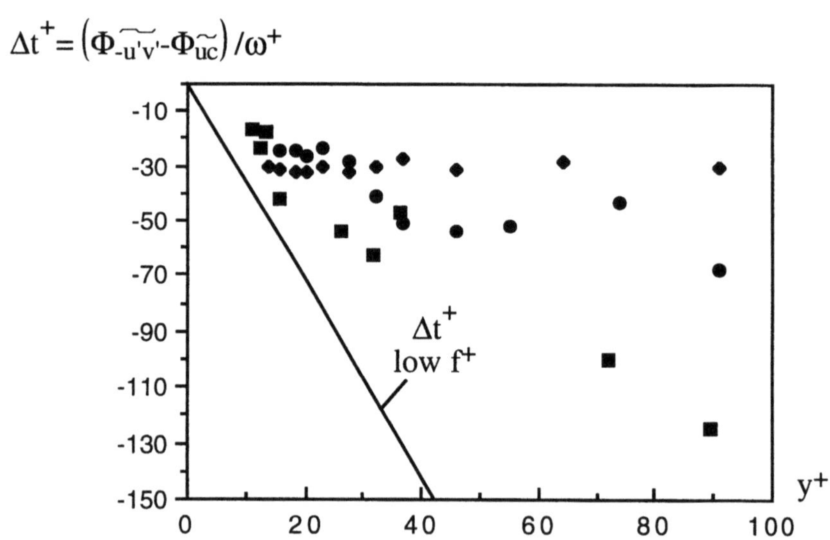

Fig. 2. Time lag of the modulation of the Reynolds shear stress with respect to the modulation of the centerline velocity; **a)** low imposed frequency range; **b)** high imposed frequency range.

a)

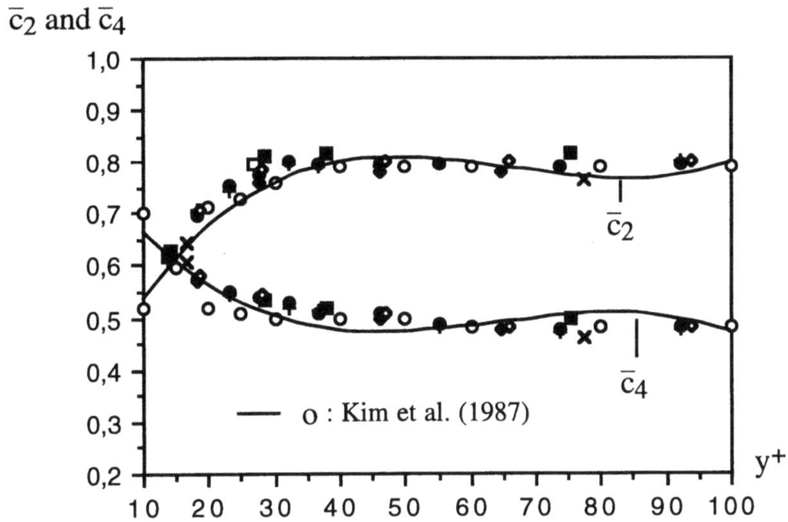

b)

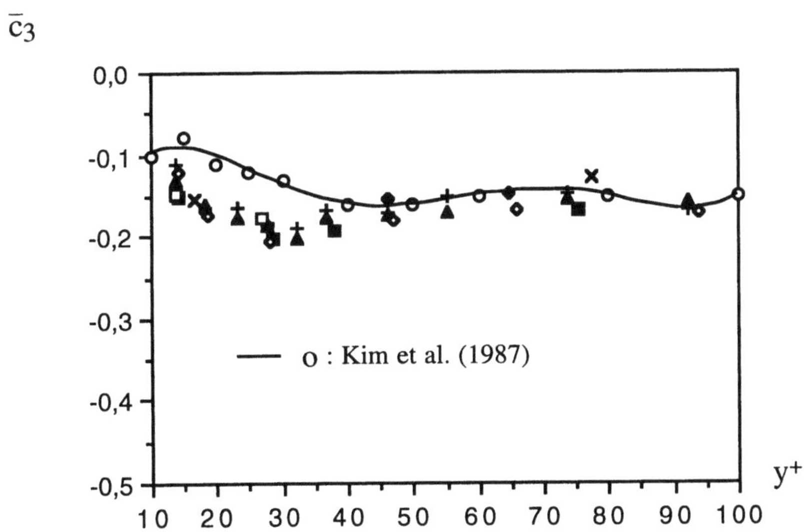

Fig. 3. Time mean quadrant contributions compared with steady flow; **a)** ejection and sweep type events; **b)** interaction type events.

($u' > 0, v' > 0$; interactions: Q1), 2 ($u' < 0, v' > 0$; ejections: Q2), 3 ($u' < 0, v' < 0$; interactions: Q3) or 4 ($u' > 0, v' < 0$; sweeps: Q4). Figures 3a and 3b show respectively the time mean contributions of quadrants 2 and 4, and of quadrant 3. The time mean contributions are insensitive to the imposed unsteadiness and the profiles of $\bar{c}_i$ compare well with previous steady flow data as for instance with those reported by Kim *et al.* (1987). There are some discrepancies in the range $15 < y^+ < 30$ between $\bar{c}_3$ and $\bar{c}_4$ reported here and the direct numerical simulation (DNS) data of Kim *et al.* (1987). These differences, however, are not significant, since they result in a contribution to $\overline{u'v'}$ by less than $\pm$ 5%. It has to be pointed out that, the satisfactory agreement between the profiles of $\bar{c}_i$ with recent steady flow data is encouraging, since serious disagreements exist between past experiments and the DNS results particularly in the buffer and viscous sublayer (Wallace *et al.* 1972; Brodkey *et al.* 1974; Kim *et al.* 1987).

The contribution of the quadrant 2 event is predominant in the larger part of the boundary layer except near the wall where the contribution of the quadrant 4 event increases progressively. The major part of $\overline{u'v'}$ is due to the contributions of Q2 and Q4 events. The interactions contribute only weakly and $\bar{c}_1 + \bar{c}_3$ hardly exceeds -0.3. Therefore, the accent here is only on $\langle c_2 \rangle$ and $\langle c_4 \rangle$.

The contributions of the quadrants to the Reynolds shear stress are firstly analyzed by separately grouping positive and negative contributions defined by:

$$\langle c_{24} \rangle = \langle c_2 \rangle + \langle c_4 \rangle \,,$$

and,

$$\langle c_{13} \rangle = \langle c_1 \rangle + \langle c_3 \rangle \,.$$

These quantities are only weakly modulated. The relative amplitude $a_{\tilde{c}_{24}}$ shown in Fig. 4 hardly exceeds 5% of the relative amplitude of the centerline velocity. A simple explanation of this behaviour may be given by considering that:

$$\langle c_{24} \rangle + \langle c_{13} \rangle = \bar{c}_{24} + A_{\tilde{c}_{24}}\, e^{i(\omega t + \Phi_{\tilde{c}_{24}})} + \bar{c}_{13} + A_{\tilde{c}_{13}}\, e^{i(\omega t + \Phi_{\tilde{c}_{13}})} = 1 \,.$$

Now, since $\bar{c}_{24} + \bar{c}_{13} = 1$, dividing this relationship by $\bar{c}_{24}$ gives:

$$a_{\tilde{c}_{24}}\, e^{i(\omega t + \Phi_{\tilde{c}_{24}})} + \frac{\bar{c}_{13}}{\bar{c}_{24}}\, a_{\tilde{c}_{13}}\, e^{i(\omega t + \Phi_{\tilde{c}_{13}})} = 0 \,.$$

Since $\bar{c}_{13}/\bar{c}_{24} \approx 0.2 \ll 1$ the second term is negligible (except when $a_{\tilde{c}_{13}}/a_{\tilde{c}_{24}} \approx 5$ which is quite unlikely) and one obtains approximately $a_{\tilde{c}_{24}} \approx 0$. Now, the small value of $a_{\tilde{c}_{24}}$ gives in return a simple relationship between the amplitude and phase of the contribution of quadrant 2 and quadrant 4 events. An analysis similar to the preceding one indicates indeed that:

$$\frac{a_{\tilde{c}_4}}{a_{\tilde{c}_2}} \approx \frac{\bar{c}_2}{\bar{c}_4} \,,$$

and,

$$\Phi_{\tilde{c}_2} - \Phi_{\tilde{c}_4} \approx 180^\circ \,.$$

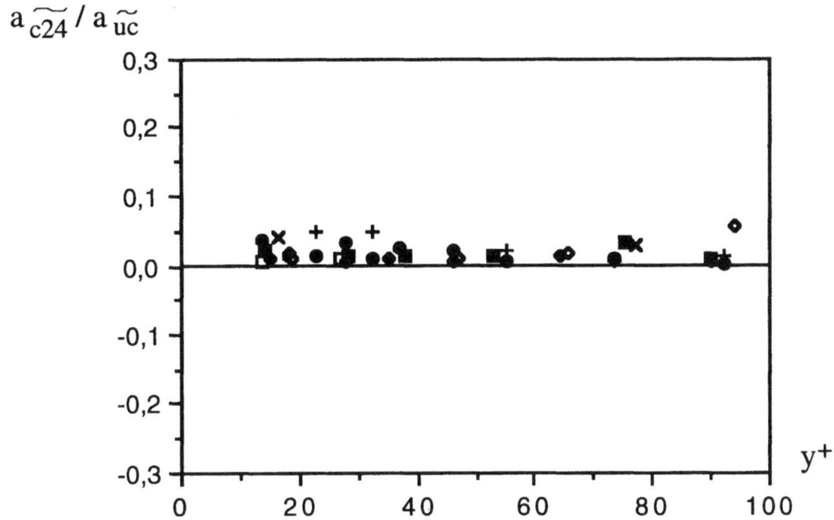

Fig. 4. Profiles of the relative amplitude of the contribution of Quadrant 2 and Quadrant 4 events.

The quasi-steady behaviour of $a_{\tilde{c}_i}$ is found easily from their profiles in the steady flow, since they only slightly depend on the Re number in the range investigated in this study. Let $\bar{c}_i$ be expressed in the form:

$$\bar{c}_i = \sum_0^N \alpha_j \, y^{+j} .$$

In the quasi-steady limit, the phase averages of the contributions are therefore related to:

$$\langle c_i \rangle = \bar{c}_i + A_{\tilde{c}_i \, \mathrm{qs}} \, \mathrm{e}^{i\left(\omega t + \Phi_{\tilde{c}_i \, \mathrm{qs}}\right)} = \sum_0^N \alpha_j \left(\frac{y}{\nu}\right)^j \langle u_\tau \rangle^j{}_{\mathrm{qs}} =$$

$$= \sum_0^N \alpha_j \left(\frac{y}{\nu}\right)^j \bar{u}_\tau^j \left(1 + a_{\tilde{u}_\tau \, \mathrm{qs}} \, \mathrm{e}^{i\left(\omega t + \Phi_{\tilde{u}_\tau \, \mathrm{qs}}\right)}\right)^j .$$

When the imposed amplitude is small, then to the first approximation this relationship becomes:

$$\bar{c}_i + A_{\tilde{c}_i \, \mathrm{qs}} \, \mathrm{e}^{i\left(\omega t + \Phi_{\tilde{c}_i \, \mathrm{qs}}\right)} \approx \sum_0^N \alpha_j \left(\frac{y}{\nu}\right)^j \bar{u}_\tau^j \, j \left(1 + a_{\tilde{u}_\tau \, \mathrm{qs}} \, \mathrm{e}^{i\left(\omega t + \Phi_{\tilde{u}_\tau \, \mathrm{qs}}\right)}\right) .$$

Consequently:

$$A_{\tilde{c}_i\,\mathrm{qs}} \approx \sum_0^N j\,\alpha_j\,y^{+j}\,a_{\tilde{u}_\tau\,\mathrm{qs}} \approx \sum_0^N j\,\alpha_j\,y^{+j}\,a_{\tilde{u}_c}\,,$$

and $\Phi_{\tilde{c}_i\,\mathrm{qs}} \approx \Phi_{\tilde{u}_c}$. That results in:

$$\frac{a_{\tilde{c}_i\,\mathrm{qs}}}{a_{\tilde{u}_c}} \approx 2\,\frac{a_{\tilde{c}_i\,\mathrm{qs}}}{a_{\tilde{\tau}\,\mathrm{qs}}} \approx \frac{\sum_0^N j\,\alpha_j\,y^{+j}}{\sum_0^N \alpha_j\,y^{+j}}\,.$$

It is seen from the above relationship that both the ratios $a_{\tilde{c}_i}/a_{\tilde{u}_c}$ and $a_{\tilde{c}_i}/a_{\tilde{\tau}}$ can be independently analyzed. The ratios $a_{\tilde{c}_i}/a_{\tilde{u}_c}$ and $a_{\tilde{c}_i}/a_{\tilde{\tau}}$ are not simply proportional (except in the low imposed frequency range wherein $a_{\tilde{\tau}}/a_{\tilde{u}_c} \approx 2$), because beyond $f^+ = 0.003$, the relative amplitude of the wall shear stress follows the viscous Stokes solution: $a_{\tilde{\tau}}/a_{\tilde{u}_c} = \sqrt{2\,\pi\,f^+}\,(\bar{U}_c/\bar{u}_\tau)$ and hence, increases with imposed frequency (Mao and Hanratty 1986; Tardu and Binder 1993).

Figures 5a and 5b show respectively the $a_{\tilde{c}_2}/a_{\tilde{u}_c}$ and $a_{\tilde{c}_2}/a_{\tilde{\tau}}$ profiles. The dashes in these figures correspond to the distributions in the quasi-steady regime computed by means of a third-order polynomial approximation. In the quasi-permanent flow and at $y^+ > 35$, $\langle\, c_2 \rangle_{\mathrm{qs}}$ is not modulated since $\bar{c}_2$ is almost constant in this region in the steady flow. Near the wall, however, where $\bar{c}_2$ decreases, the ratio $a_{\tilde{c}_2\,\mathrm{qs}}/a_{\tilde{u}_c}$ increases up to 0.3. Figure 5a shows that the profile corresponding to $l_s^+ = 42$ coincides quite well with the quasi-steady one with only a slight departure at the upper part of the inner layer where $a_{\tilde{c}_2}/a_{\tilde{u}_c}$ is about 0.2 and therefore hardly significant.

The unsteadiness rapidly affects the redistribution of the turbulent shear stress as f^+ increases from the quasi-steady limit and $\langle\, c_2 \,\rangle$ becomes strongly modulated at $l_s^+ = 15$. The increase of $a_{\tilde{c}_2}/a_{\tilde{u}_c}$ is followed by a subsequent decrease in the high imposed frequency range. The ratio $a_{\tilde{c}_2}/a_{\tilde{\tau}}$ regroups the data better at $l_s^+ < 8$ compared with $a_{\tilde{c}_2}/a_{\tilde{u}_c}$ as is seen in Fig. 5b. The progressive decrease of the $\langle\, c_2 \,\rangle$ modulation at $l_s^+ < 8$ is also more conspicuous when $a_{\tilde{c}_2}$ is scaled with $a_{\tilde{\tau}}$.

The dependence of the modulation of the contributions on the imposed frequency is more evident in Fig. 6a, which recapitulates $a_{\tilde{c}_2}/a_{\tilde{\tau}}$ and $a_{\tilde{c}_4}/a_{\tilde{\tau}}$ measured at $y^+ \approx 15$. The sharp increase of the relative amplitudes of $\langle\, c_2 \,\rangle$ and $\langle\, c_4 \,\rangle$ near $f^+ = 0.0014$ ($l_s^+ = 15$) is striking in this figure. The subsequent decrease towards higher imposed frequencies takes place as if the quadrant redistribution or the unsteady organization of $\langle\, u'v' \,\rangle$ "returns" back to its quasi-steady behaviour. The phase shifts $\Phi_{\tilde{c}_2} - \Phi_{\tilde{\tau}}$ and $\Phi_{\tilde{c}_4} - \Phi_{\tilde{\tau}}$ shown in Fig. 6b confirm also this suggestion. The change in $\Phi_{\tilde{c}_2} - \Phi_{\tilde{\tau}}$ from approximately 180° at $f^+ = 0.0014$ to zero in the high frequency regime is as rapid as the decrease of $a_{\tilde{c}_2}/a_{\tilde{\tau}}$ in the same range of f^+. Such rapid modifications of the response of the turbulent quantities are common in unsteady wall flows. A typical example is the sharp decrease of the modulation of the wall shear stress intensity from the quasi-steady regime to $f^+ = 0.004$ (Mao and Hanratty 1986; Tardu and Binder 1993) or the decrease of $\Delta t^+ = \left(\Phi_{\widetilde{-u'v'}} - \Phi_{\tilde{u}_c}\right)/\omega^+$, when f^+ varies from $f^+ = 0.005$ to $f^+ = 0.002$ (Fig. 2b).

a)

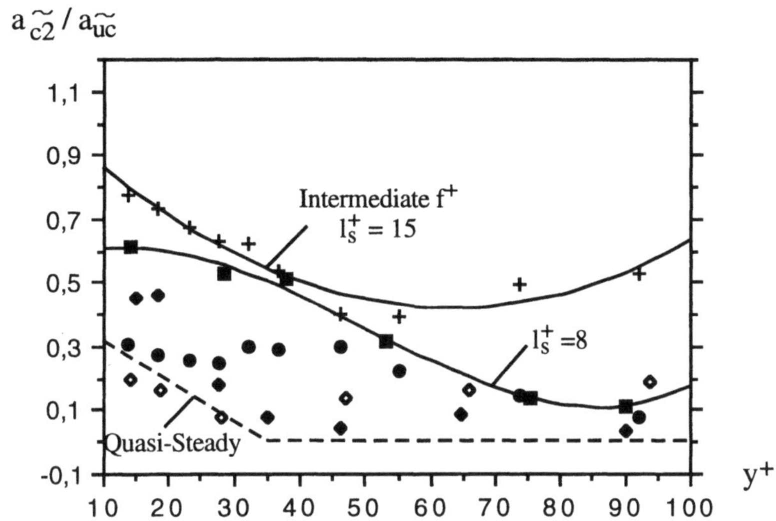

b)

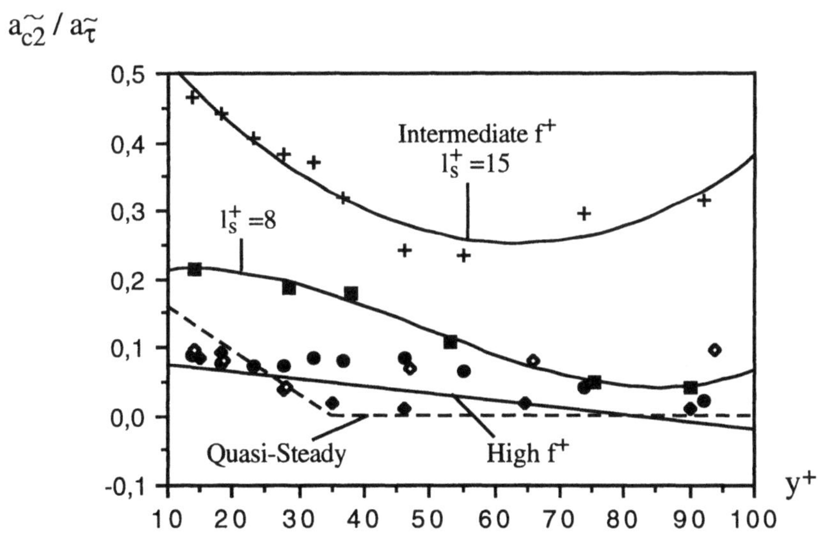

Fig. 5. Profiles of the relative amplitude of the contribution of Quadrant 2 events with respect to **a**) centerline velocity; **b**) wall shear stress. For legend see Table 1.

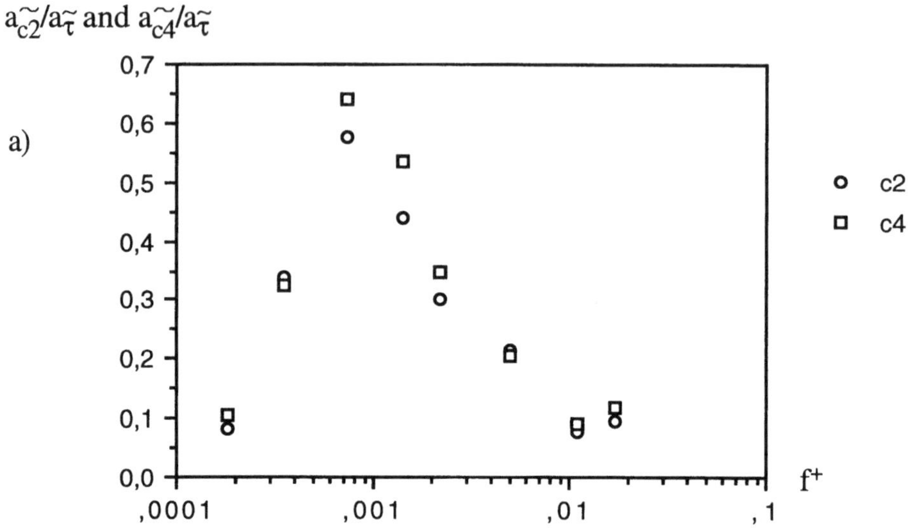

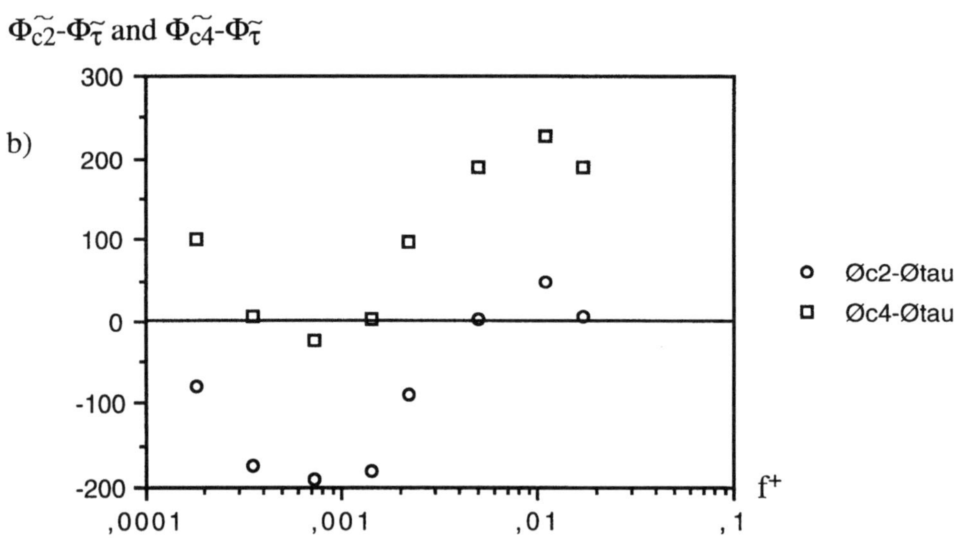

Fig. 6. Modulation characteristics of Quadrant 2 and Quadrant 4 contributions at $y^+ = 15$ vs. the imposed frequency; **a**) relative amplitude; **b**) phase shift.

3.3 Modulation of Ejection Frequency

One should be careful in the adaptation of the ejection detection schemes to the unsteady flow. In a steady channel flow, the detection function $D(t)$ which identifies an ejection by the quadrant 2 technique is:

$$D(t) = 1 \quad \text{if} \quad u'(t) < 0\,;\ v'(t) > 0 \quad \text{and}$$

$$u'v'(t) < -\bar{H}_2 \sqrt{\overline{u'u'}}\,\sqrt{\overline{v'v'}}\,.$$

The threshold used in the quadrant scheme is usually defined as:

$$\bar{H}_2 = -\frac{\overline{(u'v')_2}}{\sqrt{\overline{u'u'}}\,\sqrt{\overline{v'v'}}}\,,$$

where subscript 2 refers to the quadrant where $\overline{(u'v')}$ is computed (Compte Bellot *et al.* 1978; Bogard 1982). More clearly $\overline{(u'v')}_i$ denotes the mean turbulent shear stress in the quadrant i, i.e., $\overline{(u'v')}_i = \sum u'v'_i/\bar{N}_i$ with $\bar{N}_i$ being the number of samples in the quadrant i. A simple relationship between the contribution $\bar{c}_i$ of the quadrant i event and the threshold $\bar{H}_i$ can be given in terms of $\bar{t}_i = \bar{N}_i/\bar{N}$, which is the percentage of time (u', v') is in the quadrant i ($\bar{N}$ is the total number of samples):

$$\bar{H}_i = \frac{-\overline{u'v'}\,\bar{c}_i}{\bar{t}_i \sqrt{\overline{u'u'}}\,\sqrt{\overline{v'v'}}}\,.$$

In a similar way, the ejections are identified in an unsteady flow by:

$$D(t) = 1 \quad \text{if} \quad u'(t) < 0\,;\ v'(t) > 0 \quad \text{and}$$

$$u'v'(t) < -\langle H\rangle_2 \sqrt{\langle u'u'\rangle}\,\sqrt{\langle v'v'\rangle}\,,$$

where the modulated threshold is:

$$\langle H\rangle_2 = -\frac{\langle (u'v')_2\rangle}{\sqrt{\langle u'u'\rangle}\,\sqrt{\langle v'v'\rangle}}\,.$$

The differences in the modulation characteristics of the thresholds and the contributions are due to the modulation of the residence time $\langle t_i \rangle$ and the unsteady correlation coefficient $-\langle u'v'\rangle/\sqrt{\langle u'u'\rangle}\,\sqrt{\langle v'v'\rangle}$. Note on the other hand, that it would not be adequate to use the lo¬g time average of the threshold $\bar{H}_2$ in the detection scheme, since otherwise the unsteady effects would not be taken into account.

The modulation characteristics of the threshold with respect to the centerline velocity are shown in Table 2 for the seven imposed frequencies investigated so far. The detection is done at $y^+ = 15$, and the results are presented as a function of the frequency parameter $l_s{}^+$. The time mean of the threshold is $\bar{H}_2 \approx 1$ as in the steady flow. The threshold is weakly modulated in the quasi-steady limit as would be expected, since in the steady flow $\bar{H}_2$ almost equals one at all positions accross the channel height (Bogard 1982). The true unsteady effects are important towards the high frequency regime where $a_{\bar{H}_2}$ is large. The error that the omitting of the modulation of $\langle H\rangle_2$ would cause is estimated to be about 20%.

Table 2. Modulation characteristics of the threshold used in the quadrant technique and of the ejection frequency $y^+ = 15$.

l_s^+	$\overline{H}$	$\frac{a_{\tilde{H}}}{a_{\tilde{u}_c}}$	$\Phi_{\tilde{H}} - \Phi_{\tilde{u}_c}$
44	0.94	0.30	25°
30	0.99	0.55	-32°
21	1.06	0.48	-17°
15	1.02	0.55	20°
12	0.94	0.65	-9°
8	0.94	0.50	-40°
4	1.09	0.79	-160°

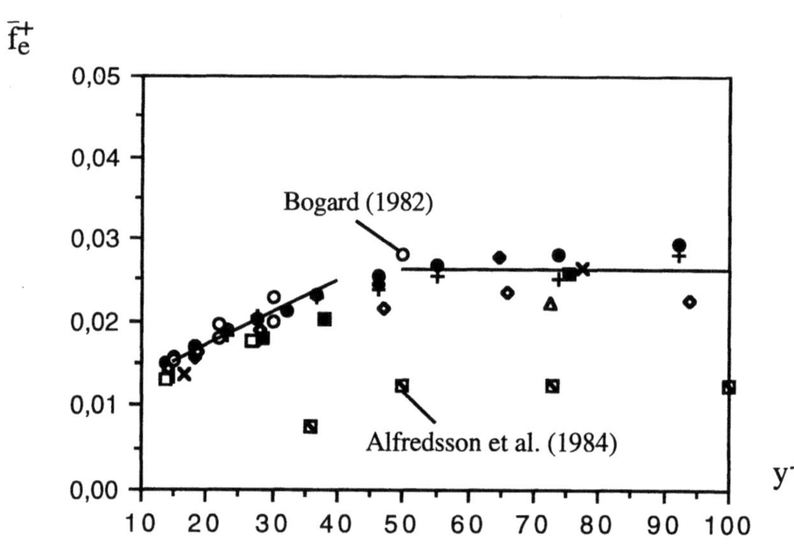

Fig. 7. Time mean ejection frequency. For legend see Table 1.

The time mean ejection frequency is not affected by the imposed unsteadiness as shown in Fig. 7. It is seen that the distribution of the ejection frequency compares well quantitively with the steady flow data of Bogard (1982), who used the same detection scheme adopted here. The unsteady flow data agrees also qualitatively well with that of Alfredsson and Johansson (1984). The quantitative disagreement is due to the differences in the threshold chosen by these authors which is 4 times larger.

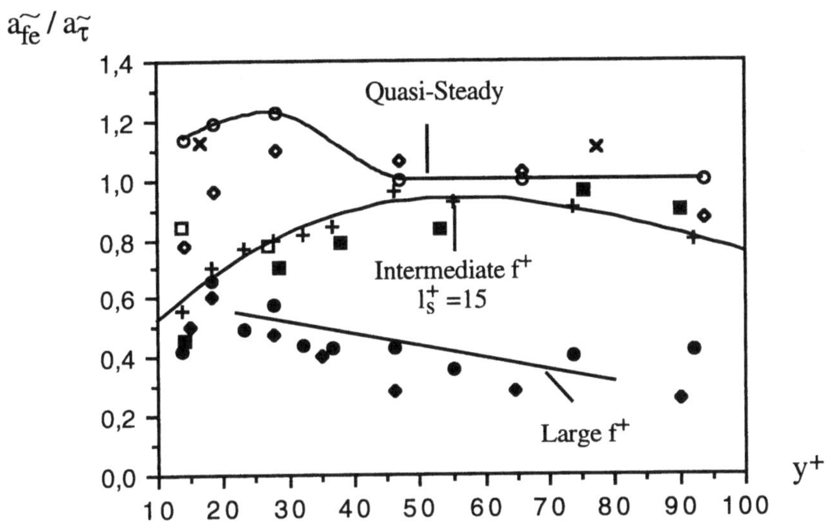

Fig. 8. Profiles of the relative amplitude of modulation of the ejection frequency. For legend see Table 1.

The amplitude and the phase shift of the modulation frequency in the quasi-steady regime may be determined by the distribution of $\bar{f}_e^+$ in a similar way to the modulation of the contributions given in the above section and the results are shown in Fig. 8. The relative amplitude of $\langle f_e \rangle$ in the quasi-steady regime is almost equal to the relative amplitude of the modulation of the wall shear stress at $y^+ > 40$ and is slightly larger than $a_{\tilde{\tau}}$ near the wall where the time mean ejection frequency decreases. It is seen in Fig. 8 that the behaviour of $\langle f_e \rangle$ is close to the quasi-steady response in both amplitude and phase shift (not shown here) at $l_s^+ = 42$ through the entire boundary layer. This confirms, in return, the validity of the inner scaling in the steady flow within the range of Re number investigated here.

In the intermediate imposed frequency range ($l_s^+ = 15$), $a_{\tilde{f}_e}/a_{\tilde{\tau}}$ is near one at $y^+ > 50$ but decreases in the buffer layer by almost 50%. This indicates that the ejection frequency can no longer follow the wall shear stress modulation near the wall once the imposed frequency is about $f^+ = 0.0014$. It is worth pointing out that the beginning of the relaxation regime corresponding to $\langle f_e \rangle$ coincides with the confining of the oscillating shear layer near the wall. Past research has indeed shown that the oscillating gradient $\partial\widetilde{u}/\partial y$ is confined to $y^+ < 2\,l_s{}^+ = 2/\sqrt{\pi\, f^+}$ once $f^+ > 0.002$ (mostly to $y^+ < l_s{}^+$, Mao and Hanratty 1986; Brereton *et al.* 1991; Tardu *et al.* 1994; the frequency limits which demark several flow regimes in unsteady flow are not universal and a Re number dependence may be suspected). The thickness of the oscillating shear layer decreases therefore, once $f^+ > 0.002$. Although $f^+ = 0.0014$ is slightly different than the limit given above, one can consider that the part of the flow corresponding to $y^+ > 40$ is almost solely under the effect of the oscillating pressure gradient $\partial\widetilde{p}/\partial x$, while the buffer layer and the viscous sublayer are exposed to both $\partial\widetilde{u}/\partial y$ and $\partial\widetilde{p}/\partial x$. The Reynolds shear stress producing eddies are consequently also forced by both an oscillating pressure gradient and an oscillating shear. The response of the ejection frequency results from the negotiation of these quantities. It is therefore suggested here, that the relaxation of the modulation of the ejection frequency at $y^+ < 40$ ($f^+ = 0.0014$) may result from a mutual cancellation of the effects of $\partial\widetilde{u}/\partial y$ and $\partial\widetilde{p}/\partial x$. Let us define a modulating pressure gradient by:

$$\widetilde{\Delta_p} = \frac{\nu}{\bar{u}_\tau^3}\left(-\frac{1}{\varrho}\frac{\partial\widetilde{p}}{\partial x}\right).$$

The amplitude of the oscillating pressure gradient is:

$$A_{\widetilde{\Delta_p}} = 2\,\pi\, f^+\, a_{\widetilde{u}_c}\,\frac{\overline{U}_c}{\bar{u}_\tau},$$

and it is in quadrature with $\langle\, U_c\,\rangle$. Since in the low and intermediate frequency range and in the viscous sublayer, the amplitude of the oscillating shear is $A_{\widetilde{\partial u^+/\partial y}} = 2\,a_{\widetilde{u}_c}$, these two effects are related by:

$$\frac{A_{\widetilde{\Delta_p}}}{A_{\widetilde{\partial u^+/\partial y^+}}} = \pi\, f^+\,\frac{\overline{U}_c}{\bar{u}_\tau},$$

near the wall. This ratio is smaller than 0.1 at $f^+ < 0.0014$ so that the oscillating shear seems to be the main mechanism that induces the oscillating ejection frequency. The oscillating pressure gradient may however reach important values during the oscillation cycle. Figure 9 shows the phase average of these quantities at $f^+ = 0.0014$. It is well established (Kline *et al.* 1967) that an adverse pressure gradient increases the ejection frequency, while the former is decreased or completely suppressed (relaminarization) in a favorable pressure gradient. The modulation of the ejection frequency is thought to be partly due to $\partial\widetilde{u}/\partial y$ (through $\widetilde{f}_{e\,\partial\widetilde{u}/\partial y}$) and partly to $\partial\widetilde{p}/\partial x$ ($\widetilde{f}_{e\,\partial\widetilde{p}/\partial x}$). It is seen that the interaction between these quantities may weaken the modulation of the ejection frequency. This crude interpretation may explain the relaxation of $\langle f_e\rangle$ near the wall, but its behaviour at $y^+ > 40$ remains unexplained. It seems that the oscillating pressure gradient acts more on the flow near the wall and that the vortical structures have a long "memory" of the oscillating shear as they develop in the low logarithmic region. Such a situation was also conjectured by Smith *et al.* (1991) who reported that "the influence (of the pressure gradient) is thought to be most important for the portions of the (vortical structure) where the vortex makes its closest approach to the surface". This is also in agreement with the fact that the viscous wall region responds faster to an imposed pressure gradient than the outer flow. Finally, this intermediate frequency corresponds to the upper limit of an "unsteady equilibrium state" reported by Finnicum (1987; p. 189). The results presented here show that the equilibrium ceases more rapidly near the wall as expected. The multiple aspects of the near-wall Reynolds shear-stress producing coherent structures (Robinson 1991; Smith *et al.* 1991; Hon and Walker 1987; Jimenez and Moin 1991; Brooke and Hanratty 1993 Hanratty *et al.* 1993 etc.) in steady canonical boundary layers may be the key to a clearer understanding of the reaction of the ejection and bursting mechanism to unsteady forcing. These points are currently under consideration.

When the imposed frequency is high ($l_s^+ < 8$), the ejection frequency modulation becomes weak almost everywhere in the boundary layer with respect to the modulation of the wall shear stress (Fig. 8). The imposed frequency $l_s^+ \approx 5$, at which the ejection regeneration mechanism can only partly respond is closely related to the time mean ejection frequency $\bar{f}_e^+ \approx 0.015$ at $y^+ = 20$ (Fig. 7). In other words, the response of the ejection regeneration mechanism is stabilized and weak in the entire boundary layer when the imposed period is of the same order of the ejection interarrival time.

In the high imposed frequency regime wherein rapid modifications of the pressure gradient take place, the flow near the wall responds only partly to $\partial\tilde{p}/\partial x$. The modulation of the pressure gradient has therefore to be replaced by an effective pressure gradient following Mao and Hanratty (1986), i.e.:

$$A_{\widetilde{\Delta}_{p\mathrm{eff}}} = \frac{A_{\widetilde{\Delta}_p}}{\sqrt{1+(2\,f^+\theta^+)^2}} \approx \frac{\pi a_{\tilde{u}_c}}{\theta^+}\frac{\overline{U}_c}{\bar{u}_\tau},$$

where θ^+ is the mean relaxation time (following the model D of Mao and Hanratty 1986). On the other hand, the amplitude of the oscillating shear depends on the imposed frequency at $y^+ < l_s^+$ as pointed out in section 3.3:

$$A_{\widetilde{\partial u^+/\partial y^+}}\, a_{\tilde{\tau}} \approx A_{\tilde{\tau}\,\mathrm{Stokes}}/\,\bar{\tau} = \sqrt{2\,\pi\,f^+}\,a_{\tilde{u}_c}\frac{\overline{U}_c}{\bar{u}_\tau} = \sqrt{\frac{2\,f^+}{\pi}}\,\theta^+ A_{\widetilde{\Delta}_{p\mathrm{eff}}}.$$

It is seen that the amplitude of the pressure gradient acting on the near wall structures becomes independent of the imposed frequency. The amplitude of the shear increases with f^+ while the thickness of the oscillating shear layer decreases leaving a larger part of the near-wall viscous region under the only effect of the pressure gradient. The near-wall vortical structures are therefore in a highly non-equilibrium environment.

The striking decrease of $a_{\tilde{f}_e}/a_{\tilde{\tau}}$ at $f^+ \approx 0.002$ and the strong dependence of the modulation of the ejection frequency upon the imposed frequency can be seen more clearly in Fig. 10a, which recapitulates the results obtained at $y^+ = 15$. The phase-shift $\Phi_{\tilde{f}_e} - \Phi_{\tilde{\tau}}$ is also maximum at this critical imposed frequency (Fig. 10b), but decreases rapidly towards zero in the high imposed frequency regime. Note that, on the contrary the modulation of the quadrant 2 and quadrant 4 contributions is significantly larger at $f^+ \approx 0.002$. It was thought that in the range $0.0008 < f^+ < 0.002$ the modulation

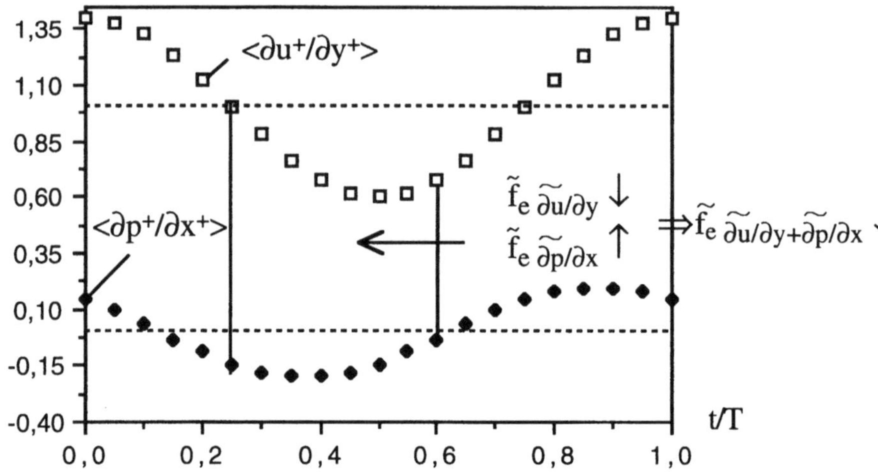

Fig. 9. Phase average of the modulation of the oscillating shear and pressure gradient at $y^+ = 15$ and $f^+ = 0.0014$. Possible combined effects on the damping of the ejection frequency modulation.

of the ejection frequency responds to an effective shear which introduces a relaxation time in a manner quite similar to the effective pressure gradient, i.e.:

$$\frac{a_{\tilde{f}_e}}{a_{\tilde{\tau}}} = \frac{1}{\sqrt{1 + \left(2\, f^+ \theta^+_{\mathrm{ej}}\right)^2}} \,,$$

and,

$$\Phi_{\tilde{f}_e} - \Phi_{\tilde{\tau}} = -\arctan\left(2\,\pi f^+ \theta^+_{\mathrm{ej}}\right) .$$

The relaxation time corresponding to the modulation of the ejection frequency is of the same order of $\theta^+ \approx 250$. Figure 10 shows that both the phase and the amplitude of the modulation of the ejection frequency are well predicted in the range $0.000\,8 < f^+ < 0.002$ with this simple model. The data indicates that the relaxation time decreases rapidly with the imposed frequency in the non-equilibrium regime $f^+ > 0.002$. The reaction of the near wall turbulence in this regime is still not well understood.

3.4 Modulation of the Contribution of Ejections to Reynolds Shear Stress

The response of the near wall turbulence to imposed unsteadiness is further analyzed by investigating the modulation characteristics of the contribution of the ejections to the quadrant 2. The former is expressed as:

$$\langle c_e \rangle = \frac{\langle \sum u'(t) v'(t)\, D(t) \rangle}{\langle \sum u' v_2' \rangle} = \frac{\langle \sum u'(t) v'(t)\, D(t) \rangle}{\langle S_2 \rangle} .$$

a)

b)

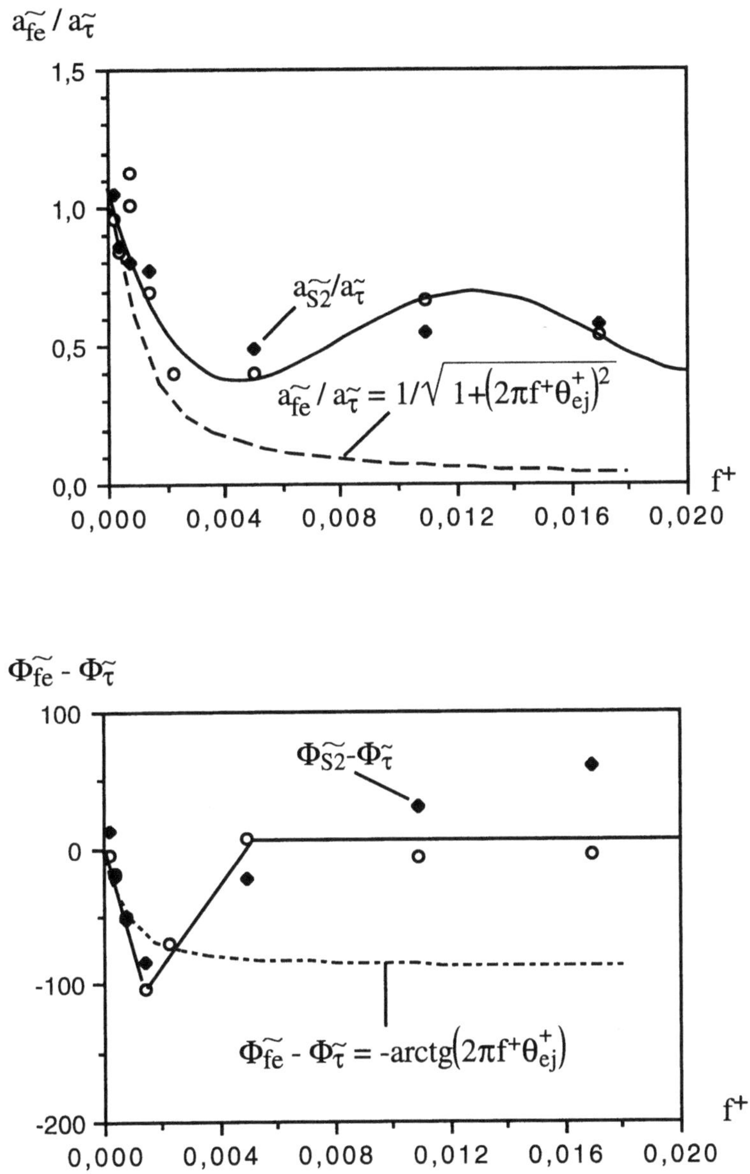

Fig. 10. Modulation characteristics of the ejection frequency at $y^+ = 15$ vs. the imposed frequency. Comparison with $\langle S_2 \rangle$.

The contribution c_e of the ejections to the second quadrant is 75% in the mean as in steady flow. The relative amplitude $a_{\tilde{c}_e}$ (not shown here) is, in return, smaller than 20% of the centerline amplitude. This shows that the modulation of the contribution of the ejections is negligible whatever the imposed frequency. Furthermore, let us define the phase average of the contribution per ejection $\langle c_E \rangle$ by:

$$\left\langle \sum u'(t)v'(t)\, D(t) \right\rangle = \langle N_e \rangle \langle c_E \rangle ,$$

where $\langle N_e \rangle$ is the phase average of the number of ejections whose relative amplitude and phase shift are the same as those of $\langle f_e \rangle$. It is important to note that, if there is considerable variation of $\langle c_E \rangle$ during the oscillation cycle, the physical information inferred from $\langle f_e \rangle$ is meaningless at least in a dynamic sense. For instance, in a situation where $\langle f_e \rangle$ is slightly but $\langle c_E \rangle$ is strongly modulated, one can not come to the conclusion that the related turbulent structure is "frozen" during the oscillation cycle by simply analyzing $a_{\tilde{f}_e}$ and $\Phi_{\tilde{f}_e}$. Fortunately, the modulation of $\langle c_E \rangle$ is negligible in the present conditions. The immediate implication is that the modulation characteristics of $\langle f_e \rangle$ coincide with those of $\langle S_2 \rangle$. Indeed since both $\langle c_e \rangle$ and $\langle c_E \rangle$ are only slightly modulated we have approximately:

$$\frac{\langle N_e \rangle \langle c_E \rangle}{\langle S_2 \rangle} \approx \frac{\langle N_e \rangle \bar{c}_E}{\langle S_2 \rangle} \approx \bar{c}_e ,$$

and therefore,

$$\langle N_e \rangle \approx \frac{\bar{c}_e}{\bar{c}_E} \langle S_2 \rangle \qquad \text{and} \qquad a_{\tilde{N}_e} = a_{\tilde{f}_e} \approx a_{\tilde{S}_2} ; \, \Phi_{\tilde{f}_e} \approx \Phi_{\tilde{S}_2} .$$

Figure 10 shows clearly that the distributions of the amplitudes and phase shifts of the ejection frequency and $\langle S_2 \rangle$ are more or less similar both quantitatively and qualitatively. This also indirectly proves that the individual contribution of the ejections is uniformly distributed during the oscillation cycle. Consequently, the modulation of the ejection frequency is an effective tool in describing the reaction of the near-wall turbulence.

4. Concluding Remarks

The results presented in this study confirm those of Tardu et Binder (1992) obtained by a single hot-film. It is clearly shown that in unsteady flows there is no simple relationship between the modulation of the ejection frequency and the modulation of the inner or the outer time scales. In such complex and nonequilibrium flows, the part of the boundary layer which is under the influence of the oscillating unsteady shear layer is about $2\, l_s^+$ and depends therefore on the imposed frequency, while the oscillating pressure gradient affects the entire boundary layer. The interaction of these two effects, combined with the complexity of the near wall relaxation mechanism make the finalizing of a complete picture of the unsteady wall flows difficult.

The grouping of the ejections into bursts are done in a way similar to Tardu and Binder (1992), but the characteristics of the modulation of the bursting frequency could not be discussed here and this part of the study will be published elsewhere. Let us simply indicate that these results also confirm our previous investigations, namely that the clusters of closed ejections (multiple ejection bursts) are governed by the modulation of the wall shear stress near the wall in the whole imposed frequency range, but the modulation of the frequency of the single ejection bursts depends strongly on l_s^+.

Acknowledgment

Financial support of the Délégation Générale de l'Armement via the Direction des Recherches, Etudes et Techniques is gratefully acknowledged.

References

Alfredsson, P.H., Johansson, A.V. (1984): On the detection of turbulence generating events. J. Fluid Mech. **139**, 325

Bogard, D.G. and Tiederman, W.G. (1986): Burst detection with a single point velocity measurements. J. Fluid Mech. **162**, 389

Bogard, D.G. (1982): "Investigation of burst structures in turbulent channel flows through simultaneous visualization and velocity measurements", *Ph.D. Thesis*, Purdue University

Brereton, G.J., Reynolds, W.C., and Jarayaman, R. (1990): Response of a turbulent boundary layer to sinusoidal free-stream unsteadiness. J. Fluid Mech. **221**, 131

Brereton, G.J., Reynolds, W.C. (1991): Dynamic response of boundary-layer turbulence to oscillatory shear. Phys. Fluids A **3** (1), 178

Brodkey, R.S., Wallace, J.M., Eckelmann, H. (1974): Some properties of truncated turbulence signals in a bounded shear flow. J. Fluid Mech. **63**, 209

Brooke, J.W., Hanratty, T.J. (1993): Origin of turbulence- producing eddies in a channel flow. Phys. Fluids A **5** (4), 1011

Compte-Bellot, G., Sabot, J., Saleh, I. (1978): "Detection of intermittent events maintaining Reynolds stress", in *Proceedings of Dynamic Flow Conference- Dynamic Measurements in Unsteady Flows*, Marseille, Baltimore, Skovlunde, Denmark, p. 213

Feng, M.Q., Tardu, S., Binder, G. (1993): "Inner region of unsteady channel flow", in *Near Wall Turbulent Flows* (R.M.C. So, C.G. Speziale and B.E. Launder eds) (Elsevier Science Publishers, Amsterdam, London, New-York, Tokyo), p. 457

Finnicum, D.S., Hanratty, T.J. (1987): Pressure gradient effects in the viscous wall region of a turbulent flow. Rep. № 7, Dept. of Chemical Engineering, University of Illinois

Finnicum, D.S., Hanratty, T.J. (1988): Influence of imposed flow oscillations on turbulence. PCH PhysicoChem. Hydrodyn. **10**, 585

Hanratty, T.J., Adrian, R.J., Liu, Z.C., Brooke, J., Papavassiliou, D., McLaughin J. (1993): "Recent results on the structure of wall turbulence", in *Proc. of 9th International Symposium on Turbulent Shear Flows,* Kyoto, Japan, p. 11-3

Hon, T.L., Walker, J.D.A (1987): An analysis of the motion and effects of hairpin vortices. AFOSR Rep. FM-11, Dept. Mechanical Engineering, Lehigh University

Kim, J., Moin, P.R., Moser, R. (1987): Turbulence statistics in fully-developed channel flow at low Reynolds number. J. Fluid Mech. **177**, 133

Kline, S.J., Reynolds, W.C., Schraub, F.A., Runstadler, P.W. (1967): The structure of turbulent boundary layers. J. Fluid Mech. **30**, 741

Lueptow, R.M., Breuer, K.S. and Haritonidis, J.H. (1988): Computer aided calibration of X probes using a look-up table. Experiments in Fluids **6**, 115

Mao, Z.X., Hanratty, T.J. (1986): Studies of wall shear stress in a turbulent pulsating pipe flow. J. Fluid Mech. **170**, 454

Robinson, S.K. (1991): The kinematics of turbulent boundary layer structure. NASA Technical Memorandum 103859.

Smith, C.R., Walker, J.D.A., Haidari, A.H., Sobrun, U. (1991): On the dynamics of near wall turbulence. Phil. Trans. Roy. Soc. **336**, 131

Tardu S., Binder, G., Blackwelder, R. (1987): "Modulation of bursting by periodic oscillations imposed on channel flow" in *Proc. of 6th International Symposium on Turbulent Shear Flows*, Toulouse, France, p. 4.5.1

Tardu, S., Binder, G. (1992): Response of bursting to imposed oscilations of high frequencies. J. Fluid Mech. In Revision

Tardu, S., Binder, G. (1993): Response of turbulence to imposed oscillations of high frequencies. Physics of Fluids-A, **5** (8), 2028

Tardu, S., Binder, G., Blackwelder, R.F. (1994): Turbulent channel flow subjected to large imposed velocity oscillations. J. Fluid Mech. In Press

Wallace, J.M., Eckelmann, H., Brodkey, R.S. (1972): The wall region in turbulent shear flow. J. Fluid Mech. **204**, 57

Computation of Oscillating Turbulent Flows at Transitional Re-Numbers

K. Hanjalić, S. Jakirlić, and I. Hadžić[1]

Lehrstuhl für Strömungsmechanik
Friedrich-Alexander Universität
Cauerstr. 4, 91058 Erlangen, Germany

Abstract

The paper describes the application of a second moment closure to the computations of oscillating boundary layers, channel- and pipe flows at transitional and higher Reynolds numbers. The model reproduced well the 'conditional turbulence' with sudden turbulence bursts and subsequent relaminarization in the oscillating boundary layers in the whole range of transitional Reynolds numbers in accord with results of direct numerical simulation and experiments. Predictions of turbulence dynamics in the outer region of a channel or pipe show also a good qualitative agreement with experimental records.

1. Introduction

Oscillating and periodic turbulent flows exhibit a number of features, seldom encountered jointly in other flows. Such are the periodic interchange of favourable and adverse pressure gradients, reversal of the mean flow and the shear stress, inflection of the mean velocity profiles. Yet oscillating boundary layer and fully developed pipe and channel flow still fall into a category of simple, self-similar flows, and are suitable for studying some aspects of the dynamics of turbulence response to the imposed periodic perturbation. More complex, but of course, of more significant practical importance, are the flow oscillations in finite-length pipes where non-similar cyclic oscillations of flow properties vary along the pipe. Of particular interest are the oscillating flows around a zero mean at transitional Re numbers where both forward and reverse transition may appear within a single cycle. The oscillations can be sinusoidal, or have some other periodic appearence, or can have a resting period between the acceleration and deceleration. In all these cases a sudden laminar-to-turbulent transition occurs around the onset of the deceleration phase. Subsequent acceleration, if strong enough, may cause relaminarization and the process will be repeated in the next cycle. Because of these features, but also because of a substantial practical relevance in engineering (e.g., rotating aerofoils and turbomachinery blades, heat exchangers attached to reciprocating engines), in the environment, (e.g.,

[1] All three authors are on leave from the University of Sarajevo, Bosnia Hercegovina

coastal waters) as well as in biology and bioengineering (flows in live organisms and their experimental simulation), oscillating flows have been the subject of extensive experimental and, more recently, of numerical investigation.

Prediction of laminar-to-turbulent (or reverse) transition has been regarded as a challenging test for computational models of turbulence. Although the statistical averaging approach - inherent in most single-point closure models - conceals the dynamics of flow disturbances, the development of instability and the actual mechanism of the transition phenomena, a turbulence model for more complex flows should be capable of reproducing a change-over from one flow regime to another at appropriate conditions. Statistical models can reproduce this transition only when either a continuous source of turbulence exists somewhere in the flow or at its edge from where the turbulence will diffuse into the rest of the flow (by-pass transition), or the laminar flow considered must contain some background turbulence, sufficiently weak not to influence laminar-like mean flow properties, but still sufficient to amplify itself when the flow deformation is imposed.

Unlike some other flows where the predicted location of the transition depends on the initial conditions, the cyclic variation of ensemble-averaged properties in a fully developed oscillating flow becomes self-similar and independent of the initial field. In comparison with monotonic acceleration or deceleration, oscillating and periodic flows are characterized by *hysteresis* of the turbulence field to the variation of external flow conditions. Moreover, different turbulence quantities exhibit different kinds of hysteresis, which vary across the flow, and can serve as an indicator of the response of turbulence to the external perturbation as well as of the ability of models to mimic such behaviour. Of course, the response of turbulence structure will vary over the spectrum, but the single-point closure can not depict such details.

Oscillating boundary layers are characterized by the Reynolds number Re_δ, based on the Stokes-layer thickness $\delta_S = \sqrt{2\nu/\omega}$ and the peak free stream velocity U_0 (where ω is the angular frequency). Stability analysis suggests $Re_\delta \approx 600$ as the laminar stability threshold. The direct numerical simulation (DNS) by Spalart and Baldwin (1989) for this Re_δ showed flow features which differ from typical laminar flows. But it was only for $Re_\delta = 800$ and 1000 that DNS showed a marked transition to turbulence around the start of the deceleration phase to be followed by laminarization during the acceleration phase. Very similar shapes of the wall-shear stress distribution over a cycle were found experimentally by Jensen *et al.* (1989) for Re_δ =760, 1000 and 1140, (Fig. 1). Hino *et al.* (1983) recorded a similar profile in a rectangular duct for Re_δ = 876.

In the case of a pipe or channel flow the critical conditions can not be defined solely by Re_δ, but depend on the ratio of the pipe diameter and Stokes-layer thickness, $\lambda = d/(2\delta_S)$. Hino *et al.* (1976) proposed a stability diagram in terms of Re_δ and λ in which they identified - in addition to regions of laminar- and distorted laminar flow - three turbulent regimes: weakly turbulent flow, conditional turbulence and fully turbulent flow. The weakly turbulent regime (hereafter denoted as 'turbulescence') is characterized by visible, but small amplitude perturbation in the mean velocity oscillogram over the whole cycle, which only marginally affects the laminar-like mean

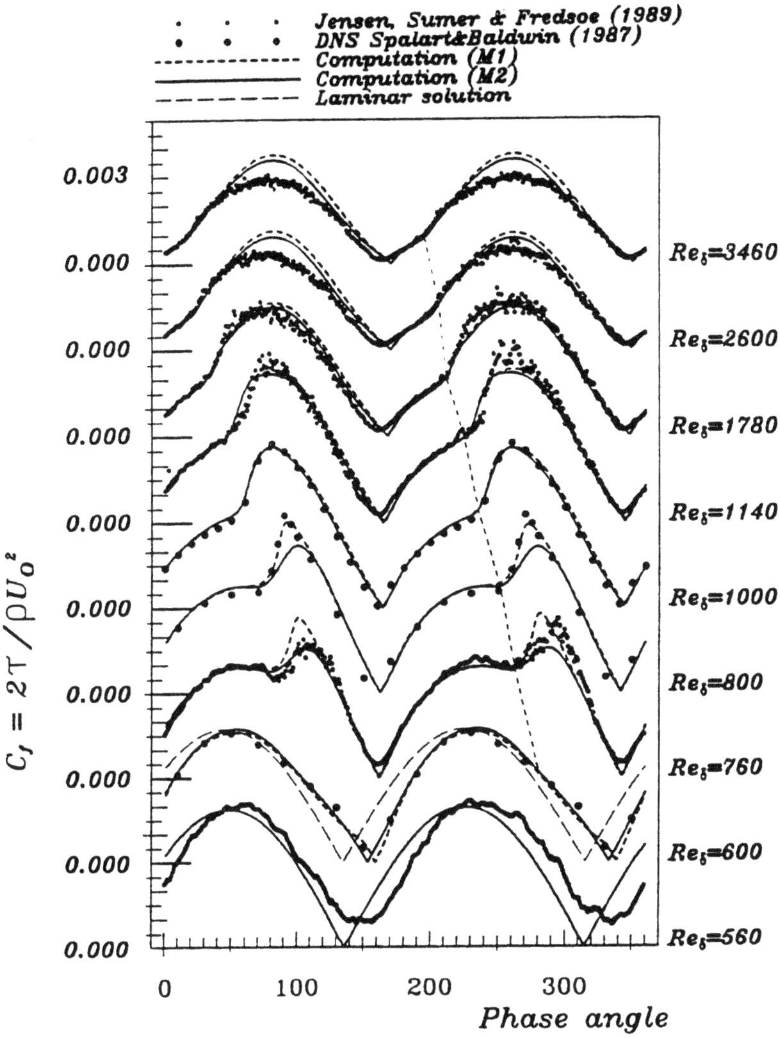

Fig. 1. Evolution of the wall shear stress in an oscillating boundary layer over a cycle for different Re-numbers; the dotted line denotes the incipience of turbulent burstes.

flow properties. The conditional turbulence denotes the pattern with pronounced turbulence bursts, sustained only over a part of the cycle. It seems more appropriate to call this regime periodically turbulent, or simply 'bursting' regime. Hino *et al.* (1976) argued that this regime will prevail always if $Re_\delta > 550$, (close to the stability criterion for the oscillating boundary layer) provided that Stokes parameter $\lambda > 1.6$. They plotted a demarcation line separating the weakly and conditionally turbulent regimes, which follows the relation $Re_m \approx 1100\lambda$, where Re_m is the pipe Reynolds number based on diameter and mean velocity amplitude. Seume (1988) identified a similar relationship expressed in terms of the Valensi number, $Va = \omega d^2/4\nu = 2\lambda^2$

for the transition Reynolds number above which a rapid increase in the measured *rms* velocity fluctuations was detected. Hino *et al.* concluded that oscillatory pipe flows are more 'stable' than unidirectional steady flows and can be turbulent only during the part of each cycle, decaying suddenly when the direction of flow is reversed. This conclusion can be extended by saying that the bursting regime in an oscillating pipe or channel flow can exist over a wide range of Reynolds numbers, its span increasing progressively as the Stokes parameter λ increases. Indeed, the experiments of Jensen *et al.* (1989) on the oscillating boundary layer were performed in a rectangular channel, but with a width which exceeded many times the Stokes penetration depth (i.e. $\lambda \approx 75$) so that the periodic (bursting) turbulence was detected over a range in which Re number varied almost by one order of magnitude.

Neither Hino *et al.* (1976) nor other authors defined the criterion above which no relaminarization will occur and the flow will remain fully turbulent over the whole cycle.

Simulation of the above described experimentally recorded events and different regimes at transitional Reynolds numbers, particularly the subtle details of the turbulence initiation, goes beyond the capabilities of one-point averaging closure models. Yet some of the features can be and have been simulated and this paper discusses some earlier and more recent results. Ha Minh *et al.* (1989) reported reasonable agreement between computations with a low-Re-number second-moment model and the DNS results of Spalart and Baldwin (1989) for a boundary layer with sinusoidal oscillation of the free stream velocity at $Re_\delta = 1000$, but reported that at $Re_\delta = 600$ the model could not maintain the turbulence. Justesen and Spalart (1990) applied two low-Reynolds-number $k - \varepsilon$ models to compute the oscillating boundary layer with 'mild' (sinusoidal) and 'steep' periodic variation of pressure gradients and reported good overall agreement with DNS data, though noting some important defficiencies, particularly in reproducing the location, appropriate slope and peak value of the sudden rise of the wall shear stress at the onset of forward transition. Shima (1993) computed the same sinusoidal flow with his version of a second-moment closure and obtained better response of the wall shear stress at the incipience of transition, but the predicted peak and subsequent attenuation are much underpredicted. Numerical computations of oscillating flows in a finite pipe, with two low-Reynolds-number $k - \varepsilon$ models have been reported by Koehler *et al.* (1992) for several combinations of Re_δ and Valensi numbers in the transition and fully turbulent regime. Close to the wall the turbulence response time is short and a fast response of stresses to the variation in mean velocity can be reasonably well reproduced even by a direct stress-strain relationship as in eddy-viscosity models, particularly at higher Reynolds numbers. However, a faithful reproduction of the dynamics of response of the turbulence field in the whole flow, particularly at transitional Reynolds numbers, requires the application of higher order models.

This paper reports on the computation of oscillating wall-parallel flows at transitional and higher Reynolds numbers by applying a version of second-moment closure model which includes models of the low-Re-number and wall-vicinity effects. The paper focuses on boundary layers with oscillating free streams and for which DNS data are available and the experimental results are more abundant than for conduit flows, but considers also few cases of oscillating channel and pipe flows. We focused our attention to the evolution of turbulence properties over the cycle and tested the model ability to reproduce the fine gradation at different Reynolds numbers within the transition regime. These, seemingly too fine details, proved to be sensitive indicators of the model ability to respond adequately to the conditions leading to laminar-turbulent transition, and influence to a great extent the ability of the model to predict the transition at appropriate Reynolds number.

2. Turbulence Model and Numerical Method

The model uses the transport equations for the Reynolds stress $\overline{u_i u_j}$ and energy dissipation ε in the following form (Hanjalić and Launder 1976):

$$\begin{aligned}\frac{\mathrm{D}\overline{u_i u_j}}{\mathrm{D}t} &= \frac{\partial}{\partial x_k}\left[\left(\nu + C_s \frac{k}{\varepsilon}\overline{u_k u_l}\right)\frac{\partial \overline{u_i u_j}}{\partial x_l}\right] \\ &\quad - \left(\overline{u_i u_k}\frac{\partial U_j}{\partial x_k} + \overline{u_j u_k}\frac{\partial U_i}{\partial x_k}\right) + \Phi_{ij} - \varepsilon_{ij},\end{aligned} \tag{1}$$

$$\begin{aligned}\frac{\mathrm{D}\varepsilon}{\mathrm{D}t} &= \frac{\partial}{\partial x_k}\left[\left(\nu + C_\varepsilon \frac{k}{\varepsilon}\overline{u_k u_l}\right)\frac{\partial \varepsilon}{\partial x_l}\right] - C_{\varepsilon_1}\frac{\varepsilon}{k}\overline{u_i u_j}\frac{\partial U_i}{\partial x_j} \\ &\quad - C_{\varepsilon_2} f_\varepsilon \frac{\varepsilon\tilde{\varepsilon}}{k} + C_{\varepsilon_3}\nu\frac{k}{\varepsilon}\overline{u_j u_k}\frac{\partial^2 U_i}{\partial x_j \partial x_l}\cdot\frac{\partial^2 U_i}{\partial x_k \partial x_l}.\end{aligned} \tag{2}$$

We focused on selecting and examining models of the pressure-strain redistribution Φ_{ij} and dissipation-rate tensor ε_{ij}, particularly their modifications to account for viscous and wall effects. These modifications depend of course on the selection of the basic high-Re-number model. Several non-linear models for Φ_{ij} have been proposed over the past few years (e.g., Launder and Tselepidakis 1993), but their full verification and expected advantages are still to be verified. For that reason, for the wall-parallel flows considered here we retained the simplest linear models for both the 'slow' and 'rapid' parts $\Phi_{ij,1}$ and $\Phi_{ij,2}$, and, consistent with this, the standard wall-reflection terms $\Phi^w_{ij,1}$ and $\Phi^w_{ij,2}$ (Gibson and Launder 1978), which proved to yield satisfactory predictions of a variety of thin shear flows:

$$\Phi_{ij,1} = -C_1 \varepsilon a_{ij}, \qquad \Phi_{ij,2} = -C_2\left(P_{ij} - \frac{2}{3}P_k\delta_{ij}\right), \tag{3}$$

$$\Phi^w_{ij,1} = C^w_1 f_w \frac{\varepsilon}{k}\left(\overline{u_k u_m} n_k n_m \delta_{ij} - \frac{3}{2}\overline{u_i u_k} n_k n_j - \frac{3}{2}\overline{u_k u_j} n_k n_i\right), \tag{4}$$

$$\Phi^w_{ij,2} = C^w_2 f_w \left(\Phi_{km,2} n_k n_m \delta_{ij} - \frac{3}{2}\Phi_{ik,2} n_k n_j - \frac{3}{2}\Phi_{kj,2} n_k n_i\right), \tag{5}$$

where n_p denotes the unit vector normal to the wall. Necessary modifications for low-Reynolds-number wall flows are commonly accomplished by modifying the coefficients C_1, C_2, C_1^w and C_2^w by damping functions in terms of suitable parameters. For this purposes we employed the turbulence Re number, $Re_t = k^2/\nu\varepsilon$, and the stress anisotropy invariants $A_2 = a_{ij}a_{ji}$, $A_3 = a_{ij}a_{jk}a_{ki}$, where $a_{ij} = \overline{u_i u_j}/k - 2/3\delta_{ij}$, and the 'eddy flatness' parameter $A = 1 - 9/8(A_2 - A_3)$. The first parameter introduces the effect of the viscosity which is unavoidable in treating the transition phenomena, whereas the second parameter is expected to account for strong stress anisotropy in the near-wall region. One of the novelties of the present modellling is the introduction of the 'flatness' parameter of the dissipative scales $E = 1-9/8(E_2-E_3)$, analogous to A, where $E_2 = e_{ij}e_{ji}$ and $E_3 = e_{ij}e_{jk}e_{ik}$ are the second and third invariants of the anisotropy of the dissipation rate tensor, i.e., $e_{ij} = \varepsilon_{ij}/\varepsilon - 2/3\delta_{ij}$. This parameter was to account for the anisotropy of the dissipating motion which persists at low Re numbers not only in the wall vicinity, but also over a considerable part of the outer flow domain. The scalar character of all of these variables makes them suitable for representing the low-Reynolds-number and wall-vicinity effects irrespective of the topology of the flows considered.

Various forms of the functional expression for the coefficients in conjunction with different levels of modelling of the pressure strain term Φ_{ij} have been tested in a number of wall-flows. In this paper we present two variants of the model which differ only slightly, both related to the simple linear model given by expressions (3) to (5). This approach was adopted, with the purpose of further testing the conventional, relatively simple, but well-explored form of the second-moment closure model in more demanding situations with very strong variation of external conditions and under the dominating influence of the low-Re-number effects. By adopting the linear models of Φ_{ij}, we followed initially the approach of Launder and Shima (1989), but modified first the coefficient C_1 to achieve the best overall performances for several non-equilibrium flows with pronounced low-Reynolds number and wall-proximity effects. The modifications were also necessitated by the form of ε-equation which differs from the equation of Launder and Shima. The first model, denoted as M1, reproduced very well the appearance of turbulent bursts over the whole range of transitional Reynolds numbers, but produced an increase in the friction factor in wall flows at high Re numbers, where the modifications were expected to fade out completely. The second, slightly different model, denoted as M2, does not have this defficiency, as discussed later. The coefficients adopted for both models of Φ_{ij} are summarized in Table 1.

The dissipation rate tensor ε_{ij} is defined as (for details, see Hanjalić and Jakirlić 1993):

$$\varepsilon_{ij} = f_s\varepsilon_{ij}^* + (1 - f_s)\frac{2}{3}\delta_{ij}\varepsilon, \tag{6}$$

$$\begin{aligned}\varepsilon_{ij}^* &= \frac{\varepsilon}{k}\left[\overline{u_i u_j} + (\overline{u_i u_k}n_j n_k + \overline{u_j u_k}n_i n_k\right.\\ &\quad \left.+\overline{u_k u_l}n_k n_l n_i n_j)f_d\right] \Big/ \left(1 + \frac{3}{2}\frac{\overline{u_p u_q}}{k}n_p n_q f_d\right) .\end{aligned} \tag{7}$$

Table 1. Specification of model coefficients.

	Model M1	Model M2
C	$2.6AA_2^{1/4}f$	$2.5AF^{1/4}f$
F	-	$\min\{0.6; A_2\}$
f	$1 - \exp\left[-\left(Re_t/150\right)\right]$	$\min\{(Re_t/150)^{3/2}; 1\}$
C_1	$C + \sqrt{A}E^2$	
C_2	$0.75A^{1/2}$	$0.80A^{1/2}$
C_1^w	$1 - 0.67C$	$1 - 0.70C$
C_2^w	$\max\{2/3 - 1/(6C_2); 0\}$	$\min(A; 0.3)$
C_{ε_3}	0.35	0.25
f_w	$\min\{k^{3/2}/(2.5\varepsilon x_n); 1\}$	$\min\{k^{3/2}/(2.5\varepsilon x_n); 1.4\}$

In comparison with earlier models, we employ the Reynolds-number function $f_d = (1 + 0.1Re_t)^{-1}$ only to suppress the wall correction of Lauder and Reynolds (1983) outside the viscous region. A new function f_s is introduced, expressed in terms of the dissipation tensor 'flatness' parameter mentioned earlier. The use of E to define the 'switch over' from isotropic to non-isotropic dissipation, when the wall is approached, was inspired by the DNS data which indicate a much higher degree of anisotropy in ε_{ij} in a plane channel - at least at the two Re numbers considered, 5600 and 13 750 - than hitherto anticipated. This behaviour can not be simulated either in terms of Re_t or A. The parameter E seems to respond adequately to the bulk Re number effect and wall influence (Hanjalić and Jakirlić 1993). Of several forms tested, the DNS results of ε_{ij} are best reproduced by the function:

$$f_s = 1 - \sqrt{A}E^2. \tag{8}$$

The dissipation equation retains the form which was originally proposed by Hanjalić and Launder (1976). The coefficients in the ε equation are: $C_\varepsilon = 0.18$, $C_{\varepsilon 1} = 1.44$, $C_{\varepsilon 2} = 1.92$, whereas $C_{\varepsilon 3}$ was adjusted to achieve best overall agreement in a variety of low-Reynolds-number flows. The only empirical function in the ε-equation is

$$f_\varepsilon = 1 - \frac{C_{\varepsilon_2} - 1.4}{C_{\varepsilon_2}} \exp\left[-\left(\frac{Re_t}{6}\right)^2\right], \tag{9}$$

which was introduced to ensure a switch from the initial to the final period of decay of isotropic turbulence at the appropriate Re_t.

The above modifications improved the reproduction of individual ε_{ij} components, satisfied better the asymptotic and limiting turbulence states in the case of vanishing Re_t or close to a solid wall where the turbulence approaches a two-component limit, and resulted in overall improvement of predictions in most of the flows considered.

The model was first tested in the computation of several simpler two-dimensional flows including the constant-pressure boundary layer and plane channel at low and moderate Re numbers, and subsequently applied to the computation of several more complex cases, including the non-equilibrium boundary layers in strong adverse and favourable pressure gradients, laminarizing two-dimensional sink flows, as well as a three-dimensional shear driven boundary layer at various pressure gradients.

The computations were performed by applying a parabolized Navier-Stokes time-dependent numerical code using the control-volume approach with 80 to 100 nodes across the flow in a collocated variable arrangement. The time-dependent pressure gradient was specified explicitly and the V-component of velocity evaluated from the continity equation. Stones strongly-implicit procedure (SIP) was applied with the selective interpolation technique of the stress equation. Typical time steps were around 1°; tests with 1/2, 1/5 and 1/10° showed insignificant differences.

3. Results and Discussion

3.1 Oscillating Boundary Layers

We consider first boundary layers on a flat wall subjected to sinusoidal free stream variation $U_\infty = U_0 \cos(\omega t)$, at several Re_δ ranging from the stability limit of 560 to fully turbulent flow at 3460. The evolution of the wall shear stress over the cycle provides a good illustration of the flow dynamics in the near-wall area, where the instability is first generated. Figure 1 compares our computations with the measurements of Jensen *et al.* (1989) and with the results of DNS by Spalart and Baldwin (1989) for a range of Reynolds numbers. At the lowest Re_δ = 560 which corresponds to the Hino *et al.* (1976) demarcation between the 'deformed laminar' and 'weakly turbulent' (turbulescent) regimes, the experimental profile looks laminar, though shifted by a small phase angle and slightly perturbed over the whole cycle in the same sense, as seen in velocity oscillograms of Hino *et al.* (1976). Our computations could not reproduce these small perturbations and yielded laminar solutions irrespective of the initial turbulence field. Similar features have been reproduced by the DNS by Spalart and Baldwin (1989) at Re_δ = 600. Although this Re number is marginally higher and the flow is still in the turbulescent regime, which is far from real turbulence, both of the models considered agreed very well with the DNS data, as is shown in Fig.1. Already at Re_δ = 760, which falls in the range of bursting ('conditionally turbulent regime') the models reproduce very well the experimentally detected sudden jump in the wall shear stress corresponding to turbulent bursts after the onset of the deceleration phase, as well as subsequent relaminarization during the acceleration. As the Reynolds-number increases, transition appears earlier and this behaviour is well predicted by the models, as indicated by a sharp bend in the curves in Fig. 1. There is, however, a distinct difference between the performances of the two models at this and slightly higher Reynolds numbers. The earlier model (M1) produces a sharp turbulent burst very similar to that recorded by experiments in the second half of the cycle, but its peak appears about 12 to 15° earlier than recorded by experiments. The second model (M2) yields a more gradual development of the turbulent burst and agrees very well with the experimental data in the first half of the cycle. The experimental scatter and the difference between the two halves of the cycle makes it difficult to judge which of the models reproduces the experiments better. However at a still slightly higher Re number of 800, for which Spalart and Baldwin performed direct numerical simulations, the model M1 gave excellent agreement with the DNS data. At higher Reynolds numbers the difference between the performances of the two models fades away, but becomes noticeable again at the two highest Reynolds numbers, where the flow becomes fully turbulent over the whole cycle. Although both models yield too high maximum wall shear stress, it is believed that the model M2 gives the realistic value and that the experimental data might be somewhat too low. This belief is partly substantiated by the fact that the model M2 gave very good reproduction of the channel flow and of boundary layers at various pressure gradients at high Reynolds numbers. Also the semi-logarithmic plot of the mean velocity profiles, shown in Fig. 2 for a range of phase angles φ at $Re_\delta = 3460$ seems to confirm the above conclusion. Excellent agreement is obtained

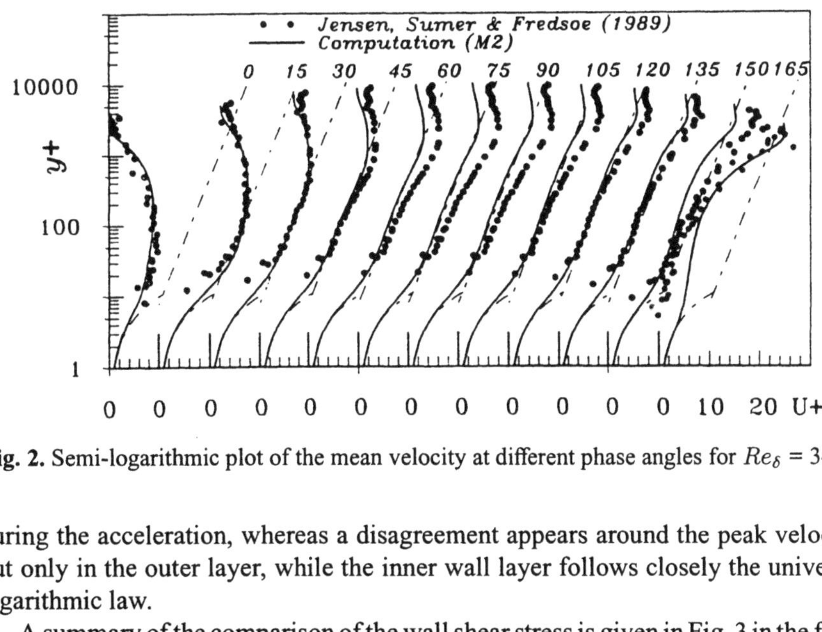

Fig. 2. Semi-logarithmic plot of the mean velocity at different phase angles for $Re_\delta = 3460$.

during the acceleration, whereas a disagreement appears around the peak velocity, but only in the outer layer, while the inner wall layer follows closely the universal logarithmic law.

A summary of the comparison of the wall shear stress is given in Fig. 3 in the form of the peak τ_w, normalized with U_0, with the DNS results of Spalart and Baldwin. The diagram confirms a very good agreement, apart from the highest Reynolds numbers, as discussed earlier. It should be noted, however, that a line drawn through the experimental points in the Reynolds-number region after transition, would exhibit a convex form mainly due to a low friction factor at the highest Re number, whereas typical experimental friction-factor curves in a channel or for boundary layer are concave. Our predicted results also show such pattern.

Figure 4 compares the phase lead of the maximum wall shear stress over the peak free-stream velocity showing a change-over from 45° for laminar to about 8° at $Re_\delta = 3460$. In a range of Reynolds numbers where the transition occurs *after* the start of the deceleration phase, τ_w exhibits two peaks: the first corresponding to

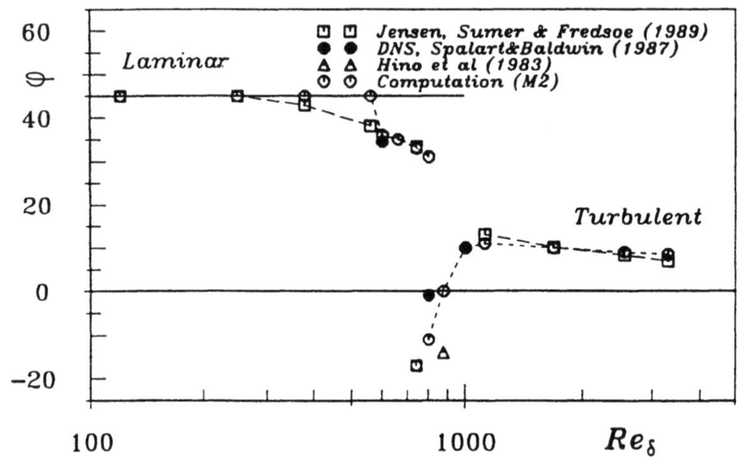

Fig. 3. Non-dimensional maximum wall shear stress (wave friction factor) versus Re_δ.

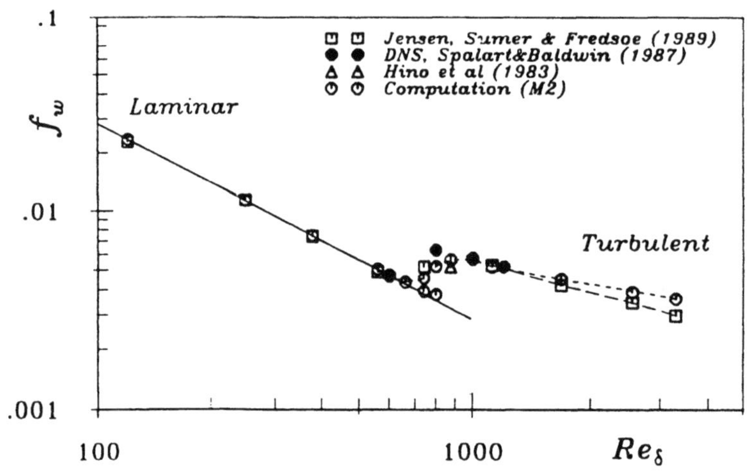

Fig. 4. Phase lead of the maximum wall shear stress with respect to the maximum free-stream velocity.

an almost laminar regime and the second, which lags behind the velocity maximum, corresponding to the turbulent burst (negative φ).

A further comparison with the DNS data of Spalart and Baldwin showed also a very good agreement of the mean velocity and all stress components at different phase angles for $Re_\delta = 1000$. For illustration the semi-logaritmic plot of the turbulent shear stress profiles $\overline{uv}$ is shown for a range of phase angles in Fig. 5. The comparison of

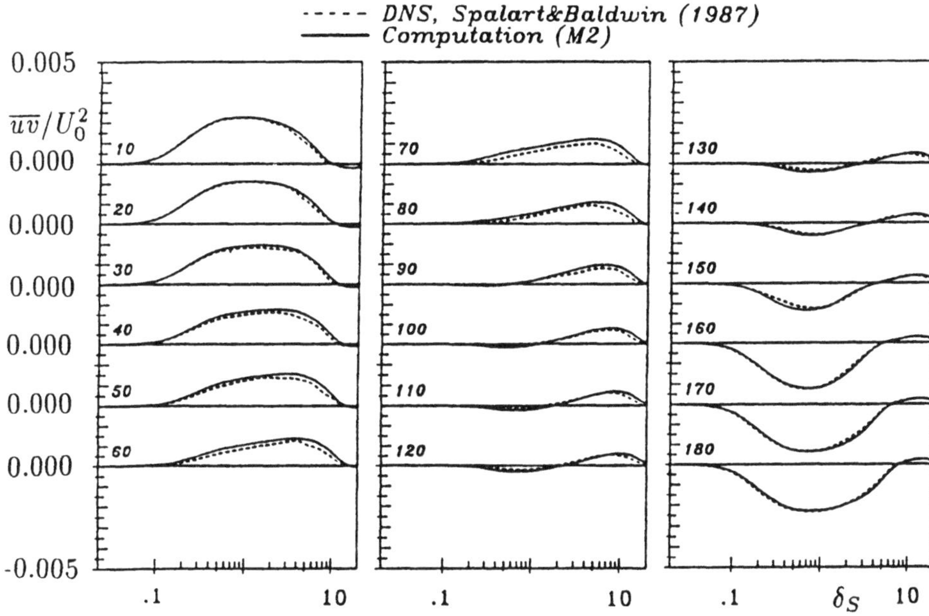

Fig. 5. Comparison of computed profiles of turbulent shear stress with DNS results at different phase angles.

the wall shear stress at $Re_\delta = 1000$ is shown in Fig. 6. Both models follow closely the DNS points, better than any earlier reported model, though they do not reproduce in full the asymmetric sharp peak. A sharper transition can be generated, e.g., by replacing the f-function in the model M1 by $1 - \exp\left[-(Re_t/150)^2\right]$ (not shown here), but this function delays transition and already at $Re_\delta = 760$ could not maintain the turbulence.

Some further illustration of the ability of the models to reproduce the dynamics of the turbulence field can be obtained from Fig. 7 where the computed time contours of all stress components are compared with those obtained from DNS. Note that the maximum boundary layer thickness on the ordinate is $20\delta_s$, corresponding to the thickness over which the data in Fig. 5 are plotted. For comparison, the DNS by Spalart and Baldwin covered δ, whereas the half channel width of the experimental flow of Jensen *et al.* was approximately $75\delta_s$. The time coordinate on the abscissa covers one cycle. Both the qualitative and quantitative comparisons show excellent agreement with the DNS results. These plots give the possibility of comparing the

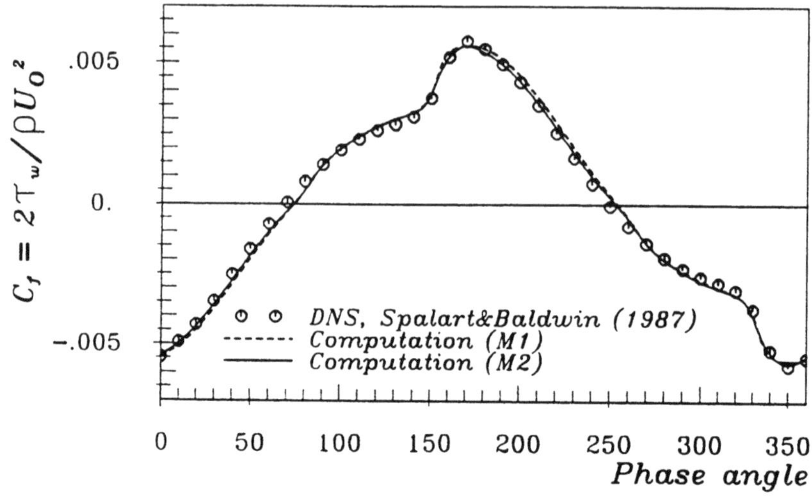

Fig. 6. Evolution of the wall shear stress at Re_δ = 1000 for 'mild' (sinusoidal) variation of the free-stream velocity.

time response and development of different stress components. In wall-parallel flows, the energy from the mean motion is transferred directly into the $\overline{u^2}$ component due to the shear deformation so that $\overline{u^2}$ reaches the maximum value earlier than any other stress component. The computations give the maximum $\overline{u^2}$ at $\varphi \approx 70°$, in excellent agreement with DNS, as indicated in Fig. 7. Other normal stresses, $\overline{v^2}$ and $\overline{w^2}$ lag behind since they receive energy solely through the redistributive action of the fluctuating pressure. Our computations yield the maximum $\overline{v^2}$ at $\approx 85°$ and $\overline{w^2}$ at $\approx 75°$ while the DNS gives $\approx 90°$ for both. Both the computed and DNS maxima of $\overline{uv}$ are at $\approx 80°$. Although the agreement for $\overline{v^2}$ and $\overline{w^2}$ is not complete, it can be concluded that the model of the pressure redistribution process (Φ_{ij}) responds well to the imposed sinusoidal variation of the free stream velocity.

We turned our attention to a more rigorous case simulated by Justesen and Spalart (1990), in which much steeper accelerations and decelerations were imposed, with quasi-steady intervals in between as shown in Fig. 8. Predicted and DNS results for the wall shear stress are compared in Fig. 9, which shows good agreement between them. More details for this case are shown in Figs. 10 and 11, where logarithmic velocity profiles, and normal and shear stress components are plotted in parallel. Excluding the positions around the zero wall shear stress (70 to 80°), the velocity plot at all $\varphi's$ agrees very well with DNS results. Turbulent stresses are also well predicted, apart from some curious bumps exhibited by the DNS results in the outer region at φ of 10 and 20°.

3.2 Oscillating Channel and Pipe Flow

If $\lambda \gg 1$, the boundary layer in a pipe or channel flow remains much thinner than the channel half-width and there is no effect of the channel Reynolds number. However,

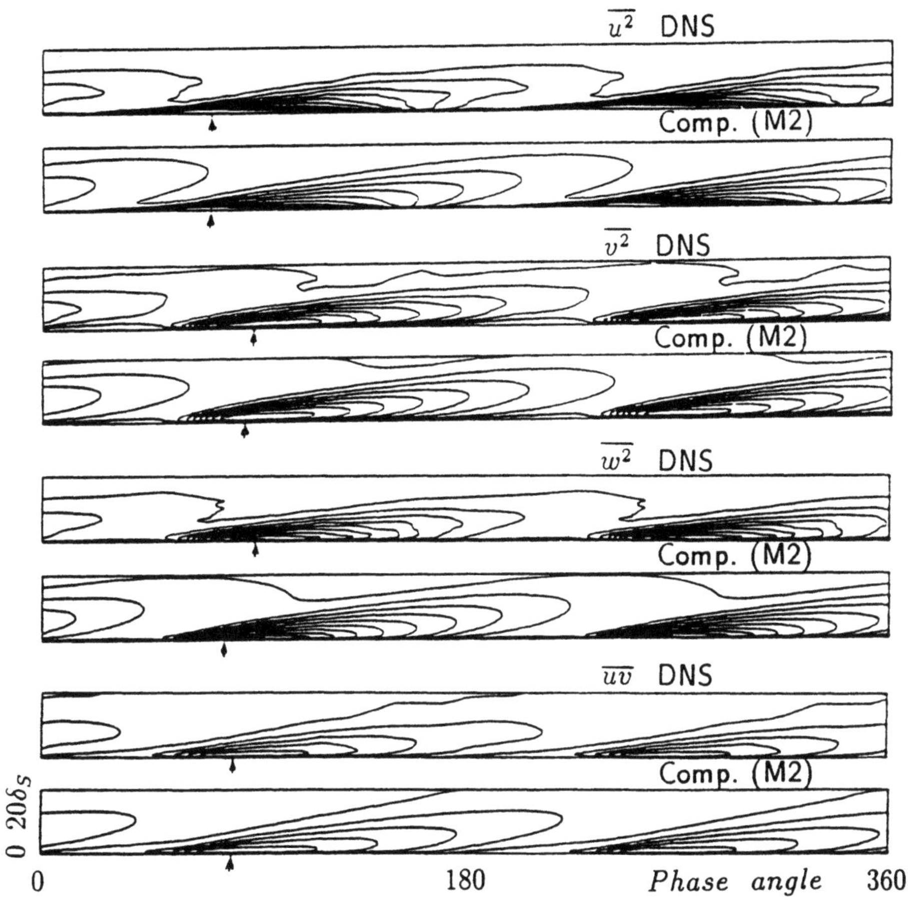

Fig. 7. Time development of contours of constant turbulent stresses at Re_δ = 1000 for 'mild' (sinusoidal) variation of free-stream velocity; phase angles of maximum values are denoted by arrows.

if the penetration depth approaches or exceeds the channel half width, the interaction of the two merging boundary layers influences the stability and turbulence pattern. Application of the same model to the fully developed oscillating channel and pipe flow reproduced the same pattern in the near-wall region at similar Re_δ, but not so well in the outer region for λ close to 1. In the absence of DNS data or detailed experimental data we present here only few results obtained for a channel and pipe flow at the conditions close to the experiments.

Figure 12 displays the computed wall shear stress and $\overline{u_i u_j}$ very close to the wall, together with the experimental data of Hino *et al.* (1983) obtained in a rectangular duct at Re_δ = 876 and λ = 12.8. At this value of λ we would expect that the flow is essentially of the boundary-layer type. The computed and measured profiles of the wall shear stress look similar, but there is a considerable phase difference. The

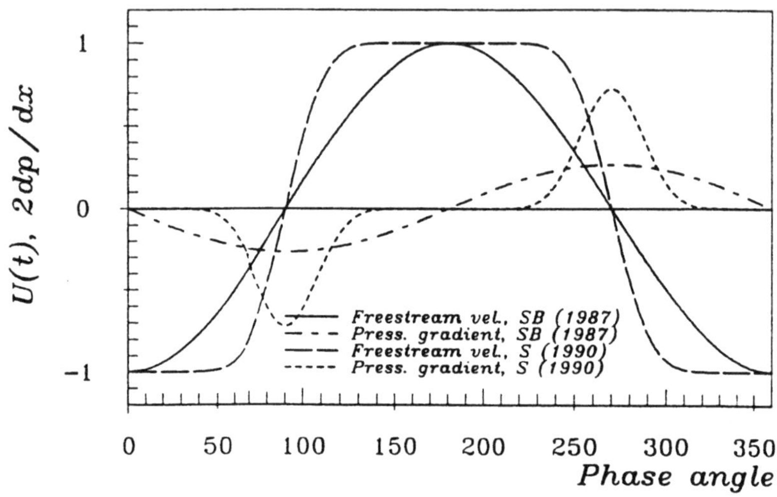

Fig. 8. Variation of the free-stream velocity and pressure gradient for 'mild' and 'steep' case.

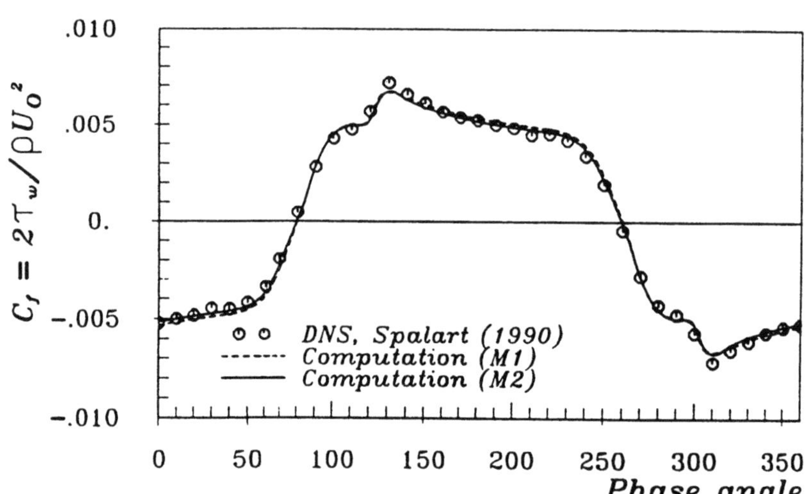

Fig. 9. Evolution of the wall shear stress at $Re_\delta = 1000$ for 'steep' variation of the free-stream velocity.

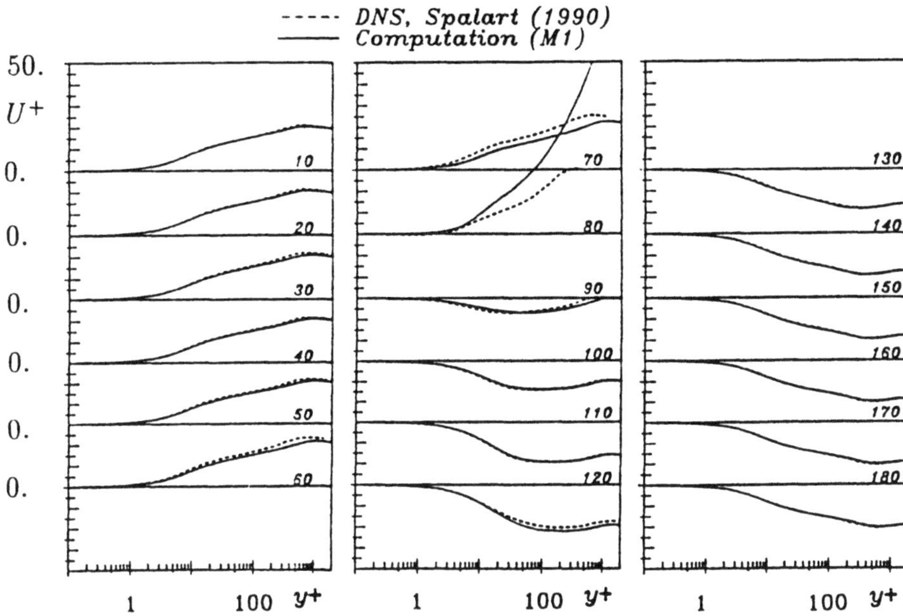

Fig. 10. Comparison of computed profiles of mean velocity with DNS results for the 'steep' case at different phase angles.

same could be said for the $\overline{u_i u_j}$, obtained at $y = 5$ mm (Note: $\delta_S = 3.9$ mm, channel width $H = 100$ mm). In the light of excellent agreement previously shown with the measurements for the same Re_δ it is difficult to explain these discrepancies. For illustration, we show in Fig. 13 the comparison between the recorded and computed evolutions of the ensemble-averaged velocity at several positions across the flow. Excluding the region very close to the wall (for y=0.05 and 1 mm), the agreement is close - both records show the transition at the start of the deceleration phase. However, the experiments indicate a relatively low level of turbulence in spite of a large channel Re number, ($Re = 2.25 \times 10^4$). Much closer agreement is shown with the results of an earlier experiment of Hino *et al.* (1976), Fig. 14, where the computed velocities at several radial positions are compared with the oscillograms obtained in a pipe, at a larger Stokes Re-number, $Re_\delta = 1530$, but at much smaller channel Re number, $Re = 5830$.

The corresponding lines are obtained at the same non-dimensional distance, normalized with the Stokes thickness δ_S. Both pictures exhibit the same sudden jump in the velocity profiles in the near-wall region and a corresponding sudden decrease in the central region. In this case the Stokes parameter $\lambda = 1.91$ and the instability and transition occur also at the pipe centerline. The phase shift of the velocity maximum is also well reproduced.

Finally we present in Fig. 15 a comparison of the histograms of the bursts in streamwise turbulent fluctuations at three radial positions, recorded by Seume (1988) and our computations.

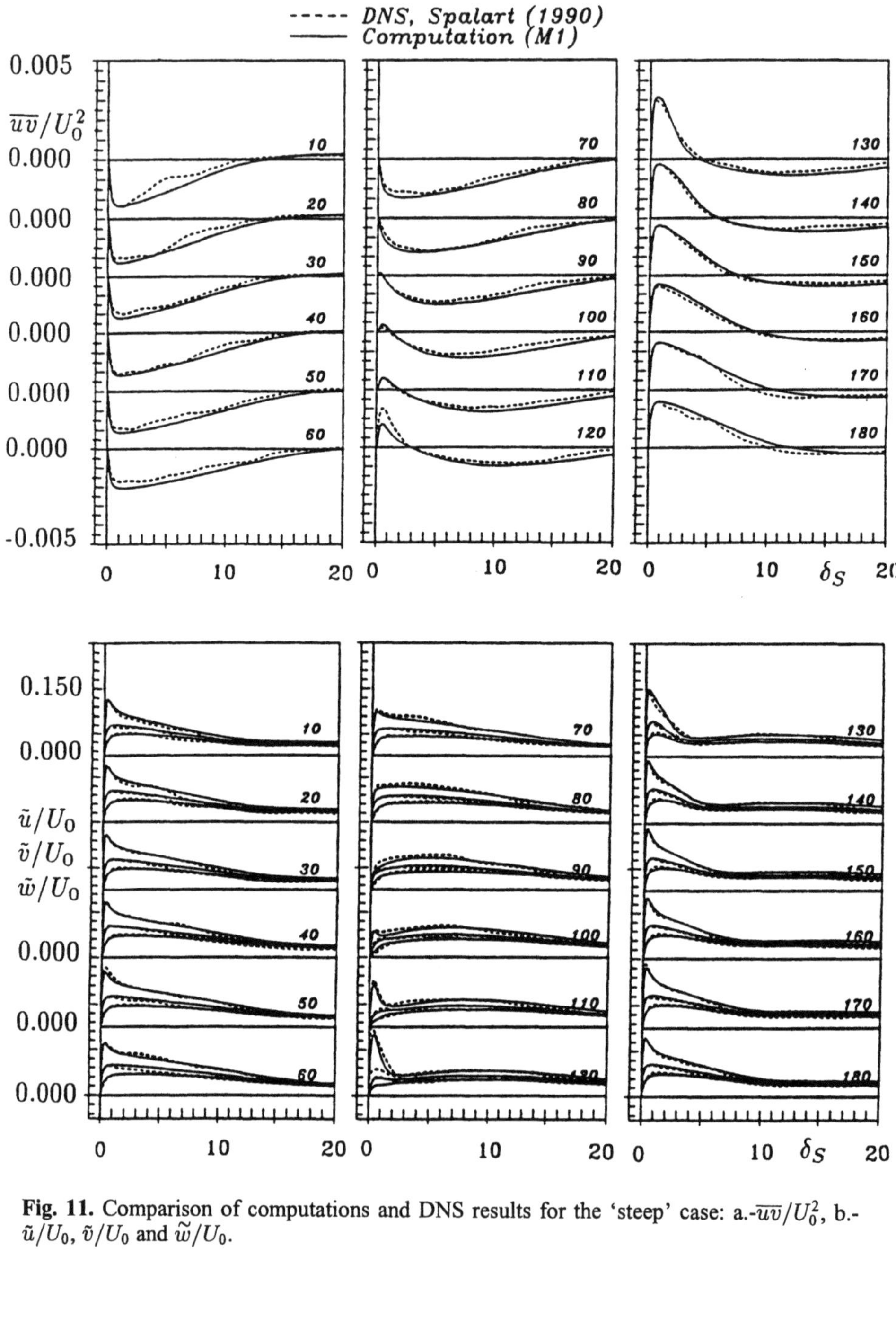

Fig. 11. Comparison of computations and DNS results for the 'steep' case: a.-$\overline{uv}/U_0^2$, b.-$\tilde{u}/U_0$, $\tilde{v}/U_0$ and $\widetilde{w}/U_0$.

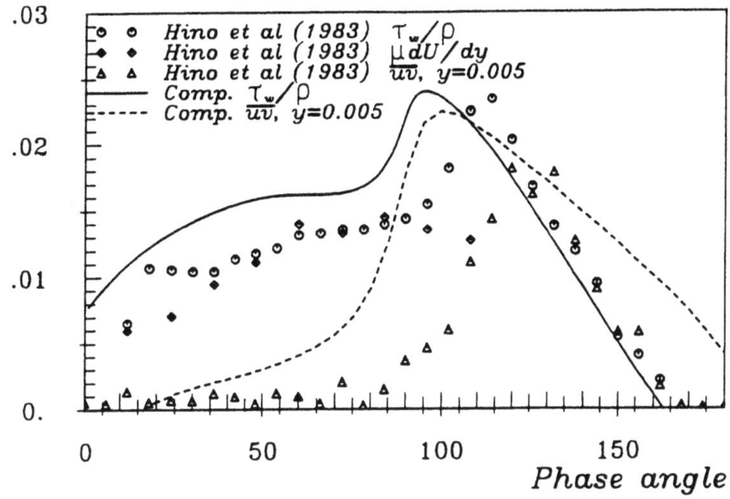

Fig. 12. Evolution of the shear stress at the wall and at a position in the flow in a plane channel.

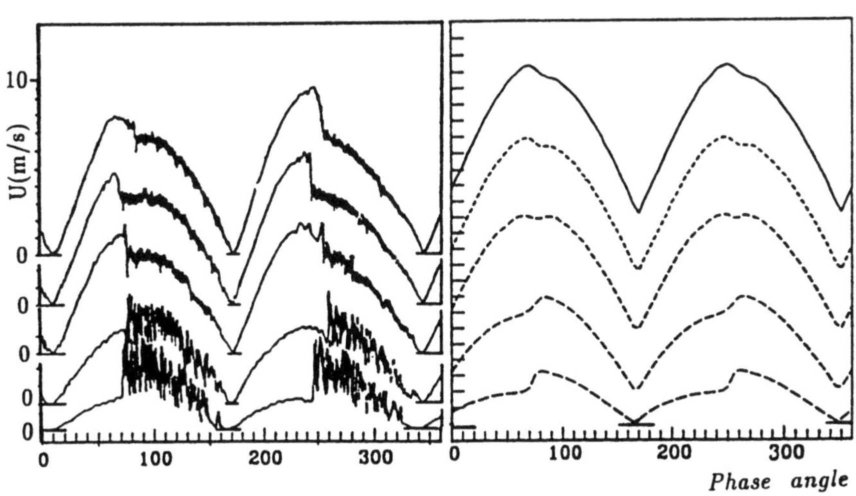

Fig. 13. Recorded (Hino et. al. 1983) and computed histograms of the mean velocity at various positions across the channel ($Re_\delta = 876$, $\lambda = 12.8$).

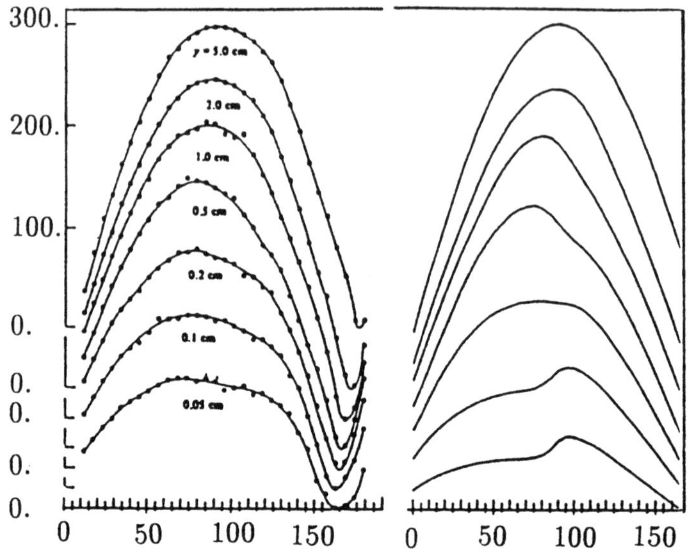

Fig. 14. Recorded (Hino et. al. 1976) and computed histograms of the mean velocity at various positions across the pipe (Re_δ = 1530, λ = 1.91).

The resemblance of the signals, both in shape and magnitude, are so striking, in particular at two positions away from the wall, that we considered presenting the results to be worthwhile, although they were obtained for frequency about one-third of that in the experiment (other parameters are the same). Computations at the exactly same conditions produced similar signals, but with smaller peaks and milder variation over a cycle. It should be noted that the experiments were carried out in a finite pipe of $L/D = 60$ and that the oscillograms were recorded at $L/D = 44$, whereas our time-marching computations takes no account of the pipe length.

4. Conclusions

Computations of oscillating boundary layers for a range of transitional Reynolds numbers, Re_δ, between 760 and 1140 show a very good agreement with DNS and experimental results. For oscillating flows in pipes and channels for the same Re_δ, the transition features are equally well reproduced if $\lambda \gg 1$ and if the duct Reynolds number is higher than the critical value, defined approximately by $Re = 1500\lambda$. It is obvious that these (and probably other) turbulence models can not simulate the subtle details of non-laminar (or perturbed laminar) pattern at the initiation of instability at the lower transition limit, as found by experiments or DNS, e.g., at $Re_\delta \approx 600$. However, the model seems to be capable of reproducing the conditionally turbulent regime with sudden bursts of turbulence at the start of deceleration phase and subsequent re-laminarization, over a range of transitional Re numbers. The model

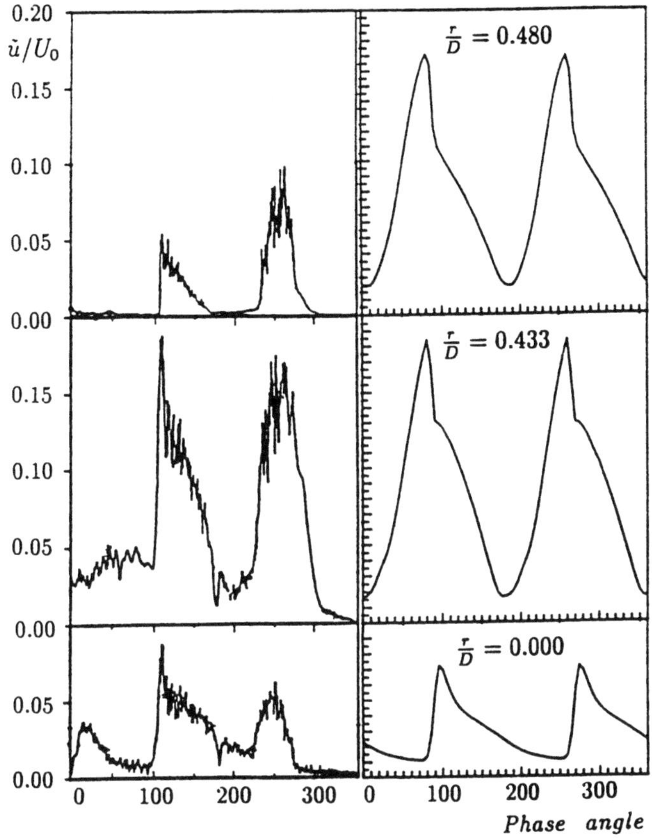

Fig. 15. Recorded (Seume 1988) and computed histograms of the streamwise velocity fluctuations at three positions across the pipe ($Re_\delta = 925, \lambda = 6.32$).

requires a minimum of 'background' turbulence to be able to reproduce its amplification and eventual transition to the fully turbulent regime. What this minimum is depends on the choice of the adjustment functions, which have in the past been tuned to satisfy the switch over from the outer turbulent to the inner viscous regimes in near-wall flows. The present model reproduces better the transition dynamics in the corresponding range of Re numbers, including the form and magnitude of the sudden bursts of turbulence quantities at various positions in the flows.

Acknowledgements

The authors acknowledge the valuable support of the Deutsche Forschungsgemeinschaft (to K. H.) and of the International Bureau of the Forschungszentrum Jülich (to S.J. and I. H.). We thank P. R. Spalart and B. L. Jensen for supplying data for model validation.

References

Gibson M.M., Launder B.E. (1978): "Ground effects on pressure fluctuations in the atmospheric boundary layer", J. Fluid Mech. **86**, 491-511

Ha Minh H., Viegas J.R., Rubesin M.W., Vandromme D.D., Spalart P.R. (1989): Physical Analysis and Second-Order Modelling of an Unsteady Turbulent Flow: The Oscillating Boundary Layer on a Flat Plate", Proc. 7th Symp. on Turbulent Shear Flows, Stanford University, Paper 11-5.

Hanjalić K., Jakirlić S. (1993): "A Model of Stress Dissipation in Second-Moment Closures", Applied Scientific Research **51**, 513-518, ed. F.T.M. Nieuwstadt, Advances in Turbulence IV, Kluwer Academic Publishers.

Hanjalić K., Launder B.E. (1976): "Contribution towards a Reynolds-stress closure for low-Reynolds-number turbulence", J. Fluid Mech. **74**, 593-610

Hino M., Sawamoto M., Takasu S. (1976): "Experiments on transition to turbulence in an oscillatory pipe flow", J. Fluid Mech. **75**, 193-207

Hino M., Kashiwayanagi M., Nakayama A., Hara T. (1983): "Experiments on the turbulence statistics and the structure of a reciprocating oscillatory flow", J. Fluid Mech. **131**, 363-400

Jensen B.L, Sumer B.M., Fredsoe J. (1989): "Turbulent oscillatory boundary layers at high Reynolds numbers", J. Fluid Mech. **206**, 265-297

Justesen P., Spalart P.R. (1990): "Two-Equation Turbulence Modeling of Oscillatory Boundary Layers", AIAA Paper 90-0496

Koehler W.J., Patankar S.V., Ibele W.E. (1991): "Numerical prediction of turbulent oscillating flow in a circular pipe". (unavailable):

Launder B.E., Reynolds W.C. (1983): "Asymptotic near wall stress dissipation rates in a turbulent flow", Phys. Fluids **26**, 1157

Launder B.E., Shima N. (1989): "Second-Moment Closure for the Near-Wall Sublayer: Development and Application", AIAA J. **27**, 1319-1325

Launder B.E., Tselepidakis D.P. (1993): "Contribution to the Modelling of Near-Wall Turbulence", Turbulent Shear Flows 8, ed. F. Durst *et al.*, Springer-Verlag, 81-96

Seume J.R. (1988): Ph.D. thesis, Univ. Minnesota.

Shima N. (1993): "Prediction of Turbulent Boundary Layers With a Second-Moment Closure: Part I - Effects of Periodic Pressure Gradient, Wall Transpiration and Free-Stream Turbulence", J. Fluid Eng. **115**, 56-69

Spalart P.R., Baldwin B.S. (1989): "Direct Simulation of a Turbulent oscillating Boundary Layer", Turbulent Shear Flows 6, ed. J. C. André *et al.*, Springer-Verlag, 417-440

Structure of Heat Transfer in the Thermal Layer Growing in a Fully Developed Turbulent Flow

Y. Nagano[1], H. Sato[1,2], and M. Tagawa[1]

[1] Department of Mechanical Engineering, Nagoya Institute of Technology, Gokiso-cho, Showa-ku, Nagoya 466, Japan
[2] On leave from Nippondenso Co., Ltd., 1-1, Showa-cho, Kariya 448, Japan

Abstract

An experimental investigation of the transport processes of heat has been made in the thermal entrance region of a fully developed turbulent pipe flow. Statistical quantities closely related to turbulent heat transfer have been measured and analyzed. Basic data presented can be used for assessing or developing a turbulence model for heat transfer. Turbulence quantities such as the temperature intermittency factor and the skewness and flatness factors of temperature fluctuations are found to have similarities in the growing thermal layer. Also, from the results of conditional analyses, the weighted probability density function of a turbulent heat flux, and the wavelet transform of velocity and temperature fluctuations, it becomes evident that the ejection-type fluid motions dominate the heat transfer in the outer intermittent region and play a key role in the evolution of the thermal boundary layer.

Nomenclature

d	pipe diameter
$F(\chi)$	flatness factor of χ, $= \overline{\hat{\chi}^4}$
Pr_t	turbulent Prandtl number
$P(\chi)$	probability density function (p.d.f.) of χ
r_0	pipe radius, $= d/2$
$S(\chi)$	skewness factor of χ, $= \overline{\hat{\chi}^3}$
T	instantaneous fluid temperature, $= \overline{T} + t$
$\overline{T}$, t	time-averaged and fluctuating temperatures
$\overline{T}_0$	fluid temperature at beginning of heated section
$\overline{T}_w$	wall temperature
$\overline{T}^+$	dimensionless temperature, $= (\overline{T}_w - \overline{T})/t_\tau$
t_τ	friction temperature
$\overline{U}_0$	bulk velocity at beginning of heated section
$\overline{U}_c$	centerline velocity
u, v	fluctuating velocities in x and y directions
u_τ	friction velocity
W_m	weighted p.d.f. of moment m [Eq. (8)]

x	streamwise distance from beginning of heated section
y	radial distance from wall
y^+	dimensionless distance, $= u_\tau y/\nu$
δ_t	thermal boundary-layer thickness at point where $(\overline{T}_w - \overline{T})/(\overline{T}_w - \overline{T}_0) = 0.99$
δ^*	displacement thickness
ν	kinematic viscosity
τ	time
χ	probability variable normalized by r.m.s. value
$\widehat{(\ )}$	normalization by r.m.s. value
$\overline{(\ )}$	time mean value
$(\ )^+$	normalization by inner variables
$\langle (\ ) \rangle_{in}$	zone-average inside instantaneous thermal layer
$\langle (\ ) \rangle_{out}$	zone-average outside instantaneous thermal layer

1. Introduction

In the thermal boundary layer growing in a fully developed turbulent pipe flow, usually called the thermal entrance region, the analogy between turbulent heat and momentum transfer disappears almost completely. Especially, in the outer region of a thermal boundary layer, the distributions of temperature-fluctuation intensities are totally dissimilar because of the distortion due to the intermittent entrainment of thermal bulges. Therefore, detailed measurements are needed to gain more insight into the fundamental mechanism of the turbulent heat transfer process. Moreover, the experimental data may be used to assess the existing turbulence models and to develop new ones.

In the fully developed thermal field, the detailed characteristics of temperature fluctuations are investigated by Krishnamoorthy and Antonia (1987), and Nagano and Hishida (1985) experimentally, and by Kasagi et al. (1992), Kim and Moin (1989) and Lyons et al. (1991), using direct numerical simulation (DNS). However, the thermal entrance region problems remain unsolved.

In the present paper, we show the measured distributions of turbulence quantities related to heat and momentum transfer, together with relevant budget profiles. To delineate the relations between the outer-region intermittency and turbulent quantities, we investigate experimentally and theoretically the conditional probability density functions (p.d.f.s) of both temperature fluctuations and their spatial derivatives. Conditional analyses, weighted p.d.f.s and the wavelet transform of velocity and temperature fluctuations give full details of heat transport processes in the thermal entrance region and reveal the essential features responsible for the development of a thermal boundary layer.

2. Experimental Apparatus

The test section was constructed from a reamed brass tube with d = 45.68 mm in diameter and 2079 mm (45.5 d) in length. The section was heated at a uniform wall temperature by saturated steam of atmospheric pressure. An unheated pipe upstream of the test section constitutes a 127d hydrodynamic entrance length. Measurements were made in an air flow for a Reynolds number of $Re = \overline{U}_0 d/\nu = 4 \times 10^4$ at representative axial locations of $1.00 \leqslant x/d \leqslant 39.89$. The thermal boundary layer thicknesses δ_t/r_0 are 0.25, 0.41 and 0.54 at x/d = 1.89, 3.89 and 5.89, respectively.

Fluctuations of velocity components, u and v, and temperature, t, were simultaneously measured with a specially devised three-wire probe technique (Hishida and Nagano 1978; 1988). The probe consisted of two hot wires, one a symmetrically bent V-shaped wire and the other a single straight one, together with a cold wire. For each set of u-, v- and t-fluctuations, the real-time sampling frequency was 32 kHz, and the number of data points per measurement was 65 536. The statistical analysis of the data was performed on a FACOM M-1800 computer system.

3. Results and Discussion

3.1 Mean Temperature

The mean temperature profiles in the thermal entrance region are shown in Fig. 1. The solid line represents the fully developed logarithmic profile (Hishida and Nagano 1979) given by

$$\overline{T}^+ = 2.17 \ln y^+ + 4.30. \qquad (1)$$

As seen in Fig. 1, the mean temperature profile in the buffer layer ($y^+ < 30$) has been almost fully developed at x/d = 1.0, but the logarithmic temperature profile in the fully turbulent region is not established even at x/d = 5.89. The mean temperature requires a large distance from a merging point of the thermal boundary layer ($x/d \simeq$ 15, where $\delta_t = d/2$) to reach the fully developed state at x/d = 39.89.

3.2 Instantaneous Signals of Turbulence Quantities

Figure 2 shows the typical instantaneous signal traces of turbulence quantities measured at x/d = 3.89 (δ_t/r_0 = 0.41). Intermittency of heat and momentum transfer in the vicinity of the wall (Fig. 2a) can be attributed to well-organized fluid motions such as ejection (u<0, v>0) and sweep (u>0, v<0) (for details, see Nagano and Hishida 1990). Near the edge of the thermal boundary layer (Fig. 2b), on the other hand, thermal turbulence quantities show another type of remarkable intermittency due to the fluctuating interface between hot and ambient cold fluids. As seen from this figure, the ejection-type outward motions dominate the transport process of heat in the outer intermittent region.

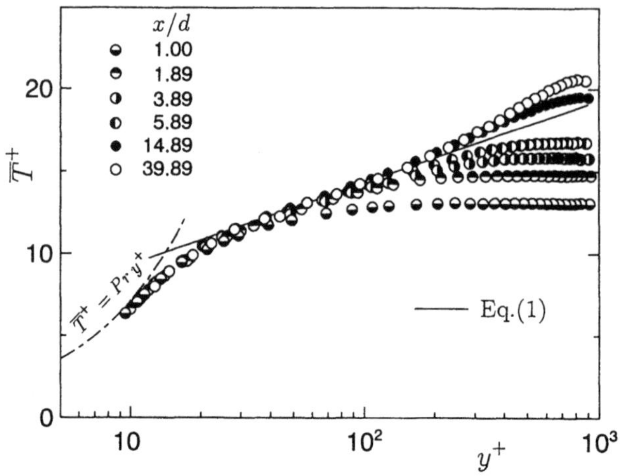

Fig. 1. Mean temperature profiles.

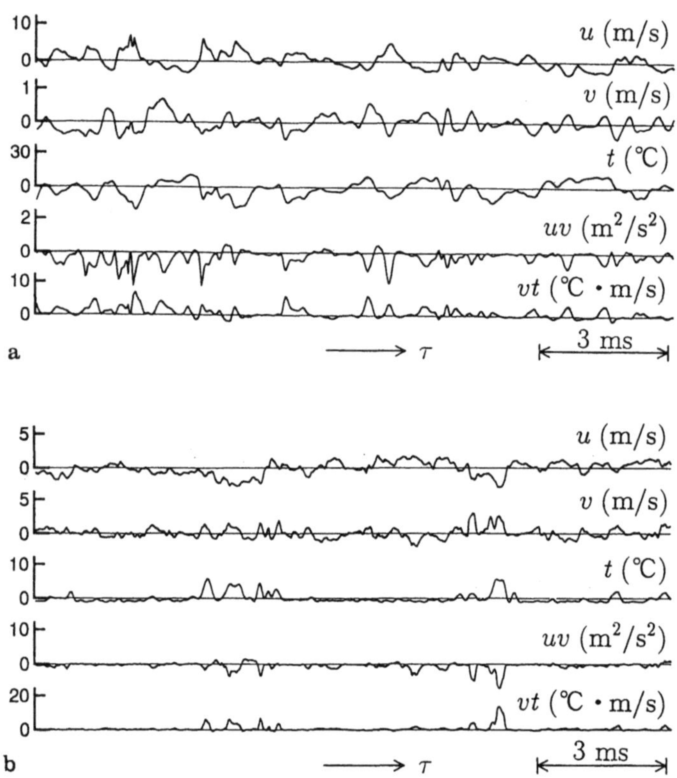

Fig. 2. Signal traces of turbulence quantities. **a)** $y^+ = 11.0$, $y/\delta_t = 0.030$; **b)** $y^+ = 363$, $y/\delta_t = 0.987$

3.3 Temperature Fluctuation and Turbulent Heat Flux

The distributions of the relative intensity of temperature fluctuations, $\sqrt{\overline{t^2}}/(\overline{T}_w - \overline{T})$, are presented as a function of x/d in Fig. 3. This figure shows that the profile develops progressively as the distance from the beginning of the heated section, x/d, becomes large. It should be noted that at the position where the thermal boundary layers reach the pipe centerline ($x/d \simeq 15$), the $\sqrt{\overline{t^2}}/(\overline{T}_w - \overline{T})$ distribution is not fully developed and grows further downstream, especially in the outer intermittent region of the thermal boundary layer.

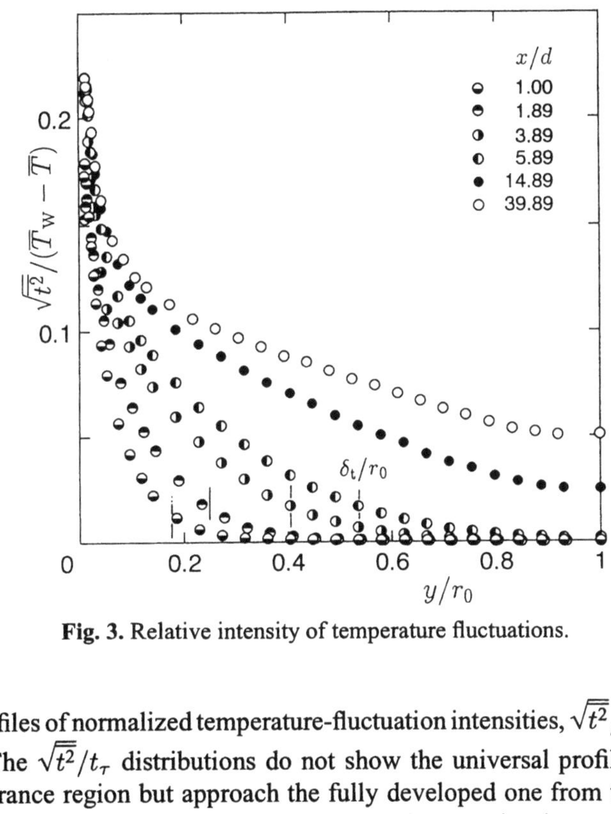

Fig. 3. Relative intensity of temperature fluctuations.

The profiles of normalized temperature-fluctuation intensities, $\sqrt{\overline{t^2}}/t_\tau$, are shown in Fig. 4. The $\sqrt{\overline{t^2}}/t_\tau$ distributions do not show the universal profile through the thermal entrance region but approach the fully developed one from the vicinity of the wall with the thermal layer developing. From Figs. 1 and 4, the mean temperature in the thermal boundary layer is found to develop more rapidly than the temperature fluctuations.

The profiles of normalized turbulent heat flux, $\overline{vt}/u_\tau t_\tau$, are presented in Fig. 5 as a function of x/d. Also included is the fully developed profile of the normalized Reynolds shear stress, $-\overline{uv}/u_\tau^2$. It can be seen that the distribution of $\overline{vt}/u_\tau t_\tau$ changes with the development of the thermal layer while that of $-\overline{uv}/u_\tau^2$ remains unchanged as a matter of course, and thus there exists no similarity between heat and momentum transfer. In the fully developed region (x/d = 39.89), there is a

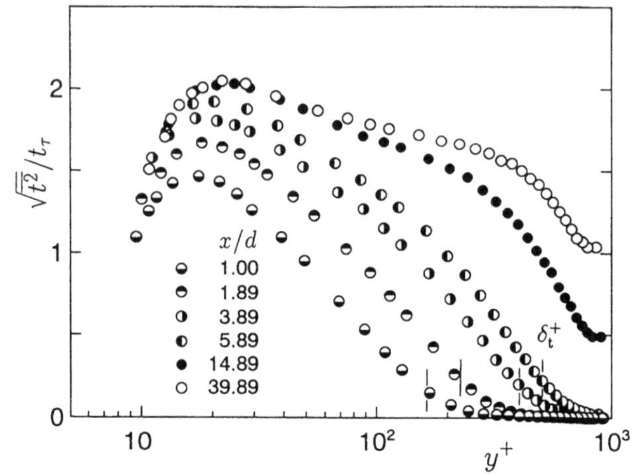

Fig. 4. R.m.s. intensity of temperature fluctuations.

close similarity between $\overline{vt}/u_\tau t_\tau$ and $-\overline{uv}/u_\tau^2$ profiles (for details, see Nagano and Tagawa 1988).

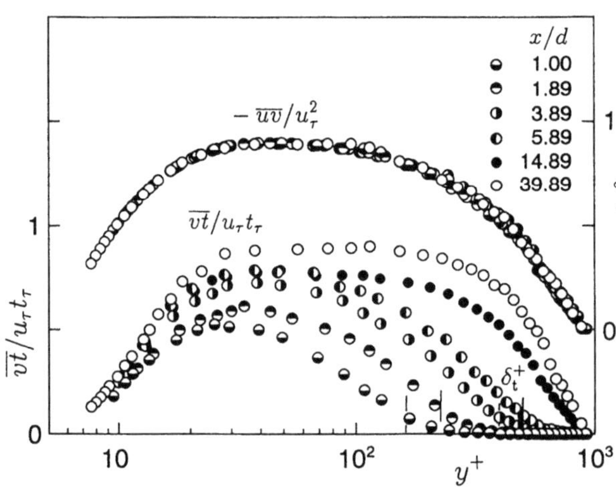

Fig. 5. Turbulent heat flux $\overline{vt}$ and Reynolds shear stress $-\overline{uv}$.

Figures 6a-c show the budgets of temperature variance $\overline{t^2}$ at $x/d = 1.0, 3.89$ and 39.89, respectively. Values of each term in the $\overline{t^2}$- and $\overline{vt}$-transport equations are obtained in the same manner as in Nagano and Hishida (1985) and are normalized by the wall variables ν, u_τ and t_τ. Temperature fluctuations in the thermal entrance region are produced mainly in the buffer layer ($y^+ < 30$) regardless of the axial position x/d, and this fact is similar to that found in the fully developed thermal layer (Nagano and Hishida 1985). The production rate peaks at about $y^+ = 13$, and the energy produced here is transferred toward and away from the wall by the turbulent diffusion, and toward the wall by the molecular diffusion. In the region $y^+ > 30$, the sum of the production and the turbulent diffusion is balanced with the dissipation.

From Fig. 6, we can investigate the effects of the thermal layer growth on the budget of $\overline{t^2}$ in detail. The main results are summarized as follows:

(i) The energy-producing region extends to a larger y^+ region and the peak value of the production rate, located at about $y^+ = 13$, becomes larger as the thermal layer develops.

(ii) The dissipation rate shows characteristics similar to those of the production, though it has the opposite sign.

(iii) The turbulent diffusion is hardly affected by the thermal layer development. This is because the turbulent diffusion is dominated almost always by the near-wall coherent structures, i.e., ejections and sweeps (Nagano and Tagawa 1991) which exist in a fully-developed velocity field regardless of the growth of the thermal layer.

(iv) The molecular diffusion keeps virtually the same distribution through the thermal entrance region, since it becomes significant only in the vicinity of the wall where the $\overline{t^2}/t_\tau^2$-profile has fully developed very quickly.

From the above observations, we may conclude that the development of the thermal layer affects mainly the production and dissipation processes of $\overline{t^2}$, whereas the diffusive transports near the wall reach a fully-developed state in the early stage of thermal layer development.

The budgets of turbulent heat flux $\overline{vt}$ at $x/d = 1.0,\ 3.89,\ 39.89$ are shown in Figs. 7a-c. In contrast to the temperature variance, the turbulent heat flux is produced over a wide portion of the wall region ($y^+ < 100$); and besides, it is mostly destructed there by the pressure scrambling interactions. The turbulent diffusion near the wall is quite large. Therefore, it is clear that turbulent diffusion also contributes much to the development of the thermal boundary layer. Even so, the near-wall development of the thermal layer is found to affect mainly the production and the destruction processes, which is also the case with the temperature variance.

3.4 Turbulent Prandtl Number

As shown in Fig. 5, the analogy between the velocity and the thermal fields disappears in the thermal entrance region. Thus, it is important to know how the turbulent Prandtl number Pr_t behaves in the thermal entrance region, since the conventional turbulence models assume a constant Pr_t value based on the Reynolds analogy. The turbulent Prandtl number Pr_t is defined by

$$Pr_t = \frac{\nu_t}{\alpha_t} = \frac{\overline{uv}\ \partial\overline{T}/\partial y}{\overline{vt}\ \partial\overline{U}/\partial y}, \tag{2}$$

where ν_t and α_t are the eddy viscosity and the eddy diffusivity for heat, respectively. The results are shown in Fig. 8, together with the experimental data by Blackwell

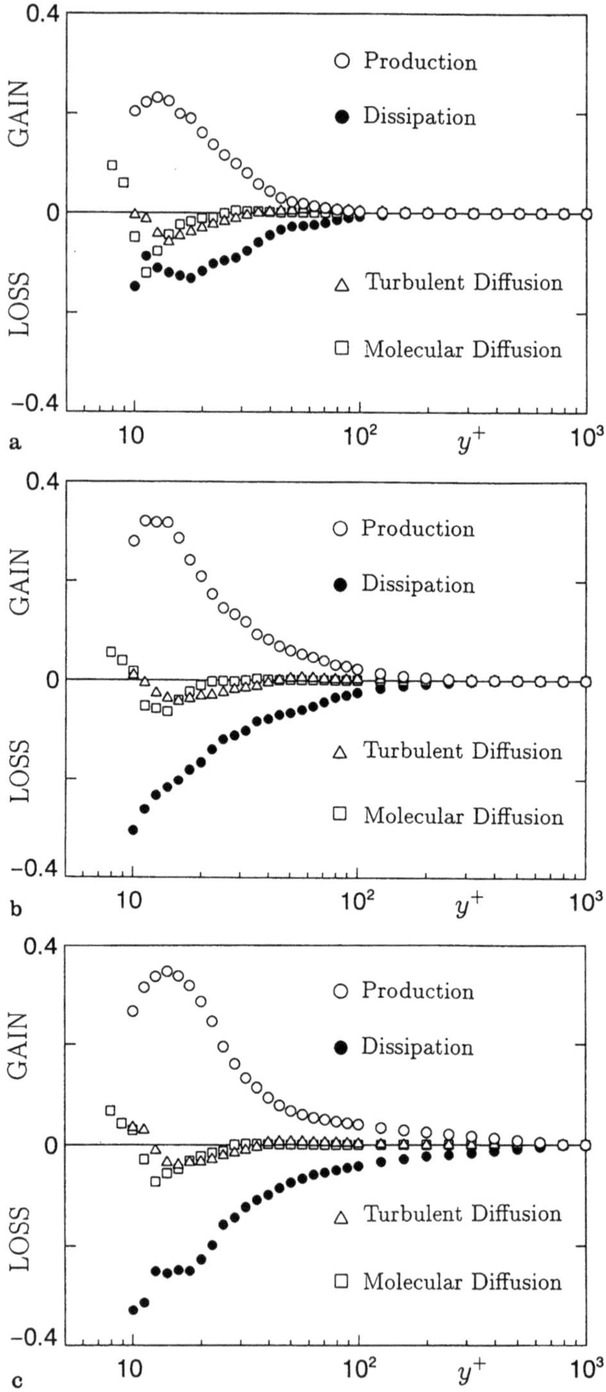

Fig. 6. Budget of temperature variance $\overline{t^2}/t_\tau^2$. **a)** $x/d = 1.0$; **b)** $x/d = 3.89$; **c)** $x/d = 39.89$

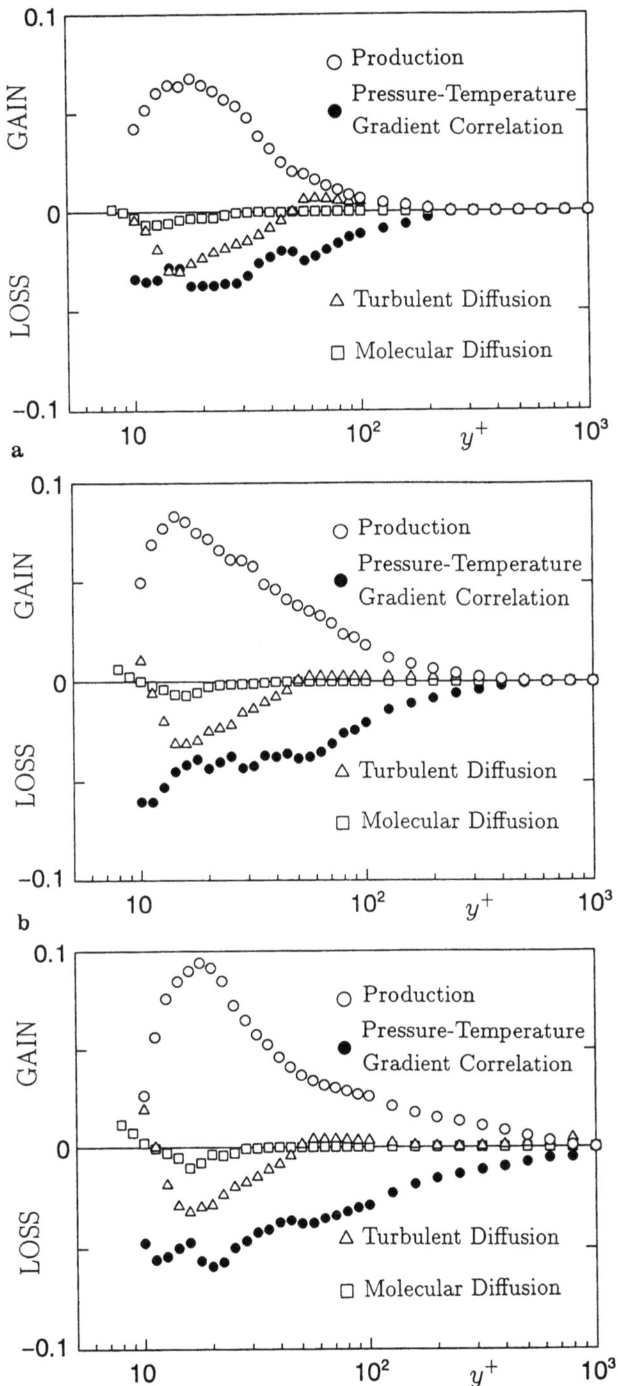

Fig. 7. Budget of turbulent heat flux $\overline{vt}/u_\tau t_\tau$. **a)** $x/d = 1.0$; **b)** $x/d = 3.89$; **c)** $x/d = 39.89$

et al. (1972). It is evident from this figure that Pr_t is not constant in the thermal boundary layer and has no universal distribution throughout the thermal entrance region. For $y^+ > 20$, the Pr_t profile does not change greatly with y^+ and approaches a fully-developed value of 0.87 as the thermal layer grows. On the other hand, in the near-wall region $y^+ < 20$, Pr_t increases steeply as the wall is approached. The overall profile of the fully-developed Pr_t distribution ($x/d = 39.89$) is in good agreement with that of Blackwell et al. (1972).

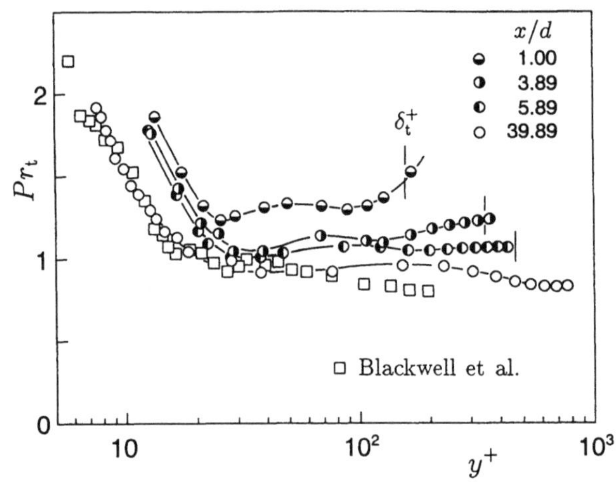

Fig. 8. Turbulent Prandtl number.

3.5 Intermittency of Thermal Field

It is important to capture the instantaneous interface between hot and ambient cold fluids in order to analyze the developing process of a thermal boundary layer. Thus, we have used a detection scheme based on the time-derivative of temperature fluctuations $(\partial t/\partial \tau)^2$ and the instantaneous temperature T in a manner similar to the technique of capturing turbulent/non-turbulent interfaces (Hedley and Keffer 1974). Using the detector functions, the probability distributions of instantaneous thermal interface, i.e., the temperature intermittency factor, were obtained as shown in Fig. 9. The abscissa is the distance from the wall normalized by δ_t. The solid line in the figure is a Gaussian probability distribution expressed by the following equation:

$$\Omega = \frac{1}{\sqrt{2\pi}\sigma} \int_y^{\infty} \exp\left[-\left(y-\overline{Y}\right)^2/2\sigma^2\right] dy, \tag{3}$$

where the averaged position of interface of a thermal boundary layer is $\overline{Y} = 0.87\,\delta_t$, and the standard deviation is $\sigma = \left[\overline{(Y-\overline{Y})^2}\right]^{1/2} = 0.28\,\delta_t$. Obviously, the instantaneous position of the interface fluctuates widely around the time-averaged position in conformity to a Gaussian p.d.f.. As a result, ambient cold fluids penetrate deeply into the near-wall hot region, say, down to 0.3 δ_t. The turbulent outcroppings of the thermal boundary layer, on the other hand, extend to about 1.5 δ_t. This intermittent phenomenon is the entrainment process leading to the development of the thermal boundary layer in a fully developed turbulent shear flow.

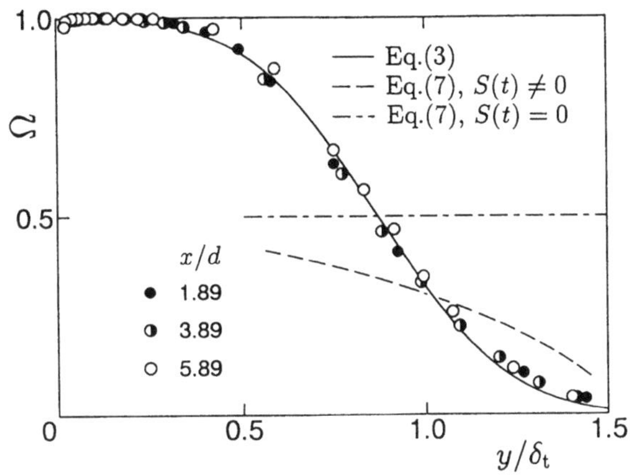

Fig. 9. Probability of the thermal boundary layer edge.

Comparing this value with that of the velocity field obtained by Kibens (1968) in a turbulent boundary layer, i.e., $\sigma = 0.15\,\delta_{99}$ (δ_{99} : boundary layer thickness), one can see that the instantaneous edge of the thermal boundary layer in the thermal entrance region fluctuates more widely than that of the turbulent boundary layer. This important phenomenon can be ascribed to the fact that the turbulent mixing exists everywhere through the thermal entrance region (both inside and outside the thermal boundary layer) because of the fully-developed velocity field. Thus, the instantaneous thermal boundary layer may be driven to fluctuate very actively and can grow more rapidly to the fully-developed state.

3.6 Probability Density Distributions of Temperature Fluctuation and Its Derivative

In an intermittent region near the edge of a thermal boundary layer, we expect that heat transport processes are complicated by the fluctuations of an instantaneous interface between hot and cold fluids. So, we have attempted to elucidate the theoretical relation between the intermittency and the statistical characteristics of t and $\partial t/\partial x$. The derivative, $\partial t/\partial x$, is obtained by converting the temporal difference into a spatial difference using Taylor's hypothesis (Antonia et al. 1980).

To express the probability density function of turbulence quantities, we have used the cumulant-discard method (Monin and Yaglom 1971) to take into account a departure from a Gaussian distribution. Using the skewness and flatness factors, $S(\chi)$ and $F(\chi)$, the one-dimensional p.d.f. is defined (Nagano and Tagawa 1988) as

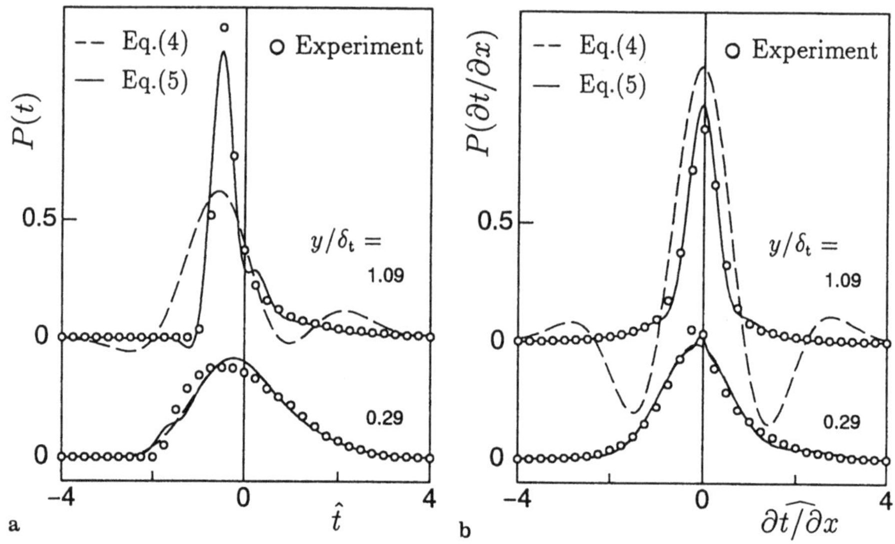

Fig. 10. Probability density distributions of $\hat{t}$ and $\widehat{\partial t/\partial x}$. **a**) $\hat{t}$; **b**) $\widehat{\partial t/\partial x}$

$$P(\chi) = P_G(\chi)\left\{1 + \frac{1}{6}S(\chi)\,(\chi^3 - 3\chi) + \frac{1}{24}[F(\chi) - 3]\,(\chi^4 - 6\chi^2 + 3)\right\}, \tag{4}$$

where $P_G(\chi)$ denotes the one-dimensional Gaussian p.d.f. given by $P_G(\chi) = \exp(-\chi^2/2)/(2\pi)^{1/2}$.

First, the p.d.f.s of t and $\partial t/\partial x$ are calculated at two representative positions, the near-wall region ($y/\delta_t = 0.29$) and the outer intermittent region of the thermal boundary layer ($y/\delta_t \simeq 1$) without taking the intermittency into consideration, and they are shown by the broken lines in Figs. 10a, b together with the experimental data. For the p.d.f. calculation of t, we have neglected the flatness-factor-affected term in the right-hand side of Eq. (4) to avoid disparate results (this treatment corresponds to the so-called Edgeworth expansion, in which the optimal combination of the coefficients in the Gram-Charlier expansion is used). As shown in Fig. 10, in the near-wall region, the calculated results of both t and $\partial t/\partial x$ agree well with the experiments. Since turbulence quantities near the wall have reached the fully-developed state in the early stage of the thermal layer development (see, e.g., Figs. 4 and 5), their statistical characteristics tend to obey the quasi-Gaussian stochastic process as seen in Nagano and Tagawa (1988). Thus, near the wall, we may use the Edgeworth expansion which is naturally applicable only to the case of slight deviation from Gaussianity. In the intermittent region, on the other hand, the calculations do not agree with the experiments at all. This is mainly due to the intermittency of the thermal field there and to the fact that the instantaneous temperature inside the thermal layer is consistently much larger than that of the ambient fluid. Therefore, we regard the p.d.f. in the intermittent region as the intermittency-factor-weighted sum of the zone-averaged p.d.f. for the inside of the thermal layer and that for the outside of the thermal layer, which yields

$$P(\chi) = \Omega P_{in}(\chi - \Delta\chi_{in}) + (1 - \Omega)P_{out}(\chi - \Delta\chi_{out}), \tag{5}$$

where P_{in} and P_{out} respectively denote the p.d.f.s for the inside and the outside of the thermal boundary layer with $\Delta\chi_{in} = \left(\langle T\rangle_{in} - \overline{T}\right) / \sqrt{\overline{t^2}}$ and $\Delta\chi_{out} = \left(\langle T\rangle_{out} - \overline{T}\right) / \sqrt{\overline{t^2}}$.

The p.d.f. distributions, $P\left(\hat{t}\right)$ and $P\left(\widehat{\partial t/\partial x}\right)$, calculated from Eqs. (4) and (5), are shown by the solid lines in Figs. 10a, b. The results are in good agreement with the experiments. Therefore, the assumption given by Eq. (5) is valid, and can be used to analyze the entrainment process of the thermal layer.

To know further details on the statistical characteristics of temperature fluctuations, we investigated the distributions of skewness and flatness factors of both t and $\partial t/\partial x$. The results are shown in Figs. 11a, b as a function of x/d together with the relevant Gaussian values. In the immediate neighborhood of the wall, the skewness factor of t becomes negative due to the dominance of the sweep-type motions of cold fluid. On the other hand, away from the wall, $S(t)$ increases monotonously to be a large positive value, since the ejection-type motions with high temperature dominate the turbulent core (Nagano and Tagawa 1988). The skewness factor of $\partial t/\partial x$, which is considered to be the measure of the fine structures of a thermal field (Sreenivasan and Antonia 1977), remains nearly at unity over the thermal boundary layer.

Flatness factors of both t and $\partial t/\partial x$ (Fig. 11b) stay at respective constant values for $y/\delta_t < 0.35$ and begin to increase as the distance from the wall becomes larger. These characteristics can be ascribed to increasingly intermittent nature of temperature fluctuations in the outer region of the thermal boundary layer. From Figs. 11a, b, it is found that both the skewness and flatness factors of t and $\partial t/\partial x$ lie in the respective unique curves independent of x/d. This is strong evidence for the possible existence of a structural similarity in the thermal entrance region.

3.7 Relation between Intermittency Factor and Temperature Skewness

From Fig. 2b, the time trace of temperature fluctuation t in the outer intermittent region can be modeled as illustrated in Fig. 12. Based on this pictorial model, we can estimate the intermittency factor Ω as follows:

$$\Omega = \frac{1}{\tau_{total}} \sum_i \Delta\tau_i \simeq \frac{1}{\tau_{total}} \sum_i \Delta\tau_i' \equiv \gamma_+(t), \tag{6}$$

where τ_{total} is the total measuring time, and $\Delta\tau_i$ and $\Delta\tau_i'$ denote the period for $T > \overline{T}_0$ and that for $t > 0$, respectively. According to Nagano and Tagawa (1988), the time fraction for $t > 0$, $\gamma_+(t)$, is given by

$$\gamma_+(t) = \int_0^\infty P\left(\hat{t}\right) d\hat{t} = 1/2 - S(t) \Big/ 6\sqrt{2\pi}\,. \tag{7}$$

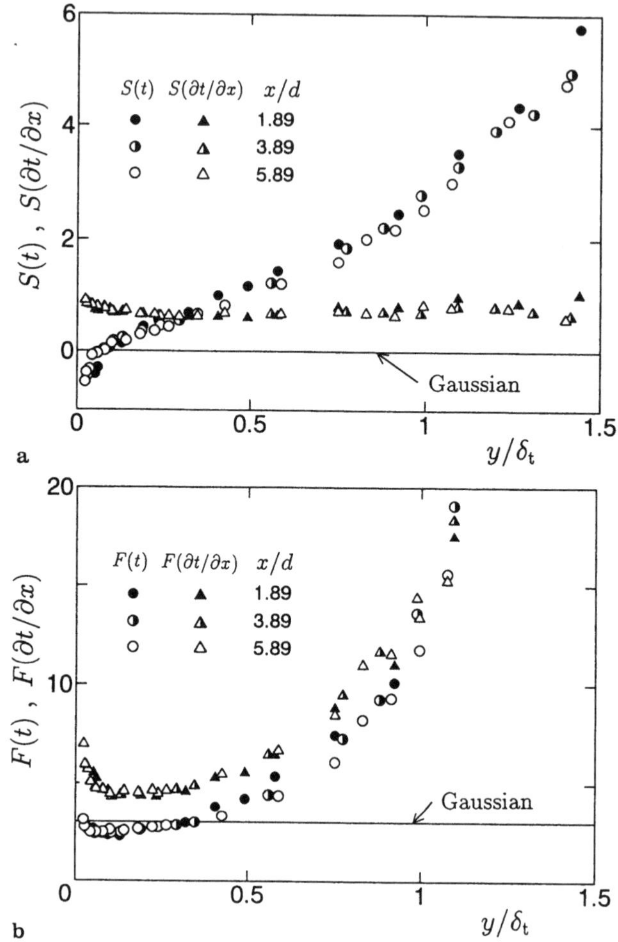

Fig. 11. Statistical quantities of t and $\partial t/\partial x$. **a)** Skewness factors; **b)** Flatness factors

Thus, we can predict the value of Ω from Eq. (6) using Eq. (7). The predictions of Ω are shown in Fig. 9 in broken and dashed lines. The broken line represents the calculations using the measured skewness factor (Fig. 11a), while the dashed line is the result calculated assuming that the p.d.f. of t is Gaussian, i.e., $S(t) = 0$. From Fig. 9, it is seen that the broken line can trace the behavior of the measurements of Ω qualitatively, although the agreement with the experiment is not quantitatively sufficient. On the other hand, the dashed line shows that the Gaussian p.d.f. of t does not represent the essential features of the turbulent structures in the outer intermittent thermal field. This means that $S(t)$ can be regarded as a primary parameter which characterizes the outer intermittent region of the thermal boundary layer. Furthermore, we may estimate the intermittency factor Ω by calculating the skewness factor of t using a turbulence model for the third-order moment $\overline{t^3}$ (see Lumley 1978; Nagano and Tagawa 1991).

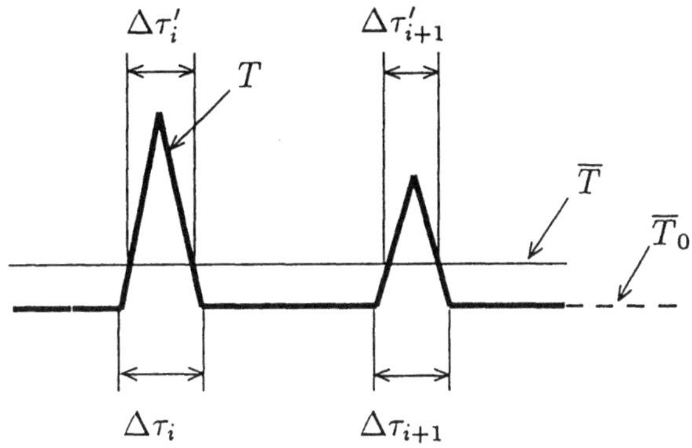

Fig. 12. Schematic model of temperature fluctuation in the outer intermittent region.

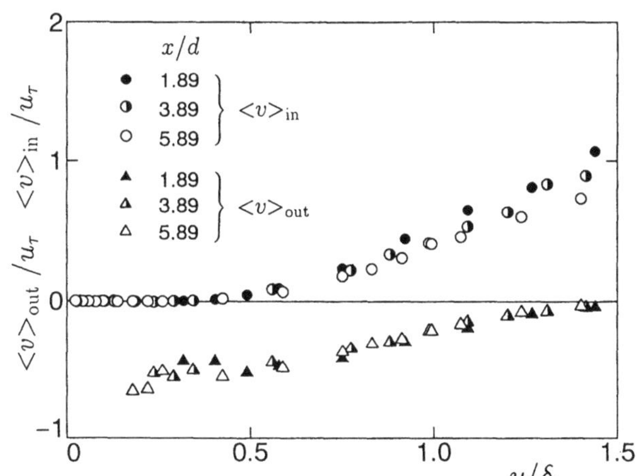

Fig. 13. Conditional averages of v-fluctuations.

3.8 Contributions of Organized Fluid Motions to Turbulent Heat Transport

Figure 13 shows the zone averages of the wall-normal velocity fluctuation v for the inside and the outside of the instantaneous thermal boundary layer, $\langle v\rangle_{in}$ and $\langle v\rangle_{out}$. Since the fluid inside the thermal layer is generally hot, $\langle v\rangle_{in}$ may correspond to the motion of a hot fluid. On the other hand, $\langle v\rangle_{out}$ is considered as a representative velocity of a cold fluid. As seen from Fig. 13, $\langle v\rangle_{in}$ is nearly zero in the near-wall region and increases monotonously with an increase of the distance from the wall. The zone average outside the thermal boundary layer, $\langle v \rangle_{\text{out}}$, on the other hand, becomes nearly zero far away from the wall and decreases gradually as the wall is approached. So, it can be concluded that the turbulent heat transport is dominated by the ejection-type motion ($v > 0$) in the outer region of the thermal boundary layer and by the sweep-type motion ($v < 0$) near the wall, respectively. Furthermore, the distributions of $\langle v \rangle_{\text{in}}$ and $\langle v \rangle_{\text{out}}$ lie in their unique curves regardless of the axial position x/d. This gives further support to the supposed similarity of the developing process of the thermal field up through the thermal entrance region.

To know in detail the quantitative contributions of each fluid motion classified in the (u, v)-plane to the turbulent heat flux $\overline{vt}$, we may use the weighted p.d.f. of $\overline{vt}$, $W_{\hat{v}\hat{t}}(\hat{u}, \hat{v})$, defined by

$$W_{\hat{v}\hat{t}}(\hat{u}, \hat{v}) = \int_{-\infty}^{\infty} \hat{v}\hat{t}\, P\left(\hat{u}, \hat{v}, \hat{t}\right) \mathrm{d}\hat{t}. \tag{8}$$

Figures 14a-c show the weighted p.d.f.s of $\overline{vt}$, $W_{\hat{v}\hat{t}}(\hat{u}, \hat{v})$, measured at three different wall-normal positions in the outer intermittent region. The interval between the contour lines is 5×10^{-3}. From these figures, it can be seen that in the outer region of the thermal boundary layer the second-quadrant (Q2) motion (i.e., ejection) is the only contributor to the production of $\overline{vt}$ independent of y/δ_{t}. However, the fourth-quadrant (Q4) motion (i.e., sweep), which is known to contribute equally to $\overline{vt}$ in the fully-developed thermal layer, does not play an important role at all, especially in the outermost region $y/\delta_{\text{t}} > 1$.

To obtain further detailed information, we have investigated the conditionally weighted p.d.f.s of $\overline{vt}$ for the inside of the instantaneous thermal boundary layer, $\langle W_{\hat{v}\hat{t}}(\hat{u}, \hat{v}) \rangle_{\text{in}}$. The results are shown in Figs. 15a-c. The distributions of $\langle W_{\hat{v}\hat{t}}(\hat{u}, \hat{v}) \rangle_{\text{in}}$ are very similar to those in the second-quadrant in Figs. 14a-c, and the contributions of the Q4-motions disappear completely.

From the comparison of $W_{\hat{v}\hat{t}}(\hat{u}, \hat{v})$ in Fig. 14 with $\langle W_{\hat{v}\hat{t}}(\hat{u}, \hat{v}) \rangle_{\text{in}}$ in Fig. 15, we may conclude that the Q2 motions (ejection-type motions) dominate the transport processes of heat in this region and that the Q4 motions (sweep-type motions) contribute little outside the instantaneous thermal boundary layer. However, as the wall is approached, the contribution of the Q4-motions to $\overline{vt}$ becomes increasingly important (not shown).

To investigate the local similarity between the thermal and velocity fields in the thermal entrance region, we apply the wavelet transform technique to the instantaneous signals of v and t. The wavelet transform has a feature that the relation between the local phenomena and turbulent structure can be observed visibly. The one-dimensional continuous wavelet transform is defined (Farge 1992) as

$$R(\tau,\, a) = a^{-1/2} \int_{-\infty}^{\infty} g^{*}\left(\frac{\tau' - \tau}{a}\right) f(\tau')\mathrm{d}\tau', \tag{9}$$

where $g(\tau)$ is an analyzing wavelet (* denotes the complex conjugate), a (> 0) is the scaling parameter, and $f(\tau)$ is an instantaneous signal to be examined. Values of

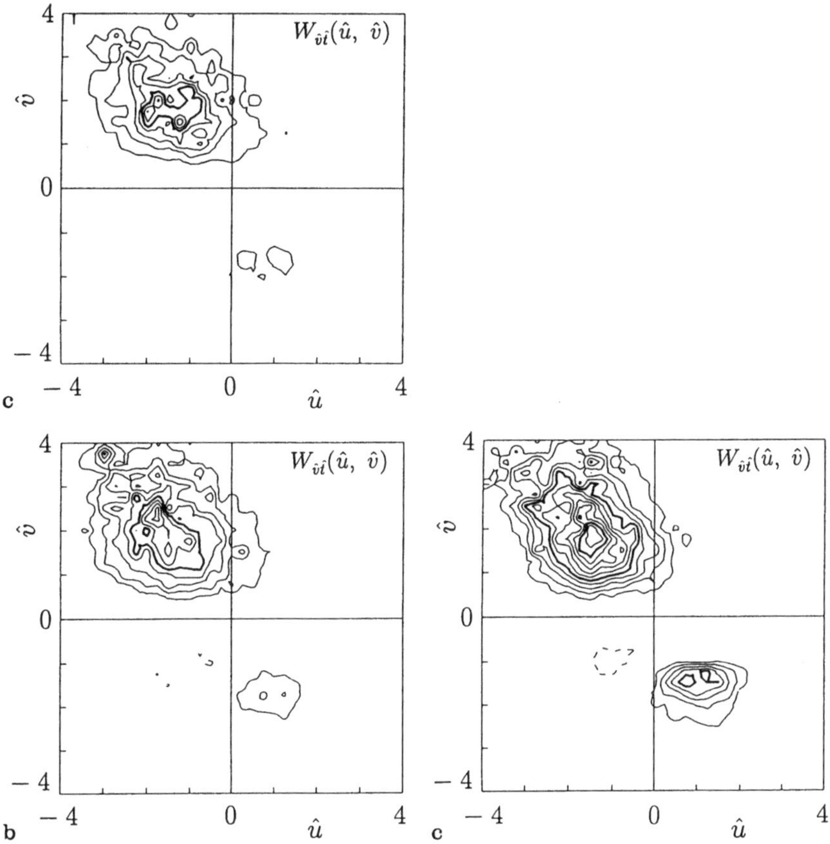

Fig. 14. Weighted p.d.f.s $W_{\hat{v}\hat{t}}(\hat{u}, \hat{v})$ (x/d = 3.89). **a)** y/δ_t = 1.31; **b)** y/δ_t = 1.09; **c)** y/δ_t = 0.773

a, τ and τ' are normalized by $\delta^* / \overline{U}_c$. As the analyzing wavelet, $g(\tau)$, we have used the second derivative of the Gaussian distribution function, i.e., "Mexican hat".

Figures 16a-c show the contour maps of $R(\tau,\ a)$ for u, v and t in the near-wall region (y^+ = 13.0, x/d = 3.89) together with the respective instantaneous signal traces. The solid and broken lines represent positive and negative values, respectively. The interval between contour lines is 8×10^{-5}. From these figures, we find that temperature fluctuations with the time scale of $a = 5 \sim 10$, which corresponds to the frequency of occurrence of the well-ordered motions (Nagano and Hishida 1990), are well synchronized with the ejection- and sweep-type motions. In addition, the strong correlation between v and t is seen in the very large time-scale regions as observed in a fully developed thermal field (Hishida and Nagano 1979).

The wavelet transforms of u, v and t in the outer intermittent region of the thermal boundary layer are shown in Figs. 17a-c. As seen from these figures, the positive t regions correspond very well to the positive v regions (ejection), while there is no

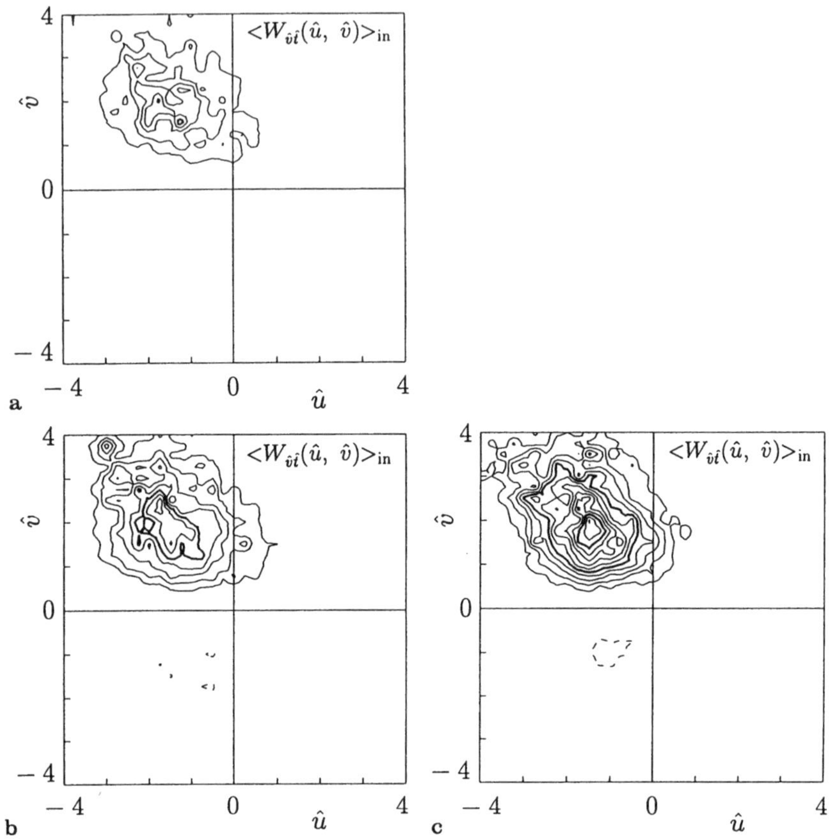

Fig. 15. Conditioned weighted p.d.f.s $\langle W_{\hat{v}\hat{t}}(\hat{u}, \hat{v})\rangle_{\text{in}}$ ($x/d = 3.89$). **a)** $y/\delta_t = 1.31$; **b)** $y/\delta_t = 1.09$; **c)** $y/\delta_t = 0.773$

correlation between the negative t region and the negative v region (sweep). This may support the results in the previous section that the ejection-type motions dominate turbulent heat transfer in the outer intermittent region of the thermal boundary layer.

4. Conclusions

From the budgets of temperature variance and turbulent heat flux, it becomes clear that $\sqrt{\overline{t^2}}$ and $\overline{vt}$ are produced mainly in $y^+ < 30$ and $y^+ < 100$, respectively, regardless of the thermal boundary layer thickness. The development of the thermal layer affects mainly the magnitude of their production and dissipation rates. It is also shown that the probability density functions of intermittent quantities (t and $\partial t/\partial x$) can be expressed by the weighted sum of the probability densities for the zones inside and outside the instantaneous thermal boundary layer.

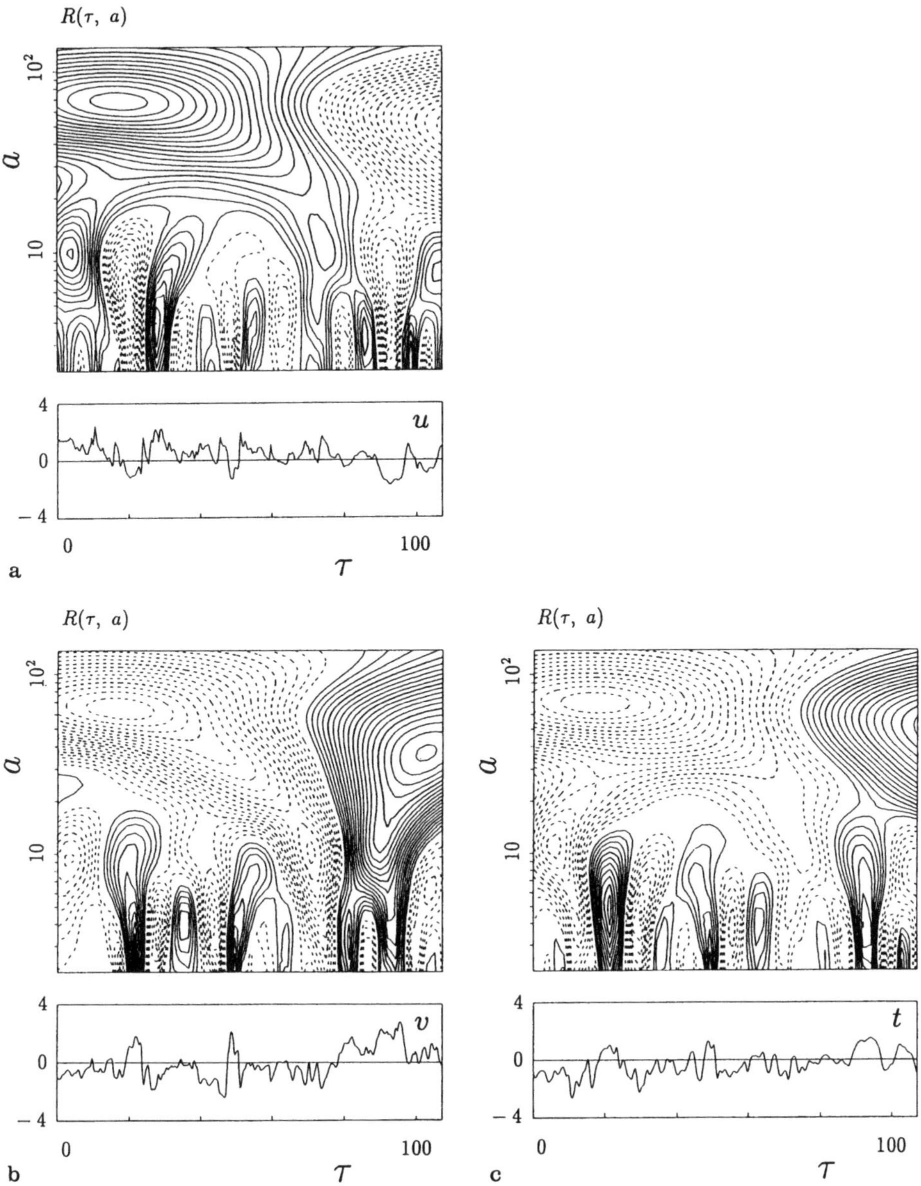

Fig. 16. Wavelet transform ($y^+ = 13.0$, $x/d = 3.89$). **a)** u; **b)** v; **c)** t

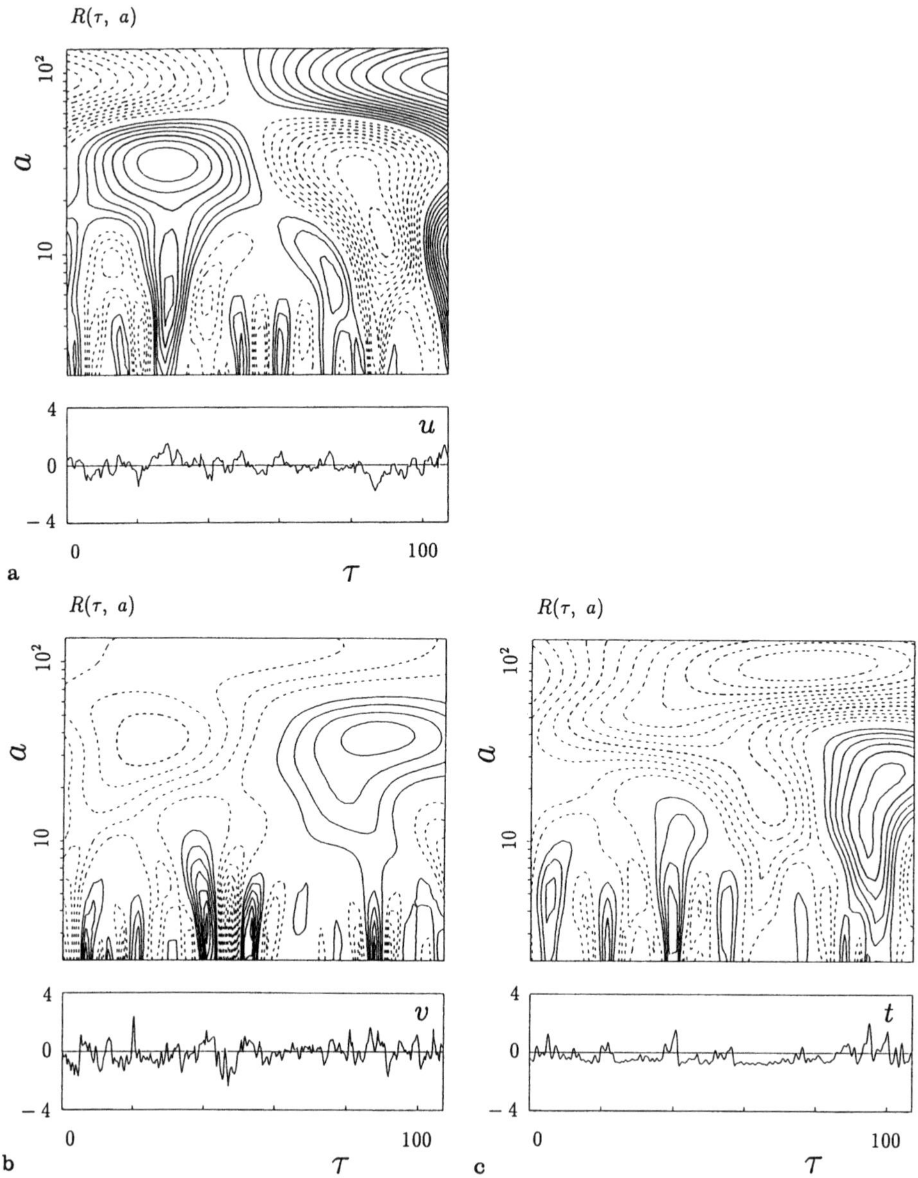

Fig. 17. Wavelet transform ($y^+ = 284.2$, $x/d = 3.89$). **a)** u; **b)** v; **c)** t

In the thermal entrance region of the fully developed turbulent pipe flow, the temperature intermittency factor and the skewness and flatness factors of temperature fluctuations show marked similarities in the thermal boundary layer. From the results of the conditional averages, the weighted p.d.f.s and the wavelet transforms of turbulence quantities, it is found that the ejection-type fluid motions dominate the heat transport processes in the outer intermittent region and play a key role in the development of the thermal boundary layer. The present data-analysis techniques provide full details on the heat transport mechanism in turbulent thermal boundary layers. The present experimental results may also be useful for critical assessment of the existing turbulence models and to develop new ones.

References

Antonia, R. A., Phan-Thien, N., Chambers, A. J. (1980): Taylor's hypothesis and the probability density functions of temporal velocity and temperature derivatives in a turbulent flow. J. Fluid Mech. **100**, 193–208

Blackwell, B. F., Kays, W. M., Moffat, R. J. (1972): The turbulent boundary layer on a porous plate: an experimental study of the heat transfer behavior with adverse pressure gradients. Rep. № HMT-16, Thermosciences Division, Dept. of Mechanical Engineering, Stanford University

Farge, M. (1992): Wavelet transforms and their applications to turbulence. Annu. Rev. Fluid Mech. **24**, 395–457

Hedley, T. B., Keffer, J. F. (1974): Turbulent/non-turbulent decisions in an intermittent flow. J. Fluid Mech. **64**, 625–644

Hishida, M., Nagano, Y. (1978): Simultaneous measurements of velocity and temperature in nonisothermal flows. Trans. ASME, J. Heat Transfer **100**, 340–345

Hishida, M., Nagano, Y. (1979): Structure of turbulent velocity and temperature fluctuations in fully developed pipe flow. Trans. ASME, J. Heat Transfer **101**, 15–22

Hishida, M., Nagano, Y. (1988): Turbulence measurements with symmetrically bent V-shaped hot-wires. part 1: principles of operation. Trans. ASME, J. Fluids Engng. **110**, 264–269

Hishida, M., Nagano, Y., Tagawa, M. (1986): "Transport processes of heat and momentum in the wall region of turbulent pipe flow", in *Proceedings of 8th International Heat Transfer Conference*, San Francisco, USA, pp. 925–930

Kasagi, N., Tomita, Y., Kuroda, A. (1992): Direct numerical simulation of passive scalar field in a turbulent channel flow. Trans. ASME, J. Heat Transfer **114**, 598–606

Kibens, V. (1968): "The intermittent region of a turbulent boundary layer", *Ph.D. Thesis*, Johns Hopkins University

Kim, J., Moin, P. (1989): "Transport of passive scalars in a turbulent channel flow", in *Turbulent Shear Flows 6* (J.-C. André, J. Cousteix, F. Durst, B. E. Launder, F. W. Schmidt, J. H. Whitelaw eds) (Springer-Verlag, Berlin, Heidelberg), pp. 85–96

Krishnamoorthy, L. V., Antonia, R. A. (1987): Temperature-dissipation measurements in a turbulent boundary layer. J. Fluid Mech. **176**, 265–281

Lumley, J. L. (1978): "Computational modeling of turbulent flows", in *Advances in Applied Mechanics* **18** (C.-S. Yih, ed) (Academic Press), pp. 123–176

Lyons, S. L., Hanratty, T. J., McLaughlin, J. B. (1991): Direct numerical simulation of passive heat transfer in a turbulent channel flow. Int. J. Heat Mass Transfer **34**, 1149–1161

Monin, A. S., Yaglom, A. M. (1971): *Statistical Fluid Mechanics*, Vol. 1, MIT Press, Cambridge

Nagano, Y., Hishida, M. (1985): "Production and dissipation of turbulent velocity and temperature fluctuations in fully developed pipe flow", in *Proceedings of 5th Symposium on Turbulent Shear Flows*, Ithaca, USA, pp. 14.19–14.24
Nagano, Y., Hishida, M. (1990): "Turbulent heat transfer associated with coherent structures near the wall", in *Near-Wall Turbulence* (S. J. Kline, N. H. Afgan, eds) (Hemisphere), pp. 568–581
Nagano, Y., Tagawa, M. (1988): Statistical characteristics of wall turbulence with a passive scalar. J. Fluid Mech. **196**, 157–185
Nagano, Y., Tagawa, M. (1991): "Turbulence model for triple velocity and scalar correlations", in *Turbulent Shear Flows 7* (F. Durst, B. E. Launder, W. C. Reynolds, F. W. Schmidt, J. H. Whitelaw eds) (Springer-Verlag, Berlin, Heidelberg), pp. 47–62
Sreenivasan, K. R., Antonia, R. A. (1977): Skewness of temperature derivatives in turbulent shear flows. Phys. Fluids **20**, 1986–1988

Structure of the Turbulent Boundary Layer and Heat Transfer Downstream of a Vortex Generator Attached to a LEBU Plate

Kyoji Inaoka and Kenjiro Suzuki

Department of Mechanical Engineering, Kyoto University, Kyoto 606-01, Japan

Abstract

An experimental study was performed of a flat plate turbulent boundary layer disturbed by insertion of a Large Eddy Break-Up (LEBU) plate with a vortex generator attached. The vortex generator was found to work well to augment the wall heat transfer over a large streamwise distance, although the LEBU plate alone causes deteoriations in both heat transfer and skin friction coefficients on the wall. To investigate the mechanism of the achieved heat transfer augmentation, time mean values and fluctuating intensities of three velocity components were measured. It is concluded that the heat transfer augmentation is mainly caused by the approach of lower temperature fluid from the free stream toward the near-wall region. This results from a downwash fluid motion induced by the longitudinal vortex. Heat transfer enhancement can also be achieved in the region where a upwash fluid motion is observed. Such enhancement is produced by the activated near-wall turbulence.

Nomenclature

C_f	skin friction coefficient
H	elevation of LEBU plate, mm
h	heat transfer coefficient, $W/(m^2K)$
L	streamwise width of LEBU plate, mm
s	height of half-delta wing vortex generator, mm
T	mean temperature, K
U, V, W	mean velocities in the x-, y- and z- directions, m/s
u, v, w	fluctuating velocities in the x-, y- and z- directions, m/s
$\overline{uv}$	turbulent shear stress, m^2/s^2
$\overline{uw}$	turbulent shear stress, m^2/s^2
$\overline{w\theta}$	turbulent heat flux, mK/s
x, y, z	streamwise, wall-normal and spanwise coordinates, see Fig. 3, mm
α	attack angle of half-delta wing vortex generator, °
δ	boundary layer thickness, mm
Θ	non-dimensional temperature, $= (T_w - T)/(T_w - T_e)$
$(\ )_e$	free stream

$(\ )_L$ LEBU plate
$(\ )_w$ wall
$(\ )_0$ undisturbed boundary layer
$(\ )'$ r.m.s. value

1. Introduction

The present study is related to an effort of seeking for an effective way to generate the dissimilarity between the turbulent momentum and heat transfer processes. In the case where a cylinder was inserted into a flat plate turbulent boundary layer, it was found that the wall heat transfer was enhanced downstream of the cylinder but the skin friction acting on the same surface was reduced (Kawaguchi et al. 1984, Suzuki et al. 1988). This feature of turbulent transport processes, i.e., the turbulent momentum and heat transfer dissimilarity in the boundary layer, is very interesting from a physical viewpoint and also important from a practical viewpoint because it suggests a possibility of improving the heat transfer performance of heat exchangers while paying less amount of pressure drop penalty. However, an increase of total momentum loss is not effectively suppressed with a cylinder. Its form drag must be added as an additional source of momentum loss and it is relatively large. Therefore, it is of practical worth to examine what is the best shape of the body to be inserted or the best way of insertion of a body in order to effectively suppress the increase of total momentum loss and to generate the dissimilarity most effectively.

As its first step, a study was previously made for the case where a Large Eddy Break-Up manipulator (LEBU plate) was inserted in a flat plate turbulent boundary layer (Suzuki et al. 1991). The LEBU plate is known to be effective in reducing the wall skin friction because it destroys the large scale eddy structures in a turbulent boundary layer (Nguyen et al. 1987, Savill and Mumford 1988) and has small form drag. Therefore, its pressure drop penalty should be small. However, it was found that no strong dissimilarity could be generated with the LEBU plate (Suzuki et al. 1991). Heat transfer was also suppressed by the LEBU plate. This was the penalty to be paid in the effort to reduce the wall skin friction. Therefore, further experimental study has been initiated to determine if the deterioration in heat transfer could be recovered by attaching some supplementary body to the LEBU plate. As a body to be attached, a half-delta wing vortex generator was first tested. The present study aims to explore if and how effectively this half-delta wing vortex generator works in generating a longitudinal vortex and in enhancing wall heat transfer.

2. Experimental Apparatus and Procedures

Experiments were performed in a low-speed wind tunnel, which was used previously by Suzuki et al. (1991), and is outlined as follows. Figure 1 shows a schematic view of the wind tunnel. A turbulent boundary layer develops on a flat plate denoted as 9, which composes the floor of the wind tunnel. Cross-sectional uniformity of the flow velocity was established at the leading edge of the flat plate by sucking the upstream boundary layer on the tunnel floor with an auxiliary blower 11. The sucking flow rate was adjusted with a controller 10. The main stream velocity was kept at 14 m/s within an accuracy of one percent. Streamwise non-uniformity of static pressure was minimized by adjusting the slope of the ceiling of the wind tunnel except for an inevitable steep change around the inserted body, i.e., a LEBU plate with a vortex generator attached. The resultant pressure non-uniformity was less than one percent of the main stream dynamic pressure.

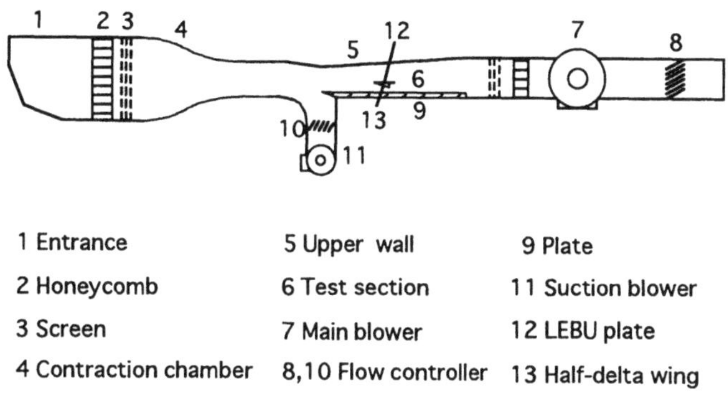

Fig. 1. Schematic view of wind tunnel.

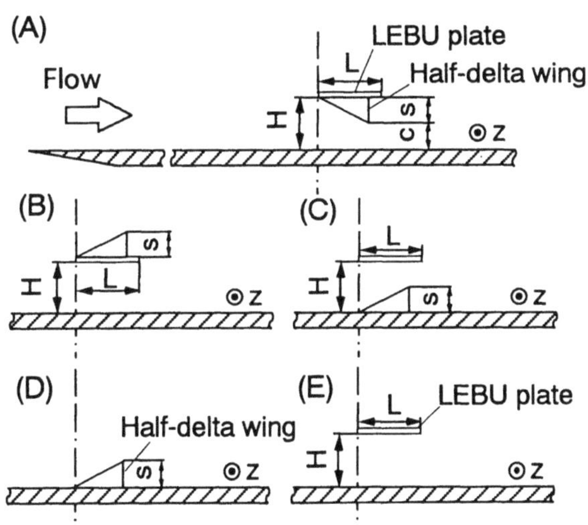

Fig. 2. Combination form of the vortex generator and the LEBU plate.

As will be described later, in the first stage of the present study, examination was made to see if the vortex generator is effective when it is combined with the LEBU

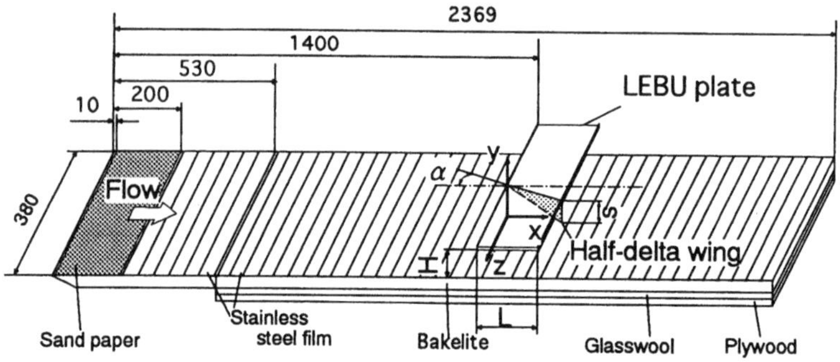

Fig. 3. Geometry of the studied flow system, coordinate system and geometric parameters.

plate, as in the case where the vortex generator is introduced alone into a boundary layer (Eibeck and Eaton 1987, Torii et al. 1991). An attempt was made to optimize geometrical arrangement of the vortex generator and the LEBU plate. Figure 2 (A) through (E) show the arrangements studied. As a result of this study, a particular type of arrangement, i.e., case A in which the vortex generator is attached to the down-facing surface of the LEBU plate, was chosen. This combination is shown schematically in Fig. 3 together with the coordinate system and several geometric parameters to be used in the following. A LEBU plate and a half-delta wing vortex generator were both made of bakelite of 1 mm thickness. The chord length of the LEBU plate, L, was kept at 28 mm, the value equal to the thickness of the approaching boundary layer, δ, since its change does not significantly alter either the value of skin friction or of heat transfer coefficients (Suzuki et al. 1991). The longer side of half-delta wing vortex generator was glued to the LEBU plate in a fashion that the leading edges of the half-delta wing and the LEBU plate were located at the same streamwise position. The LEBU plate was mounted normal to the flow direction and parallel to the flat plate at a streamwise station of 1400 mm downstream from the leading edge of the flat plate. The thickness of temperature boundary layer almost coincided with the velocity boundary layer thickness at that position. Study was made on the influence of elevation of the LEBU plate H, height of the vortex generator s and the angle of attack of the vortex generator α. Both of s and α were changed in three steps, i.e., $s = 5$, 10, 15 mm, and $\alpha = 10, 20, 30°$. The elevation H was also changed from 7 mm to 24 mm in less than six steps within the geometrically possible range. A preliminary study was made for the two cases where the angle of attack α of the vortex generator takes positive and negative values. The obtained data were found essentially the same, i.e., symmetric with respect to the plane of $z = 0$.

To measure the local heat transfer coefficient, the surface of the bakelite flat plate was covered by thin stainless steel strips of 20 μm thickness, leaving an upstream part (from 10 mm to 210 mm from the leading edge) uncovered. This part was covered by a sandpaper to promote the laminar to turbulent transition of the boundary layer. The stainless steel strips glued on the flat plate surface were heated at almost uniform heat flux (about 1 kW/m^2) with an electric current through them. In order to measure the streamwise and spanwise distributions of surface temperature, 360 alumel-chromel thermocouples of 100 μm diameter were attached to the back surface of the stainless steel strips. About 60 of them were arranged closely along the center line of the flat plate, $z = 0$. A number of thermocouples of the same thickness were also allocated spanwise at eight different streamwise positions of $x/\delta = -4.6$, 0.9, 2.7, 4.5 ,8.1, 15.4, 20.9, and 30.0. Non-uniformity of the local wall heat flux due to the heat conduction loss toward the back surface of the flat plate was taken into account. The heat loss was evaluated by making use of one-dimensional heat conduction analysis. The surface temperature of metal sheets and the back surface of the flat plate were substituted into the heat conduction equation. For this use, another 60 thermocouples of the same thickness were also attached to the back surface of the flat plate. For the evaluation of radiative heat loss, the emissivity of the stainless steel surface was assumed to be 0.2. The streamwise Nusselt number distribution measured for a flat plate boundary layer was confirmed to agree well with the empirical equation given by Johnson and Rubesin (1949), over a whole range of streamwise position of -500 mm $\leqslant x \leqslant$ 840 mm, if the virtual origin of the boundary layer is assumed at the position of 163 mm downstream of the leading edge of the flat plate.

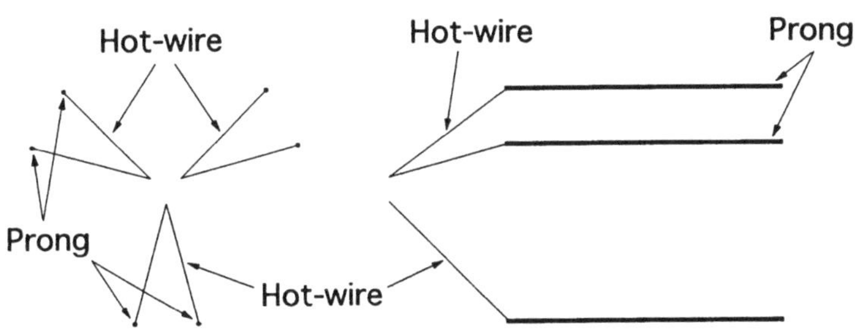

Fig. 4. Configuration of V-shaped triplet hot-wire probe.

To investigate the heat transfer mechanism, measurement of the three components of averaged velocity was made in some particular cases by using a five-hole pitot yaw probe having the size of 3 mm in diameter. Five-hole pressures in reference to atmospheric pressure were measured by electronic pressure gages. The output signal from each of the pressure gages were digitized on line at the sampling rate of 500 Hz and recorded over about fifteen seconds on a magnetic tape recorder. Furthermore, three components of fluctuating velocity were measured with hot-wire anemometry. Two types of hot-wire probes were used for this purpose, i.e., a single I-shaped hot-wire probe and a V-shaped triplet hot-wire probe both home-made following the paper by Hishida and Nagano (1988). Measurement of streamwise velocity with the former probe is of higher accuracy than with the latter probe. Thus, the data obtained with the former probe were used for evaluating the three velocity components obtained by the latter probe. Each hot-wire sensor was made of a tungsten wire of 5 μm in diameter with copper plated ends. Figure 4 shows the geometry of the V-shaped triplet hot-wire probe. The vertex angle of each V-shaped hot-wire was set equal to about 40° and pitch angles with respect to the free stream direction were set equal to about 30°. The output signals from the three V-shaped hot-wires were digitized at the sampling rate of 10 kHz for each signal and recorded over about twenty seconds on a magnetic tape recorder. Data processing of the replayed signals was performed later by a computer at the Kyoto University Data Processing Center. Transformation of the signals from anemometers into the three velocity components was made by using the calibration relationship between the output voltage and the effective velocity established for each sensor prior to the experiments. Measurement of mean fluid temperature was additionally made by traversing a thin thermocouple of 100 μm diameter in the y-direction at several streamwise and spanwise positions.

3. Results and Discussions

3.1 Dissimilarity Caused by the LEBU Plate

As a basis of the following discussions, the effect of insertion of the LEBU plate alone (case E) is first mentioned. Figure 5 shows the streamwise distributions of the heat transfer coefficient h_L and the wall skin friction coefficient C_{fL} for several different elevations of the LEBU plate, H. The results are normalized by the counterparts for an undisturbed flat plate boundary layer. As previously reported (Suzuki et al. 1991), the increase of H makes the depression of h_L shallower, its depressed region wider and its minimum position further downstream. In the figure, the heat transfer coefficient as well as the skin friction coefficient is observed to decrease downstream of the LEBU plate. Thus, no significant dissimilarity is generated by the LEBU plate.

In order to analyze the mechanism of the dissimilarity between the momentum and heat transfer from a structural view point of turbulence, a quadrant analysis was applied to the fluctuating velocity signals when a cylinder was inserted into a turbulent boundary layer (Suzuki et al. 1988). It was concluded that the intensification of interactive fluid motions observed at the positions close to the wall plays an important role in generating the strong dissimilarity. As a result of similar analysis applied to the case where the LEBU plate was introduced into the boundary layer, intensification of interactive fluid motions was found to be insignificant while the contributions from the ejection and sweep events to the Reynolds shear stress were reduced at such positions near the wall. Consequently, the insertion of the LEBU plate led to the reduction of near-wall turbulence level and also of heat transfer coefficient. With the LEBU plate, weak dissimilarity was still found to occur just downstream the LEBU plate. This is caused by the changes in the flow and thermal fields behind the LEBU plate. Just downstream the LEBU plate, the flow velocity is reduced near the wall because of the drag exerted by the LEBU plate, but the thermal field is not altered noticeably in that region.

3.2 Heat Transfer Augmentation by Half-delta Wing

Figure 6 shows how the spanwise distribution of heat transfer coefficient develops along the streamwise direction for cases A and B. The same half-delta wing vortex

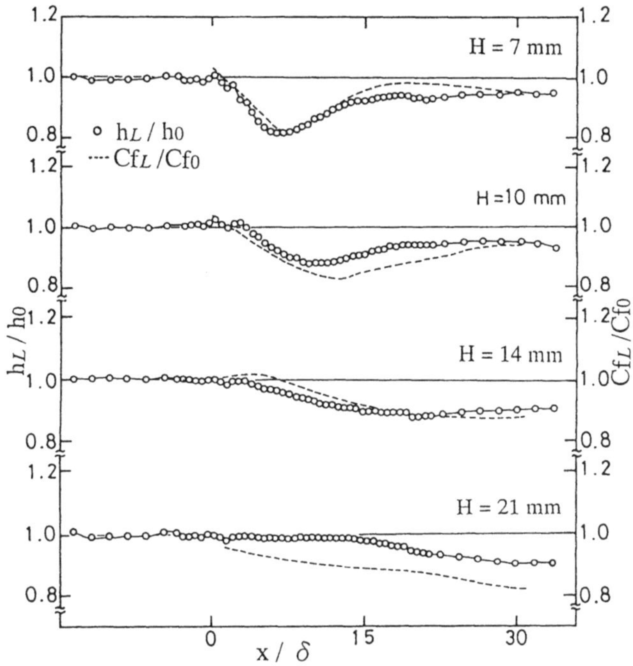

Fig. 5. Streamwise distribution of local heat transfer coefficient (in case E).

generator was used in both cases. The heat transfer coefficient is normalized by its value for an undisturbed flat plate boundary layer. Here, circle with an arrow head and dotted horizontal lines shown in the figure respectively indicate the rotating direction of the longitudinal vortex in a cross-stream section and the Nusselt number obtained for case E, where only LEBU plate is inserted in the boundary layer. The hatched area shows the amount of heat transfer augmentation caused by attaching the half-delta wing to the LEBU plate. As seen in the figure, the depression of heat transfer caused by insertion of the LEBU plate is recovered and the heat transfer augmentation is brought about over a large streamwise and spanwise extent. At several downstream positions, the width of the augmentation region is more than twice larger than the thickness of the turbulent boundary layer. This demonstrates that the effectiveness of the longitudinal vortex in enhancing the wall heat transfer is not lowered even if the half-delta wing is combined with the LEBU plate which can eliminate the large-scale vortex structures in the boundary layer. The peak value and spanwise distribution of heat transfer coefficient do not vary so much in the streamwise direction at $x/\delta \geqslant 8.1$ (this region is called the developed region hereafter), but they show substantial dependence upon the streamwise distance in the region of $x/\delta \leqslant 8.1$ (the initial region, hereafter). The following discussion will be concentrated on the results of case A, because the augmentation of heat transfer does not depend noticeably on how the vortex generator is combined with the LEBU plate (cases A through D in Fig. 2). It is of practical importance to notice that there is no necessity to directly attach the vortex generator to the heated surface in case A.

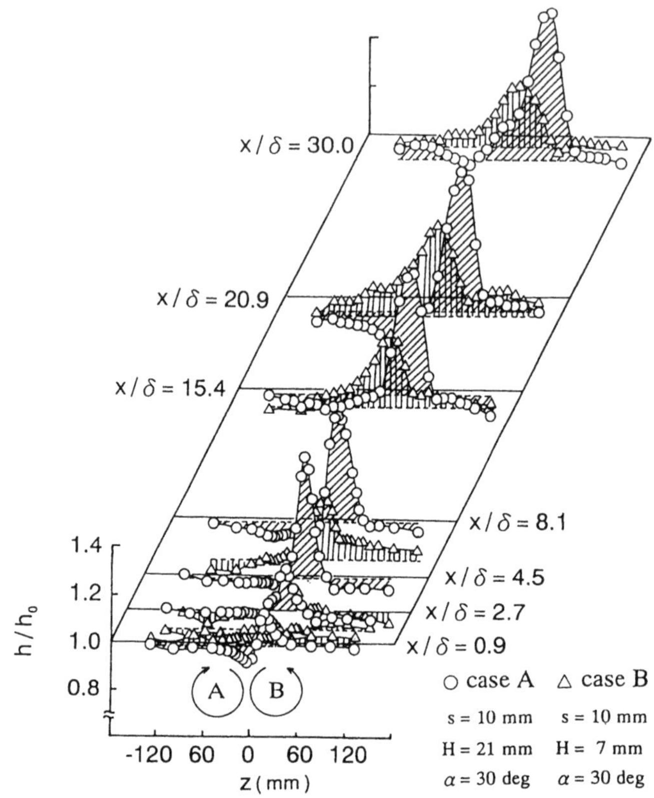

Fig. 6. Spanwise distribution of local heat transfer coefficient (in cases A and B).

Within the present range of geometrical parameters, H was not a factor significantly affecting the effectiveness in enhancing the heat transfer, but s and α affect the degree of heat transfer enhancement significantly (Inaoka et al. 1992). This is related to the fact that the magnitude of vorticity is controlled by s and the cross-sectional size of the vortex is controlled by s and α. It is concluded that more effective heat transfer enhancement is generated both in the initial and developed regions with larger values of s and α. Figure 7 shows an example of the heat transfer coefficient obtained for a few cases of different height of the half-delta wing s.

3.3 Time-averaged Flow and Thermal Structures

In order to grasp the mechanism of the above mentioned heat transfer augmentation, time-averaged values of three velocity signals, U, V and W, and of temperature in the boundary layer were measured at several positions in the case of $H = 21$ mm, $s = 15$ mm and $\alpha = 30°$. For this purpose, a five-hole pitot yaw-meter and thermocouples were used. Figures 8 through 10 show the results obtained in three cross-sections at different streamwise positions, i.e., at $x/\delta = 2.7$ in the initial region, and at $x/\delta = 8.1$

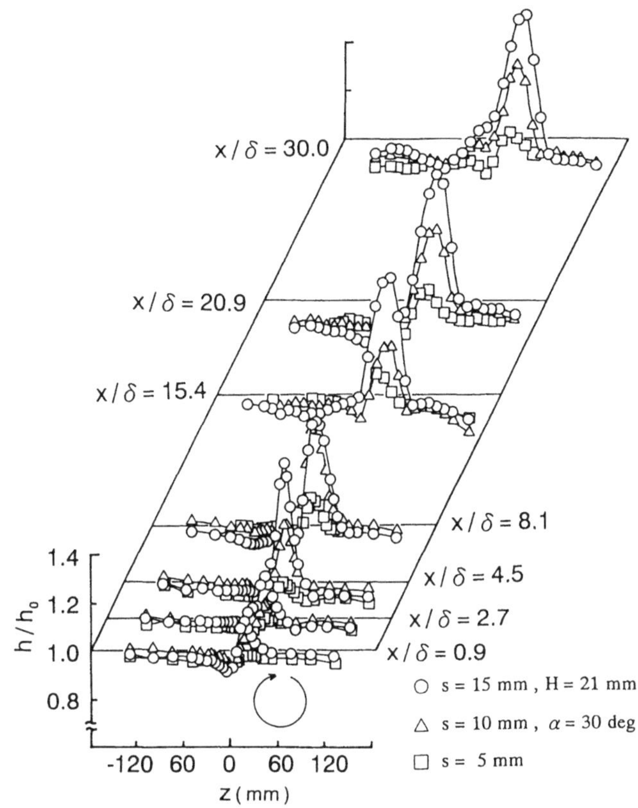

Fig. 7. Spanwise distribution of local heat transfer coefficient (influence of s).

and 30.0 in the developed region. The results plotted are the V-W velocity vector maps of the secondary flow and contour maps of streamwise velocity U, vorticity and non-dimensional temperature Θ. Thick lines and circles on the abscissa indicate the spanwise area where the heat transfer augmentation is attained and the position of peak Nusselt number, respectively. It is clearly found in the figures that the both of flow and thermal boundary layers have three-dimensional structures and that the influence of the vortex generator on the averaged velocity field is continuously observed over a long streamwise distance. Especially, in the V-W velocity vector diagrams in Figs. 9 and 10, downwash fluid motion from the free stream region to the near-wall region as well as upwash fluid motion away from the wall can be seen, i.e., there exists the longitudinal vortex induced by the vortex generator. High velocity and low temperature fluid in the main stream, which is brought into the boundary layer by the action of downwash fluid motion, penetrates deeper and wider into the boundary layer as the fluid flow downstream. This fluid motion in the initial region matches the streamwise change of the peak value of the heat transfer very well. Although strong vorticity can no longer be observed at the position of x/δ = 30.0 in Fig. 10, there still exists a region occupied by higher velocity and lower temperature fluid near the wall. The spanwise extension of such region corresponds well to the counterpart of the region where the heat transfer augmentation is obtained.

3.4 Turbulence Intensities

In order to further analyze the heat transfer augmentation mechanism, the V-shaped triplet hot-wire probe was introduced into the inner layer of the boundary layer at x/δ = 30.0. The I-shaped hot-wire probe was also used at x/δ = 8.1 and 30.0. Figure 11 shows the cross-stream distributions of turbulence intensities u', v' and w' measured at x/δ = 30.0.

First, it is observed that the distributions of u', v' and w' over the whole thickness of boundary layer differ from the counterparts for the undisturbed boundary layer. Every distribution is of the shape strongly distorted by the action of the longitudinal vortex as much as the averaged velocity field observed above. It is also found that the intensification of turbulence is not very significant in the near-wall region. This is quite different from the case where a cylinder was inserted into the boundary layer (Kawaguchi et al. 1984, Suzuki et al. 1988) and is rather similar to the case where a LEBU plate alone was inserted into the boundary layer (Suzuki et al. 1991). The spanwise non-uniformity of turbulence intensity does not correlate very well with the distribution of heat transfer augmentation. This is discussed again later. As is also observed in the figure, u', v' and w' distributions at the same z position are similar in shape, but depend seriously upon the spanwise location. Thus, attention is paid only to the cross-sectional distribution of u', the intensity of streamwise velocity fluctuation, to further examine the effect of the longitudinal vortex both on the turbulent boundary layer structures and wall heat transfer.

Figure 12 shows the contour map of u' measured with the I-shaped probe at x/δ = 8.1. In the figure, temperature contours are replotted from Fig. 9. At this position, protrusion of low turbulence region from the main stream into the boundary layer is observed. Particularly at the positions of 20 mm $\leqslant z \leqslant$ 40 mm, the turbulence level near the wall is lower in the present case than in the undisturbed case. As is seen in the figure, this region corresponds well to the region where lower temperature fluid exists near the wall. This region almost coincides the region where higher velocity fluid exists near the wall as seen if compared with the results in Fig. 9. Thus, this reduction of near-wall turbulence level should have resulted from the downwash motion of the higher velocity, lower temperature and low turbulence intensity fluid from the free stream or from the upper region of the turbulent boundary layer toward the near-wall region. Because the spanwise positions where heat transfer augmentation is achieved corresponds to this region, it is concluded that the mechanism of heat transfer augmentation is mainly due to the supply of cold fresh fluid from the free stream to the near-wall region. It is not directly related to the change in the near-wall turbulence intensities. Since the downwash fluid motion produces the penetration of high velocity fluid to the region close to the wall, the wall skin friction may not be reduced effectively, that is, strong dissimilarity may not be generated.

Figure 13 shows the contour map of the turbulence intensity u' measured by the I-shaped probe at x/δ = 30.0. At this position, a similar feature as that at x/δ = 8.1

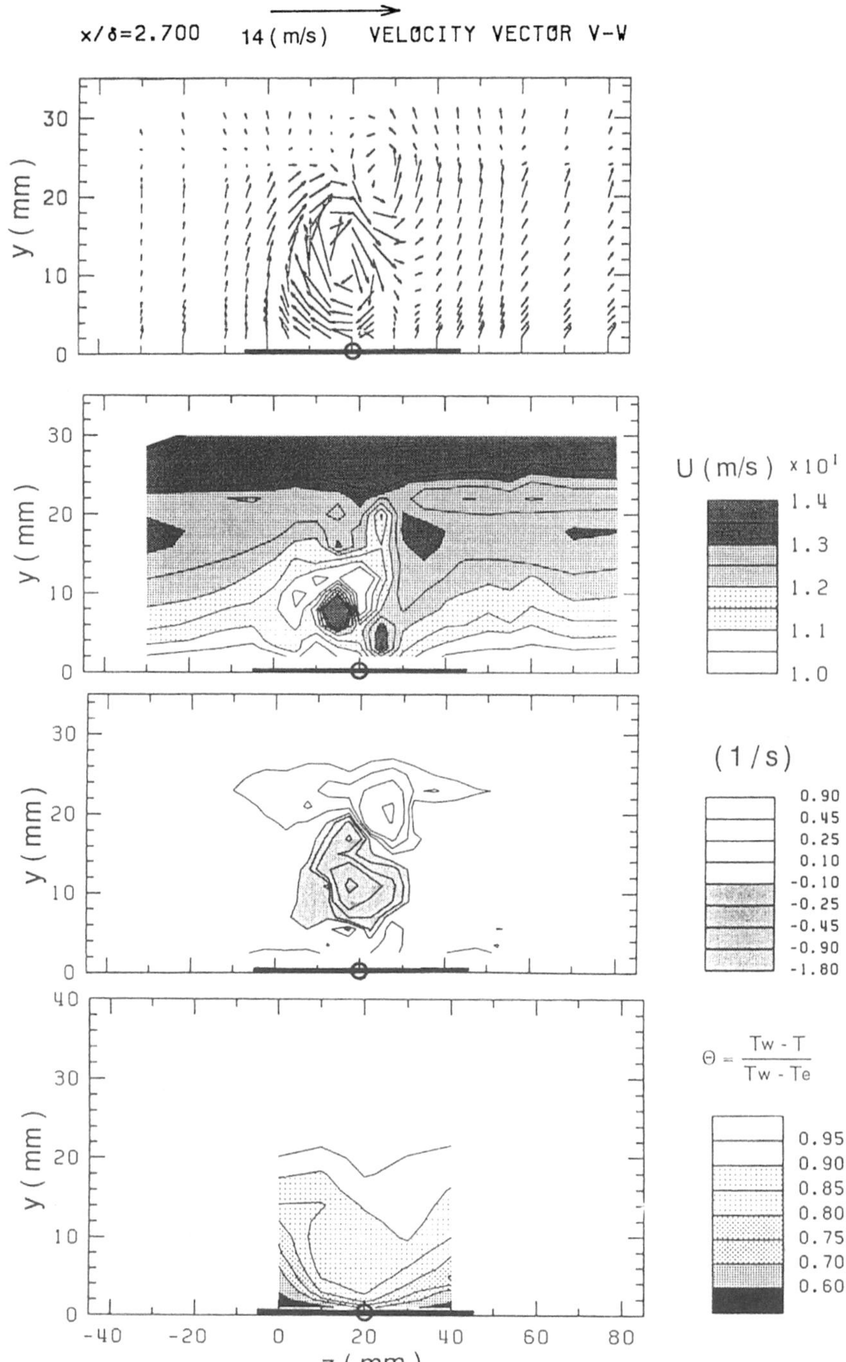

Fig. 8. Cross-sectional distribution of averaged velocity vector V-W, contour map of stream-wise velocity, contour map of vorticity, contour map of temperature (measured at $x/\delta = 2.7$).

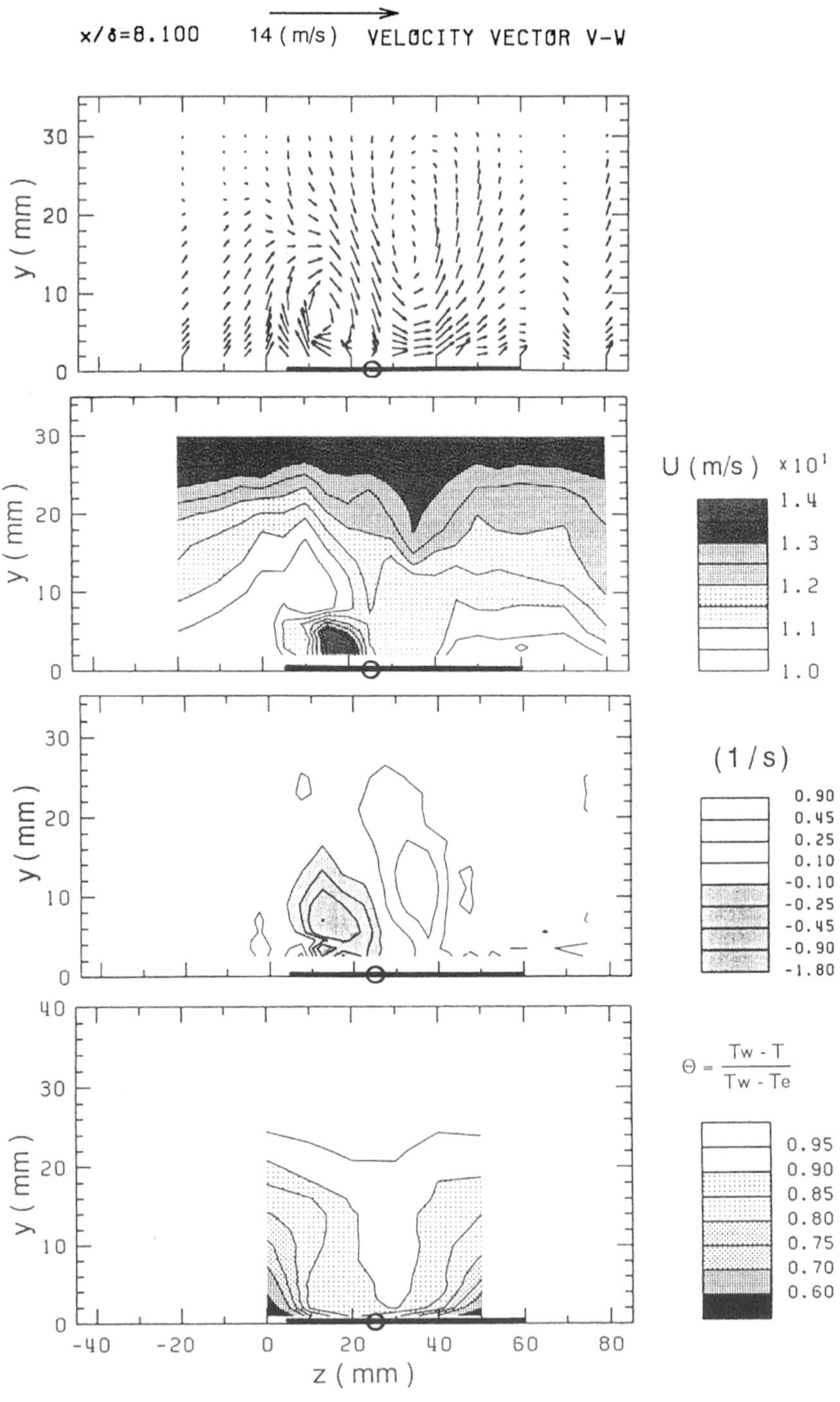

Fig. 9. Cross-sectional distribution of averaged velocity vector V-W, contour map of stream-wise velocity, contour map of vorticity, contour map of temperature (measured at $x/\delta = 8.1$).

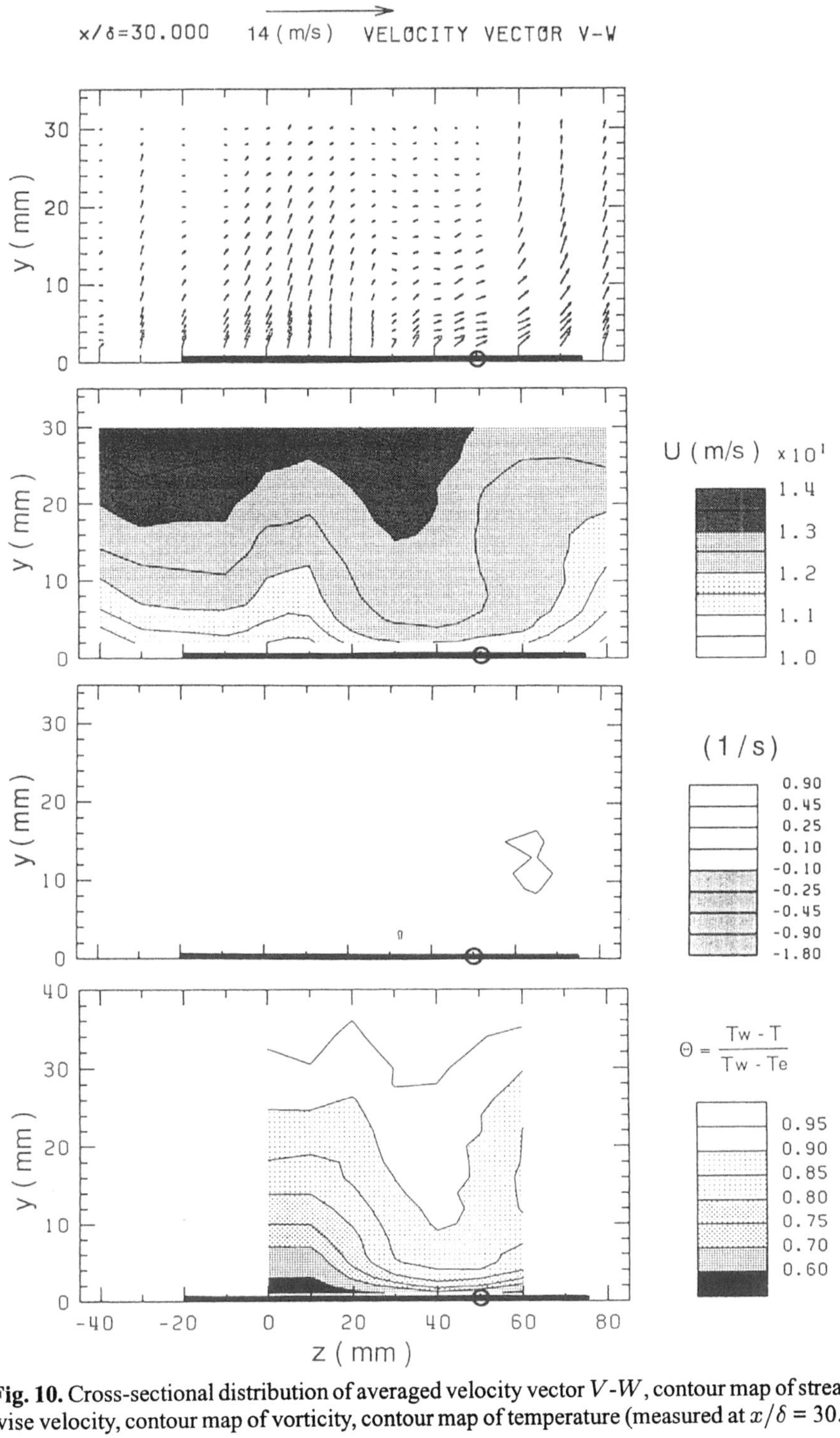

Fig. 10. Cross-sectional distribution of averaged velocity vector V-W, contour map of streamwise velocity, contour map of vorticity, contour map of temperature (measured at x/δ = 30.0).

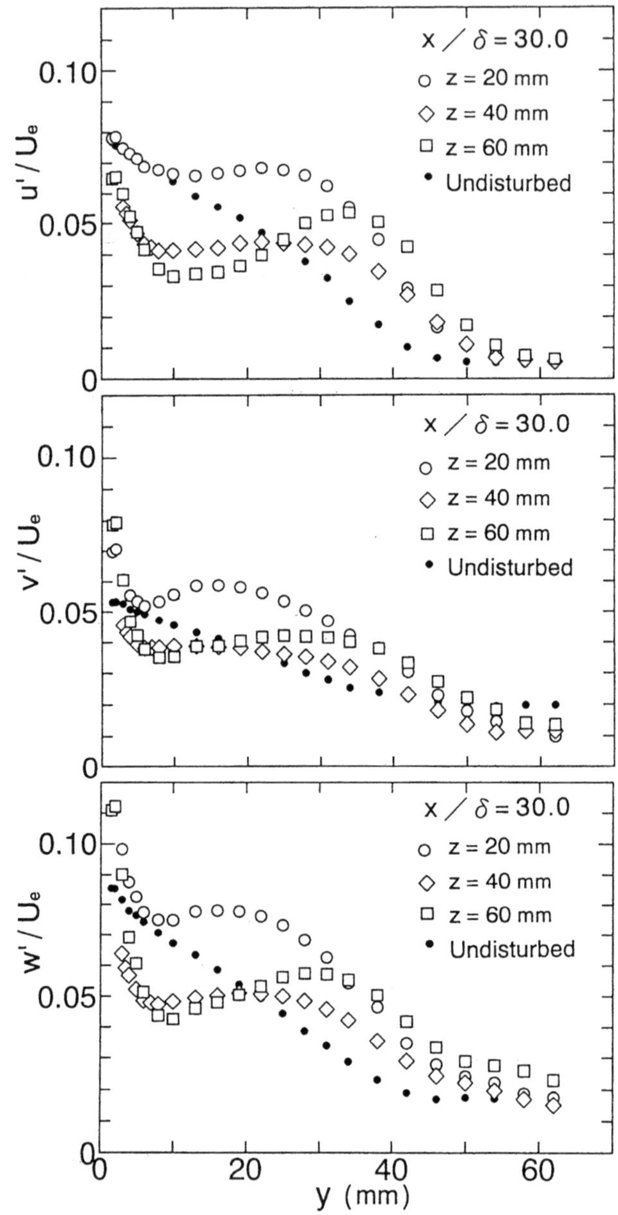

Fig. 11. Cross-streamwise distributions of turbulence intensities, $u'/U_e, v'/U_e$ and w'/U_e.

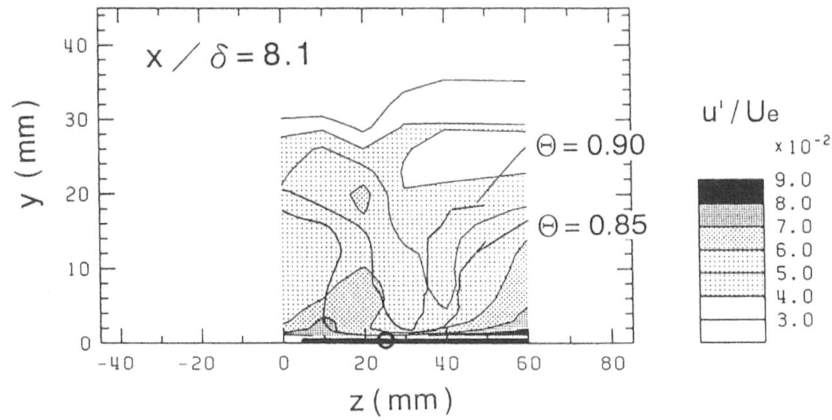

Fig. 12. Contour map of the streamwise turbulence intensity, u' (measured at $x/\delta = 8.1$).

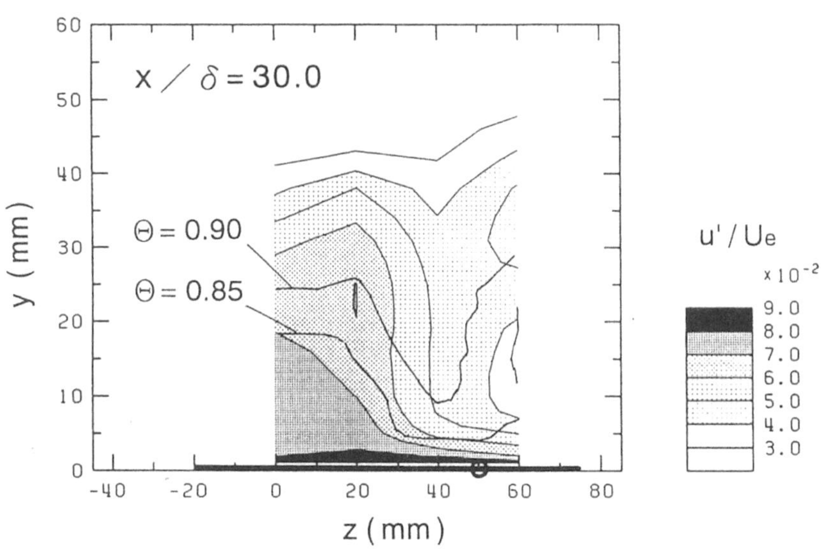

Fig. 13. Contour map of the streamwise turbulence intensity, u' (measured at $x/\delta = 30.0$).

is observed, i.e., a deep indent of streamwise velocity, temperature and turbulence intensity contours is observed in the spanwise region of 30 mm $\leqslant z \leqslant$ 60 mm, except that such an indent appears at positions a little shifted toward positive direction of z, compared with that observed at $x/\delta = 8.1$. Since the vorticity is already low in intensity at this position (see Fig. 10), indent of the contours has been gradually established by the downwash fluid motion caused by the longitudinal vortex at the upstream positions. Spanwise shift of the indent of the contours toward the positive z direction may be caused by the deflection of the longitudinal vortex in that direction, i.e., the longitudinal vortex lies with a small angle with respect to the x-axis. The region where heat transfer augmentation is attained at $x/\delta = 30.0$ is also expanded to the negative direction compared with that at $x/\delta = 8.1$. Its expansion to the negative direction may be related to an increase in the turbulence intensity near the wall. In the region of 0 mm $\leqslant z \leqslant$10 mm, for instance, heat transfer enhancement is achieved. However, the approach of low temperature fluid close to the wall is not noticeable there, since the wallward bulge of temperature contour appears only in the region of $z \geqslant$ 20 mm. On the other hand, high turbulence intensity region near the wall is thicker there than in the region $z \geqslant$ 20 mm. This thicker high turbulence intensity region is supposed to be produced by the upwash fluid motion of intense turbulence fluid otherwise existing near the wall.

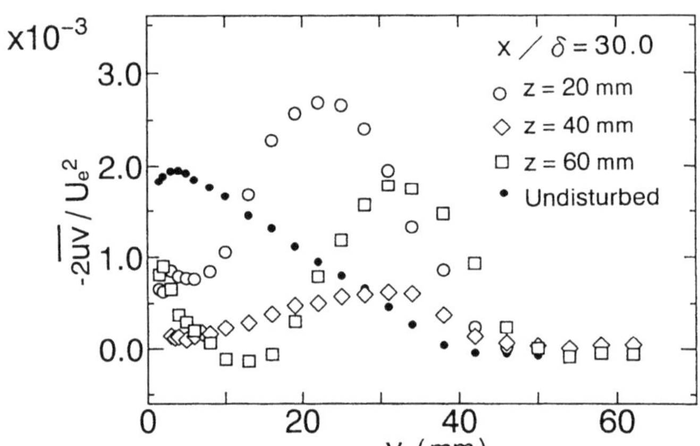

Fig. 14. Cross-streamwise distribution of Reynolds shear stress, $-2\overline{uv}/U_e^2$ (measured at x/δ = 30.0).

Finally the measured data of Reynolds shear stress are presented. Figures 14 and 15 show the cross-stream distributions of two components of Reynolds shear stress, $-\overline{uv}$ and $-\overline{uw}$. It is observed that both distributions are distorted by the influence of the inserted body, and that their shapes significantly vary from one spanwise location to another. Particularly, $-\overline{uw}$, which should be zero in the undisturbed case, shows a large finite value in the disturbed boundary layer. At z = 20 mm, it shows noticeable positive values over a whole thickness of boundary layer. Such positive value of

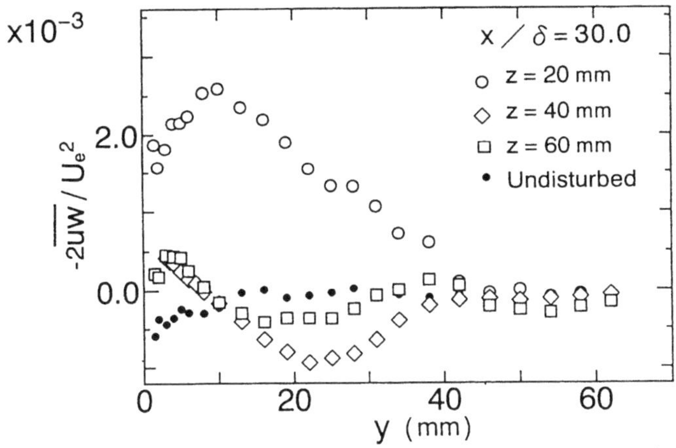

Fig. 15. Cross-streamwise distribution of Reynolds shear stress, $-2\overline{uw}/U_e^2$ (measured at $x/\delta = 30.0$).

$-\overline{uw}$ may suggest non-zero value of $\overline{w\theta}$, which produces the spanwise turbulent transport of heat, therefore, also contributes to the enhancement of heat transfer.

4. Concluding Remarks

Heat transfer data were obtained for a flat plate turbulent boundary layer disturbed by a vortex generator attached to a large eddy break-up manipulator (LEBU plate). It was found that the deterioration in heat transfer caused by the insertion of a LEBU plate could be recovered by attaching the vortex generator to it. The vortex generator was found to work well to augment the wall heat transfer over a large streamwise distance. To investigate the mechanism of this recovery of heat transfer or heat transfer enhancement, further experimental studies were performed on the velocity and the temperature fields, such as, mean velocities, mean temperature and fluctuating velocities by using a five-hole pitot yaw meter, thermocouples and hot-wire anemometry, respectively. Judging from the measured three components of mean and fluctuating velocities, it was concluded that the main cause of the heat transfer augmentation was the changes in the mean velocity and temperature fields produced by the action of the longitudinal vortex generator. That is, an approach of higher velocity, lower temperature and low turbulence fluid from the free stream to the near-wall region is produced by the downwash fluid motion accompanied by the longitudinal vortex and plays a predominant effect on heat transfer characteristics. It was also found that intensification of the near-wall turbulence may play some role on heat transfer enhancement, particularly, in the upwash fluid region.

Acknowledgement

The present study was partially supported by a Grant-in-Aid on Priority-Area Research № 043203108 of the Ministry of Education, Science and Culture, Japan. The present experiments were conducted assisted by Mr. Kazuyuki Suzuki, a former post-graduate student of Kyoto University. The authors extend their acknowledgement also to Dr. Hiroshi Suzuki, Associate Professor of Hiroshima University for his valuable discussions.

References

Kawaguchi,Y., Matumori,Y. and Suzuki,K., (1984): "Structural study of momentum and heat transfer in the near-wall region of a disturbed turbulent boundary layer", Proc. 9th Biennial Symp. on Turbulence. pp. 28.1–28.10

Suzuki, H., Suzuki, K. and Sato, T., (1988): "Dissimilarity between heat transfer and momentum transfer in a turbulent boundary layer disturbed by a cylinder", Int. J. Heat Mass Transfer. Vol. 31 № 2, pp. 259–265

Suzuki, K., Suzuki, H., Inaoka, K. and Kigawa, H., (1991): "Heat transfer in a turbulent boundary layer with an insertion of a LEBU plate", 8th Symp. on Turbulent Shear Flows, pp. 24.5.1–24.5.6

Nguyen, V.D., Savill, A.M. and Westpal, R.V., (1987): "Skin friction measurements following manipulation of a turbulent boundary layer", AIAA Journal, 25, pp. 498–500

Savill, A.M. and Mumford, J.C., (1988): "Manipulation of turbulent boundary layers by outerlayer devices: Skin friction and flow-visualization results", J. Fluid Mechanics, 191, pp. 389–418

Eibeck, P.A. and Eaton, J.K., (1987): "Heat transfer effects of a longitudinal vortex embedded in a turbulent boundary layer", Journal of Heat Transfer, Vol. 109, № 1, pp. 16–24

Torii, K., Yanagihara, I.J. and Nagai, Y., (1991): "Heat transfer enhancement by vortex generators", Proc. ASME/JSME Thermal Engineering, pp. 77–83

Johnson, H.A. and Rubesin, M.M., (1949): "Aerodynamic heating and convection heat transfer - Summary of literature survey", Trans. ASME, 71, pp. 447–456

Hishida, M. and Nagano, Y., (1988): "Turbulence Measurements of with symmetrically bent V-shaped hot-wires. Part 1: Principles of operation. Journal of Fluids Engineering", Vol. 110, pp. 264–269

Inaoka, K., Suzuki, K., Hagiwara, Y. Suzuki, K. and Suzuki, H., (1992): "Heat transfer augmentation in a turbulent boundary layer disturbed by means of a vortex generator" Heat Transfer Japanese Research, Vol. 21, № 7, pp. 721–735

Part IV

Combustion and Recirculating Flows

Introductory Remarks

Derek Bradley

Department of Mechanical Engineering, University of Leeds,
Leeds LS2 9JT, England

Abstract

This brief introductory paper summarises the salient coverage in the subsequent ones. It covers:

(i) the development of shear layers and recirculation zones, under conditions of steady and explosive combustion,
(ii) the structure of recirculation zones and the role of counter-gradient diffusion, and
(iii) some aspects of turbulent premixed flames, including mathematical modelling.

1. Development of Shear Layers and Recirculation Zones

Recirculating flows are of great importance in combustion and occur both behind bluff bodies and within fully containing boundaries. In addition, recirculation can be induced by swirling flow, as it is in industrial burners, and ever since the early Whittle gas turbine (Watson and Clarke 1947), in gas turbine combustion chambers also. The swirling component of momentum induces a flow divergence that creates recirculation, whilst the swirl velocity is an extra source of momentum and kinetic energy, able to modify the flow pattern and enhance the transport processes. Dissipation of the swirl kinetic energy enhances the mixing processes. Recirculation zones can control the shape of a flame, increase the mean volumetric heat release rate, and extend stability over a wider range of flow rates. A review of combusting swirling flows has been presented by Weber and Dugué (1992). Each of the present papers demonstrates important facets of combustion in recirculating flows and these are summarised briefly in this introductory paper. The development of a shear layer and recirculation zone just behind an obstacle in non-swirling flow is shown by Fig. 3 of the second paper. In this study of transient flame growth, by Lindstedt and Sakthitharan, the gases initially are at rest and then accelerate, during an explosion, to a velocity of about 50 $m\,s^{-1}$ at the plane of the obstacle. Laser-doppler velocimetry and spark schlieren photography trace how an initially laminar flame evolves through laminar instability, to a turbulent, recirculating flame as it passes over an obstacle. The third paper reports a fundamental study by Li, Brown, and

Bilger of a slowly reacting mixing layer. The reactants, nitric oxide and ozone, are introduced into separate streams and their concentrations measured in the mixing layer. The Damköhler number is about 1.6. The importance of this study arises from the interplay of the rates of mixing and reaction. It relates to the turbulent frequency at which fluctuations in specie concentration might be affected by chemical reaction. Chemiluminescent analysers give a continual record of concentration, with a frequency response of up to 20 Hz. Power spectral density functions of velocity fluctuations are employed widely to characterise turbulence: the lower frequency range is associated with the energy containing eddies, and the higher range is associated with dissipation. Spectra of fluctuating concentrations of reactants are less common. The spectra presented in Chapter 3 show a high coherence of the two species over most of the frequency range, contrary to the expectation that coherence would be lost at the frequency where reaction loss balances spectral transfer. This raises many problems and, clearly, further research is necessary.

2. Recirculation Zone Structure

Researches in the aerospace era on reacting recirculation zones behind bluff bodies lead to the bounded, "well-stirred", spherical reactor of Longwell and Weiss (1955). This was followed by the jet-stirred, conical-hemispherical, reactor of Spalding (Jenkins *et al.* 1967). Interpretations of the experimental measurements of blow-out points in spherical, Longwell reactors assumed a homogeneous structure within them, to yield global reaction rates for both fuel disappearance (Clarke *et al.* 1965) and carbon dioxide formation (Hottel *et al.* 1965). Subsequent measurements of gas temperatures showed the structure to be non-homogeneous, particularly at the higher reaction rates (Williams *et al.* 1969). In addition, measurements in conical, jet-stirred, reactors and the application of mathematical models, revealed spatially distributed reaction zones and yielded lower global reaction rates (Bradley *et al.* 1977). It was becoming clear that reaction and aerodynamics in recirculation zones are strongly coupled. Although the blow-out data were useful, the rate data, unlike those for laminar flames, could not readily be generalised. Because of the strong non-linearity of reaction rates with temperature, both the measurement and prediction of temperature pdfs became imperative. Initially, compensated thermocouples were employed (Lockwood and Odidi 1975) to measure the fluctuating temperatures. Heitor *et al.* (1985) correlated laser doppler velocimetry signals with those from compensated thermocouples. Difficulties arise because the thermocouple time constant is a function of both the instantaneous temperature and velocity and because, even at moderate Reynolds numbers, the signal to noise ratio is too low, rendering accurate compensation almost impossible (Bradley *et al.* 1989). There is no such disadvantage with laser thermometry, by either the CARS or Rayleigh scattering techniques, although the finite size of the measuring volume limits the spatial resolution. Rayleigh scattering and laser-doppler velocimetry are employed by Ferrao and Heitor to measure temperature- velocity correlations, indicative of the turbulent heat flux, in baffle-stabilized flames. This is reported in the fourth paper. At the lower values of Reynolds number and Karlovitz stretch factor the temperature distribution at a point is close to bimodal across the reacting shear layer. At the higher values of these parameters it becomes gaußian. A consequence of the large temperature fluctuations in turbulent flames is the important phenomenon of counter- gradient diffusion. This arises because small pockets of low density gas are accelerated more rapidly by the mean pressure gradient than are pockets of high density. This generates a net flux of higher temperature gas towards the burnt products, contrary to the predicted direction of the gradient transport formulation, usually associated with the $k - \varepsilon$ model of turbulence. The experimental demonstration of this provides the main thrust of this paper. The mean pressure gradient normal to the boundary of the reacting recirculation zone changes along the boundary. This, in combination with counter-gradient diffusion, directs the turbulent heat flux vectors along the isotherms, rather than normal to them, as would be expected with gradient diffusion. The different accelerations, arising from different densities, also are the cause of the Taylor-Markstein instability, discussed by Lindstedt and Sakthitharan in the earlier paper on explosion phenomena.

3. Turbulent Premixed Flames

In the fifth paper, Mantel, Borghi, and Picart present a new closure procedure to obtain the turbulent flux of the progress variable, c. They create a transport equation for its correlation with velocity, in which their Eq. (6) models the pressure gradient term. This is part of a numerical simulation of the propagation of a one-dimensional, premixed, turbulent, flame. Interestingly, the predictions of turbulent burning velocity are similar to those obtained with the gradient transport formulation, but the predicted flame brush thicknesses differ. The present author and colleagues, with a different model that employed computed heat release rate profiles from stretched laminar flame sub models, have found the same effect. They attributed it to the small magnitude of the turbulent flux relative to the chemical source and convective terms, even though the pressure gradient term is dominant in the turbulent flux equation (Bradley *et al.* 1994). The other important transport equation in this fifth paper is that for the dissipation rate, $\bar{\varepsilon}_c$, of c, which is directly related to the mean reaction rate. The case of fast chemistry is analysed, in which the chemical reaction time is short in comparison with the aerodynamic, or eddy, time. On the Borghi diagram in Fig. 2 of the fourth paper this is in the régime of high Damköhler number and low Karlovitz stretch factor: that of the wrinkled, continuous, laminar, flame sheet. A rich variety of flamelet models, the underlying assumption of which is that a turbulent flame is an array of laminar flames, is currently in use to describe premixed combustion. The approach of Bray (1990) and co-workers is to express the mean volumetric chemical source term in terms of mass burning rate, w, as $w = \varrho_u u_\ell \Sigma I_0$. Here ϱ_u is the unburned gas density, u_ℓ is the laminar burning velocity, Σ is the laminar flamelet surface area per unit volume, and I_0 represents the influence of flame stretch upon u_ℓ. Elsewhere, Borghi (1990) has shown that Σ can be correlated with $\bar{\varepsilon}_c$. Another approach in this *genre* is that of Peters and co-workers (Wirth and Peters

1992) who employ the equation of motion of a scalar field, G, to represent the flame surface. Duclos *et al.* (1993) present the transport equation for Σ and compare values of the turbulent burning velocities predicted from five flamelet models with those suggested by the Leeds group's correlation of experimental values. An alternative to deriving Σ from its transport equation is to derive it experimentally (Bray *et al.* 1984). It is to be anticipated that the greater the value of Σ, the greater the frequency at which flamelets cross a small measuring volume. The greater this crossing frequency, the smaller the integral length scale of the near-square wave that represents the crossings of the contour of a given mean value of c. It has been found that Σ for a mean value of c is, indeed, inversely proportional to such a length scale (Bray 1990). The derivation of w is the central problem addressed by Rousselle-Mounaïm and Gökalp in the sixth paper. They report experimental measurements of flamelet crossing times at a given point within the flame brush, as well as of the orientation of the flamelets. A 6.5 W continuous laser source is employed to induce Rayleigh scattering, from which the instantaneous density changes at a point are inferred, at a sampling rate of up to 1 MHz. The focused waist diameter of the beam is 0.08 mm and the scattered light is collected through variable diameter pinholes. Variation of pinhole diameter changes the measuring volume to allow optimisation of the signal. From the measurements of crossings, an expression, Eq. 8, is obtained for the mean mass rate of production of burned gas. This is compared with an earlier expression suggested by Bray.

References

Borghi, R.: Combust. Flame **80**, 304 (1990).
Bradley, D., Chin, S. B., Draper, M. S., and Hankinson, G.: Sixteenth Symposium (International) on Combustion, p. 1571, The Combustion Institute, 1977.
Bradley, D., Gaskell, P. H., and Gu, X. J.: Combust. Flame **96**, 221 (1994).
Bradley, D., Lau, A. K. C., and Missaghi, M.: Combust. Sci. and Tech. **64**, 119 (1989).
Bray, K. N. C., Libby, P. A., and Moss, J. B.: Combust. Sci. Tech. **41**, 143 (1984).
Bray, K. N. C.: Proc. R. Soc. Lond. A**431**, 315 (1990).
Clarke, A. E., Odgers, J., Stringer, F. W., and Harrison, A. J.: Tenth Symposium (International) on Combustion, p. 1151, The Combustion Institute, 1965.
Duclos, J. M., Veynante, D., and Poinsot, T.: Combust. Flame **95**, 101 (1993).
Heitor, M. V., Taylor, A. M. K. P. , and Whitelaw, J. H.: Experiments in Fluids **3**, 323 (1985).
Hottel, H. C., Williams, G. C., Nerheim, N. M., and Schneider, G. R.: Tenth Symposium (International) on Combustion, p. 111, The Combustion Institute, 1965.
Jenkins, D. R., Spalding, D. B., and Yumlu, V.: Eleventh Symposium (International) on Combustion, p. 779, The Combustion Institute, 1967.
Lockwood, F. C., and Odidi, A. O. O.: Fifteenth Symposium (International) on Combustion, p. 561, The Combustion Institute, 1975.
Longwell, J. P. , and Weiss, M. A.: Ind. Eng. Chem. **47**, 1634 (1955).
Watson, E. A., and Clarke, J. S.: J. Inst. Fuel **21**, 2 (1947).
Weber, R., and Dugué, J.: Prog. Energy Combust. Sci. **18**, 349 (1992).
Williams, G. C., Hottel, H. C., and Morgan, A. L.: Twelfth Symposium (International) on Combustion, p. 913, The Combustion Institute, 1969.
Wirth, M., and Peters, N.: Twenty-Fourth Symposium (International) on Combustion, p. 493, The Combustion Institute, 1992.

Transient Flame Growth in a Developing Shear Layer

R.P. Lindstedt and V. Sakthitharan

Mechanical Engineering Department,
Imperial College of Science, Technology & Medicine,
Exhibition Road, London SW7 2BX, U.K.

Abstract

The present work is concerned with further understanding the transition of a laminar flame to a turbulent deflagration in a transient shear layer. An experimental study of the transient interaction of a premixed stoichiometric methane-air flame with a developing shear layer and recirculation zone formed by an obstacle placed normal to the axis of a long closed rectangular channel has been performed. The interaction was visualized by spark schlieren photography. Velocity measurements were obtained by LDA used in a standard dual-beam forward scatter system. Silicone oil droplets were used as seeding particles to obtain the mean and rms velocities in the reactant gas. Significant new information has been obtained on the transition of an essentially laminar flame to a strongly turbulent deflagration in a developing shear layer. The results show a complex interaction of the accelerating flame with the wall and obstacle boundary layers as well as with the recirculation zone behind the obstacle. It is shown that the onset of flame instabilities leading to a strongly turbulent flame occurs along the obstacle induced shear layer.

1. Introduction

Stationary bluff body stabilized flames at high turbulence Reynolds numbers have been comprehensively investigated by several authors, e.g. Taylor and Whitelaw (1984), Heitor et al. (1987). These studies have resulted in a much improved understanding of the effects of combustion on the flow field. Transient flames accelerated by the use of bluff bodies have also been studied, e.g. Lindstedt and Michels (1989), but detailed measurements of flow field characteristics have not been made. The lack of detailed information is associated with the complexity of the flow field which results in measurement difficulties. There is, therefore, a lack of understanding of the physical processes involved. The absence of any detailed measurements of flame interactions with transient shear layers has also necessarily meant that the accuracy of modelling techniques applied to such flows remains uncertain. In view of its practical applications this lack of a proper understanding of flows of this type is of considerable concern. In particular, the transient interaction of flames with obstacles is a determining factor in the severity of turbulent gas explosions. The purpose of the present study is, therefore, to address the lack of qualitative understanding as well as to provide quantitative information which is suitable for the further development of modelling techniques.

The current investigation focuses mainly on the transition of a laminar flame to a turbulent deflagration in a developing shear layer and its interaction with the recirculation zone formed by a flat plate obstacle. The phenomena occurring are of significant fundamental and practical importance and as such require detailed attention. It is shown that the instabilities leading to a strongly turbulent flame occur along the obstacle induced shear layer and the resulting highly turbulent flame subsequently travels preferentially along the wall boundary layer leading to the generation of an increasingly turbulent flame structure with a large flame area parallel to the tube axis. It is also shown that large distortions in the obstacle shear layer structure occur as a result of the subsequent increase in volume production downstream of the obstacle. The final stage of the propagation is characterized by a violent explosion occurring when the flame interacts with the recirculation zone behind the obstacle.

2. Experimental Details

The flame tube used consists of five interchangeable rectangular sections (72 mm × 34 mm) each 1.825 m in length and two window sections each 0.385 m in length. The latter were placed at the beginning of the flame tube with an obstacle 36 mm in height placed 0.478 m from the ignition source. The obstacle was made of a 5 mm thick steel plate. The large L/D ratio (= 137) of this configuration allowed the investigation of flame-obstacle interactions without the interference of the waves reflected off the non-ignition end plate. Stoichiometric methane-air mixtures were prepared using CP grade methane and 99% dry air. Mixture preparation was done in an initially evacuated ($p < 5$ kPa) flame tube used as a partial pressure mixing vessel. The partial pressures were measured by a pressure transducer. The gas mixture was subsequently circulated in a closed loop system for a sufficiently long period to ensure homogeneity.

The basic instrumentation includes six PCB pressure transducers and sixteen coaxial ionization probes. The pressure records were stored in a six channel transient recorder and the ionization probes with integrated circuits (Lindstedt 1984) were connected to 24-bit 10 MHz counters for the measurement of flame arrival times. Spark schlieren photography was used to elucidate the flame propagation process. A standard parallel beam optical arrangement composed of lenses was used with an argon-jet spark light source whose duration was typically 500 ns. The parallel beam of approximately 80 mm diameter was passed across the flame tube and focused on a schlieren stop. An aperture of suitable size was employed as a schlieren stop to increase the sensitivity of the system and to ensure that density variations in all directions were recorded. A 35 mm single-lens reflex camera was used with appropriate filters and black and white film of suitable sensitivity to secure correct exposure of photographs. A time-delay system similar to the one used by Markstein (1956, 1964) was implemented by selecting an ionization probe as a trigger device and by the introduction of a suitable time delay between the arrival of the flame at the ionization probe and the trigger of the light source. Thus, a series of photographs was constructed at different positions around the obstacle through the selection of ionisation of probes at different positions and the variations of time-delay. More details of the schlieren set-up used can be found elsewhere (Lindstedt and Sakthitharan 1993a).

A standard dual-beam forward scatter laser Doppler anemometer was used to obtain gas particle velocities around the obstacle. The anemometer consisted of a 1.5 W argon-ion laser operating in single-line mode at 514.5 nm wavelength. The transmitting optics consisted of a beam splitter with a 600 mm focal length lens. The high focal length was chosen in order to increase the fringe spacing and hence the maximum velocity that could be measured. A fixed frequency shift of 40 MHz was applied to one of the beams via a Bragg cell to eliminate directional ambiguity. The arrangement permitted the measurement of positive velocities up to 400 $m\,s^{-1}$ and negative velocities up to 150 $m\,s^{-1}$. A list of optical characteristics of the present arrangement is shown in Table 1.

Table 1. Principal characteristics of LDA optical arrangement.

Wavelenght of 1.5 W Ar - ion Laser	514.5 nm
Laser e^{-2} Intensity Diameter	1.5 mm
Focal Length of Focusing Lens	600 mm
Measured Half Angle of Intersection	2.94°
Fringe Spacing	5.02 μm
Calculated Number of Fringes	52
Calculated Diameter of Measuring Volume	0.262 mm
Calculated Length of Measuring Volume	5.110 mm

The receiving optics consisted of lenses and a photomultiplier placed at an angle with a pinhole of 0.1 mm diameter controlling the amount of scattered light received. The photomultiplier signals were processed by a frequency counter (TSI 1990c; TSI Inc.). The counter was set to examine first 16 cycles (2^4), compare the measured frequencies over 10 and 16 cycles and accept the results with a difference less than 1%. The reason for using a 16 cycle setting on the counter is discussed below. The counter was digitally interfaced to the microcomputer via a purpose-built interface card. This card has FIFO (First-In-First-Out) memory of 8 kilobytes to store the validated 16-bit digital frequency signals and 24-bit counter values generated by the card's own 1 MHz timer. The data was transferred to the computer using DMA (Direct Memory Access) co-ordinated by the software. The transfer of three 16-bit words per measurement coupled with 16-bit DMA transfer and sufficiently large FIFO memory ensured the high data transfer rate essential for measurement in high velocity flows. Silicone oil droplets of 1 μm nominal size were generated by the use of an atomizer placed in the circulation loop of the flame tube. During the introduction of seeding particles the gas mixture was allowed to pass through the atomizer for about sixty seconds at the end of the gas mixing. Afterwards, a short period of time was allowed for the gas mixture to become quiescent before the ignition was triggered. Usage of silicone oil aerosol facilitated the measurement of conditionally averaged velocities in the unburnt gas (Cheng 1984). Proper synchronization of triggering various devices and the ignition source was accomplished through a software routine written in assembly language.

3. Results and Discussion

3.1 Preliminaries

In preliminary work reported elsewhere (Lindstedt and Sakthitharan 1993b), a large number of experiments were performed such that at least six sets of flame arrival time and pressure records were available at each port within the region under investigation. From these data the average flame arrival time at each port was estimated. The fact that the flame arrival time itself was very reproducible with a standard deviation of less than 3% can be used to adjust the time scale of each experiment in the present study to get a mean time scale. This was achieved by adding an offset to the time scale of the experiment concerned. The offset was chosen to be the difference between the mean flame arrival time and the actual flame arrival time at a port located at 1.076 m from the ignition point. In the present paper, schlieren photographs taken at two different positions are presented. However, account is taken of the knowledge obtained from photographs taken at other positions as well (Lindstedt and Sakthitharan 1993a). The two camera positions and the points at which the gas velocities were measured are shown in Fig. 1.

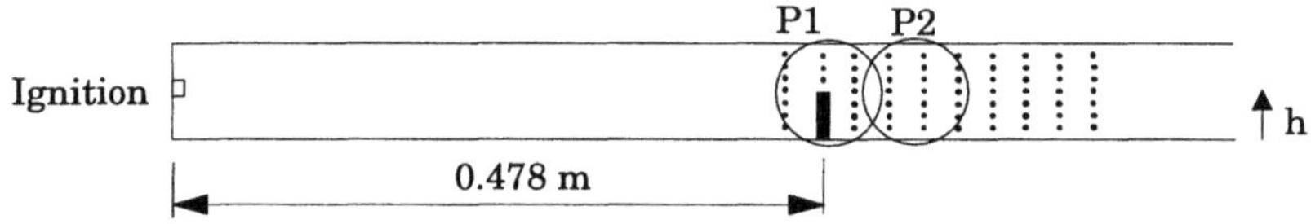

Fig. 1. Obstacle configuration marked with positions at which photographs were taken and the points at which LDA measurements were made. Height h has its origin at the bottom wall and increases upwards.

Position P1 is chosen to capture the flame propagation over the obstacle, which is located at 0.478 m from the ignition point. The marker for the position P2 is placed at the bottom wall at 0.558 m from the ignition point. Velocity components along the axial direction (u) and along the vertical direction (v) were measured at sixty-six points above and beyond the obstacle as shown in Fig. 1. All these points lie on the vertical plane passing through the axis of the flame tube. The points were separated by 9 mm in the vertical direction and by 25 mm in the horizontal direction. Thus, the points were located at 63, 54, 45, 36, 27, 18 and 9 mm from the bottom wall along the vertical planes normal to the axis at 0.453, 0.478, 0.503, 0.528, 0.553, 0.578, 0.603, 0.628, 0.653 and 0.678 m from the ignition point. At the position of the obstacle measurements were made at the top three points.

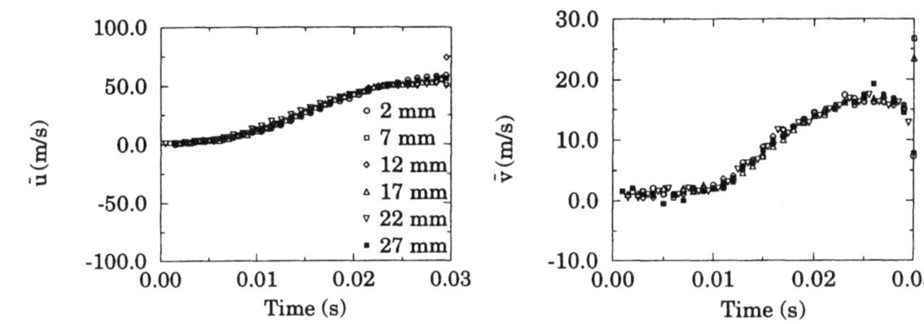

(a) At 0.478 m

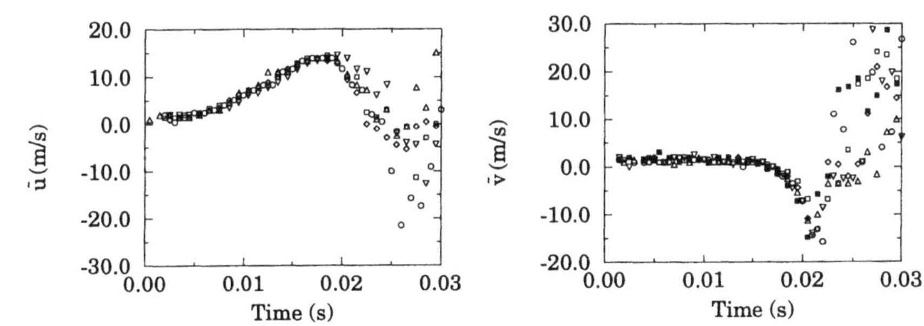

(b) At 0.578 m

Fig. 2. Test for two-dimensionality of flow: ũ and ṽ profiles at different positions across the flame tube. (**a**) At 0.478 m and $h = 54$ mm (**b**) At 0.578 m and $h = 18$ mm

In the analysis of the LDA data, overlapping time windows of 500 μs duration were used in an ensemble averaging technique for the estimation of mean and rms velocities. During ensemble averaging the time in individual experiments was converted to an average time frame by adding an offset as explained earlier. At least three experiments were performed for each velocity component at each measurement point. Data obtained from each experiment at a point were ensemble averaged separately and the mean velocity was given by the weighted-average of all means. The standard deviation was estimated using the *pooled sample variance* and an estimate of cycle-to-cycle variation was computed from the *between-sample variance*. The statistical uncertainty due to the finite number of measurements within each time window was evaluated at the 95% confidence level (Yanta 1973). Statistical errors of 10% or less in mean velocity and of 30% or less in rms velocity were accepted. Performing at least three experiments for each component of velocity at each measurement point was found to fulfill the statistical error requirement for most of time windows.

The two-dimensionality of the flow was examined at two different locations: one 18 mm above the obstacle (0.478 m, 54 mm) and the other 18 mm above the bottom wall and 100 mm from the obstacle (0.578 m, 18 mm). The ensemble averaged mean velocities $\tilde{u}$ and $\tilde{v}$ across the tube (2 mm, 7 mm, 12 mm, 17 mm, 22 mm, 27 mm from the side wall) at these locations are shown in Fig. 2a–b. These two regions were chosen to be representative of a mainly laminar region (0.478 m, 54 mm) and a recirculation zone (0.578 m, 18 mm) — see below. As can be seen from Fig. 2a–b, the flow is mainly two-dimensional with the wall boundary layer effects being confined to a small region from either side wall of the flame tube. It is, therefore, reasonable to assume that the flow around the obstacle in the present study is mainly two-dimensional and the side wall effects can be ignored for most part of the transient flow concerned.

3.2 Shear Layer Growth

Figures 3–5 show velocity profiles at 0.478, 0.528, 0.578, 0.628 and 0.678 m after 0.015, 0.0175, 0.020, 0.0225, 0.025 and 0.0275 s. The shear layer growth is clearly seen in these figures. At 0.015 s (Fig. 3a), along the 0.478 m plane $\tilde{u}$ is naturally higher closer to the obstacle than at the top two points and all $\tilde{u}$ velocities are positive. The shear layer at 0.015 s bends over and terminates just beyond 0.528 m where $\tilde{u}$ reduces from 18.7 m s^{-1} to about 2.8 m s^{-1}. After this plane the $\tilde{u}$ profile is almost flat. Strong positive $\tilde{v}$ is present close to the obstacle at 0.478 m as expected due to the presence of the *vena contracta*. The maximum positive $\tilde{v}$ of 12.0 m s^{-1} occurs at h = 45 mm. The $\tilde{v}$ velocities at 0.528 m are negative indicating that the centre of the recirculation zone is behind this plane. The maximum negative $\tilde{v}$ is about −9.0 m s^{-1} occurring at h = 36 mm. At 0.578 m and beyond $\tilde{v}$ velocities are almost zero, making the flow parallel to the tube axis.

At 0.0175 s (Fig. 3b), $\tilde{u}$ velocities at the top three points are high at 35.9 m s^{-1} at 0.478 m. The shear layer has now extended beyond 0.528 m, along which plane $\tilde{u}$ decreases from 37.8 m s^{-1} to zero below 45 mm. Along this plane at 9 mm and 18 mm both $\tilde{u}$ and $\tilde{v}$ are negative and this means that the recirculation zone has moved further down. This has, in essence, pushed the shear layer up and the shear layer swiftly moves down behind the recirculation zone between 0.528 m and 0.578 m. Beyond 0.578 m the $\tilde{u}$ profile is almost flat. The v-component closer to the obstacle at 0.478 m continues to be high and the maximum positive $\tilde{v}$ is about 18.3 m s^{-1} at h = 45 mm. The $\tilde{v}$ velocities at 0.528 m are all negative with a maximum negative value of −14.7 m s^{-1} at 36 mm. Beyond 0.578 m $\tilde{v}$ velocities are zero and therefore the flow is parallel to the tube axis.

At 0.020 s (Fig. 4a), the shear layer growth has reached the 0.578 m plane, as the recirculation zone moves further down the tube. The highest $\tilde{u}$ is observed at 0.528 m plane close to the top wall and is about 54.0 m s^{-1} at 54 mm with the minimum of −28.3 m s^{-1} occurring at 9 mm. Beyond 0.578 m the $\tilde{u}$ profiles are flat. Strong positive $\tilde{v}$ velocities still persist just above the obstacle with a maximum of

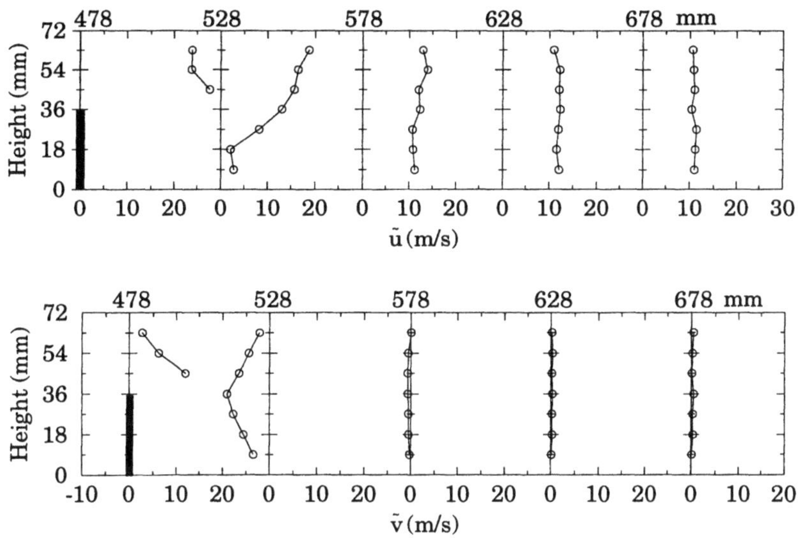

(a) At 0.015 s

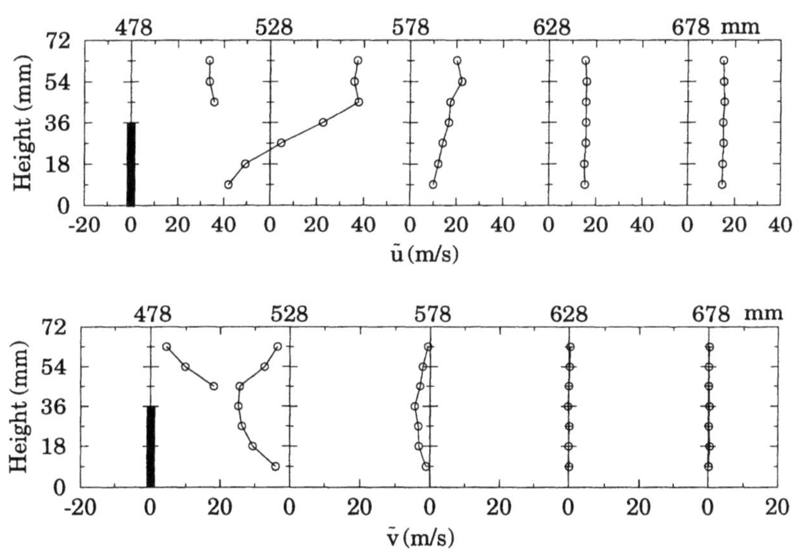

(b) At 0.0175 s

Fig. 3. ũ and ṽ profiles at different times at different positions along the flame tube: at 0.478 m, 0.528 m, 0.578 m, 0.628 m and 0.678 m: (**a**) at 0.015 s (**b**) at 0.0175 s.

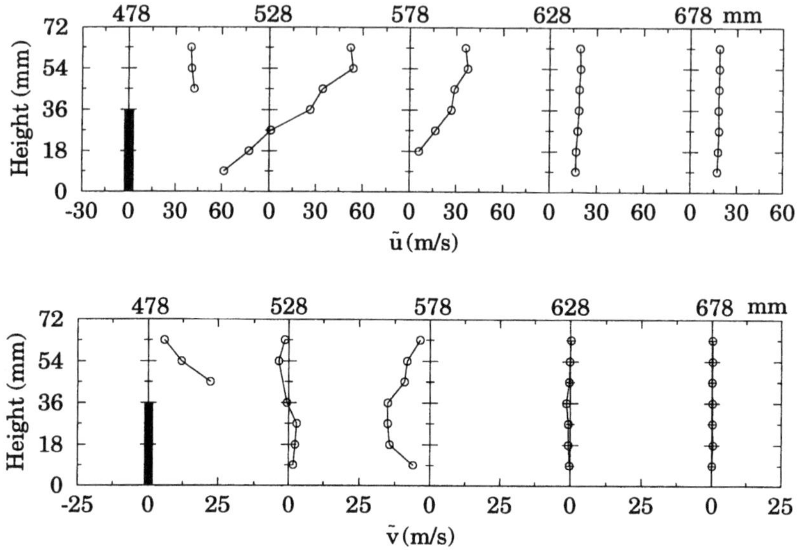

(a) At 0.020 s

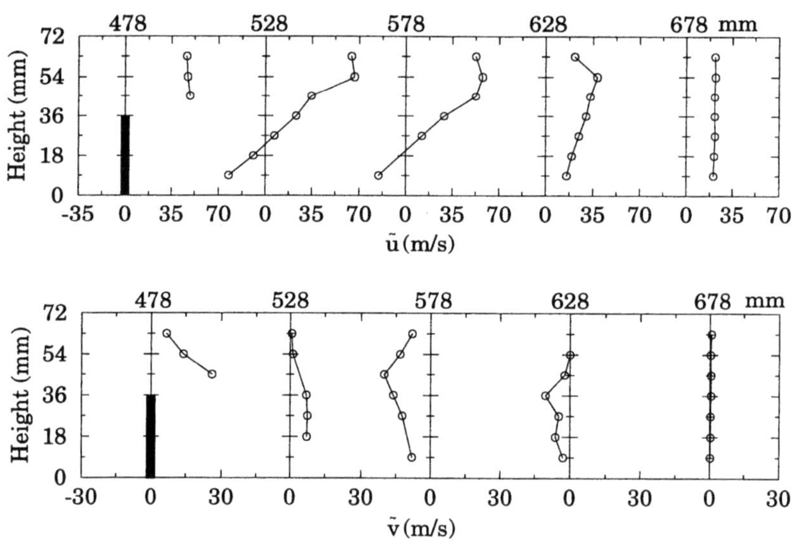

(b) At 0.0225 s

Fig. 4. ũ and ṽ profiles at different times at different positions along the flame tube: at 0.478 m, 0.528 m, 0.578 m, 0.628 m and 0.678 mm: (**a**) at 0.020 s (**b**) at 0.0225 s.

22.3 m s^{-1} at 45 mm. At the 0.528 m plane the $\tilde{v}$ velocities are very small and beyond the 0.578 m plane they are almost zero indicating that the flow beyond 0.578 m is parallel to the tube axis. The maximum negative $\tilde{v}$ component is found to be about −15.0 m s^{-1} occurring at 36 mm and 27 mm along the 0.578 m plane. At 0.0225 s (Fig. 4b), the shear layer has reached the 0.628 m plane. However, as can be seen from Fig. 4b, the $\tilde{u}$ velocity at h = 63 mm is smaller than at the next point down (h = 54 mm). This is possibly the starting point of flow separation which can clearly be seen to continue in the subsequent figures. The maximum positive $\tilde{u}$ is about 66.8 m s^{-1} at 54 mm and maximum negative $\tilde{u}$ about −27.5 m s^{-1} at 9 mm along the 0.528 m plane. The $\tilde{u}$ profile at 0.678 m is still flat. The recirculation zone now extends up to the 0.628 m plane as indicated by the negative $\tilde{v}$ velocities. However, the flow continues to be parallel at the 0.678 m plane. The maximum $\tilde{v}$ of 26.2 m s^{-1} occurs at 45 mm along the 0.478 m plane and the minimum of −20.1 m s^{-1} occurs at 45 mm along the 0.578 m plane.

As the recirculation zone extends further down the tube (Fig. 5a), the shear layer is also pushed further down and this causes the flow separation to extend to the 0.678 m plane at 0.025 s. The maximum positive $\tilde{u}$ is about 72.7 m s^{-1} occurring at 54 mm and maximum negative is about −26.8 m s^{-1} at 9 mm along the 0.528 m plane. At the 0.528 m plane the positive $\tilde{v}$ velocity within the recirculation zone shows that the centre of the recirculation zone has moved past it. Strong positive $\tilde{v}$ velocities are still found above the obstacle with a maximum of 28.0 m s^{-1} at 45 mm. The maximum negative $\tilde{v}$ velocity is about −27.9 m s^{-1} at 27 mm along the 0.628 m plane.

At 0.0275 s above the obstacle at 0.478 m the $\tilde{u}$ and $\tilde{v}$ have hardly changed their pattern. Almost flat $\tilde{u}$ profiles and strong positive $\tilde{v}$ velocities close to the obstacle can be seen in Fig. 5b. The $\tilde{u}$ velocity profile at 0.628 m has now become parallel from the top down to h = 45 mm. It follows that the flow separation point has also moved down along the tube and the separation can still be observed at the 0.678 m plane. The highest $\tilde{u}$ is about 74.2 m s^{-1} at 63 mm along the 0.578 m plane and the minimum of −20.0 m s^{-1} occurs at 9 mm along the 0.528 m plane. The maximum and minimum values of $\tilde{v}$ are respectively 28.0 m s^{-1} at 45 mm along the 0.478 m plane and −23.2 m s^{-1} at 45 mm along the 0.628 mm plane. The recirculation zone development is discussed in detail below.

3.3 Flow Characteristics

The velocity evolution with time along the vertical planes at 0.453 m and 0.553 m from the ignition point are shown in Fig. 6a–b. Each velocity component is shown separately. The $\tilde{u}$ velocity in front of the obstacle (shown in Fig. 6a) increases almost uniformly until the flame arrives. The only exception is at h = 9 mm where it becomes negative after 0.030 s as the eddies are formed in the wake in front of the obstacle. The $\tilde{v}$ velocity profile along the vertical line is almost symmetrical along the central point with the maximum velocity at h = 36 mm.

At 0.553 m plane (Fig. 6b) the $\tilde{u}$ and $\tilde{v}$ velocities are nearly uniform until 0.013 s. The $\tilde{u}$ velocities reach a maximum of 80.0 m s^{-1} along the top two points at 0.0305 s

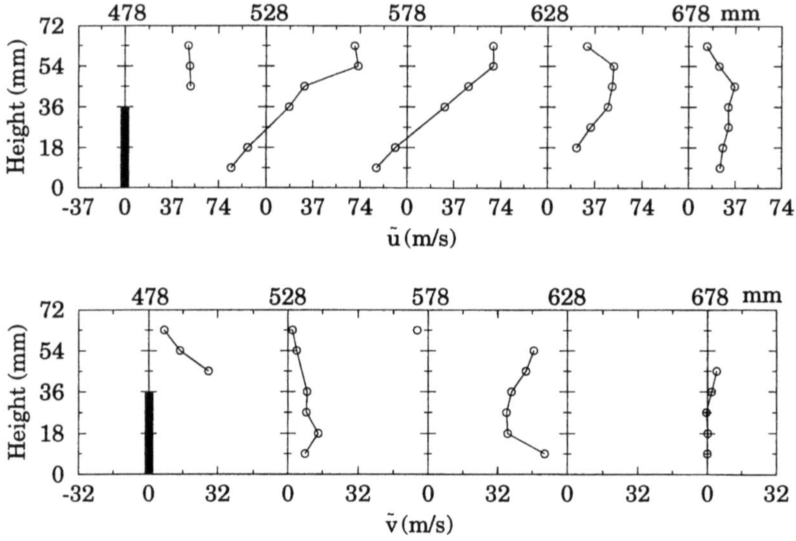

(a) At 0.025 s

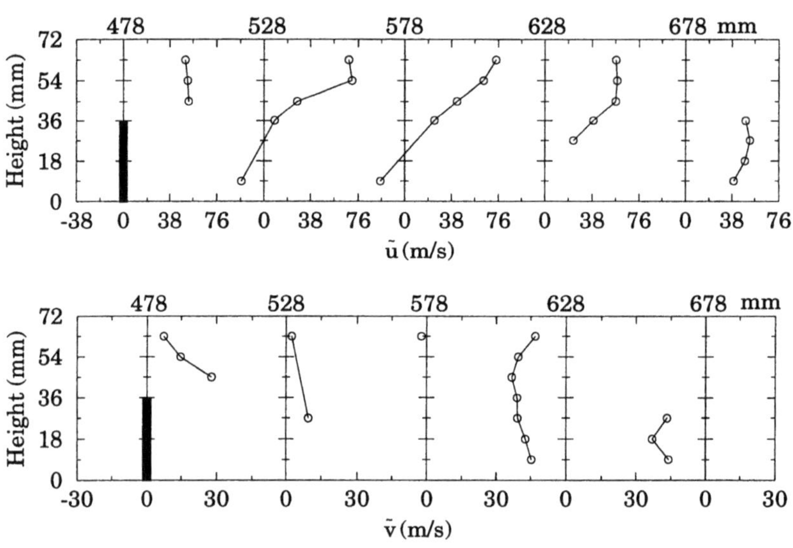

(b) At 0.0275 s

Fig. 5. ũ and ṽ profiles at different times at different positions along the flame tube: at 0.478 m, 0.528 m, 0.578 m, 0.628 m and 0.678 m: (**a**) at 0.025 s (**b**) at 0.0275 s.

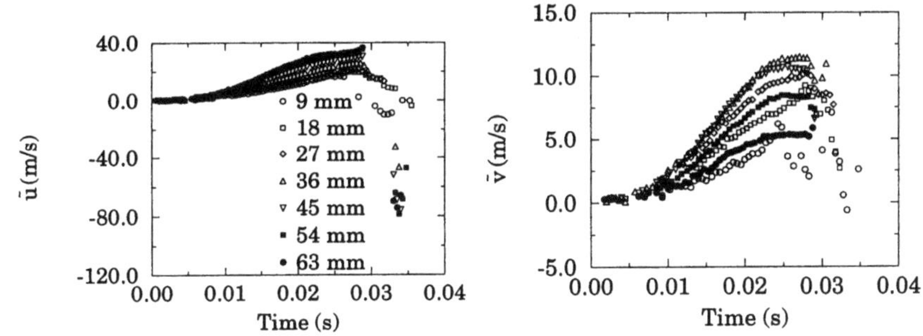

(a) At 0.453 m

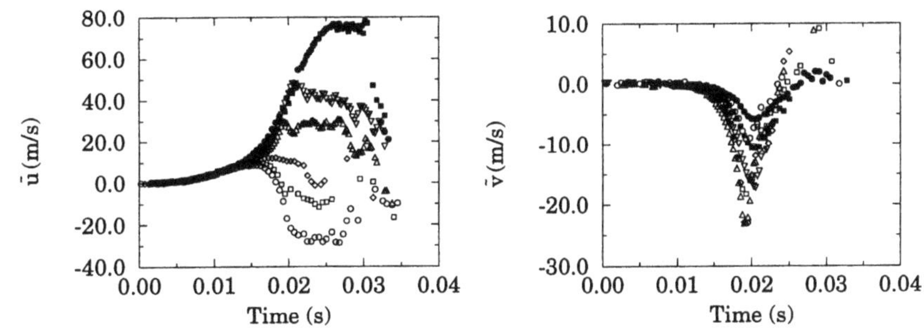

(b) At 0.553 m

Fig. 6. Variations of ũ and ṽ with time at two different positions along the flame tube. (**a**) At 0.453 m (**b**) At 0.553 m

while those at the bottom two points turn negative as the recirculation zone extends to this plane. It is interesting to note that there are oscillations in ũ velocities which are not equally as strong in the ṽ velocities. These oscillations occur at h = 36 mm and at h = 45 mm between 0.020 s and 0.030 s. It is during this period that the shear layer develops between these two points along the 0.553 m plane (Fig. 4–5). Hence, there is high turbulence generation taking place within this region. In addition, the turbulent eddies shed from the top of the obstacle travel along the shear layer. The acceleration of the flow and the normal velocity gradient along the top half of the flame tube may stretch the eddies in the axial direction. This could explain why the oscillations are not observed as strongly in the v-direction. However, contributions from intermittency cannot be ruled out.

Since the mean velocities were obtained by ensemble averaging within time windows of 500 μs, these oscillations could correspond to the energy carrying low frequency turbulent eddies. In fact, these oscillations are observed along other planes too within the region of shear layer growth. In the absence of such velocity oscillations at other places it is reasonable to conclude that a time window of 500 μs is a good compromise between losing low frequency turbulence and increasing the error due

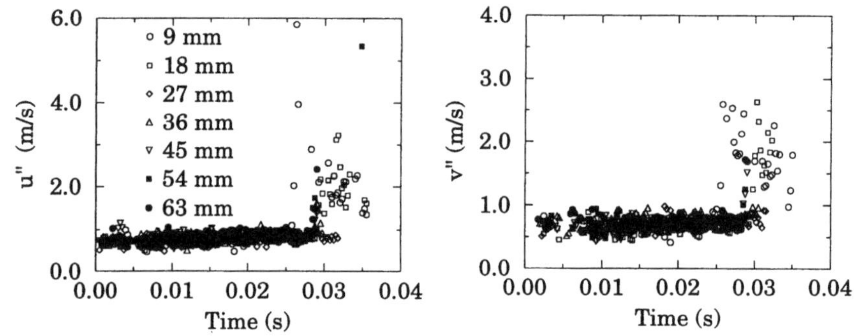

(a) At 0.453 m

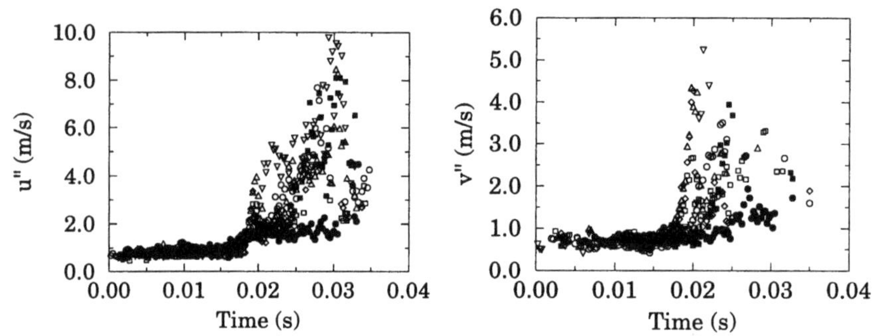

(b) At 0.553 m

Fig. 7. Variations of u″ and v″ with time at two different positions along the flame tube. (**a**) At 0.453 m (**b**) At 0.553 m

to velocity gradients within a time window. Figures 7a–b show the rms velocity variations u″ and v″ with time along the vertical planes located at 0.453 m and 0.553 m. At the 0.453 m plane where the flow is known to be laminar, u″ and v″ are of the order of 0.75 $m\,s^{-1}$. This is the error involved in the LDA measurement and is mainly caused by the TSI counter analysing a Doppler signal with a high frequency. For the present set-up, errors in the counting process for a 40 MHz signal are ± 0.5 $m\,s^{-1}$. This is the main reason for using 16 cycle setting on TSI counter. A further increase in the number of cycles will certainly reduce the error introduced by the counter, but it will inevitably result in erroneous measurement due to the limit imposed by the number of fringes within the control volume (52 in the present set-up).

At 0.453 m the rms velocities increase after 0.025 s at the lowest two points as eddies develop within the wake formed in front of the obstacle. At the 0.553 m plane the flow becomes turbulent after about 0.020 s. Until then u″ and v″ are of the order of the error. The rms velocities are high at h = 36 mm and h = 45 mm, which lie within the developing shear layer region. Isotropic turbulence is prevalent within the

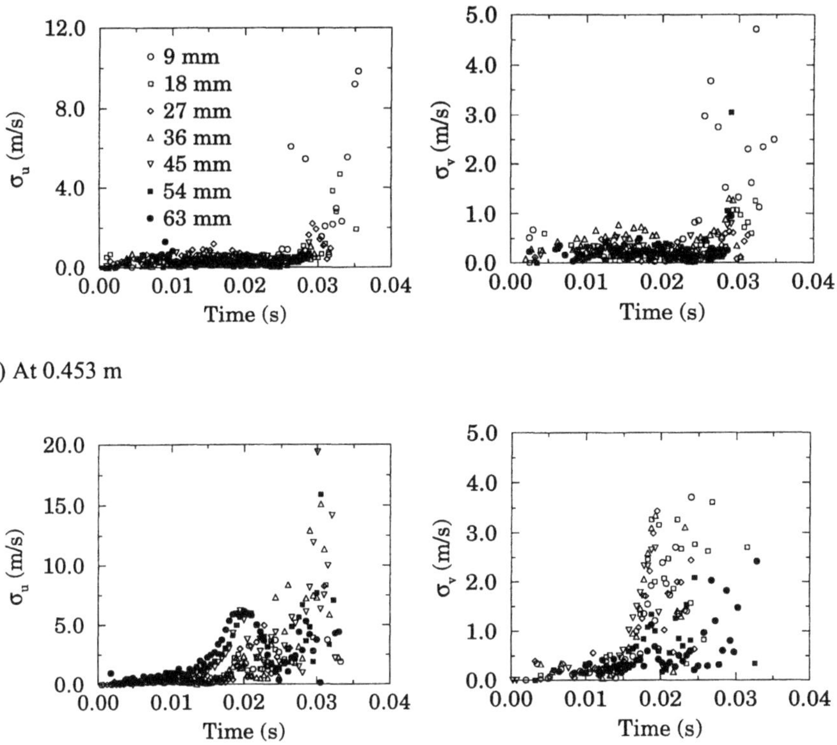

(a) At 0.453 m

(b) At 0.553 m

Fig. 8. Between-sample variations: Variations of σ_u and σ_v with time at two different positions along the flame tube. (**a**) At 0.453 m (**b**) At 0.553 m

recirculation zone. The rms velocities reach a maximum of $u'' = 9.5\ m\,s^{-1}$ and $v'' = 5.2\ m\,s^{-1}$ at h = 45 mm. Within the recirculation zone the rms velocities are typically around $3\ m\,s^{-1}$.

Between-sample variations based on between-sample variance are shown in Figures 8a–b. At the 0.453 m plane where the flow is mainly laminar flow for most part of the measurement time, the cycle-to-cycle variation is very small. Typically σ_u and σ_v are less than $1\ m\,s^{-1}$. Similar observations can be made at the 0.553 m plane (Fig 8b) where the cycle-to-cycle variation is small until the flow becomes turbulent. Within the turbulent region the between-sample variations increase significantly and, not unexpectedly, are of the order of the rms velocities. As it is clear from the figures that the between-sample variations are not greater than rms velocities it follows that the cycle-to-cycle variations excluding turbulence are indeed small.

The normal Reynolds stresses u''^2 and v''^2 are shown in Fig. 9–11 and are relatively small until 0.020 s. At 0.015 s (Fig. 9a), normal stresses are found along the developing shear layer over the recirculation zone. The u''^2 stresses are high just above the recirculation zone along the 0.503 m plane. The v''^2 stresses are high

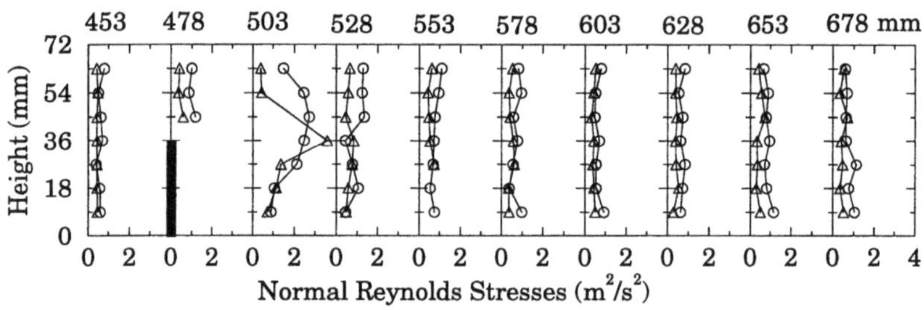

(a) At 0.015 s

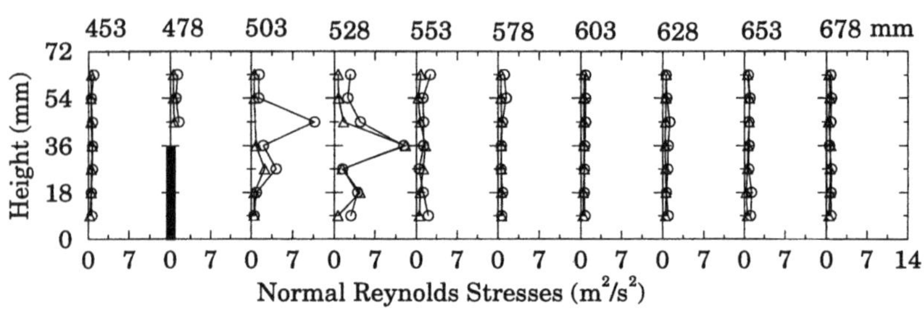

(b) At 0.0175 s

Fig. 9. Normal Reynolds stresses u''^2 and v''^2 profiles at different times. Circles show u''^2 and triangles show v''^2. (**a**) At 0.015 s (**b**) At 0.0175 s

where the mean velocity gradient becomes negative above the recirculation zone. At 0.0175 s (Fig. 9b) as the shear layer further develops, high Reynolds stresses are found along this shear layer.

At 0.020 s and 0.0225 s (Fig. 10a–b), the u''^2 stresses still continue to be high all along the shear layer region and the v''^2 stresses are high behind the recirculation zone where there is strong negative velocity gradient. This anisotropic behaviour of turbulence is particularly evident at 0.020 s (Fig. 10a). At 0.025 s (Fig. 11a) the u''^2 and v''^2 stresses continue to be high along the shear layer. It should be noted that the v''^2 stresses are not shown at several points as the v-velocities are not statistically supported at those locations. In addition, the u''^2 stresses are notably high within the wake formed in the region where the flow separates from the top wall. The initiation of the flow separation can be seen at the 603 mm and 628 mm planes in Fig. 10b. It should be noted that v-velocities are not statistically supported in this region. Further, at 0.025 s and 0.0275 s (Fig. 11a–b), the v-velocities are not statistically supported at most points, particularly within the recirculation zone and hence the nature of the turbulence is as yet unknown.

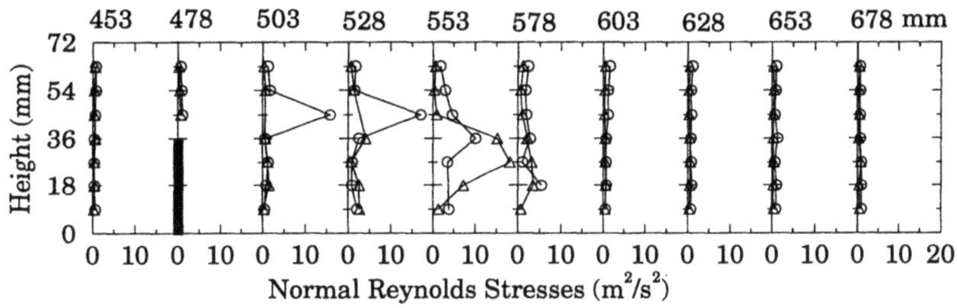

(a) At 0.020 s

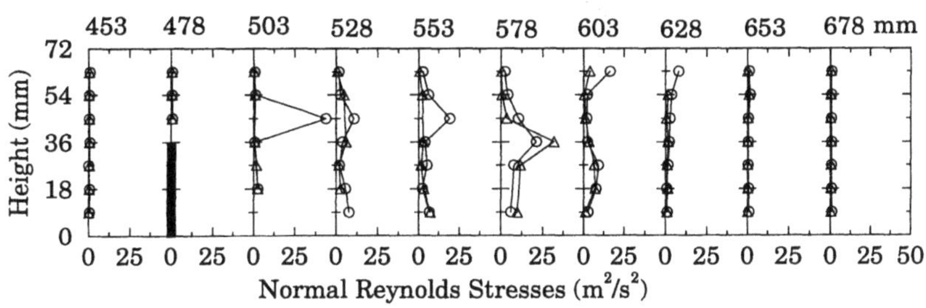

(b) At 0.0225 s

Fig. 10. Normal Reynolds stresses u''^2 and v''^2 profiles at different times. Circles show u''^2 and triangles show v''^2. (**a**) At 0.020 ms (**b**) At 0.0225 ms

3.4 Flame Interactions with Recirculation Zone

The velocity vector diagrams in Fig. 12–13 show statistically supported vectors within selected time windows. The obstacle is represented by a line in the far left and the flame propagation takes place from left to right. The instantaneous photographs taken at positions P1 and P2 are shown in Fig. 14a–f. The estimated time between the ignition of the gas mixture and triggering of the spark light source is given for each photograph. The camera position is also given in the caption of each photograph. A scale is marked along these photographs to give an impression of the physical size. Again, the flame propagation is from left to right.

At 0.015 s (Fig. 12a), the velocities at the top points located along the vertical plane at 0.478 m show an upward inclination in their directions due to the presence of a *vena contracta* and the development of the shear layer. The vortex formed behind the obstacle is still relatively small in size and its centre is located between 0.503 m and 0.528 m. The directions of velocity vectors along the 0.528 m plane show that the whole of the recirculation zone is confined in front of that plane. After this time, however, the recirculation zone begins to expand and it occupies a region up to 0.528 m at 0.0175 s (Fig. 12b). At 0.020 s the centre of the vortex has moved to 0.528 m. By this time, larger velocities are found almost everywhere in the top half of the flame tube above the shear layer (Fig. 12c). Convection of the vortex continues as it extends up to 0.578 m at about 0.0225 s (Fig. 12d) and the centre of

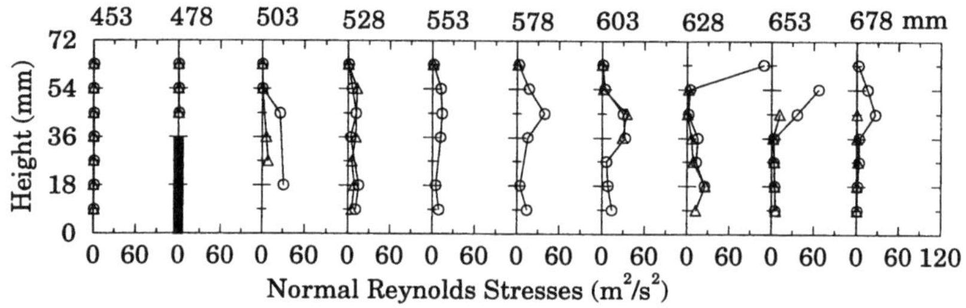

(a) At 0.025 s

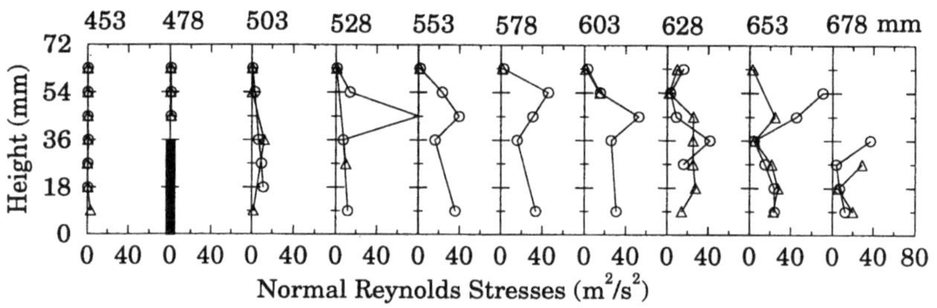

(b) At 0.0275 s

Fig. 11. Normal Reynolds stresses u''^2 and v''^2 contours at different times. Circles show u''^2 and triangles show v''^2. (**a**) At 0.025 s (**b**) At 0.0275 s

the vortex has moved to a point between 0.528 m and 0.553 m. Around this time a flow separation appears to have been initiated at the top of 0.628 m plane (see Fig. 4b as well). At 0.026 s (Fig. 13b) it appears that the flow separation point has moved further along the top wall. By the time the flame arrives at the obstacle it encounters a recirculation zone, a wake in front of the recirculation zone and a developing shear layer above the recirculation zone.

When the flame propagates over the obstacle at 0.0292 s, a laminar mode propagation still continues to dominate. As expected, slow burning is observed in the boundary layer formed along the top wall above the obstacle. The flame attains its maximum stretched state while maintaining a laminar mode, with the flow field dictating its shape. As the velocity along the top half above the obstacle is high, the flame propagation mainly takes place over the top half of the region above the obstacle. The flame thickness at the parabolic front is of the order of 1 mm (Fig. 14a), while the flame starts thickening along the shear layer developed over the obstacle. Even though the flame remains essentially laminar, its thickness along the shear layer reaches values between 3 and 4 times larger than that at the leading edge of the flame front. At this stage, due to the presence of the recirculation zone whose centre is now at about 0.578 m, the shear layer extends beyond a distance of 0.628 m from the ignition point.

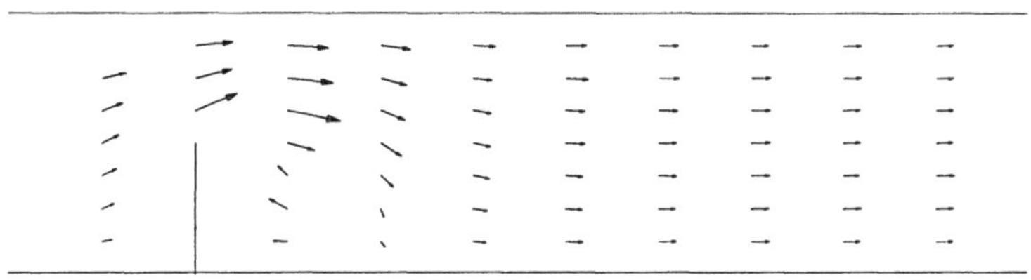

(a) At 0.015 s

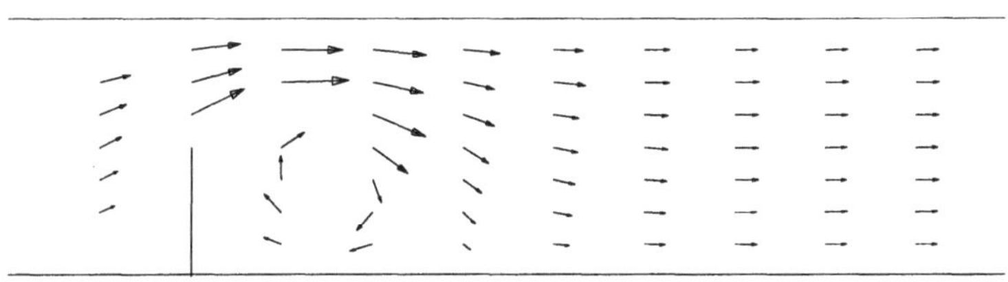

(b) At 0.0175 s

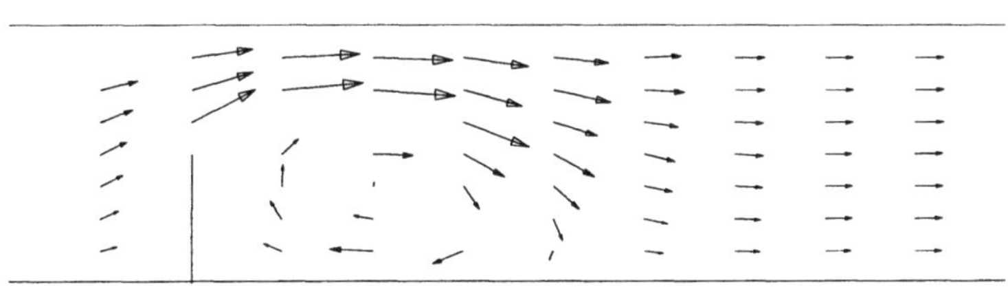

(c) At 0.020 s

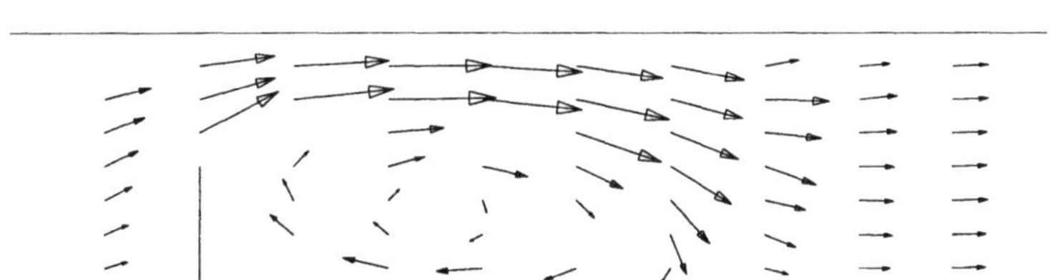

(d) At 0.0225 s

Fig. 12. Velocity vector diagrams at different times showing the evolution and convection of the recirculation zone. It also provides information on the shear layer growth.

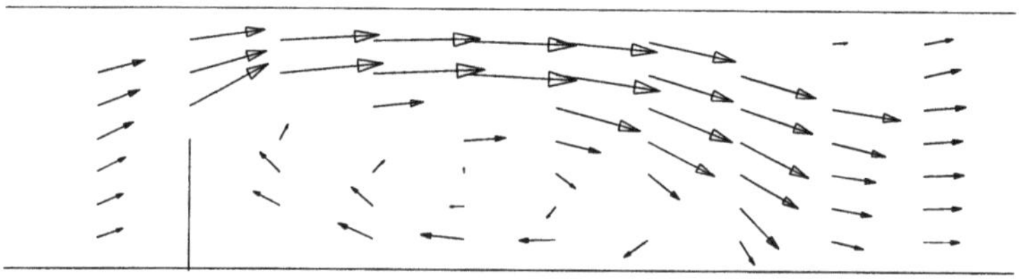

(a) At 0.0245 s

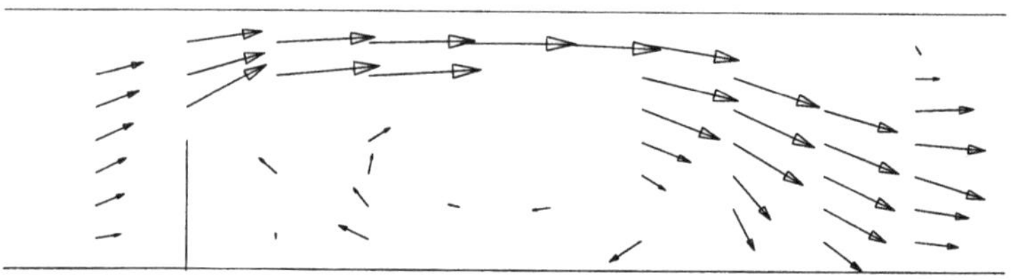

(b) At 0.026 s

Fig. 13. Velocity vector diagrams at different times showing the evolution and convection of the recirculation zone.

At about 0.0303 s after the ignition of the gas mixture, instabilities are observed at the flame front as the flame curls into the recirculation zone. Turbulence in this region is quite high due to the presence of the shear layer. The flame front is at 0.547 m and the instabilities are found to initiate at this point. It is these instabilities that contribute to the transition of the flame from laminar to turbulent propagation. Within the next 500 μs the flame front becomes turbulent. It is, however, important to note that at this stage the transformed flame front is no more than a wrinkled laminar flame. The wrinkling process of the flame front continues from its initiation point (0.547 m) backwards and travels towards the obstacle. It reaches the region above the obstacle at around 0.0317 s (Fig. 14c). The thickness of the flame burning along the shear layer above the obstacle is now of the order of 6.5 mm. The flame front continues to propagate mainly along the region closer to the top wall but as a turbulent flame brush.

By the time the flame along the shear layer becomes a wrinkled laminar flame (at about 0.0317 s), the rapid burning into the recirculation zone begins at the point where the flame has curled in (at 0.521 m). In this phase of flame propagation, the flame becomes fully turbulent with its characteristic small-scale wrinkles and the burning occurs in a large area simultaneously. The rapid burning of the turbulent flame in the recirculation zone prevents the wrapping of the flame around the centre of this zone. Instead, the burning takes place throughout the region at once. Because of the backward movement of the flame as it burns into the recirculation zone, thermal

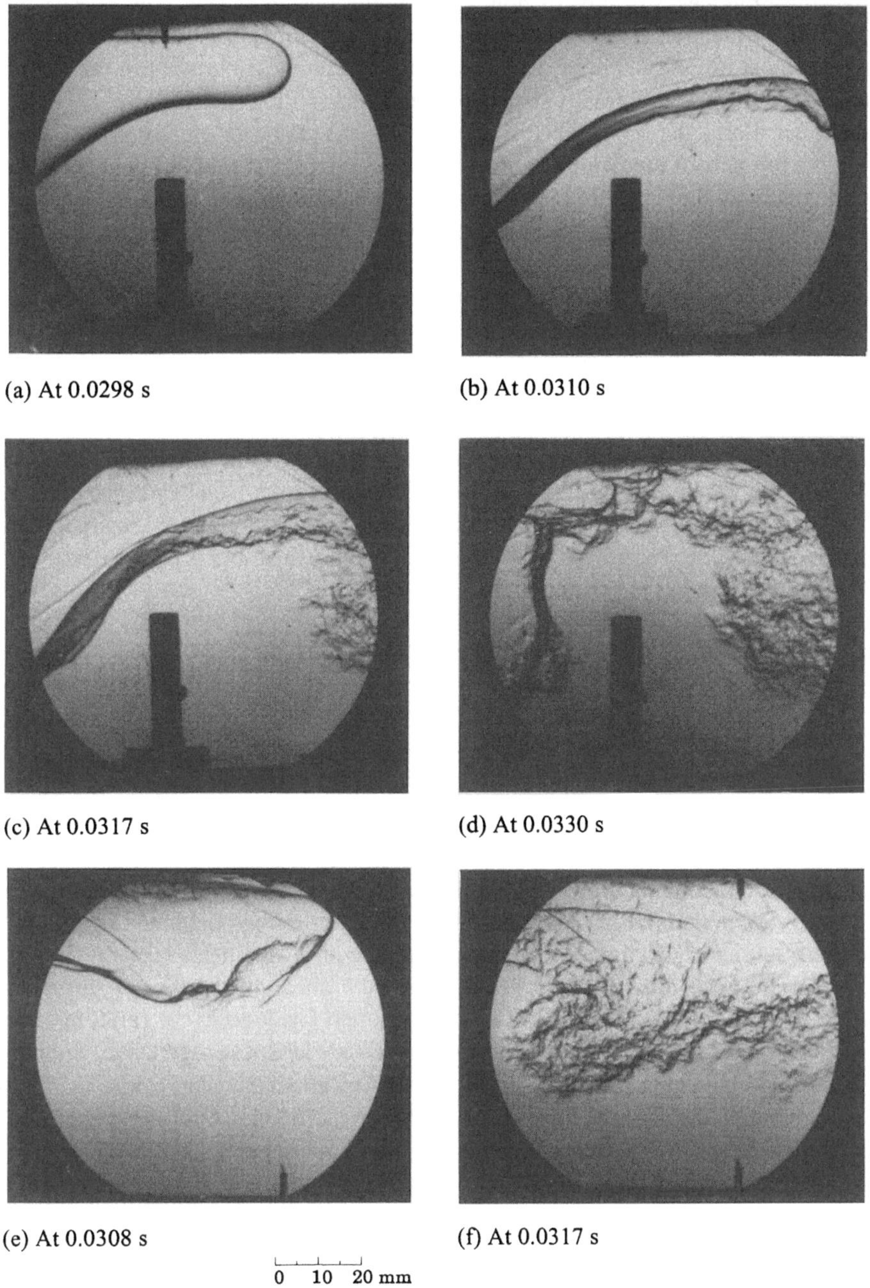

Fig. 14. Instantaneous photographs taken at position P1 after (**a**) 0.0298 s, (**b**) 0.0310 s, (**c**) 0.0317 s, (**d**) 0.0330 s and at position P2 after (**e**) 0.0308 s and (**f**) 0.0317 s.

expansion produces a backward flow towards the obstacle. This flow pushes the flame front, which remains in the shear layer for a long time, back into the burnt gas. This occurs at around 0.0329 s after the ignition. Complete burn-out in the region surrounding the obstacle finally occurs by 0.0356 s.

Further along the tube, the leading edge of the flame originally propagating along the top half of the flame tube reaches the centre of the tube at about 0.0319 s at a distance of 0.678 m from the ignition point (200 mm behind the obstacle). Afterwards, a rapid propagation along the centre is observed. The burn-out in the region around 0.678 m from the ignition finishes by 0.0330 s whereas the complete burn-out surrounding the obstacle goes on until 0.0360 s. It is the latter that contributes to the maximum pressure observed at all ports located around the obstacle. This also explains why the pressure traces contain a peak occurring for a considerable period (Lindstedt and Sakthitharan 1993b). As it has been pointed out the centre of the recirculation zone is at about 0.578 m from the ignition point when the flame arrives over the obstacle. Since the flame curls into the zone at 0.521 m, it actually breaks in between the obstacle and the centre of the recirculation zone. The flame also curls in behind the recirculation zone at 0.678 m. Intense burning begins and the recirculation zone is squeezed between these two turbulent flame brushes, and is finally destroyed before the unburnt gas is consumed.

4. Conclusion

An experimental study of the transient interaction of a premixed stoichiometric methane-air flame with a developing shear layer and recirculation gone formed by an obstacle placed in its path has been conducted. Mean gas velocities along and perpendicular to the flame axis were measured using a standard laser Doppler anemometer. Silicone oil droplets were used as seeding particles to obtain the mean and rms velocities in the reactant gas. Flow visualization was achieved by the use of a spark schlieren photographic system coupled with a time-delay system. Results show that the three-dimensional effects of the flow ahead of the flame can be neglected and the flow is essentially two-dimensional. Cycle-to-cycle variations are small in the present work and hence the velocity characteristics are reliable. Velocity measurements have provided significant information about the characteristics of the accelerated flow ahead of the flame front. The recirculation zone formed behind the obstacle has been observed to convect along the flame tube. The shear layer continues to develop over the recirculation zone and it is extended further away from the obstacle by the convection of the recirculation zone. As the flow develops, flow separation appears along the top wall and moves along with the convection of the recirculation zone. The transition of an essentially laminar flame to a strongly turbulent deflagration occurs with the onset of flame instabilities induced in a developing shear layer along the obstacle. Strong turbulence generation is observed along the developing shear layer and is found to be one of the contributory factors to the instabilities. The flame is found to engulf the recirculation zone from the front and behind.

Large distortions in the structure of the shear layer over the obstacle are observed as a consequence of the rapid increase in volume production behind the obstacle.

Acknowledgement

The authors wish to gratefully acknowledge the help given by Prof. J. H. Whitelaw, Dr. A. M. K. P. Taylor, Dr. C. Vafidis, Dr. A. Hatziapostolou and Dr. I. Hardalupas of Imperial College and the financial support of British Gas plc. The authors would also like to record their gratitude to Mr. J. R. Laker and Mr. I. Wright for their technical assistance.

References

Cheng, R. K. (1984): "Conditional Sampling of Turbulence Intensities and Reynolds Stress in Premixed Turbulent Flames", Combust. Sci. Tech., **41**, pp. 109–142

Heitor, M. V., Taylor, A. M. K. P., Whitelaw, J. H. (1987): "The Interaction of Turbulence and Pressure Gradients in a Baffle-Stabilized Premixed Flame", J. Fluid Mechanics, **181**, pp. 387–413

Lindstedt, R. P. (1984): "Deflagration to Detonation Transition in Mixtures Containing LNG/LPG Constituents", Ph.D. Thesis, University of London

Lindstedt, R. P., Michels, H. J. (1989): "Deflagration to Detonation Transitions and Strong Deflagrations in Alkane and Alkene Air Mixtures", Combust. Flame, **76**, pp. 169–181

Lindstedt, R. P., Sakthitharan, V. (1993a): "A Study of Interactions of Transient Flames with Single and Multiple Baffle-type Obstacles", Submitted to Proc. R. Soc. Lond.

Lindstedt, R. P., Sakthitharan, V. (1993b): "Stoichiometric and Obstacle Configuration Effects in Strongly Accelerated Methane and Propane Flames", In preparation

Markstein, G. H. (1956): "A Shock-tube Study of Flame Front–Pressure Wave Interaction", in Proc. 6th Symp. (Int.) on Combustion (The Combustion Institute, Pittsburgh), pp. 387–398

Markstein, G. H. (1964): "Non-steady Flame Propagation", Pergamon Press, Oxford

Taylor, A. M. K. P., Whitelaw, J. H. (1984): "Velocity Characteristics in the Turbulent Near Wakes of Confined Axisymmetric Bluff Bodies", J. Fluid Mechanics, **139**, pp. 391–416

Yanta, W. J. (1973): "Turbulence Measurements with a Laser Doppler Velocimeter", Naval Ordinance Laboratory Report, NOLTR 73-94

Spectral Measurement of Reactive and Conserved Scalars in a Turbulent Reactive-Scalar-Mixing Layer

J. D. Li, R. J. Brown, and R. W. Bilger

Department of Mechanical Engineering,
The University of Sydney, N.S.W., 2006 Australia

Abstract

Experimental results for the various spectra of scalars in a turbulent reactive scalar mixing layer have been presented. The main conclusions which can be drawn from these are: a) in the center of the mixing layer, the coherence of the two chemical species is high from very low frequency to the high frequency where most of the energy dissipation occurs; b) the results do not support the suggestion that the two chemical species will lose their coherence at the frequency where the reaction loss balances the spectral transfer; c) the theory proposed by Kosaly (1992) for the coherence between the equilibrium limits of the two chemical species in a homogeneous scalar field has been supported; and d) when the difference between the local mean mixture fraction and the stoichiometric mixture fraction is large, only the species which is locally 'rich' is strongly correlated with the mixture fraction while the species which is locally 'lean' has only a weak correlation with the mixture fraction.

1. Introduction

The study of turbulent reacting flow is important in many engineering problems, but our understanding of it is far from complete. One of the difficulties in understanding turbulent reacting flow is that turbulence involves many length and/or time scales. For a turbulent reacting flow with a finite chemical reaction rate another time scale is involved. This scale may be small relative to the time scale of the large-scale motions but large in comparison with the time scale of the fine-scale motions, particularly when the Reynolds number is high. The widely used equilibrium limit approximation assumes that the chemical reaction rate is infinite (Bilger, Saetran and Krishnamoorthy, 1991). This approximation might be reasonable for understanding the interaction between the large scale motions and the chemical reaction but is questionable for the interaction between the small scale motions of turbulence and that of chemistry. One way to study the effect of the turbulent motions of different scales on the chemical reaction is through spectral measurement and so far not much work has been done on this due to the difficulties of making measurement in turbulent reacting flows.

In the work presented here a turbulent reactive scalar mixing layer is formed in grid turbulence by introducing two streams of air into a Turbulent Smog Chamber

(TSC). Dilute nitric oxide (NO) is introduced into one stream and dilute ozone (O_3) into the other stream as reactive species. The reactants are separated before entering the TSC and allowed to mix freely after the grid. The mixture fraction, a conserved scalar, can be defined as

$$F = \frac{\Gamma_{NO} - \Gamma_{O_3} + \Gamma_{O_3 2}}{\Gamma_{NO_1} + \Gamma_{O_3 2}}. \tag{1}$$

Here Γ_{NO} is the NO concentration, Γ_{O_3} the O_3 concentration, Γ_{NO_1} the NO concentration in stream 1 and $\Gamma_{O_3 2}$ is the O_3 concentration in stream 2. Limiting cases of fast chemistry (equilibrium limit) and slow chemistry (frozen limit) can be expressed in terms of the mixture fraction (conserved scalar theory, Bilger et al., 1991),

$$\Gamma^e_{NO} = (\Gamma_{NO_1} + \Gamma_{O_3 2})(F - F_s)H(F - F_s), \tag{2}$$

$$\Gamma^e_{O_3} = (\Gamma_{NO_1} + \Gamma_{O_3 2})(F_s - F)H(F_s - F), \tag{3}$$

$$\Gamma^0_{NO} = F\Gamma_{NO_1}, \quad \Gamma^0_{O_3} = (1 - F)\Gamma_{O_3 2} \tag{4}$$

where superscript e stands for equilibrium limit and 0 for frozen limit, H is the Heaviside function and F_s is the stoichiometric mixture fraction. Saetran et al. (1989), Bilger, Krishnamoorthy and Saetran (1989), Bilger et al. (1991) and Li, Brown and Bilger (1992) have measured the various statistics of NO, O_3 and the mixture fraction. Bilger et al. (1991) measured the spectra of Γ_{NO} and Γ_{O_3}, their coherence and phase shift in the frequency domain at the center of the mixing layer. They find that at low to moderate frequency the two species are strongly correlated and 180° out of phase. At high frequency the two species are incoherent with random phase shift. Kosaly (1992) has constructed a theory to explain this.

In this paper, we present the measured spectra of the mixture fraction, the two species and their equilibrium limits (the spectra for the frozen limit is the same as that for mixture fraction as can be seen from (4)) in the turbulent reactive scalar mixing layer at two different positions across the flow. We also present the coherence and the phase shift of the two reactive species. The implication of the results for modelling is discussed.

2. Experimental Set-Up

Bilger et al. (1991) and Li and Bilger (1993) describe the experimental set-up in detail. We will give a brief description here. Figure 1 shows the schematic diagram of the experiment. The two flows with the same mean velocity and two constant reactant concentrations (Γ_{NO_1} and $\Gamma_{O_3 2}$) are introduced into the working section respectively. The turbulence is generated by a grid which is mounted at the beginning of the working section. The cylindrical working section is 8 m long and 2.8 m in diameter. The nominal mean velocity of this experiment is 0.5 m/s, giving the Reynolds number $Re = UM/\nu = 10\,700$. Here U is the streamwise mean velocity, $M = 320$ mm the mesh spacing and ν is the kinematic viscosity. The Chemiluminescent Analyzer

(CLA) developed by Mudford & Bilger (1983) has been used to measure the two chemical concentrations. Bilger et al. (1991) suggest that the frequency response of the CLA is 23 Hz. A further check finds that the frequency response of the system used then is much less than this. For this experiment the CLA has been modified. The modifications include reducing the volume of the reaction cell in front of the photomultiplier tube from 63 cm^3 to 33 cm^3, introducing the sampled gas and the gas for reactant-in-excess directly into the mixing cell, increasing the capacity of the vacuum pump and keeping the length of the sampling tube to a minimum (3.2 m). The frequency response of the CLA after the modification has been checked and is about 20 Hz. The noise of the system starts to dominate the signal from 30 Hz. The present experimental set-up is a compromise between the need for having a known turbulence field at a reasonably high Reynolds number and for achieving enough spatial and temporal resolution for the CLA to measure the species concentrations. Because of this the experiment is limited to $x/M < 21$.

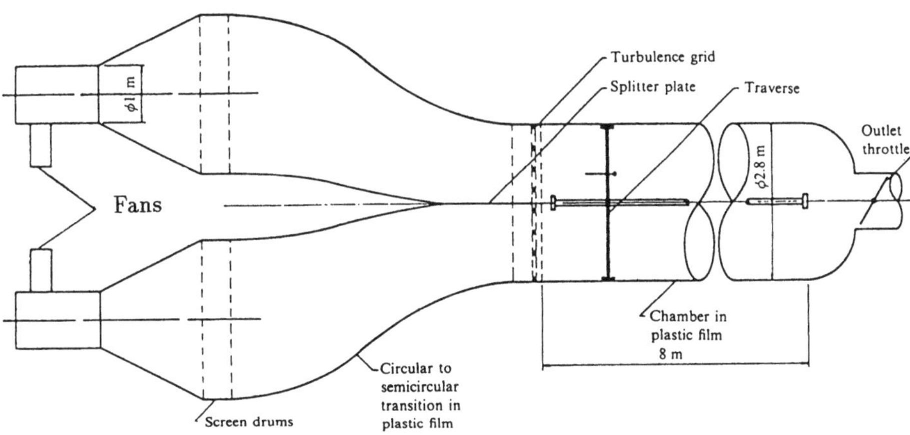

Fig. 1. Schematic diagram of experimental facility.

The NO calibration is done using a gas from a cylinder with known concentration diluted with precision flow meters. The O_3 calibration is determined indirectly by the titration method of Post & Kewley (1978). A X-wire probe with a constant temperature Hot-Wire Anemometer (HWA) is used to measure the two velocity components. One of the parameters which characterizes the time scales of both turbulence and chemical reaction is the Damköhler number. In this paper, this is defined as $N_D = \tau_{\text{flow}}/\tau_{\text{chem}} = kM(\Gamma_{\text{NO}_1} + \Gamma_{\text{O}_3{}_2})/U$. Here τ_{flow} and τ_{chem} are the turbulent and chemical time scales respectively. The experimental results given in this paper are measured at a Damköhler number of about 1.6. Simultaneous measurements of two components of velocity and two reactants are made at points across the flow field and at different stations downstream of the grid. The sampling tube, which has a sonic nozzle at its inlet, is located 18 mm above the hot-wire probe. It has been found that with this separation, the measured flow field is not interfered by the tube. The CLA samples gas at 140 std. cm^3/s through a 3.2 m Teflon tube of diameter 2.8 mm. The delay time in the tube is approximately 63 ms. The signals of the four channels (two CLA, two hot-wires) are sampled at 128 Hz using a 12 bit A/D converter. The total number of samples is 32 768, giving a sampling time of 256 seconds. The data processing is performed on a high speed PC486. From the measured concentrations of the two species the mixture fraction, the frozen limits and equilibrium limits of the two species can be determined. A Fast Fourier Transform (FFT) algorithm is used to process the time series data and the spectra, the coherence and the phase shift can be calculated (Bendat and Piersol, 1980). The spectra presented in this paper have been normalized such that the area under the curve is unity so that the variance distribution of the different quantities in the frequency domain can be compared.

3. Experimental Results and Discussion

The flow field after the grid is a near-field grid turbulence. The mean velocity and the rms of the velocity fluctuations across the flow have been checked. The profiles are fairly uniform except at the edges of the flow where a boundary layer is developing. The flow is quite anisotropic with $u'/v' \approx 1.8$ in the test section and with no tendency towards isotropy being shown. Here u' and v' are the rms values of the streamwise and traverse velocity fluctuation components. The experimental data presented here is at $x/M = 15$. The various statistics of the two species and the mixture fraction have been given in Bilger et al. (1991) and Li et al. (1992). The results are for the modified flow conditions of Li et al. (1992) and do not show the anomalous results present in the Bilger et al. (1991) data. Here only the various spectra of the scalars, the coherence and the phase shift are given.

Figure 2 shows the various measured spectra of scalars at $y/\delta \approx 0$ and $x/M = 15$ where the local mean mixture fraction is 0.56. Here δ is the mixing layer thickness which is defined as the distance from where the mean mixture fraction is 0.1 to where it is 0.9. Figure 2 shows that the spectra for Γ_{NO}, Γ_{O_3}, the mixture fraction and the equilibrium limits of the two species are the same from the very low frequency to about 10 Hz. The difference in the frequency range beyond 10 Hz is small. Also shown in figure 2 is the -5/3 law and it can be seen that an inertial subrange exists from about 2 Hz to 8 Hz. Figure 3 shows the pre-multiplied spectra where $\widetilde{S}_{\Gamma_{NO}}(\phi)$ is the normalized spectra for Γ_{NO}. The results show that the variance bearing eddies at $x/M = 15$ in the present experiment are at about 1 Hz. This is the same as that for the turbulent energy containing eddies. The pre-multiplied spectra for other quantities have not been shown since they are similar as can be inferred from figure 2. Figure 4 shows the coherence between Γ_{NO} and Γ_{O_3} and figure 5 shows the phase shift between the two species. The coherence achieves high value at low frequency and decreases slowly from 1 Hz to 20 Hz. The coherence decreases dramatically at about 20 Hz and becomes very small beyond 30 Hz. The local correlation coefficient R_{NO,O_3} is -0.81 and it can be seen that it arises almost entirely from $\phi < 20$ Hz. Bilger et al. (1991) show that the two species become incoherent at about 10 Hz. This is thought

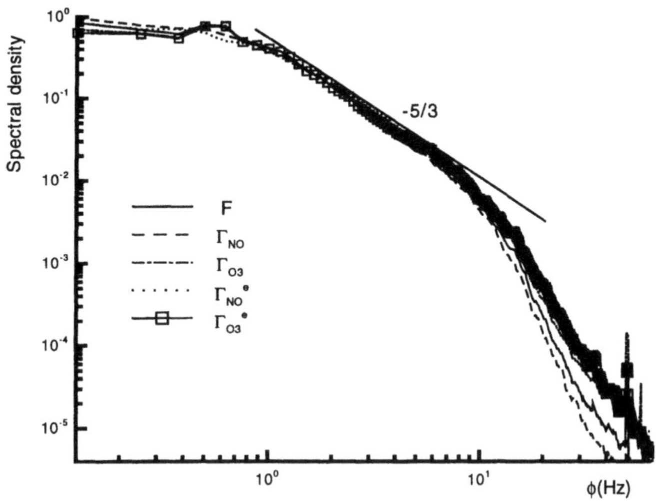

Fig. 2. The various spectra at $y/\delta \approx 0$.

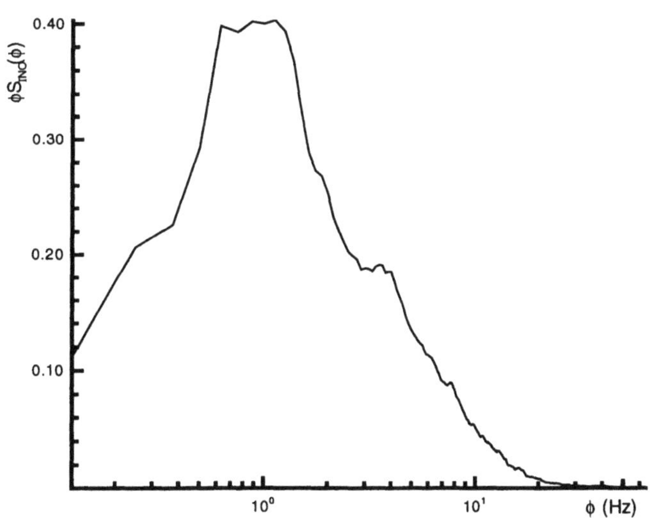

Fig. 3. The pre-multiplied spectrum of Γ_{NO}.

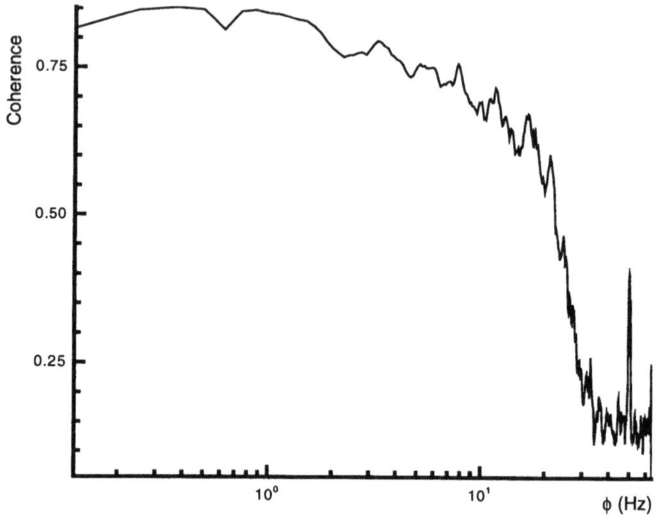

Fig. 4. The coherence between Γ_{NO} and $\Gamma_{\mathrm{O_3}}$ at $y/\delta \approx 0$.

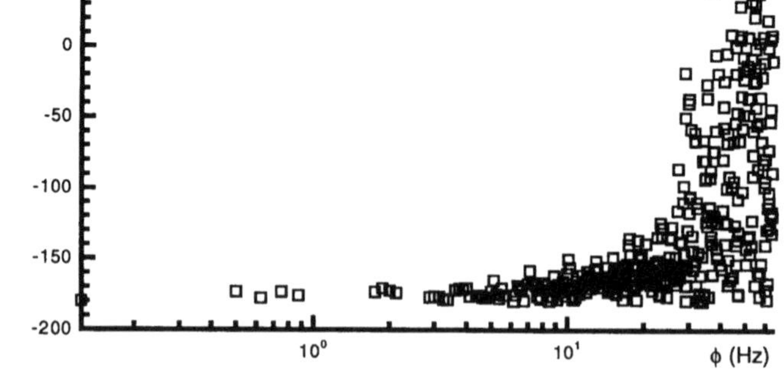

Fig. 5. The phase shift between Γ_{NO} and $\Gamma_{\mathrm{O_3}}$ at $y/\delta \approx 0$.

to be due to the low frequency response of the CLA used. Figure 5 shows that the phase shift is about 180° at low frequency and slowly decreases from 5 Hz to 30 Hz. Beyond 30 Hz the phase shift becomes random and this is consistent with the loss of coherence shown in figure 4. Since the frequency response of the CLA is about 20 Hz, it is difficult to conclude whether this loss of coherence at high frequency is due to the attenuation effect of the CLA or is due to the genuine loss of coherence between the two species. The small value of coherence beyond 30 Hz is due to noise. Because the phase shift is random in this frequency region the contribution to the overall correlation from $\phi > 30$ Hz is negligible.

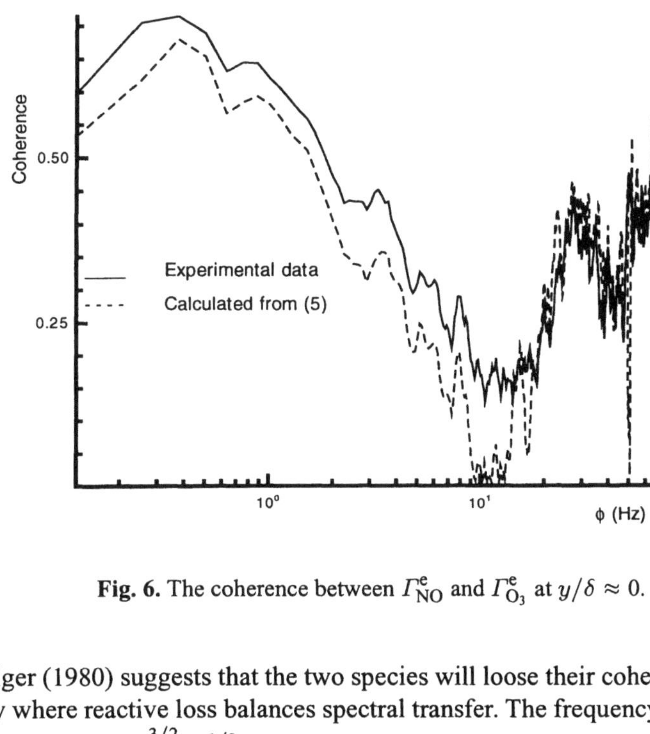

Fig. 6. The coherence between Γ^{e}_{NO} and $\Gamma^{e}_{O_3}$ at $y/\delta \approx 0$.

Bilger (1980) suggests that the two species will loose their coherence at the frequency where reactive loss balances spectral transfer. The frequency corresponding to this is $\phi_c = UK_r^{3/2}\varepsilon^{-1/2}/2\pi$. Here K_r is the rate constant for reaction and ε is the turbulent energy dissipation. Using the present experimental results it can be estimated that $\phi_c \approx 4$ Hz. Thus the present experimental data and that of Bilger et al. (1991) do not support this suggestion. The Kolmogorov length scale, η, of the present experiment is about 2.4 mm and its corresponding frequency is 210 Hz. Warhaft and Lumley (1978) and Li and Bilger (1993) show that the maximum dissipation spectrum for conserved scalar occurs at about 30η and this corresponds to 7 Hz in the present experiment. It is believed that the same conclusion can also be applied to

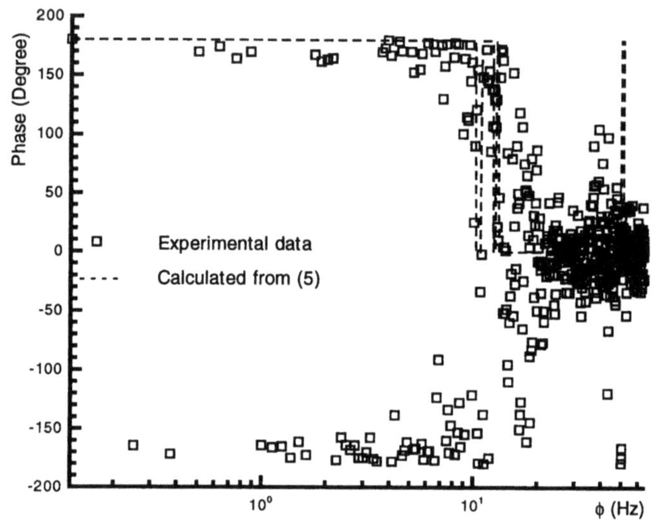

Fig. 7. The phase shift between Γ^e_{NO} and $\Gamma^e_{O_3}$ at $y/\delta \approx 0$.

reactive scalars. Figures 4 and 5 show that the two species correlate with each other even in the region where most of the dissipation for scalar variance occurs.

Figure 6 shows the coherence between Γ^e_{NO} and $\Gamma^e_{O_3}$ and figure 7 shows the phase shift between them. In comparison with figure 4, the coherence decreases quickly from 0.3 Hz to 10 Hz and becomes very small at about 10 Hz. The phase shift also becomes random at around 10 Hz. Beyond 20 Hz the coherence between the equilibrium limits of the two species increases. Also the phase shift decreases and becomes in phase. The correlation coefficient R_{NO^e,O_3^e} is -0.53 and this mainly comes from $\phi < 10$ Hz. The present results show that at the equilibrium limit the coherence between the two species is much less than that at finite Damköhler number. Since in the frozen limit the coherence between the two species is 1 and the phase shift is 180°, figures 4 and 6 show that the coherence of Γ_{NO} and Γ_{O_3} is between the frozen and equilibrium limits in the frequency domain.

Based on the experimental data of Bilger et al. (1991), Kosaly (1993) develops a theory on the auto- and cross-spectra of reactant fluctuations in a homogeneous scalar field. By assuming the pdf of the mixture fraction to be Gaussian with $\Gamma_{NO_1} \approx \Gamma_{O_{3\,2}}$, he derives

$$\frac{S^e_{NO,O_3}(\phi)}{S^e_{NO}(\phi)\cdot S^e_{O_3}(\phi)^{1/2}} = 1 - \frac{2}{1 + 0.36\frac{\tilde{S}_{|f|}(\phi)}{\tilde{S}_f(\phi)}} \tag{5}$$

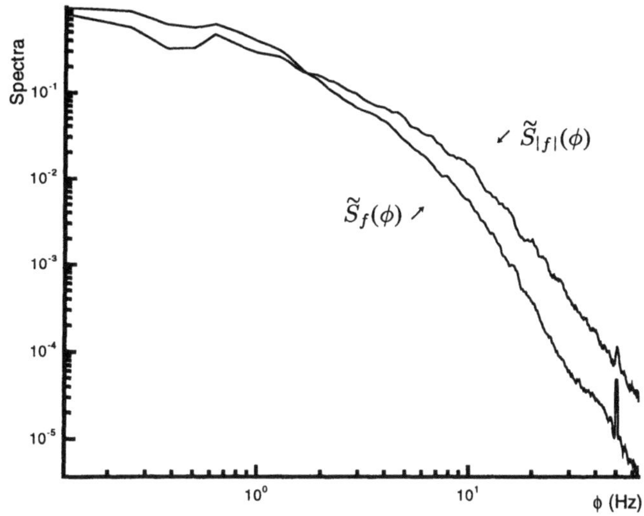

Fig. 8. The spectra of the mixture fraction and the absolute value of its fluctuation at $y/\delta \approx 0$.

for the equilibrium limits. Here S^{e}_{NO,O_3} is the co-spectrum of Γ^{e}_{NO} and $\Gamma^{e}_{O_3}$, $\widetilde{S}_f$ and $\widetilde{S}_{|f|}$ are the spectra of the mixture fraction and the absolute value of mixture fraction fluctuations. The tilde on top of the spectra means that the spectra have been normalized such that the area under the spectral curve is unity. The Γ_{NO_1} and $\Gamma_{O_3 2}$ in the present experiment are 2.8 and 3.0 parts per million (ppm), which are very close to the equality requirement for the theory of Kosaly (1993). Figure 8 shows the spectra for the mixture fraction and the absolute value of mixture fraction fluctuations. At low frequency the spectrum for mixture fraction is larger than that for its absolute value. At high frequency the spectrum for $\widetilde{S}_{|f|}$ decreases slower than that for $\widetilde{S}_f$ and this results in the decrease in the coherence of the two equilibrium limits at high frequency according to (5). The two curves cross each other at 1.5 Hz. In figures 6 and 7 we also show the coherence and phase shift calculated from (5) using the spectra for mixture fraction and its absolute value as that shown in figure 8. The coherence is the absolute value of the co-spectrum and the phase shift has been determined following the suggestion of Kosaly (1993), i.e. when the co-spectrum is less than zero the phase shift is 180° otherwise the phase shift is zero.

In comparison with the experimental data, it can be seen that the trend for the coherence predicted by (5) is correct. The coherence predicted from (5) is less than that from the experimental data for $\phi < 10$ Hz and is the same for $\phi > 30$ Hz. Figure 7 shows that the phase shift calculated from (5) in general agrees with the experimental results. In figure 6 the increase in coherence beyond 10 Hz is due to the way the coherence is calculated (Bendat and Piersol, 1971) and figure 7 shows that Γ^{e}_{NO} and $\Gamma^{e}_{O_3}$ are in phase when $\phi > 10$ Hz. Kosaly (1993) mentions that the coherence could increase after it drops to zero in the high frequency region. The present results support this but the physical explanation for it is not yet clear. Kosaly (1993) also predicts that $R_{NO^e,O_3^e} = -0.47$ and the coherence is greater than 0.47 at zero frequency. These agree well with the present experimental results.

Although not shown here it is found that at the center of the mixing layer the coherence between the mixture fraction and the equilibrium limits of both the nitric oxide and ozone are high for all the frequencies. Also it is found that the phase shift between the mixture fraction and Γ_{NO}^e are close to 0° while those between the mixture fraction and $\Gamma_{O_3}^e$ are 180°.

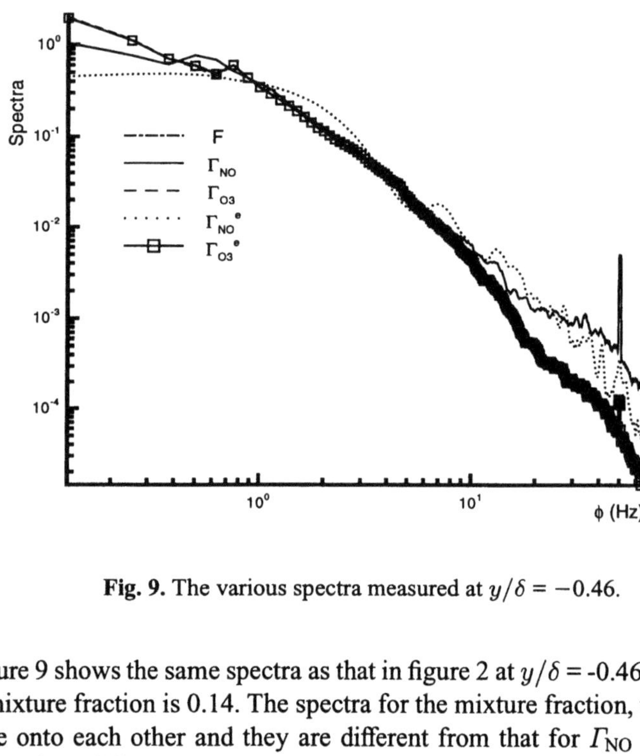

Fig. 9. The various spectra measured at $y/\delta = -0.46$.

Figure 9 shows the same spectra as that in figure 2 at y/δ = -0.46 where the local mean mixture fraction is 0.14. The spectra for the mixture fraction, the Γ_{O_3} and $\Gamma_{O_3}^e$ collapse onto each other and they are different from that for Γ_{NO} and Γ_{NO}^e which are themselves different at low frequencies. The results at $y/\delta > 0$ are not shown here since the problem is antisymmetrical, i.e. replacing the subscript NO with O_3 the conclusions will be the same. At high frequency it is thought that the slower rate of decreasing in the spectra for Γ_{NO} and Γ_{NO}^e is due to the noise from the CLA. At y/δ = -0.46 the signal level for Γ_{NO} is low and the noise starts to dominate the sampled signal at lower frequency than that at $y/\delta \approx 0.0$. This can also be seen if we compare figure 3 with figure 10 which is the pre-multiplied spectra for Γ_{NO} at y/δ = -0.46. In general we would expect that the contribution to the variance is the same

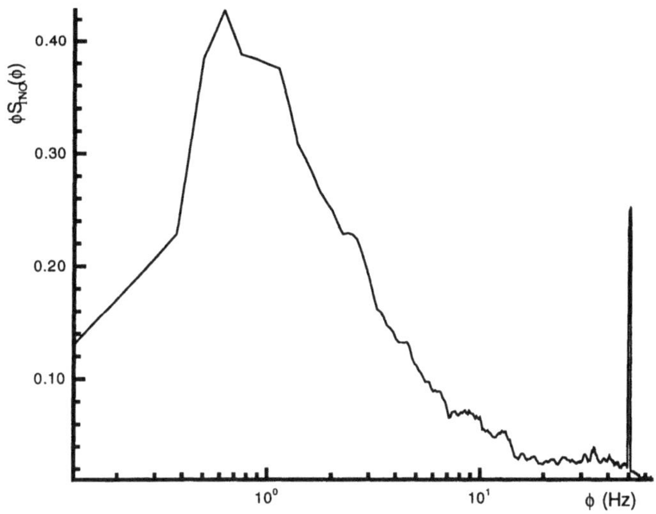

Fig. 10. The pre-multiplied spectrum of Γ_{NO} at $y/\delta = -0.46$.

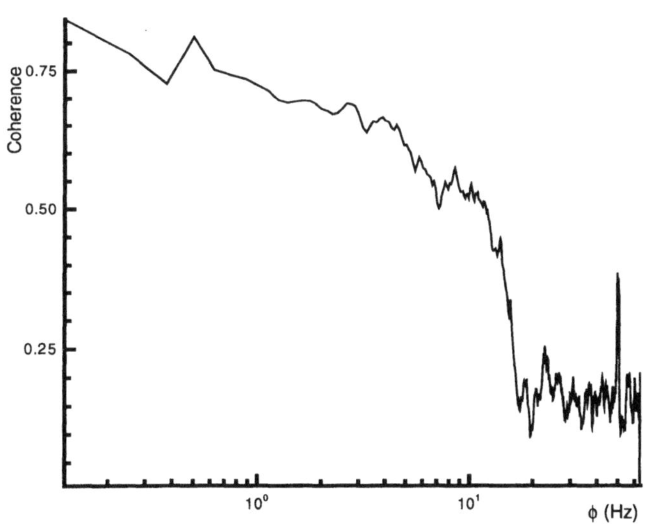

Fig. 11. The coherence between Γ_{NO} and Γ_{O_3} at $y/\delta = -0.46$.

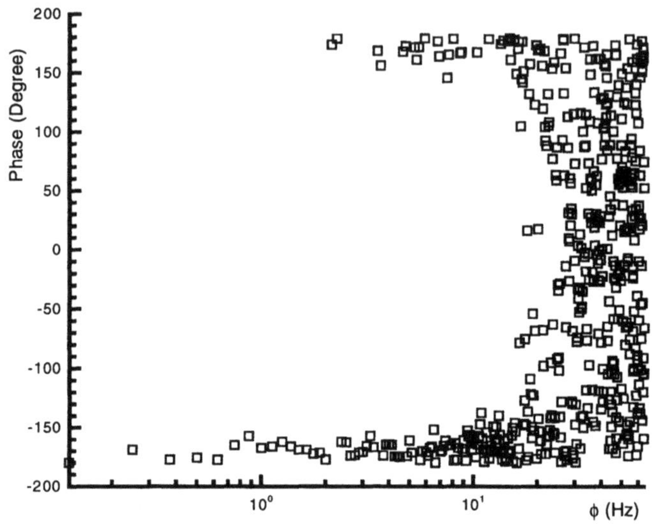

Fig. 12. The phase shift between Γ_{NO} and Γ_{O_3} at $y/\delta = -0.46$.

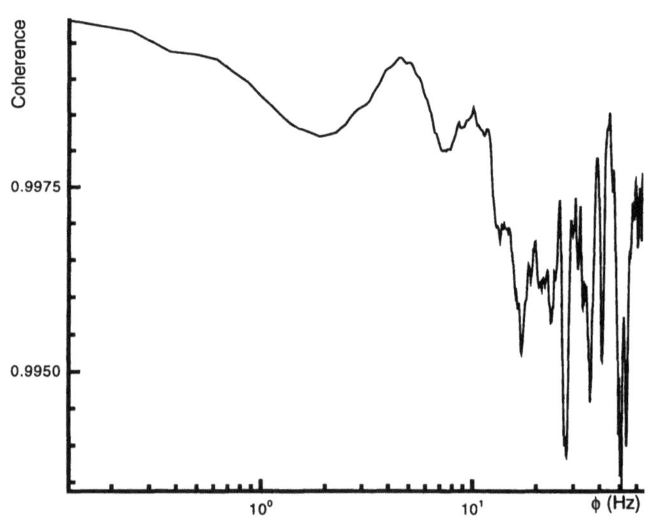

Fig. 13. The coherence between $\Gamma^{e}_{O_3}$ and mixture fraction at $y/\delta = -0.46$.

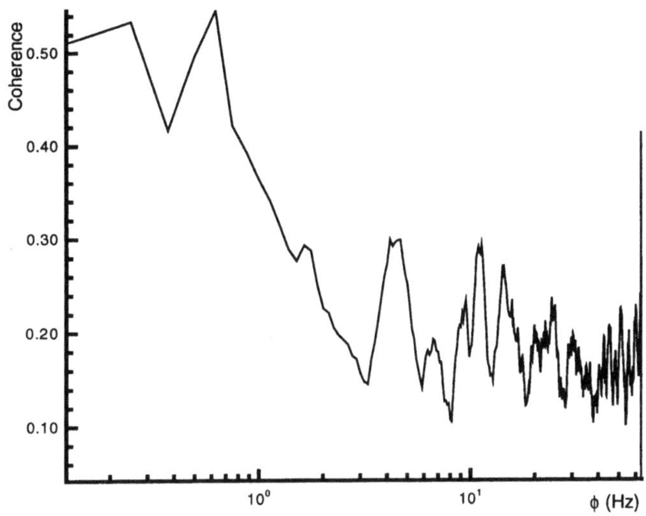

Fig. 14. The coherence between Γ^{e}_{NO} and mixture fraction at $y/\delta = -0.46$.

from variance bearing eddies of small scales irrespective the cross-stream position. This result could be improved by using a new system for measuring the species concentration with higher frequency response and better signal to noise ratio than the existing one. We need a system which not only has an overall high signal to noise ratio but also this high ratio should prevail in the high frequency region. A system based on photon counting with these characteristics is under development. Figure 11 shows the coherence between Γ_{NO} and Γ_{O_3} and figure 12 shows the corresponding phase shift at $y/\delta = -0.46$. The coherence decreases from very low frequency to about 10 Hz, then drops suddenly. The phase shift becomes random at 20 Hz. As explained above we cannot conclude whether this loss of coherence is due to the system noise or is due to the genuine loss of coherence.

Unlike that shown in figure 2, the data in figure 9 show that the spectra for Γ_{NO} and Γ^{e}_{NO} do not collapse onto each other. Figure 13 shows the coherence between $\Gamma^{e}_{O_3}$ and the mixture fraction. It can be seen that the coherence is very high for all the frequencies. The phase shift (not shown here) is 180° for all the frequencies. Figure 14 shows the coherence between Γ^{e}_{NO} and the mixture fraction and it can be seen that the coherence is low and exists only at low frequency. These results show that at y/δ = -0.46, $\Gamma^{e}_{O_3}$ is strongly correlated with the mixture fraction. For Γ^{e}_{NO} this correlation does not exist any more. Since the stoichiometric mixture fraction in the present experiment is 0.52, the conclusion can be drawn from this is that when the local mean mixture fraction is much less than the stoichiometric mixture fraction the correlation between Γ_{NO} and the mixture fraction is weak. By the antisymmetrical nature of the problem, we conclude that when the local mean mixture fraction is much higher than the stoichiometric mixture fraction only weak correlation exists between the mixture fraction and Γ_{O_3}.

4. Conclusions

The experimental results for the spectra of the two chemical species, their equilibrium limits and the mixture fraction in the turbulent reactive scalar mixing layer at two cross stream positions have been presented. The results at the center of the mixing layer show that the two species have strong coherence and this strong coherence prevails in part of the frequency domain. It is surprising that the coherence is still very high even in the frequency range where most of the scalar dissipation occurs. The two species loose their coherence only in the high frequency region. Since the frequency response of the CLA used to measure the species concentration is only 20 Hz we cannot conclude whether this loss of coherence at high frequency is due to the attenuation effect of the CLA or is due to the genuine loss of coherence of the two species. A system with a higher frequency response and better signal to noise ratio is needed to verify this. We need a system which not only has a high signal to noise ratio in the overall sense but also at high frequency. This high frequency probably should correspond to the Kolmogorov frequency. The present experimental results and that of Bilger et al. (1991) do not support the suggestion that the two species lose their coherence at the frequency where the reaction loss balances the spectral transfer. Corrsin (1961, 1964) considers this balance for a first-order chemical reaction. Extension of this idea to the second order reaction is needed. The phase shift between the two species shows similar behavior as the coherence. In general when the coherence is high, the two species are 180° out of phase. When the coherence is low, the phase shift becomes random. On the other hand the coherence between the equilibrium limits of the two chemical species show a high coherence only at very low frequency and this coherence decrease quickly with increasing frequency. At about 10 Hz, where the two species are still strongly correlated, the coherence for the two equilibrium limits drops to the level which corresponds to noise. The present experimental results at the center of the mixing layer support the theory of Kosaly (1992) about the coherence between the equilibrium limits in a reactive homogeneous scalar field, i.e. the loss of coherence for the two equilibrium limits is due to the fact that at high frequency the spectrum for mixture fraction fluctuations decreases faster than the spectrum for its absolute value.

The experimental results at the edge of the scalar mixing layer, where the difference between the local mean mixture fraction and the stoichiometric mixture fraction is large, show that the species which is 'rich' locally correlates well with the mixture fraction. On the other hand the species which is locally 'lean' only correlates weakly with the mixture fraction.

Acknowledgments

This project is supported by the Australian Research Council.

References

Bendat, J.S. and Piersol, A.G. 1980 *Engineering Application of Correlation and Spectral Analysis*. John Wiley and Sons, New York.

Bilger, R.W. 1980 Turbulent flows with nonpremixed reactants. In *Turbulent Reacting Flows* (ed. P. A. Libby and F. A. Williams), pp. 65-113. Springer.

Bilger, R.W., Krishnamoorthy, L.V. and Saetran, L.R. 1989 Use of conditional averaging to obtain biogeochemical reaction rates and stoichiometries under turbulent mixing. *10th Australasian Fluid Mechanics Conf.*, University of Melbourne, Australia, pp 4.33-4.36.

Bilger, R.W., Saetran, L.R. and Krishnamoorthy, L.V. 1991 Reaction in a scalar mixing layer. *J. Fluid Mech.* **233**, 211-242.

Corrsin, S. 1961 The reactant concentration spectrum in turbulent mixing with a first order reaction. *J. Fluid Mech.* **11**, 407-416.

Corrsin, S. 1964 Further consideration of Onsager's cascade model for turbulent spectra. *Phys. Fluids* **7**, 1156-1159.

Kosaly, G. 1993 Auto and Cross-spectra of spectra of reactant fluctuations. *J. Fluid Mech.* **246**, 489-502.

Li, J.D., Brown, R.J. and Bilger, R.W. 1992 Experimental study of a scalarmixing layer using reactive and passive scalars. *11th Australasian Fluid Mechanics Conf.*, The University of Tasmania, 1992, pp. 159-162.

Li, J.D. and Bilger, R.W. 1993 Dissipation of conserved and reactive scalars in a turbulent reactive-scalar mixing layer. Submitted for publication.

Mudford, N.R. and Bilger, R.W. 1983 A facility for the study of nonequilibrium chemistry in an isothermal turbulent flow. *Eighth Australasian Fluid Mechanics Conf.*, University of Newcastle, N.S.W., Australia, 7c

Saetran, L.R., Honnery, D.R., Starner, S.H. and Bilger, R.W. 1989 Scalar mixing layer in grid turbulence with transport of passive and reactive species. In *Turbulent Shear Flow 6* (ed. J.-C. Andre, J. Costeix, F. Durst, B. Launder, F. Schmidt and J. Whitelaw), pp. 109-118. Springer.

Warhaft, Z. and Lumley, J.L. 1978 An experimental study of the decay of temperature fluctuations in grid turbulence. *J. Fluid Mech.*, **88**. 695-684.

Turbulent Mixing and Non-gradient Diffusion in Baffle-stabilized Flames

P. Ferrão and M. V. Heitor

Department of Mechanical Engineering, Instituto Superior Técnico, Technical University of Lisbon, Av. Rovisco Pais, 1096 Lisboa, Codex – Portugal

Abstract

The interaction of turbulence and pressure gradients in baffle-stabilized premixed flames is studied making use of simultaneous measurements of time-resolved velocity and temperature obtained by laser-Doppler velocimetry and laser-Rayleigh scattering. The results reveal the existence of large zone in the flames characterized by non-gradient turbulent scalar fluxes, the extent of which is particularly influenced by the magnitude of the mean pressure gradient associated with the streamline curvature. This is associated with the streamline curvature and is not directly influenced by the flame structure.

1. Introduction

In turbulent flows, temperature-velocity correlations represent turbulent transport of energy and appear in the averaged balance equations for momentum and turbulent stresses (e.g., Heitor et al., 1993). These correlations need model approximations in order to address the closure problem (Jones and Whitelaw, 1985; Peters, 1993), which is experimentally considered in this paper with reference to a recirculating flame. Turbulent mixing in flames is altered by the accompanying heat release and can, as a consequence, be qualitatively different from that occurring in non-reacting flows: examples include counter-gradient diffusion (e.g., Libby and Bray, 1981; Bray *et al.*, 1985) in either confined non-premixed swirling flames, Takagi *et al.* (1985), or in unconfined premixed flames, Heitor *et al.* (1987), and flame generated turbulence, Gunther and Wittmer (1981), Stärner and Bilger (1986). An understanding of mixing in reacting flows is, therefore, technically important since it would provide guidelines for the design of combusting equipments. Also, the strain associated with turbulent mixing may alter the rate of chemical reaction and knowledge of this interaction is important to allow the control and the physical simulation of the practically relevant phenomena of flame extinction and lift-off. This work is aimed at achieving these objectives in laboratory recirculating flames and involves the measurement of the temperature-velocity correlations in baffle-stabilized flames for different combustion regimes.

The work follows that of Ferrão and Heitor (1992), which has assessed the extent to which a laser-Doppler velocimeter, LDV, may be used for the analysis of turbulent heat transfer in strongly sheared disc-stabilized propane air-flames through its combination with laser Rayleigh scattering. The present results quantify the processes of non-gradient diffusion as a function of the magnitude of the adverse pressure gradient to which the flames are subject, and are used to discuss the extent to which the "thin flame" model of burning represents the aerothermochemistry of premixed flames with practical interest.

The remainder of the paper includes five sections. The experimental method and the flames studied are briefly described in the next two sections, respectively. Sample results obtained from measurements of the turbulent fluxes across the reacting layer are then presented and discussed, together with the analysis of the terms in the conservation equations for the scalar fluxes. The main conclusions are summarized in the last section.

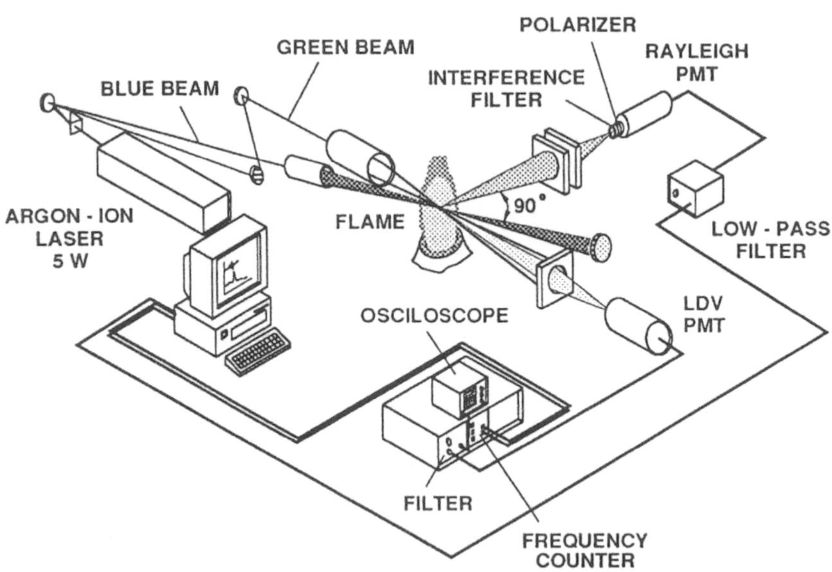

Fig. 1. Schematic diagram of the instrumentation used for the combined LDV/Rayleigh scattering system.

2. Experimental Method

The instrumentation used throughout this work is schematically shown in Fig. 1 and consists in a combined LDV/Rayleigh Scattering system based on a single laser light source (5 W argon-ion laser), the details of which can be found in Ferrão (1993).

The velocimeter was based on the green light (514.5 nm) of the laser and was operated in the dual-beam, forward-scatter mode with sensitivity to the flow direction provided by a rotating diffraction grating. The calculated dimensions of the measuring volume at the e^{-2} intensity locations were 606 μm and 44 μm. The Rayleigh scattering system was operated from the blue line (488 nm) of the same laser source, which was made prior to pass through a 5:1 beam expander. The light converged in a beam waist of 50 μm diameter, and was collected at 90° from the laser beam direction with a magnification of 1 and passed through a slit of 1 mm length. The collected light was filtered by a 1 nm bandwidth interference filter and a polarizer in order to increase the signal to noise ratio.

Rayleigh scattering has been used in a number of turbulent premixed and non-premixed flames, although the Rayleigh scattered power depends on both the total number density and the chemical composition. In the course of this work, a calibration procedure has been implemented through comparison with thermocouple measurements. A linear relationship has been found between the Rayleigh signal and the reciprocal of temperature in agreement with the work of Rajan *et al.* (1984) and Shepherd and Daily (1984).

The spatial resolution of the Rayleigh system in the direction of the temperature gradients is about 50 μm, which is one order of magnitude smaller than the typical flame thickness and, therefore, adequate for the purposes of the work.

The temporal resolution of the system depends on the integration time associated with the active filter used for processing the Rayleigh signals and was quantified to be 50 μs. This value associated with the typical flame propagation velocities found in the present flames (i.e., up to 30 m/s) gives rise to a path length of the same order of magnitude of the laminar flame thickness. Therefore, for the combustion regimes considered here (see below), the system is capable of resolving the temperature fluctuations associated with the passage of the flame front through the measuring volume. This is confirmed by the measured temperature distributions, which include instantaneous values close to either adiabatic or room temperature, which otherwise should be averaged as a consequence of the measurement of burnt and unburnt gases during the integration time.

3. The Flames Studied

The flames are open to the atmosphere and are only partially premixed, but offer the advantage of easy access. The annular bulk velocities at the trailing edge of the disc were varied between 42.4 m/s (Flame A) and 19.5 m/s (Flame B), resulting in Reynolds numbers based on the disc diameter of 1.4×10^5 and 0.7×10^5, respectively. For the experiments reported here the equivalence ratio was kept in the range 0.55–0.60 to allow analysis of the structure of disc-stabilized flames for different combustion regimes, Williams (1985), Peters (1988). Fig. 2 identifies the regimes considered and is important to evaluate the nature of the recirculating flames under analysis here.

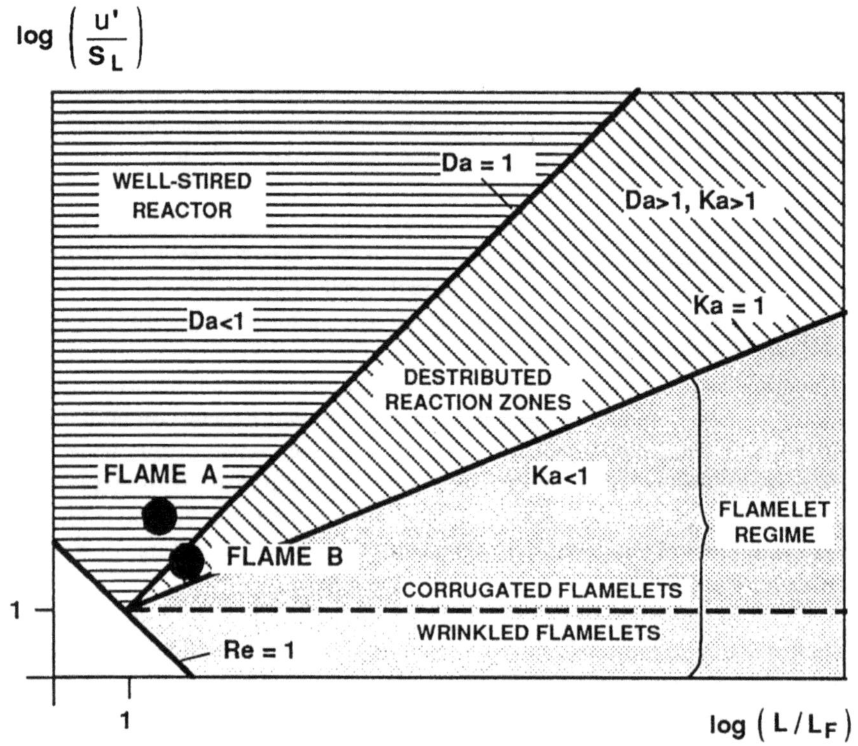

Fig. 2. Regimes of premixed turbulent combustion (e.g., Borghi, 1985) with identification of present operating conditions (see text).

This is because in most situations of practical interest combustion is far from chemical equilibrium and the flames are governed by the local interaction of non-equilibrium chemistry and turbulence. For the particular case of the two flames considered here, analysis of the diagram of Fig. 2 suggests that the thickness of the reaction surfaces, relative to the length scale of the pockets of fully burned reactants and products in the reaction zone, e.g., Bray *et al.* (1985), is not negligibly narrow. Either the flames do not follow the so-called "laminar flamelet model", and exhibit considerably high turbulence levels leading, respectively, the "well-stirred reactor" regime (Flame A) and to flames characterized by a Kolmogorov microscale smaller than the laminar flame thickness (Flame B). The consequent implications of these different regimes for the nature of the aerothermochemistry of baffle stabilized flames is quantified in the following paragraphs.

4. Experimental Results

Previous analysis has shown that the aerodynamics of the present flames is that typical of baffle-stabilized flows and exhibit regions of reverse, accelerating and near uniform velocity. The rear stagnation point, which is associated with a pressure maximum as a result of the converging of streamlines towards the centreline, occurs at $x/D = 2.2$ for $Re = 1.4 \times 10^5$ (Flame A). However, the length of the recirculation zone decreases when the Reynolds number is decreased, with the rear stagnation point located at $x/D = 1.8$ for $Re = 0.7 \times 10^5$ (Flame B). The isotherms are highly curved for the two conditions studied and reveal non-planar flames oblique to the oncoming reactants. For these conditions, the reaction regions occur outside the locus of the mean separation streamlines, which are curved along their length. This curvature varies with the Reynolds number considered and imposes mean velocity effects on the turbulence field in a way which depends on the level of interaction between the gradients of mean pressure typical of the present flames and the associated density fluctuations.

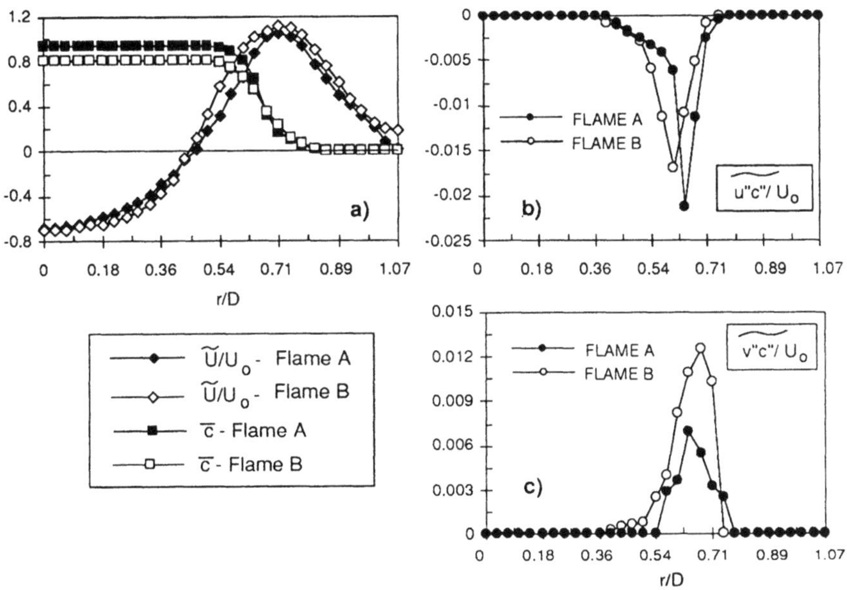

Fig. 3. Radial profiles of velocity and temperature characteristics across the maximum width of the reaction zone for Flame A ($x/D = 1.27$) and Flame B ($x/D = 1.16$).
a) mean axial velocity, $\widetilde{U}/U_0$, and reaction progress variable, $\overline{C}$
b) turbulent axial heat flux, $\widetilde{u''c''}/U_0$
c) turbulent radial heat flux, $\widetilde{v''c''}/V_0$

The detailed velocity and scalar characteristics obtained along the present flames are presented elsewhere. Here attention is focused on the zone which characterizes the maximum width of the recirculation zone of the two flames considered (i.e., $x/D = 1.27$ and 1.16 respectively for Flame A and B) and Fig. 3a) shows the related radial profiles of mean axial velocity and temperature. The latter are represented in terms of the reaction progress variable, $C = (T - T_0)/(T_{ad} - T_0)$, which reaches 0.9 in the centre of flame A and 0.8 in Flame B due to, the comparatively larger amount of entrained cold air for the conditions characterized by the lowest Reynolds number. The maximum non-dimensional reverse velocity is equal to $-0.70\ U_0$ for the two flames considered and, if we consider the dimensional values, this shows the comparatively larger adverse pressure gradient to which Flame A is subject.

Figures 3b) and c) show measured values of turbulent axial and radial heat fluxes respectively, which are restricted to the thin zone along the shear layer where the radial gradients of mean temperature is large. These quantities represent the exchange rate of reactants responsible for the phenomena of flame stabilization and the results show that the turbulent axial fluxes, Fig. 3b), are higher than the radial values, Fig. 3c), with increased differences for Flame A. In particular for this condition, the turbulent heat transfer is essentially directed along the isotherms, rather than normal to them, as would be expected from gradient-transport models of the kind used in non-reacting flows. It should be noted that the radial fluxes are always positive, as expected in a recirculating flame. The flame is then established by the heat transfer between the hot products and the cold reactants with the sign of the radial heat flux in qualitative agreement with gradient-transport models $\left(\widetilde{v''c''} = -\nu_c \partial\widetilde{C}/\partial r\right)$, where ν_c, is a scalar turbulent thermal diffusivity. Similar behaviour has been observed in other turbulent premixed flames (Heitor *et al.*, 1987) and has also been predicted analytically (Bray *et al.*, 1985), and is expected to be due to the interaction between the gradients of mean pressure typical of the present flow and the large density fluctuations that occur in the flames. However, it should be noted that the "non-gradient" nature of the transport of heat is considerably attenuated with the decrease in the Reynolds number and this is justified below.

Although the radial distributions of time-averaged temperature shown in Fig. 3 for the two conditions studied here are qualitatively similar, analysis of Fig. 2 suggests that the structure of the flames is different and the related experimental evidence is provided by the joint probability-density functions of temperature and velocity, Fig. 4.

While Flame B is characterized by near bimodal temperature distributions across the reacting shear layer with significant probability for intermediate products, the distributions obtained in Flame A exhibit typically the near-Gaußian form characteristic of the "well-stirred" combustion regime. It is clear that the absolute values of the probability distributions are influenced by the shot noise characteristics of the photomultiplier used for the Rayleigh scattering measurements, but the qualitative trends shown in Fig. 4 are not affected by experimental accuracy. The burning conditions associated with Flame B are characteristic of a regime where the flame fronts become thick and bigger than the Kolmogorov length scale. Local extinction in the inner reaction layer is expected to occur due to straining caused by the small eddies. The probability of the occurrence of intermediate products correspond, therefore, to the ratio of the volume of the instantaneous flame fronts relative to the total reacting volume, which grows by increasing the Reynolds number.

The turbulence levels associated with the conditions of Flame A lead to turbulent time scales smaller than the chemical time scales and, therefore, the reacting shear layer is characterized by large temperature and velocity fluctuations with reaction

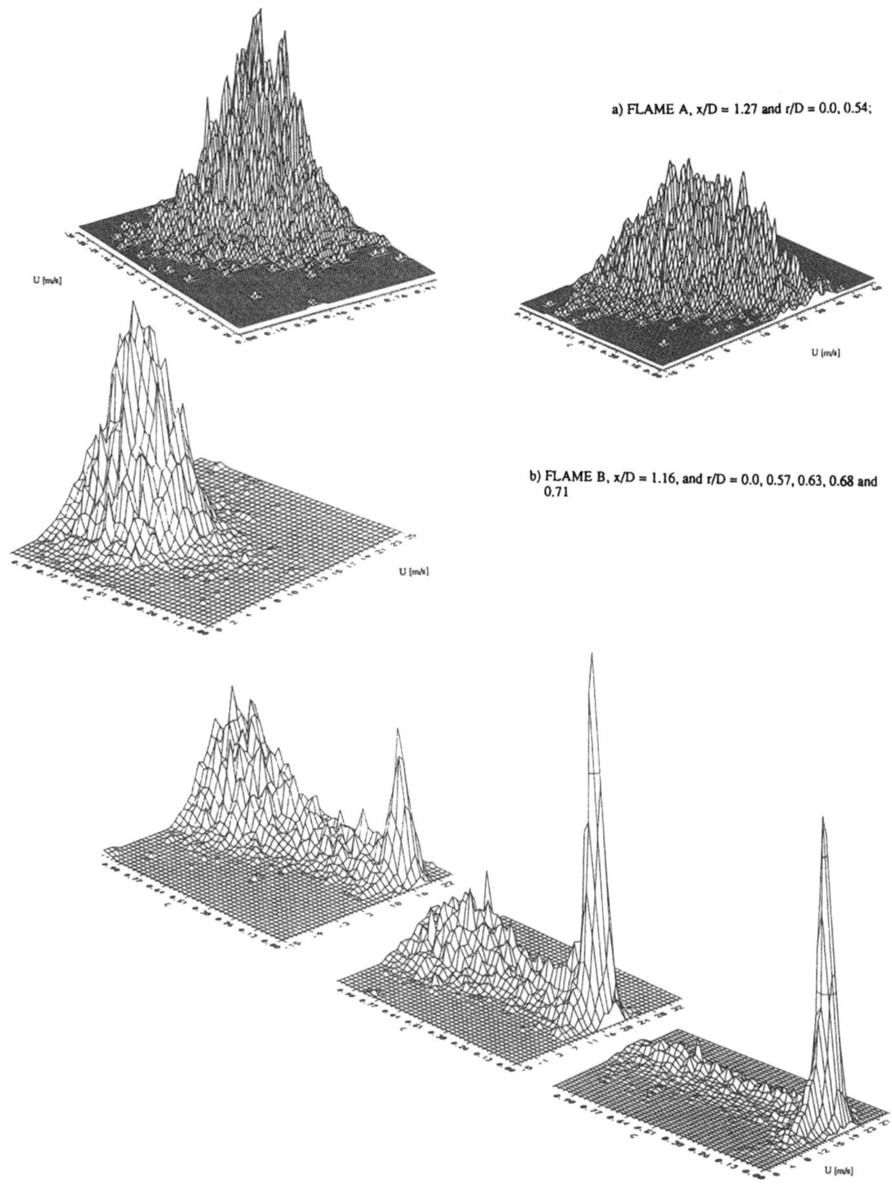

Fig. 4. Joint probability density functions of axial velocity and progress reaction variable across the reacting shear layer for two different operating conditions.
a) Flame A, $x/D = 1.27$ and $r/D = 0.0, 0.54$;
b) Flame B, $x/D = 1.16$, and $r/D = 0.0, 0.57, 0.63, 0.68$ and 0.71.

rates varying from zero (i.e., local extinction) to typical laminar values, Borghi (1988). It should be noted that the near-Gaußian distributions of Fig. 4a) exhibit values between ambient ($C = 0$) and adiabatic ($C = 1$) temperatures and, therefore, are not expected to be influenced by lack of temporal resolution of the Rayleigh system.

The results of Fig. 4 also show that, in general, lower velocities are associated with higher temperature values characteristic of the hot products of combustion. Again, this observation agrees with that of Heitor *et al.* (1987) and confirms the evidence of the preferential deceleration of the light products of combustion relative to the cold reactants due to the interaction between the mean adverse pressure gradient typical of the present recirculating flow and the large density fluctuations which characterize both flames.

5. Discussion

The previous section has shown that the turbulent heat flux vectors in premixed recirculating flames are directed along the isotherms, rather than normal to them, in a way which is particularly influenced by the magnitude of the mean adverse pressure gradient to which the flames are subject to. In addition, if the present results are transformed into streamline co-ordinates, the net turbulent scalar transport occurs mainly along the streamlines as a consequence of the highly oblique flames considered. The results suggest that this "non-gradient" nature of heat transport is not directly influenced by the flame structure and occurs even in the "well-stirred" combusting regime.

The axial gradients of $\widetilde{u''c''}$ are substantially smaller than the radial gradients of $\widetilde{v''c''}$, and hence the turbulent transport of heat is principally in the radial and not in the axial direction. However, the effect of the radial positive mean pressure gradient does not prevail in the momentum balance, since it has not preferentially deflected the light products towards the center of the flame. In general, the analysis suggests that gradients of mean pressure can affect the transport of turbulent heat flux in premixed flames and this implies the use of modeled transport equations, rather than the turbulent viscosity hypothesis, if the values of $\widetilde{u_i''c''}$ are to be calculated.

A remarkable observation is that the non-dimensional radial heat flux increases for Flame B and, therefore, the extent to which the non-gradient diffusion of heat influences the flame decreases by decreasing the Reynolds number. This is a direct consequence of the magnitude of the turbulence/combustion interactions and is explained through the relative importance of the extra terms in the equations for the transport of the scalar fluxes, which involve the mean pressure gradient and can act as sources or sinks of the rate of heat transfer.

The relative influence of the extra terms in the equations for the transport of the turbulent axial and radial heat fluxes is represented in Figs. 5 and 6, respectively. The results are the clear evidence that these terms dominate the conservation of axial heat flux (i.e., $-\overline{C''}\partial\overline{P}/\partial x$), with an increased influence for Flame A.

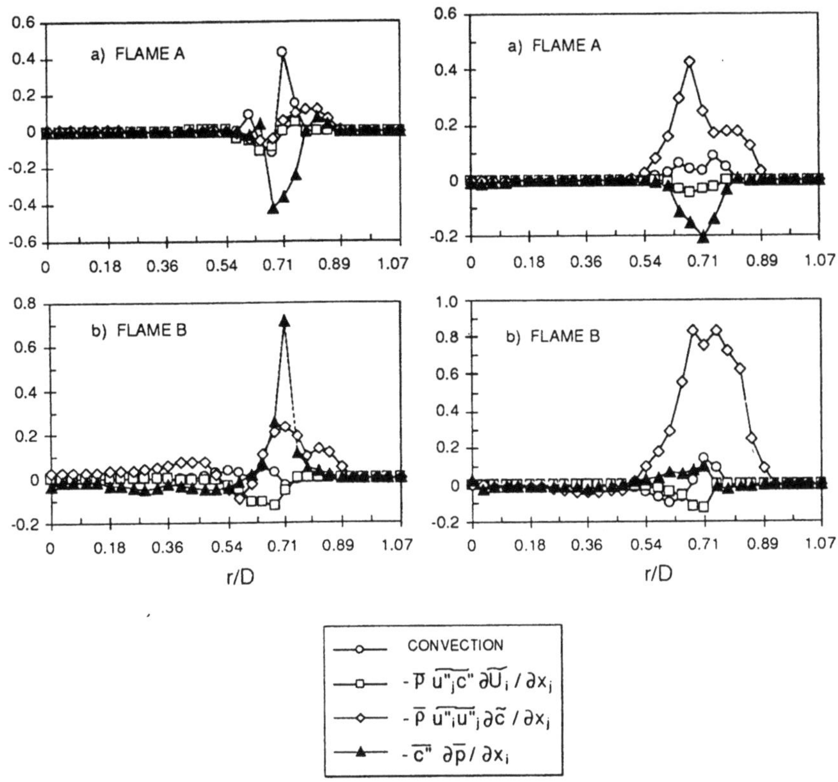

Fig. 5. Radial profiles of the convection and production terms in the conservation equation for the axial turbulent heat flux:
a) Flame A; $x/D = 1.27$;
b) Flame B; $x/D = 1.16$.

Fig. 6. Radial profiles of the convection and production terms in the conservation equation for the radial turbulent heat flux:
a) Flame A; $x/D = 1.27$;
b) Flame B; $x/D = 1.16$.

On the other hand, for the transport of radial flux in Flame A the extra source term is similar in magnitude to production through the interaction of Reynolds stresses with the spatial gradient of the mean value of the reaction progress variable, but is of the opposite sign. The turbulent heat flux results from a balance between two large, but competing, source terms. The observation is strongly dependent on the absolute value of the pressure gradient and for Flame B the extra source term in the conservation of radial flux is negligible.

A practical implication of the previous analysis is the downstream deflection of the turbulent heat flux vectors in recirculating flames, which influences the mechanism of flame stabilization and depends upon the magnitude of the adverse pressure gradients. The conditions for which the flame extinguishes are probably associated with values of $\widetilde{v''c''}$, the representing counter-gradient heat transfer into the recirculation zone. In general, the present results bring new experimental evidence of the magnitude of turbulence/combustion interactions in strongly sheared flames, in that the mean streamline directions are considerably affected by the Reynolds number of the flow considered and, consequently, the magnitude of the terms involving the mean pressure gradients in the conservation of the turbulent heat fluxes.

6. Conclusions

Simultaneous measurements of time-resolved velocity and temperature have been obtained by laser-Doppler velocimetry and laser-Rayleigh scattering in the near wake of premixed recirculating flames of propane and air. The experiments encompass two different flow conditions associated with different combustion regimes and Reynolds numbers. The results show that the interaction between gradients of mean pressure and the large density fluctuations that occur in these flames may represent the largest source terms in the transport equations for the turbulent heat fluxes, in a way which is particularly influenced by the magnitude of the mean adverse pressure gradient. These observations are related to the streamline curvature and are shown to occur even for a flame structure characterized by near-Gaußian probability distributions of the temperature fluctuations, which are typical of the"well-stirred" regime of combustion.

Acknowledgements

The assistance of Messrs. Carlos Carvalho and Jorge Coelho in the preparation of this paper is gratefully appreciated. Financial support has been provided by the SCIENCE programme of the DGXII of the Commission of the European Communities.

References

Borghi, R. 1985. Recent Advances in the Aerospace Sciences, Corrado Casci, Eds., Plennum Press Corporation, 117–138.

Borghi, R. 1988. Prog. Energy Combust. Scie. **14**, 245–292.

Bray, K.N.C., Libby, P.A. and Moss, J.H. 1985. Comb. Flame **61**, 87–102.

Dibble, R.W. and Hollenbach, R.E. 1981. *18th Symp. (Intl) on Combustion*, The Combustion Institute, 1489–1499.

Ferrão, P. and Heitor, M.V. 1992-a. Proc. 6th Intl. Symp. on Appl. of Laser Tech. to Fluid Mechanics, Lisbon, 20–23 July 1992.

Ferrão, P. and Heitor, M.V. 1992-b. Combusting Flow Diagnostics (Eds. D.F.G. Durão *et al.*). Kluwer Academic Publ. Nato/ASI series, 169–231.

Gunther, R. and Wittmer, V. 1981. Eighteenth Symposium (International) on Combustion, 961, The Combustion Institute.

Ferrão, P. 1993. Ph.D. Thesis, Technical University of Lisbon (In Portuguese).

Heitor, M.V., Stärner, S.H., Taylor, A.M.K.P. and Whitelaw, J.H. 1993. *Instrumentation for flows with combustion.* (Ed. A.M.K.P. Taylor), Academic Press, 113–250.

Heitor, M.V., Taylor, A.M.K.P. and Whitelaw, J.H. 1987. J. Fluid Mechanics **181**, 387–413.

Jones, W.P. and Whitelaw, J.H. 1985. *20th Symp. (Intl.) on Combustion*, The Combustion Institute, invited paper.
Libby, P. and Bray, K.N.C. 1981. AIAA J. **19**, 205–213.
Namer, I. and Schefer, R.W. 1985. Exp. in Fluids **3**, 1–9.
Peters, N. 1988. *21st Symp. (Intl.) on Combustion*, The Combustion Institute, 1231–1256.
Peters, N. 1993. Proc. Joint Meeting of the British and German Sections of the Combustion Institute, 26–33, 29 March–2 April.
Rajan, S., Smith, J.R. and Rambach, G.D. 1984. Comb. Flame **57**, 95–107.
Shepherd, I.G. and Daily, J.W. 1984, The Combustion Institute, Western States (USA), W98/Cl, paper № 84-15.
Stärner, S.H. and Bilger, R.W. 1986. *21st Symp. (Intl.) on Combustion*, The Combustion Institute, 1569–1577.
Takagi, T., Okamoto, T., Taji, M. and Nakasuji, Y. 1985. *20th Symp. (Intl.) on Combustion*, The Combustion Institute, 251–258.
Williams, F.A. 1985. *Combustion Theory*, The Benjamin / Cummings Publ. Conf.

Turbulent Premixed Flame Propagation Revisited Results with a New Model

T. Mantel[1], R. Borghi[2], and A. Picart[2]

[1] Renault, Dir. de la recherche, 9-11 Av. du 18 Juin 1940
92500 Rueil Malmaison, France
[2] Université de Rouen, U.R.A. C.N.R.S. 230, CORIA
76821 Mont-Saint-Aignan, Cedex, France

Abstract

Numerical simulations of the propagation of a one-dimensional turbulent flame in a premixed medium with constant density and given turbulence properties are presented. The key points of this academic problem are to propose realistic models for the mean reaction rate and for the turbulent flux which are of equal importance in this kind of problem. In the case of very fast chemistry, the mean reaction rate and the surface flame density are directly proportional to the dissipation rate of the progress variable c. The model proposed here for the mean reaction rate is based on the transport equation for the dissipation rate. Concerning the turbulent flux of c, the problem is treated in different ways. The transport equation for the correlation $\overline{u'c'}$ (for which a new modelling is proposed) is studied and is compared with the classical closure which consists of using a turbulent diffusion coefficient. The phenomenon of turbulent diffusion is also represented by random walks of fluid particles using a Monte Carlo simulation. The importance of the turbulent flux is principally evident in the transient period of the flame propagation and more on the "flame brush thickness" than on the propagation velocity. Comparisons of our simulation with experimental results concerning the turbulent flame velocity show a satisfying behaviour of the model.

1. Introduction

The propagation of a one dimensional turbulent flame in a premixed medium is a very classical problem in the field of turbulent combustion. However, the complete solution is not known and the approximate models proposed so far do not address all the aspects of the problem. The first studies concerning this flame were dealing only with the prediction of the velocity of propagation (Damköhler (1948), Schelkin (1956)). However, the existence of an intrinsical "turbulent speed of propagation" is not yet theoretically demonstrated. Many formulas have been proposed, and numerous experimental results often contradictory have been published. On the other hand, the mean thickness of the flame, often called the "flame brush thickness", has been recognized as very large, as compared to the laminar flame thickness, and almost no theoretical approach allows its calculation (with the exception of Talantov *et al.* (1969)). It can be also seen in experiments that this thickness evolves when the flame propagates from the ignition point, or similarly, when we are considering several cross sections, downstream from the flame holder, of one stabilized oblique flame. This can be clearly seen in numerical experiments, from direct numerical simulations now available (for instance El Tahry *et al.* (1991)), or from laser sheet visualizations in Bunsen burners (for instance Dumont *et al.* (1992)). The numerical simulations of Saïd and Borghi (1988) emphasize also this feature, which explain probably the experimental scatter concerning turbulent flames measurements, at least partly.

Premixed turbulent flames display different structures depending on the values of the ratios $\frac{u'}{U_L}$ and $\frac{l_t}{\delta_L}$, where u' and l_t are characteristic turbulent velocity and integral length scale, U_L and δ_L are laminar flame speed and laminar flame thickness.

We are interested here in a calculation model that could describe the mean properties of the flame, including the turbulent propagation speed, but not only this; also the mean thickness of the turbulent flame, the distribution of mean temperature (for instance) across the flame brush, and eventually, the time distribution from ignition to a steady propagation are of interest. These problems can be addressed using balance equations for statistical mean values of the relevant aerothermochemical variables, together with models for the turbulent diffusion fluxes and the mean reaction rate. There are at least three major problems for getting such equations:

i the modelling of the mean reaction rate
ii the modelling of the diffusion fluxes
iii and the modelling of the turbulence itself, including the so-called "flame generated turbulence".

We consider here only the two first problem. For the first step, although the flame generated turbulence is a very important issue, our model is a constant density model and we will assume the turbulence properties as given quantities. This renders the problem almost entirely academic, but allows to separate the effects from different causes.

Here, only the case of very fast chemistry is studied, i.e. the Damköhler number $\left(\text{defined by } Da = \frac{U_L l_t}{\delta_L u'}\right)$ is very large. As the chemistry is supposed to be very fast with respect to any aerodynamic time, the flame brush appears instantaneously as wrinkled flamelets. In the regime of "wrinkled flames", the characteristic turbulence velocity u' may be comparable to the laminar flame speed U_L, or larger than U_L (but not too large if we want to keep the flame as "wrinkled").

For such a flame, the stretching of the flamelets has been proposed as a very important phenomenon, in particular by Bradley and the Leeds group (Abdel-Gayed *et al.* (1984-1987)). But stretching leads to local extinctions of flamelets when the chemical time scale is not too small with respect to the Kolmogorov time scale, which represents a "local stretching time scale" in a turbulent flow, and this occurs only at the boundary of the domain of the "wrinkled flames", when the chemistry is not fast enough.

A more important phenomenon in our case of very fast chemistry is the curvature of the flamelet surface, provoked by the turbulent movements but strongly influenced by the propagation speed of the laminar flamelet elements. The total flamelet surface area in a given volume is controlled by such a phenomenon, and the ratio $\frac{u'}{U_L}$ is the relevant non-dimensional parameter that has to control it. The curvature of flamelet surface, depending on this parameter, can be readily seen on experiments (Mantzaras *et al.* (1989)), or in numerical simulation (Saïd and Borghi (1988)).

For wrinkled flames, it is usual to define a "progress variable" (as in Bray and Moss (1977)) to which all the aerothermochemical variables are related. In the case of adiabaticity, low velocity and Lewis number equal to one, it can be shown that this is justified.

The model that we propose here is based on a balance equation for $\overline{\varepsilon_c}$, the dissipation rate of this progress variable c, as it has been first proposed in Mantel and Borghi (1991). The dissipation rate $\overline{\varepsilon_c}$ is directly related to the mean reaction rate $\overline{w}$ (see Bray and Moss (1977)) and to the mean flamelet surface area per unit of volume Σ (see Borghi (1990)) in the regime of wrinkled flames. The model is improved here by studying the diffusion flux of the progress variable $\overline{u_i'c'}$. In order to incorporate in the modelling the influence of combustion, an equation for $\overline{u_i'c'}$ with a new modelling is studied. An attempt to built a Lagrangian model for the turbulent diffusion is also presented.

2. The Theoretical Model with the Scalar Dissipation Equation

2.1 The Scalar Dissipation Equation

As it is said in the introduction, most of the recent developments in turbulent combustion modelling are led in the so-called flamelet regime. The goal of these models is to give a realistic expression to the mean reaction rate. Bray and co-workers (see for instance Libby and Bray (1980), Bray (1985), Bray and Libby (1986)) express the instantaneous reaction rate by mean of a library of flamelet and a frequency which describes the local interaction with the turbulent flow field. In another way, the mean reaction rate can be directly connected to the surface flame density Σ. Σ can be given by the transport equation of the coherent flame model initially proposed by Marble and Broadwell (1977) and developed elsewhere (Candel and Poinsot (1990), Maistret *et al.* (1989), Pope (1988), El Thary *et al.* (1991)).

As Bray and Moss (1977) have shown for $\overline{w}$ and Borghi (1990) for Σ, these two quantities are directly linked to the dissipation rate of the progress variable $\overline{\varepsilon_c}$. Usually, this quantity is assumed to be proportional to a characteristic frequency of the turbulence $\overline{\varepsilon_c} \propto \frac{\overline{c'^2}}{\tau_t}$. However, even in the case of a non reactive medium, experiments have shown that this assumption is not always satisfactory. This is highly doubtful for a reactive scalar as it is shown in Picart *et al.* (1987). Another possibility is to use a transport equation for $\overline{\varepsilon_c}$ as it was proposed by Zeman and Lumley (1976) for a non reactive scalar. In a recent study, we have proposed to calculate $\overline{\varepsilon_c}$ directly by using its transport equation in the case of a species where c is the reaction progress variable (see Mantel and Borghi (1991)). In this equation, closures are proposed in order to take into account the presence of flamelets and their propagation in the fluid. A qualitative application of this model to the case of the propagation of a spherical flame has shown its capability to represent this kind of problem.

The exact equation for $\overline{\varepsilon_c}$ is (with constant density and constant diffusivity):

$$\begin{array}{cccccc} \dfrac{\partial\overline{\varepsilon_c}}{\partial t} + \overline{U_k}\dfrac{\partial\overline{\varepsilon_c}}{\partial x_k} = & -\dfrac{\partial\overline{u'_k\varepsilon_c}}{\partial x_k} + d\dfrac{\partial^2\overline{\varepsilon_c}}{\partial x_k^2} & - & 2d\overline{c'_{,k}c'_{,i}}\dfrac{\partial\overline{U_k}}{\partial x_i} & - 2d\,\overline{c_{,k}}\,\overline{c'_{,i}\dfrac{\partial u'_k}{\partial x_i}} & \\ (I) \quad (II) & (III) \quad (IV) & & (V) & (VI) & \\ & -2d\overline{u'_kc'_{,i}}\dfrac{\partial\overline{c_{,i}}}{\partial x_k} - 2d\overline{c'_{,k}c'_{,i}\dfrac{\partial u'_k}{\partial x_i}} & - & 2d^2\overline{\dfrac{\partial c'_{,i}\partial c'_{,i}}{\partial x_k\partial x_k}} & - 2d\overline{c'_{,i}\dfrac{\partial \dot{w}'}{\partial x_i}} & (1) \\ & (VI-bis) \quad (VII) & & (VIII) & (IX) & \end{array}$$

In the case of high Reynolds numbers and Damköhler numbers, it has been shown in Mantel and Borghi (1991) that the reaction term (IX) is compensated by some of the components of the second derivative term (VIII). On the other hand, it has been necessary to model the two terms (VII) and (VIII) separately contrarily to the usual practice, because combustion plays differently on (VII) and (VIII). Indeed, with the model of Zeman and Lumley (1976), it is not possible to get the propagation of a flame (see Mantel and Borghi (1991)).

The tentative model equation is:

$$\begin{aligned} \frac{\partial\overline{\varepsilon_c}}{\partial t} + \left(\overline{U_k} - \overline{U_{L_k}}\right)\frac{\partial\overline{\varepsilon_c}}{\partial x_k} &= \frac{\partial}{\partial x_k}\left(\frac{\nu_t}{\sigma_{\varepsilon_c}}\frac{\partial\overline{\varepsilon_c}}{\partial x_k}\right) + C_{pU}\frac{\overline{\varepsilon_c}}{\overline{k}}\frac{\nu_t}{\sigma_t}\frac{\partial\overline{U_k}}{\partial x_i}\frac{\partial\overline{U_k}}{\partial x_i} \\ +C_{p_c}\frac{\overline{\varepsilon}}{\overline{k}}\frac{\nu_t}{\sigma_c}\frac{\partial\overline{c}}{\partial x_k}\frac{\partial\overline{c}}{\partial x_k} &+ \alpha_0 Re_t^{1/2}\overline{\varepsilon_c}\frac{\overline{\varepsilon}}{\overline{k}} - \beta_0 Re_t^{1/2}\frac{\overline{\varepsilon_c}^2}{\overline{c'^2}}\left(1 + C_u\frac{U_L}{u'}\right)^{-2\gamma} \end{aligned} \tag{2}$$

This equation is written assuming that the turbulence is given with a $k-\varepsilon$ model, and so $\nu_t = c_\mu\dfrac{\overline{k}^2}{\overline{\varepsilon}}$ and $Re_t = \dfrac{\overline{k}^2}{\nu\overline{\varepsilon}}$. $\sigma_{\varepsilon_{ct}}$, σ_t, C_{pU}, C_{p_c}, α_0, β_0, C_u and γ are modelling constants. The terms proportional to $\dfrac{\partial\overline{U_k}}{\partial x_i}$ and $\dfrac{\partial\overline{c}}{\partial x_k}$ are classical terms. Their meaning can be discussed (see Antonia and Browne (1983) and Mantel and Borghi (1991)), but they will be of very little importance here. The two last terms which correspond to the terms (VII) and (VIII) in equation (1) represent respectively the production of local gradients by instantaneous strain and the dissipation due to the local curvature of the flamelets, depending of the parameter $\dfrac{u'}{U_L}$.

The quantity $\overline{U_{L_k}}$ is the mean value of the projection of the instantaneous velocities of propagation of the flamelets in the direction of the k axis. So, it depends on the distribution of the orientations of the flamelets. For this term, a closure is needed, and:

$$\overline{U_{L_k}} = \frac{U_L}{1 + C_0 \dfrac{u'}{U_L}} \frac{\overline{c_{,k}}}{\left(\overline{c_{,j}}\,\overline{c_{,j}}\right)^{1/2}} \tag{3}$$

is proposed in Mantel and Borghi (1991), because the flamelets orientation are thought to be isotropically distributed for $u' \gg U_L$. The constant C_0 has to be determined precisely. This term is negligible for large $\dfrac{u'}{U_L}$ and acts probably only when $u' < U_L$. However, if $u' \ll U_L$, the equation (2) may be not well suited because the Reynolds number is not large and this point has to be discussed further.

We will use here the following set of constants as follows: $\sigma_{\varepsilon_c} = 1.0$, $\sigma_t = 1.0$, $C_{p_U} = 1.0$, $C_{p_c} = 1.0$, $\alpha_0 = 0.9$, $\beta_0 = 1.25$, $C_u = 1.0$ and $\gamma = 0.5$.

2.2 The Turbulent Flux

For the turbulent flux of c, we propose to use the transport equation for $\overline{u_i'c'}$ which is written before any modelling (with constant density):

$$\frac{\partial \overline{u_i'c'}}{\partial t} + \frac{\partial}{\partial x_j}\overline{U_j}\,\overline{u_i'c'} = -\frac{\partial}{\partial x_j}\left(\overline{u_i'u_j'c'}\right) - \overline{u_i'u_j'}\frac{\partial \overline{c}}{\partial x_j} - \overline{u_i'c'}\frac{\partial \overline{U_i}}{\partial x_j} + \overline{\frac{c'}{\overline{\varrho}}\frac{\partial p}{\partial x_i}} + \overline{u_i'\dot{w}'} \tag{4}$$

The first term of the RHS of (4) represents a turbulent diffusion flux of $\overline{u_i'c'}$ and is classically closed using a turbulent diffusion coefficient:

$$\overline{u_i'u_j'c'} = -\frac{\nu_t}{\sigma_c}\frac{\partial \overline{u_i'c'}}{\partial x_j} \tag{5}$$

The fourth term of the RHS in which appears the pressure is known to be responsible of the dissipation of $\overline{u_i'c'}$. Classical models close this term using a characteristic time scale of turbulence. Here, we proposed to introduce a characteristic time scale of the scalar field τ_c which can be defined from $\overline{\varepsilon_c}$ and $\overline{c'^2}$:

$$\overline{\frac{c'}{\overline{\varrho}}\frac{\partial p}{\partial x_i}} = -C_D \overline{u_i'c'}\frac{\overline{\varepsilon_c}}{\overline{c'^2}} \tag{6}$$

where C_D is a constant which must be determined by comparison with experimental results.

The last term of (4) due to chemical reactions has been studied by Borghi and Dutoya (1978). The authors have deduced from the study of Dopazo and O'Brien (1974) a linear dependence between the conditional probability of U with c fixed and the value of c itself. Finally, this term can be written:

$$\overline{u_i'\dot{w}'} = \overline{u_i'c'}\,b_0\frac{\overline{\dot{w}}}{\overline{c'^2}} \tag{7}$$

where b_0 depends on the instantaneous chemical law $\dot{w}(c)$:

$$b_0 = \frac{\int_0^1 c\dot{w}(c)dc}{\int_0^c \dot{w}(c)dc} \tag{8}$$

(here $b_0 \simeq 0.2$)

Then, the modelled equation for $\overline{u_i'c'}$ can be expressed by:

$$\begin{aligned} \frac{\partial \overline{u_i'c'}}{\partial t} + \frac{\partial}{\partial x_j}\overline{U_j}\overline{u_i'c'} &= \frac{\partial}{\partial x_j}\left(d_t \frac{\partial \overline{u_i'c'}}{\partial x_j}\right) - \overline{u_i'u_j'}\frac{\partial \overline{c}}{\partial x_j} - \overline{u_i'c'}\frac{\partial \overline{U_i}}{\partial x_j} \\ &\quad - C_D \frac{\overline{u_i'c'}}{\tau_c} + \overline{u_i'c'} b_0 \frac{\overline{\dot{w}}}{\overline{\overline{c'^2}}} \end{aligned} \tag{9}$$

The term $\overline{\frac{c'}{\overline{\varrho}} \frac{\partial p}{\partial x_i}}$ which appears in the equation (4) influences directly the characteristics of the turbulent flame, its velocity U_T and its thickness δ_T. Thus, the constant C_D can be fitted on experimental results available in the literature. We will see that 2.2 is a convenient value for C_D.

3. Tests of the Model

3.1 The Set of Equations

One can write the global set of equations to be solved for a turbulent flame statistically one dimensional, in a given turbulent field and without convection:

$$\frac{\partial \overline{c}}{\partial t} = -\frac{\partial \overline{u_i'c'}}{\partial x} + \overline{\dot{w}} \tag{10}$$

$$\begin{aligned} \frac{\partial \overline{\dot{w}}}{\partial t} &= \frac{\nu_t}{\sigma_c}\frac{\partial^2 \overline{\dot{w}}}{\partial x^2} - \frac{C_{P_Y}}{\frac{1}{2} - b_0}\frac{\overline{u_i'c'}}{\overline{k}}\frac{\partial \overline{c}}{\partial x} + \alpha_0 Re_t^{1/2} \overline{\dot{w}} \frac{\overline{\varepsilon}}{\overline{k}} \\ &\quad - \beta_0 Re_t^{1/2}\left(\frac{1}{2} - b_0\right)\frac{\overline{\dot{w}}^2}{\overline{c}(1-\overline{c})}\left(1 + C_u \frac{U_L}{u'}\right)^{-2\gamma} \end{aligned} \tag{11}$$

In fact, this last equation is obtained from the equation for $\overline{\varepsilon_c}$ by noting from Bray and Moss (1977) and from Borghi and Dutoya (1978) that $\overline{\dot{w}}$ and $\overline{\varepsilon_c}$ are linked by:

$$\overline{\dot{w}} = \frac{\overline{\varepsilon_c}}{\frac{1}{2} - b_0} \tag{12}$$

(we can use an equation for $\overline{\dot{w}}$ or for $\overline{\varepsilon_c}$ as well).

For the turbulent flux of c, we propose to compare the two following models:

$$\overline{u_i'c'} = -\frac{\nu_t}{\sigma_c}\frac{\partial \overline{c}}{\partial x_i} \tag{13}$$

$$\frac{\partial \overline{u_i'c'}}{\partial t} = \frac{\nu_t}{\sigma_c}\frac{\partial^2 \overline{u_i'c'}}{\partial x^2} - \overline{k}\frac{\partial \overline{c}}{\partial x} - \overline{u_i'c'}\frac{\overline{\dot{w}}\left(\frac{1}{2} - b_0\right)}{\overline{c}(1-\overline{c})} + \overline{u_i'c'}b_0\frac{\overline{\dot{w}}}{\overline{c'^2}} \tag{14}$$

We prefer to solve the unsteady equations with no velocity of the gases instead of steady equations with the velocity of the gases as an eigenvalue for two reasons: firstly, it may be expected some problem of non-uniqueness of the eigenvalue (see Hakberg and Gosman (1984)) and secondly, we are interested by the transient behaviour before the steady propagation.

The characteristics of the turbulence ν_t and Re_t are taken as constants.

3.2 Results on the Structure of the Flame Brush

The figure 1 shows the profiles of $1-\overline{c}$ during the propagation of the flame. Figure 2 shows the profiles of $\overline{u'c'}$ and figure 3 the profiles of $\overline{\varepsilon_c}$. The numerical values are

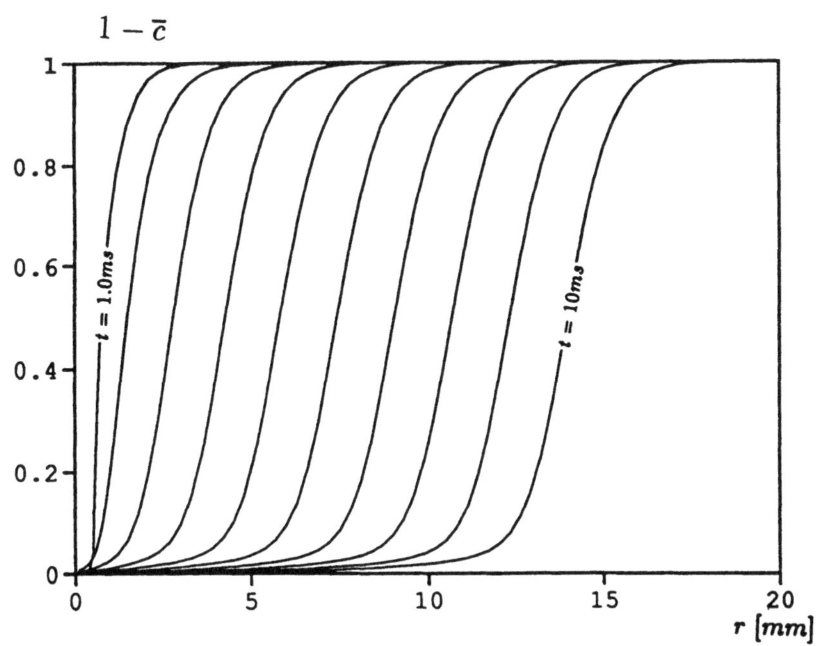

Fig. 1. Profiles of $1-\overline{c}$ during the propagation of the flame (1 ms between each profile).

$u' = 1$ m/s, $l_t = 1$ mm for a mixture of propane and air at equivalence ratio 0.9. That gives $\varepsilon = 1.84\ 10^{-3}$ m^{2}s^{-3} and $\frac{u'}{U_L} = 3.37$.

Ignition is produced by forcing the reaction rate during 1 ms; the ignition period itself is not of interest here, and we see that the propagation becomes steady after some 4 ms. The results concerning the velocity of propagation are insensitive to the exact ignition rate after 1 ms.

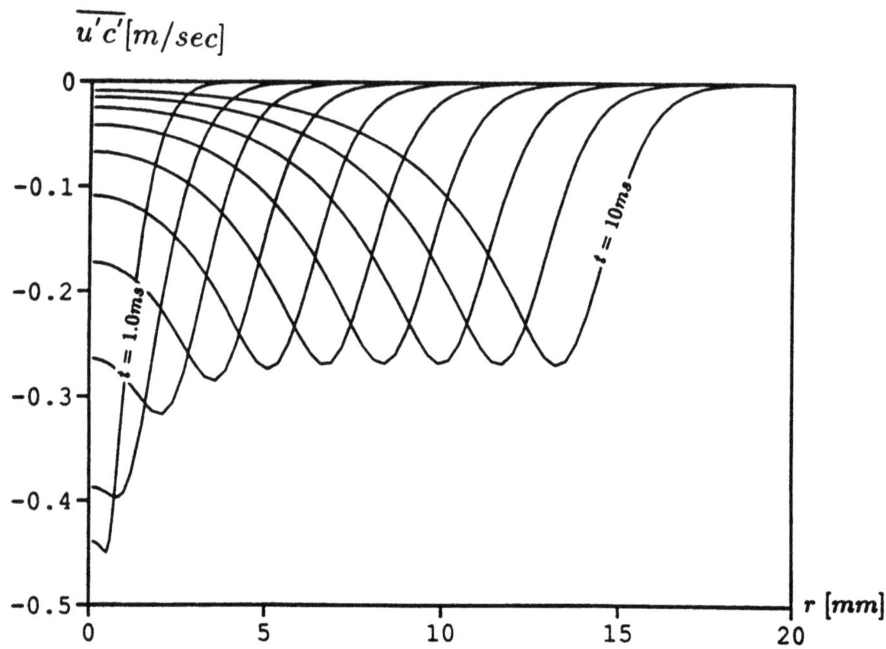

Fig. 2. Profiles of $\overline{u'c'}$ during the propagation of the flame (1 ms between each profile).

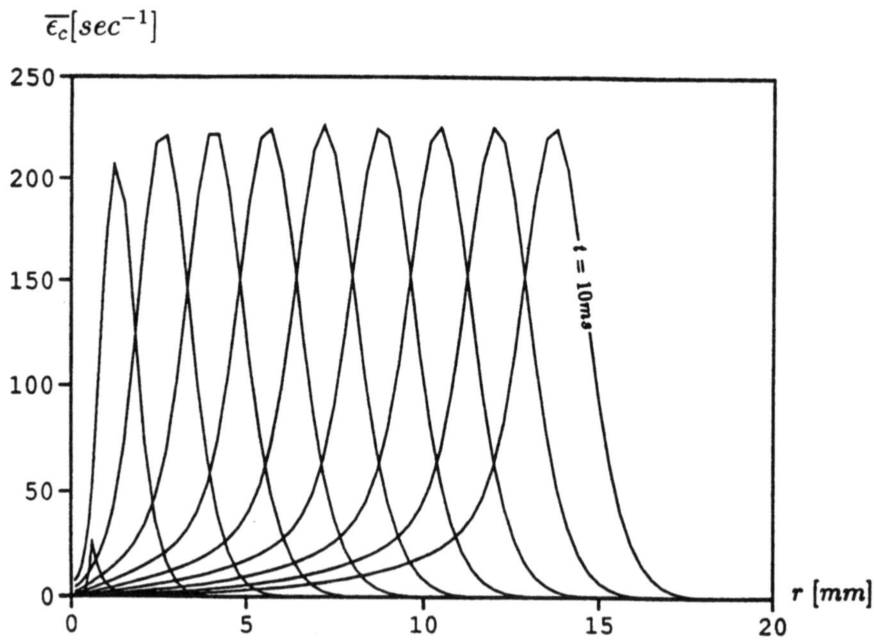

Fig. 3. Profiles of $\overline{\varepsilon_c}$ during the propagation of the flame (1 ms between each profile).

We see clearly on figure 1 that a steady propagation is obtained, with a constant velocity. This result was also obtained with the classical Eddy Break-up model but was not obtained with the $\overline{\varepsilon_c}$ equation of Zeman and Lumley (1976) (see Mantel and Borghi (1991)).

The comparison with the use of a diffusion coefficient is shown on figure 4. The flame propagates, but the initial period is different. If we define an effective Schmidt number σ_c defined as:

$$\sigma_c = -\frac{\nu_t}{\overline{u'c'}} \frac{\partial \overline{c}}{\partial x} \tag{15}$$

we obtain the profiles of figure 5. Finally, it can be noted that the differences are not so large for this case except at the beginning.

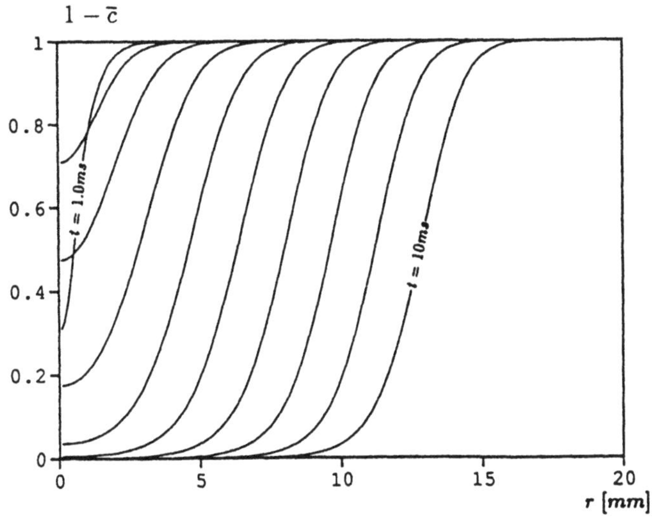

Fig. 4. Profiles of $1-\overline{c}$ during the propagation of the flame using a turbulent diffusion coefficient (1 ms between each profile).

4. Study of the Flame Velocity and Flame Thickness

The turbulent flame velocity in the steady propagation, and the turbulent flame brush thickness can be studied in function of $\frac{u'}{U_L}$.

4.1 The Turbulent Flame Velocity

The turbulent flame velocity satisfies roughly a proportionality with u' for large (but not too large $\frac{u'}{U_L}$). A recent theoretical work of Pocheau proposes, for finite $\frac{u'}{U_L}$ the

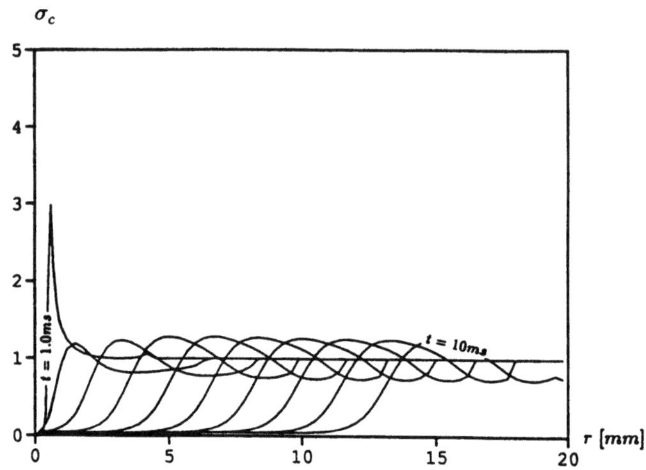

Fig. 5. Profiles of σ_c during the propagation of the flame.

following law:

$$U_T^{\alpha} = U_L^{\alpha} + \beta u'^{\alpha} \tag{16}$$

where $\alpha = 2.0$ and $\beta = 2.25$ (see Pocheau (1992-1993)).

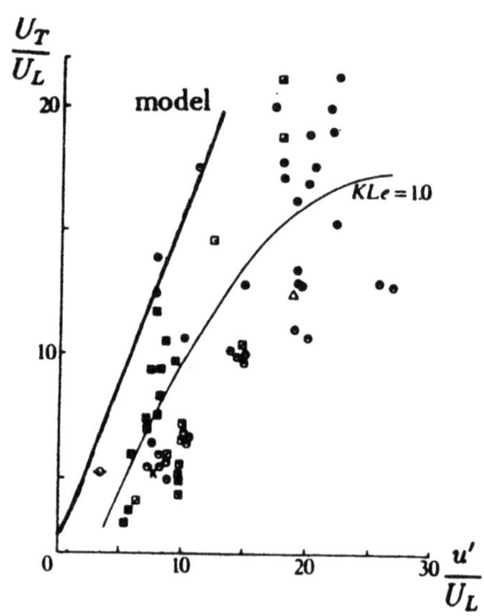

Fig. 6. Turbulent flame velocity given by the model and comparison with the data of Pocheau (1992-1993).

Figure 6 shows the comparison of our calculations with the theoretical formula, and with the experimental results of Pocheau in the range of $0 < \frac{u'}{U_L} < 2.5$. The agreement was obtained for $C_D = 2.2$. This constant has the most critical influence on U_T, once b_0, α_0, β_0 are fixed as in Mantel and Borghi (1991). The agreement with the model using a diffusion coefficient is good also, with a value of the constant C_μ (equal to 0.09). The constant C_u, which appears in the equation (2) for $\overline{\varepsilon_c}$, has a poor influence on U_T (varying C_u from 0 to 5 varies $\frac{U_T}{U_L}$ from 4.9 to 5.3 at $\frac{u'}{U_L} = 3.37$).

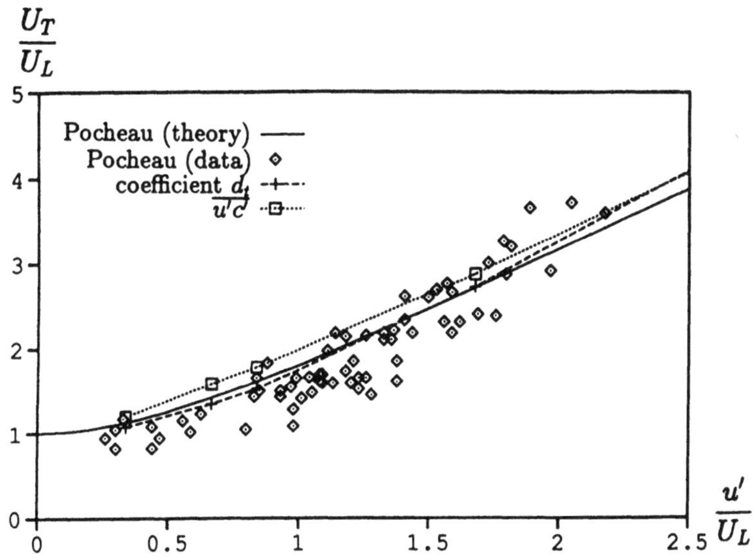

Fig. 7. Turbulent flame velocity given by the model and comparison with the data of Bradley *et al.* (1992).

Figure 7 shows a comparison with the collection of experimental results of Bradley and his group, in the range where no "bending" of the results appears (see Bradley *et al.* (1992)).

4.2 The Flame Brush Thickness

The flame brush thickness has not been studied in detail, except in the work of Talantov *et al.* (1969). However, it is a quantity of very practical interest for the design of burners.

The constant C_u appears as very important for predicting the flame thickness δ_T: a variation from 0 to 5 leads to a decrease of $\frac{\delta_T}{l_t}$ from 2.5 to 1.5.

This can be explained by the fact that the last term of equation (2) represents the destruction of $\overline{\varepsilon_c}$ by curvature effects. In the model, it has been proposed for the local radius of curvature of flamelets (Mantel and Borghi (1991)):

$$\lambda_c^f = \lambda_c \left(1 + C_u \frac{U_L}{u'}\right)^\gamma \tag{17}$$

where λ_c is the equivalent of the Taylor microscale for the non-reactive scalar field.

Thus, when $C_u \rightarrow 0$, the radius of curvature λ_c^f decreases, the flame appears more disturbed and more wrinkled. This case then leads to a thickening of the flame brush. On the contrary, when C_u increases, the radius of curvature also increases and the flame appears less wrinkled. This effect leads to a reduction of the flame brush thickness.

Figure 8 shows the transient evolution of δ_T when the flame is reaching steady propagation. It is seen that the flame brush takes some time to get a steady thickness. This time is much more than the ignition time and even than the turbulence time scale τ_t. It depends principally on $\frac{u'}{U_L}$.

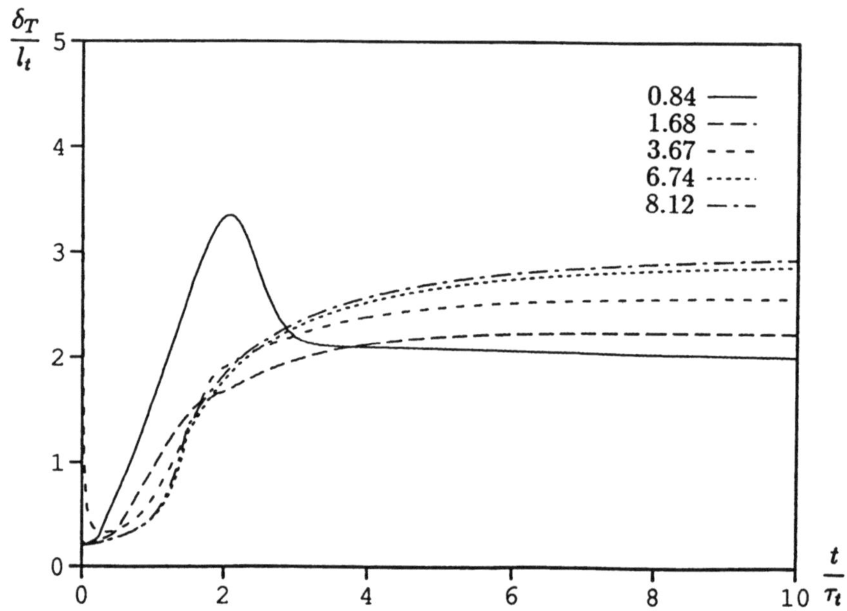

Fig. 8. Transient evolution of the flame brush thickness for different values of the ratio $\frac{u'}{U_L}$.

$\frac{U_T}{U_L}$ reaches a steady value within a smaller delay, always less than $3\tau_t$.

The figures 9 and 10 show the evolution of the flame brush thickness δ_T in function of respectively the ratio $\frac{u'}{U_L}$, the Reynolds number Re_t, for a steady propagating flame and with the two models used for the correlation $\overline{u_i'c'}$.

We note a difference on these curves between the values for δ_T given by the two models. Figure 6 also gives the difference between the two models concerning the turbulent flame velocity over a large range of $\frac{u'}{U_L}$.

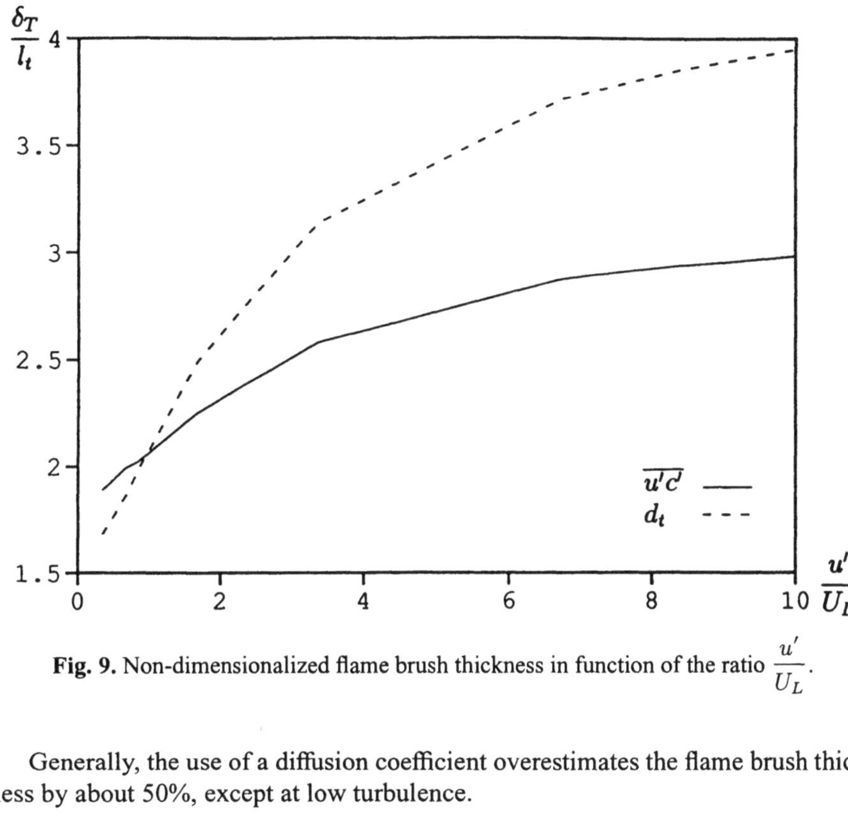

Fig. 9. Non-dimensionalized flame brush thickness in function of the ratio $\frac{u'}{U_L}$.

Generally, the use of a diffusion coefficient overestimates the flame brush thickness by about 50%, except at low turbulence.

5. Study of a Lagrangian Version of the Model

The phenomenon of turbulent diffusion can also be represented by following random walks of fluid particles (with the same point of view as G.I. Taylor (1915)). This avoids considering $\overline{u'c'}$, and avoids consequently any need of new closure assumptions for that.

We have also built a "Monte-Carlo" simulation of this type, following fluid particles (about 50 000); the turbulent dispersion has been simulated by sampling random velocities with given *rms* u' and a given spatial correlation whose mean is l_t. The velocity field is correlated during the characteristic turbulent time τ_t. During this period of time, all the particles keep the same fluctuating velocity and, after τ_t, the particles have a new fluctuating velocity randomly selected in order to obtain

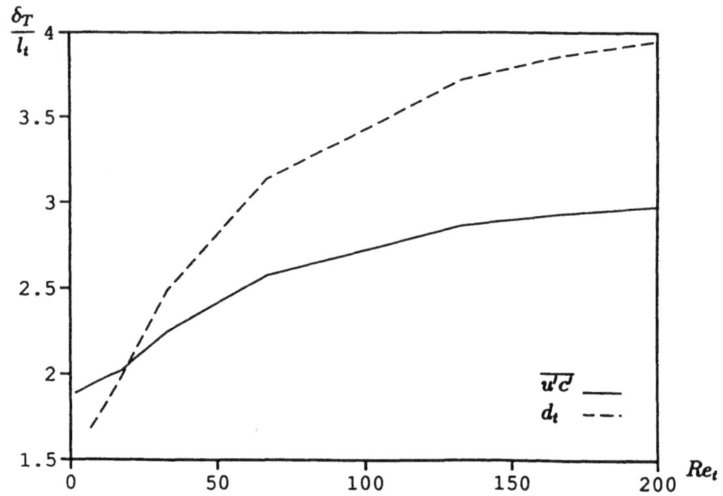

Fig. 10. Non-dimensionalized flame brush thickness in function of the Reynolds number Re_t.

a good agreement between the dispersion of this mixing model and the dispersion given by a $k-\varepsilon$ model $\left(\nu_t = C_\mu \frac{k^2}{\varepsilon}\right)$. The figure 11 shows the evolution of the mean square of the thickness of an Heaviside function due to the dispersion of the model. This evolution is close to the evolution given by the $k - \varepsilon$ model and we can see

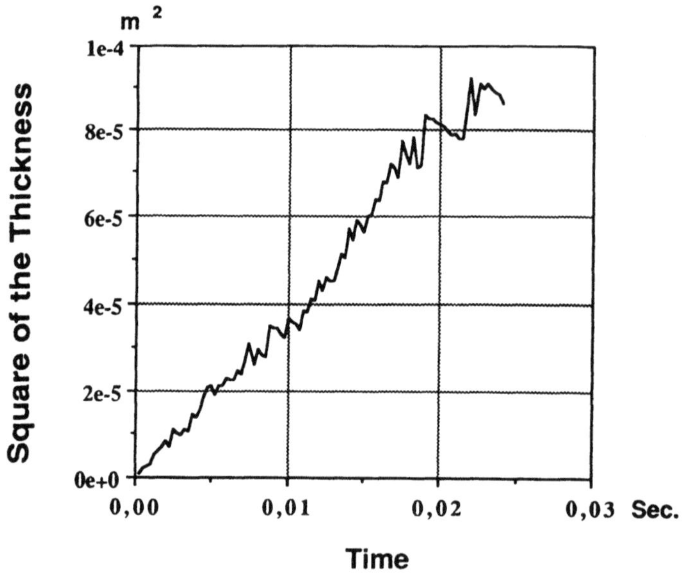

Fig. 11. Evolution of the mean square of the thickness versus time.

that this evolution is linear as predicted by Taylor's theory (1921) of the dispersion. But, to obtain a good agreement, it was necessary to increase the displacement of the fluid particles by a factor of 2.

When the chemistry is very fast, the combustion of the particles is controlled by their mutual mixing. This mixing, at small scales, is simulated as in PDF models by interactions with neighboring particles, with a frequency of mixing related to $C_{ex}\frac{\varepsilon}{k}$. The flame speed velocity U_T is very sensitive to this mixing.

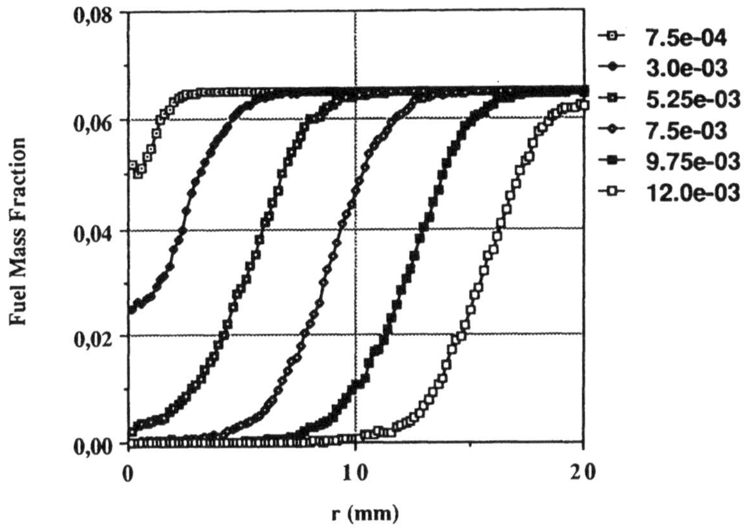

Fig. 12. Shape of the fuel mass fraction (u'=1.15 m/s, from 0.75 to 12 ms).

On the figure 12, the shape of the fuel mass fraction is plotted for different times after the ignition (from 0.75 to 12 ms). We have assumed a spherical symmetry around the area of ignition, so r is the radius of the sphere. With this symmetry, the study can be reduced to a one-dimensional phenomenon. On the curves between 5.25 and 12 ms, the flame brush is quite constant and the profiles are very similar. The figure 13 shows the shape of the mean reaction rate of the fuel for the same time and during the same computation. On the two first curves, the influence of the energy of ignition is still present. For the last four curves, the shape is preserved.

The figure 14 and the figure 15 are the same as the figures 12 and 13 except that the *rms* value of the fluctuating velocity is different (3.3 m/s instead of 1.15 ms for the two previous figures). The characteristic turbulent time is also identical but not the integral length scale. Compared to the two previous figures, it is possible to remark that the reaction zone propagates faster, that the flame brush is more important and that the mean reaction rate is higher.

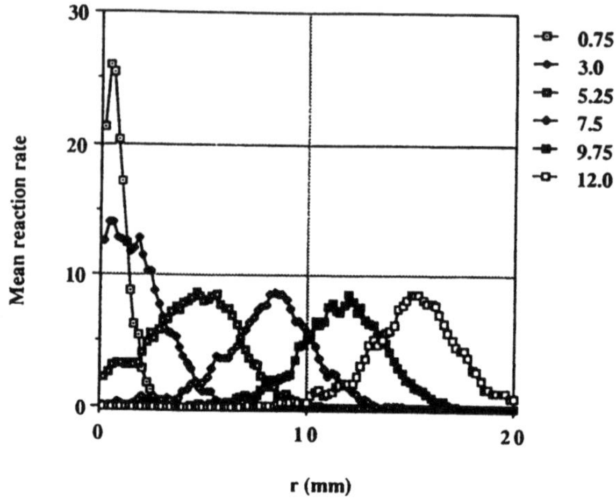

Fig. 13. Shape of the mean reaction rate (u'=1.15 m/s, from 0.75 to 12 ms).

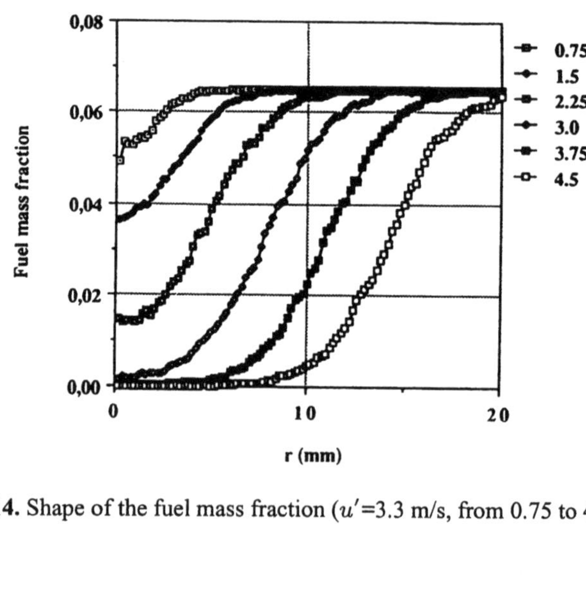

Fig. 14. Shape of the fuel mass fraction (u'=3.3 m/s, from 0.75 to 4.5 ms).

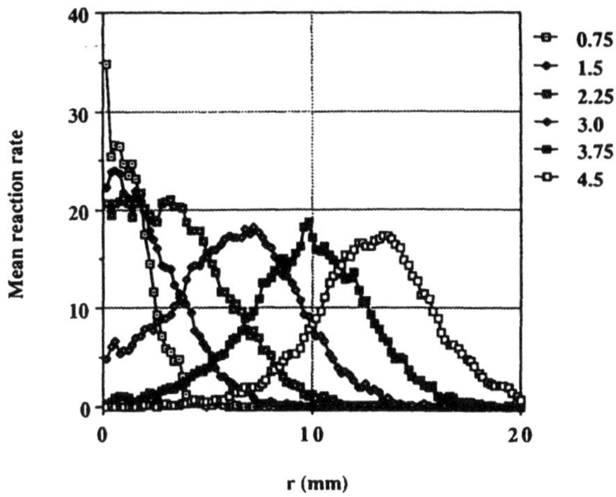

Fig. 15. Shape of the mean reaction rate (u'=3.3 m/s, from 0.75 to 4.5 ms).

On the figure 16 is plotted the ratio $\frac{U_T}{U_L}$ as a function of $\frac{u'}{U_L}$ for the 3 different values of u' (u' =1.15, 3.3 and 9 m/s). The evolution is linear and the slope is approximatively 1.5. This value is rather slow but can be adjusted with the constant C_{ex} = 4. The increase of this constant will increase the slope of the curve.

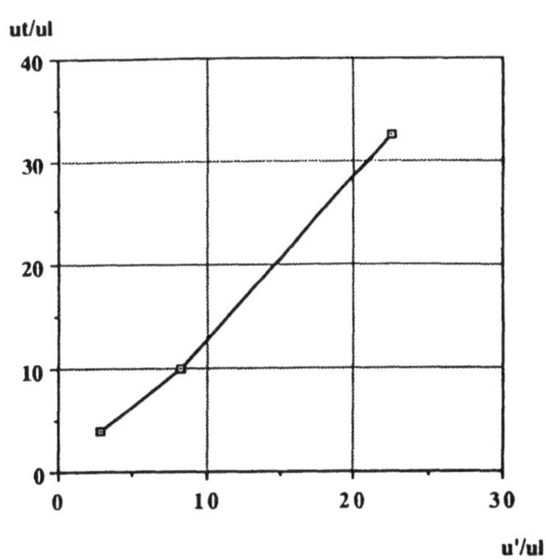

Fig. 16. Turbulent flame velocity given by the Lagrangian model.

The figure 17 presents the ratio $\frac{\delta_T}{l_t}$ versus the ratio $\frac{u'}{U_L}$. As the fluctuations of the velocity are increasing, the integral length scale is increasing, since the characteristic turbulent time is kept constant, and we can see the flame brush thickness δ_T increases faster than the integral scale.

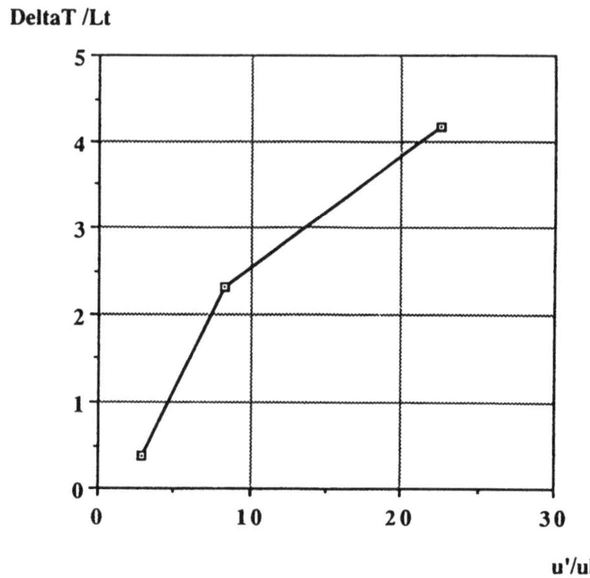

Fig. 17. Flame brush thickness given by the Lagrangian model.

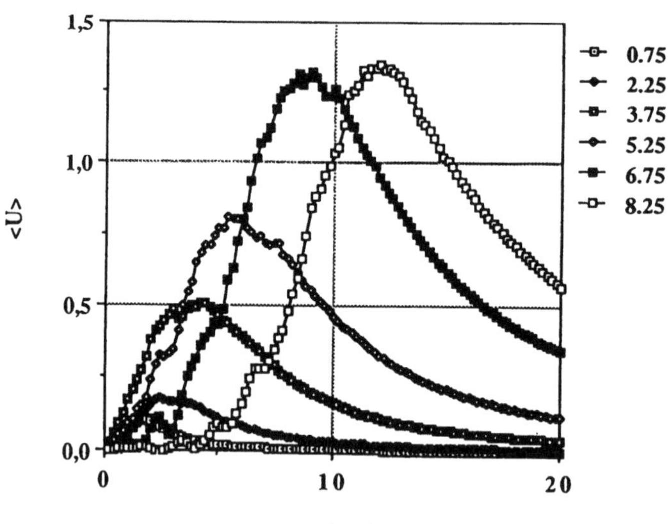

Fig. 18. Shape of the mean velocity of the fluid due to heat release.

The effect of the heat release has been omitted in all the results presented above in this section. If we take this effect into account, we can plot on the figure 18 (for $u' = 1.15$ m/s), the mean velocity of the fluid due to the expansion of the gases at different times. If we compare with figure 13, we can observe that the velocity is maximum just after the propagation of the reaction zone. The effect of the heat release is to increase the velocity of propagation of the reaction zone.

6. Conclusion

For the propagation of a turbulent premixed flame, particular attention has been paid on the modelling of the mean reaction rate and the turbulent flux of the progress variable. In order to determine the mean reaction rate, a transport equation for the scalar dissipation rate proposed by Mantel and Borghi (1991) has been used. The turbulent flux $\overline{u'c'}$ has been also calculated by using a transport equation for which a new modelling is done. We have seen that the modelling of $\overline{u'c'}$ acts more on the transient period of propagation of the flame front and more on the flame brush thickness δ_T than on the turbulent flame velocity U_T. A set of constants for the model has been found in order to obtain realistic results. This set of constants is however not definitely fixed. Other tests have to be done to assess the other constants with other experiments, in particular non reacting flows could be studied for adjusting C_{p_U}, C_{p_c}, α_0 and β_0. The interest of

the Lagrangian version of the model is that it does not need any additional closure for the $\overline{u'c'}$ equation. Due to the rough estimate of the temporal correlation, it was necessary to introduce a new constant to obtain, without reaction the same diffusion as the $k-\varepsilon$ model. The comparison of the Lagrangian model with the Eulerian model will be very helpful to assess the possibilities and the limits of the model.

References

Abdel-Gayed R.G., Al-Khishali K.J. and Bradley D., "Turbulent Burning Velocities and Flame Straining in Explosions", *Proc. R. Soc. Lond.*, **A391**, pp 393-414, 1984

Abdel-Gayed R.G., Bradley D. and Lawes M., "Turbulent Burning Velocities: a General Correlation in Terms of Straining Rates", *Proc. R. Soc. Lond.*, **A414**, pp 389-413, 1987

Antonia R.A. and Browne L.W.R., "The Destruction of Temperature Fluctuations in a Turbulent Plane jet", *Journal of Fluid Mechanics*, **134**, pp 67-83, 1983

Borghi R., "Turbulent Premixed Combustion: Further Discussions on the Scales of Fluctuations", *Combustion and Flame*, **80**, pp 304-312, 1990

Borghi R. and Dutoya D., "On the Scale of the Fluctuations in Turbulent Combustion", *Seventeenth Symposium (International) on Combustion*, pp 235-244, 1978

Bradley D., Lau A.K.C. and Lawes M., "Flame Stretch Rate as a Determinant of Turbulent Burning Velocity", *Phil. Trans. R. Soc. Lond.*, **338**, pp 359-387, 1992

Bray K.N.C., "Laminar Flamelet Modelling of Turbulent Combustion", *Lecture Notes in Physics*, **241**, pp 3-19, *Springer Verlag*, 1985

Bray K.N.C. and Libby P.A., "Passage Times and Flamelet Crossing Frequencies in Premixed Turbulent Combustion", *Combustion, Science and Technology*, **47**, pp 253-274, 1986

Bray K.N.C. and Moss J.B., "A Unified Statistical Model of the Premixed Turbulent Flame", *Acta Astronautica*, **4**, pp 291-319, 1977

Candel S.M. and Poinsot T.J., "Flame Stretch and the Balance Equation for the Flame Area", *Combustion, Science and Technology*, **70**, pp 1-15, 1990

Dopazo C. and O'Brien E.E., "A Probabilistic Approach to the Autoignition of Reaction in Turbulent Mixtures", *Acta Astronautica*, **1**, pp 1239-1255, 1974

Dumont J.P., Durox D. and Borghi R.,, "Experimental Study of the Mean Reaction Rate in a Turbulent Premixed Flame', *Combustion, Science and Technology*, **83**, pp 1-33, 1992

El Tahry S.H., Rutland C.J. and Ferziger J.H., "Structure and Propagation Speeds of Turbulent Premixed Flames - A Numerical Study", *Combustion and Flame*, **83**, pp 155-173, 1991

Hakberg B. and Gosman A.D., "Analytical Determination of Turbulent Flame Speed from Combustion Models", *Twentieth Symposium (International) on Combustion*, pp 225-232, 1984

Libby P.A. and Bray K.N.C., "Implication of the Laminar Flamelets Model in Premixed Turbulent Combustion", *Combustion and Flame*, **39**, pp 33-41, 1980

Maistret E., Darabiha N., Poinsot T., Veynante D., Lacas F., Candel S. and Esposito E., "Recent Development in the Coherent Flamelet Description of Turbulent Combustion", *Third International Conference on Numerical Combustion*, pp 157-188, 1989

Mantel T. and Borghi R., "A New Model of Premixed Wrinkled Flame Propagation Based on a Scalar Dissipation Equation", *Thirteenth International Colloquium on the Dynamics of Explosion and Reactive Systems*, Nagoya University, 28 July - 2 August 1991

Mantzaras J., Felton P.G. and Bracco F.V., "Fractals and Turbulent Premixed Engine Flames", *Combustion and Flame*, **77**, pp 295-310, 1989

Marble F.E. and Broadwell J.E., "The Coherent Flame Model for Turbulent Chemical Reactions", *Project Squid Headquarters, Chaffee Hall, Purdue University, Technical Report TRW-9-PV*, 1977

Picart A., Borghi R. and Chollet J.P., "Numerical Simulation of Turbulent Reactive Flows", *Sixth Symposium on Turbulent Shear Flows*, pp 20.5.1-20.5.5, 1987

Pocheau A., "Invariance d'Echelles en Combustion Turbulente", *C.R. Acad. Sci. Paris*, **315-II**, pp 21-27, 1992

Pocheau A., "Front Propagation in a Turbulent Medium", *to appear in Europhysics Letters*, 1993

Pope S.B., "Evolution of Surfaces in Turbulence", *International Journal of Engineering Sciences*, **26**, pp 445-469, 1988

Said R. and Borghi R., "A Simulation with a 'Cellular Automaton' for Turbulent Combustion Modelling", *Twenty-Second Symposium (International) on Combustion*, pp 569-577, 1988

Talantov A., Ermolaev V.M., Zotin V.K. and Petrov E.A., *Combustion, Explosion and Shock Waves*, **5-1**, pp 73-75, 1969

Zeman O. and Lumley J.L., "Modelling Buoyancy Driven Mixed Layers", *Journal of the Atmospheric Sciences*, **33**, pp 1974-1988, 1976

A Description of Turbulent Premixed Flames Based on Flamelet Time Statistics

Christine Mounaïm-Rousselle and Iskender Gökalp

Laboratoire de Combustion et Systèmes Réactifs,
Centre National de la Recherche Scientifique, 45071 Orléans cédex 2, France

Abstract

Laser-induced Rayleigh light scattering technique is used to determine relevant time scales in turbulent premixed flames in the flamelet regime. The major achievement of this work is the experimental determination of flamelet transit times at a given point within the flame brush. This information coupled with the usual flamelet description assumptions allows an estimate of the local mean reaction rate to be made. The statistics of instantaneous flamelet orientations are also deduced. The technique is implemented in open turbulent methane-air premixed flames stabilized on the rim of a Bunsen type burner. The variation versus the progress variable of the estimated mean reaction rate is compared with the predictions of the Bray-Moss-Champion-Libby model.

1. Introduction

Recent developments in the understanding of turbulent burning in premixed flames emphasize both the temporal and spatial statistics of instantaneous flame fronts within the average flame brush (Bray, 1990; Cant *et al.*, 1990, Gouldin, 1987; Poinsot *et al.*, 1991; Shepherd and Ashurt, 1992). On the other hand, these two types of information are used to estimate the mean reaction rate in turbulent premixed flames (Saïd and Borghi, 1988; Bray *et al.*, 1988), which is a quantity of crucial importance, from both basic and applied perspectives, and which is replacing progressively the turbulent propagation velocity concept. The development of reliable non-intrusive diagnostics and robust numerical simulation models have both contributed to the recent progress in the analysis of turbulent premixed flames.

In harmony with this evolution, we have developed an experimental technique which, combined with a theoretical analysis, yields relevant information on the temporal and spatial statistics and on the mean reaction rates in turbulent premixed flames (Boukhalfa and Gökalp, 1988; Deschamps, 1990; Deschamps *et al.*, 1992; Mounaïm-Rousselle *et al.*, 1993). The theoretical analysis relies on the laminar flamelet model as developed by Bray *et al.* (1988) and which assumes, that, at high turbulent Reynolds number and high Damköhler number, a turbulent premixed flame consists of a collection of strained flame fronts which retain the global structure of laminar premixed flames.

Three important features of the work presented here are the following: (1) the time and space statistics and the mean reaction rate in turbulent premixed flames are obtained with a single diagnostic, namely the one-point laser-induced Rayleigh scattering technique; (2) the information obtained is directly relevant for theoretical and numerical modelling purposes; and (3) the diagnostic used and the analysis developed are relevant not only for laboratory type flames but also for turbulent premixed flames in technological systems.

The present paper illustrates these features. First, the essential characteristics of the one-point Rayleigh scattering system are described. This is followed by the details of the analysis which enables to deduce the time and space statistics of the instantaneous flame fronts and the local mean reaction rate. As an application, highly turbulent large Bunsen type conical flames are explored. All the relevant time statistics of instantaneous flame fronts are determined. It is also demonstrated that the time statistics can provide relevant spatial information, such as the instantaneous flame front orientation distributions. Finally, the mean reaction rate estimated with the present approach is compared to the predictions of the Bray-Moss-Champion-Libby (BMCL) model.

2. Laser-Induced Rayleigh Light Scattering System

Laser-induced Rayleigh scattering technique allows the local instantaneous density (or the statistics of the presence and passages of instantaneous flame fronts) to be determined in a turbulent premixed flame brush (Boukhalfa and Gökalp, 1988; Dechamps and Gökalp, 1990; Deschamps, 1990; Deschamps *et al.*, 1992; Mounaïm-Rousselle *et al.*, 1993). The present Rayleigh scattering optical set-up uses the blue line (488 nm, 6.5 W) of a 20 W Argon-ion laser as the light source. The laser beam is focused down by a 106 mm focal length lens to form a waist diameter of approximately 0.08 mm. The scattered light is collected at 90° by a Hamamatsu R647-04 photomultiplier through variable diameter pin holes, ranging between 0.15 mm and 0.03 mm (0.15 mm is used for this study). The variation of the pin hole diameter from 0.15 mm to 0.03 mm changes the measuring volume from $0.4 \cdot 10^{-3}$ mm^3 to $0.008 \cdot 10^{-3}$ mm^3. The combination of high laser power and variable measuring volume allows optimization of the quality of the Rayleigh signal according to flame and turbulence conditions. The data acquisition system can collect 10^6 samples per measuring point with a sampling rate as high as 1 MHz. The results presented in this paper are obtained with a sampling rate of 100 kHz. The Rayleigh light scattering apparatus is fully described in Deschamps (1990) and Deschamps *et al.* (1992).

The Rayleigh scattering technique is applied here to the study of Bunsen type open conical turbulent premixed methane/air flames. These flames are stabilized by an annular pilot flame at the rim of a 25 mm inner diameter burner. Several combinations of turbulence and equivalence ratio conditions (producing a large spectrum of

Damköhler number, Karlovitz number and turbulent Reynolds number values) are investigated (Deschamps, 1990; Deschamps *et al.*, 1992).

For the purpose of this paper, one particular set of data corresponding to fixed turbulence conditions and to four equivalence ratios ϕ (0.7; 0.8; 0.9; 1.0) are analyzed. Turbulence is produced by a perforated plate with a mesh size of 7 mm and a hole diameter of 5 mm, placed 50 mm upstream of the burner exit section. It provides, for a cold flow velocity U_r = 5 m/s, 10% turbulence intensity and a longitudinal integral length scale of 3 mm. The results discussed in this paper correspond to measurements on the centerline (or the mean symmetry axis) of the flame brush, as this particular location allows an easy way to follow a mean streamline. Flamelet statistics are determined versus the mean progress variable $\overline{c}$. It should be noted that for a constant turbulence intensity in the reactants, decreasing ϕ from 1.0 to 0.7 increases the u'/S_L ratio approximately from 1 to 2 under the present conditions. Results corresponding to flames with a lower cold flow velocity (U_r = 2 m/s) are introduced when appropriate for comparison purposes.

3. Data Analysis

An example of the Rayleigh signal obtained with a sampling rate of 100 kHz on the axis of an open conical turbulent premixed methane/air flame (mean exit velocity of the reactants U_r = 5 m/s; equivalence ratio ϕ = 1) is presented on Fig. 1. It corresponds to a position where the mean progress variable $\overline{c}$ is approximately equal to 0.5, i.e. to the middle of the average flame brush. The light intensity $I^\star$ is normalized as:

$$I^\star = \frac{I - \overline{I_P}}{\overline{I_R} - \overline{I_P}}$$

where $\overline{I_R}$ and $\overline{I_P}$ are respectively the Rayleigh light intensity corresponding to the reactants and to the products. These are mean values determined from the pdf of the light intensity distribution.

The same figure also shows the three different threshold settings used for data analysis. The flamelet crossing frequency ν and the mean progress variable $\overline{c}$ are deduced from the mean passage times of pockets of reactants and products, $\overline{t_r}$ and $\overline{t_p}$ (or from the mean arrival times between two consecutive flame fronts) with a single threshold set at $I^\star$ = 0.5, as follows (Boukhalfa and Gökalp, 1989; Deschamps *et al.*, 1992):

$$\nu = \frac{2}{\overline{t_r} + \overline{t_p}} \quad \text{and} \quad \overline{c} = \frac{\overline{t_p}}{\overline{t_r} + \overline{t_p}} \tag{1}$$

This flamelet counting procedure includes then some partial crossings, such as the middle one on Fig. 1.

The major achievement of the Rayleigh scattering set-up we developed is that it allows the determination of transit times of instantaneous flame fronts from reactants to products, t_{rp}, and from products to reactants, t_{pr} (Deschamps and Gökalp, 1990;

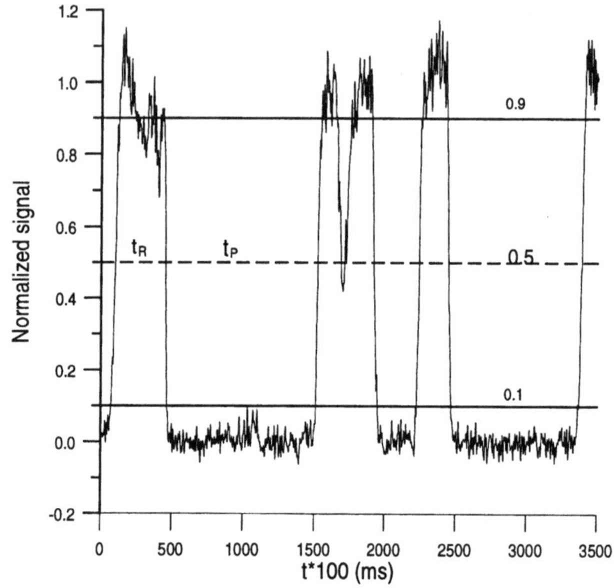

Fig. 1. Rayleigh signal ($\overline{c} = 0.5$, $\phi = 1$, $U_r = 5$ m/s).

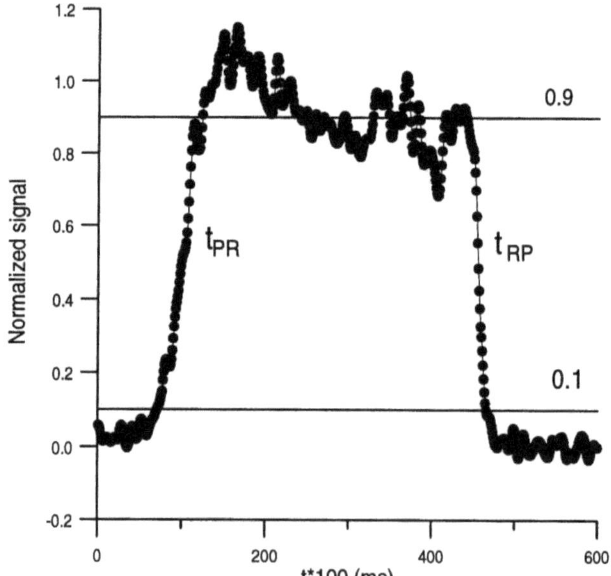

Fig. 2. Rayleigh signal ($\overline{c} = 0.5$, $\phi = 1$, $U_r = 5$ m/s).

Deschamps *et al.*, 1992). For this purpose, the time interval between the level changes of the Rayleigh signal should be determined. Fig. 2 presents an enlarged portion of the normalized Rayleigh light intensity signal, where the actually sampled data points are shown. The quality of this signal, i.e. its high temporal resolution, provides enough confidence for the correct determination of t_{rp} and t_{pr}.

A single threshold is of no use here as we have to mark the beginning and the end of the instantaneous flame front transits either from reactants to products or vice versa. A double threshold setting is then necessary. The setting of the two thresholds must correspond to an optimum: if they are too close to $I^\star = 1$ and $I^\star = 0$, even slightly partial crossings will be eliminated; this will reduce the number of crossing events and thus diminish the statistical representativeness of the signal. Also, except for $\bar{c}$ close to 0.5, this too tight threshold setting may introduce important differences between the $\bar{c}$ values calculated with single and double thresholds. On the other hand, if the thresholds are too close to $I^\star = 0.5$, the transit times will be underestimated. In the present work, after several tests we used two thresholds set at $I^\star = 0.1$ and 0.9.

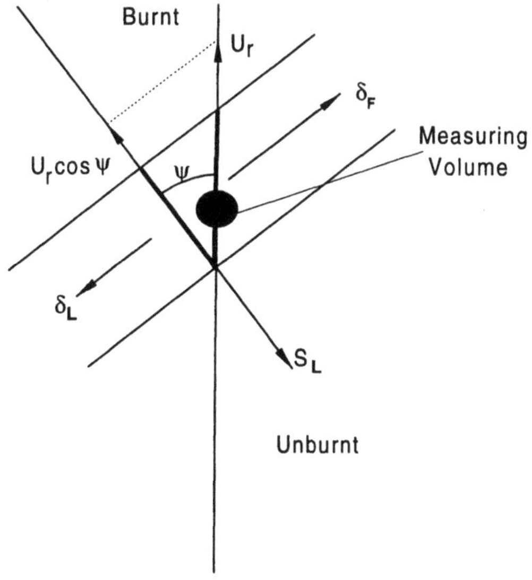

Fig. 3. Schematic representation of the relationships between δ_L, δ_F, S_L, U_r and ψ.

Fig. 3 shows schematically the relationships between the flamelet thickness (δ_L) normal to an element of the flame surface, the effective flamelet thickness (δ_F) along an instantaneous streamline, the effective flamelet normal propagation velocity (S_L), the reactants convection velocity (U_r) just upstream of the flame front along a streamline and the orientation angle (ψ) between the normal to a flamelet surface element and the streamline direction. The following expressions can be derived to relate instantaneously these parameters to each other. In these expressions d is the dimension of the measurement volume that should be taken into account.

$$t_{rp} = \frac{\delta_L + d}{U_r \cos\psi_{rp} + S_L} \quad \text{and} \quad t_{pr} = \frac{\delta_L + d}{U_r \cos\psi_{pr} - S_L} . \tag{2}$$

Following the BMCL flamelet model as reviewed in Bray (1990), the local mean reaction rate can be expressed as $\overline{w} = \overline{w_F}\nu$, where $\overline{w_F}$ is the average mass of products created per unit volume of reactants during a flamelet crossing. This last quantity can be modelled as

$$\overline{w_F} = \varrho_r \overline{S_L}\, \overline{1/V_n} \tag{3}$$

In this relation, $\overline{S_L}$, the effective mean laminar propagation velocity is different from S_L^0 , which is the propagation velocity for a planar unstretched laminar flame. It can be written, as in Bray (1990), $\overline{S_L} = S_L^0 \overline{I_0} \cdot \overline{I_0}$ is a parameter which takes into account the reduction of $\overline{S_L}$ due to mean and turbulent stretch and curvature effects. $\overline{V_n}$ represents an average flame transit velocity such that $\delta_L \overline{1/V_n}$ is the mean transit time of a laminar flamelet. $\overline{V_n}$ is calculated as $1/2\left(\overline{V_{n_{rp}}} + \overline{V_{n_{pr}}}\right)$ where:

$$\overline{V_{n_{rp}}} = U_r \cos\psi_{rp} + S_L \quad \text{and} \quad \overline{V_{n_{pr}}} = U_r \cos\psi_{pr} - S_L \tag{4}$$

The above relation for $\overline{w_F}$ expresses the fact that a single flamelet can convert the total amount of fresh gases per unit volume (ϱ_r) to combustion products only if $\overline{V_n} \leqslant \overline{S_L}$. This is best expressed by the ratio $\overline{w_F}/\varrho_r = \overline{S_L}\,\overline{1/V_n} = \overline{t_F}/\overline{t_c}$, where $\overline{t_c}$ is the characteristic chemical time expressed as $\delta_L/\overline{S_L}$, and $\overline{t_F}$ is constructed as $1/2\left(\overline{t_{rp}} + \overline{t_{pr}}\right) = \delta_L \overline{1/V_n}$. Consequently, a unit volume of unburnt gases should encounter several instantaneous flame fronts in order to be converted totally into burnt gases. The competition between ν and the ratio $\overline{t_F}/\overline{t_c}$ will then determine the local burning rate, as follows: $\overline{w_F}/\varrho_r = \left(\overline{t_F}/\overline{t_c}\right)\nu$. It is also clear that from the combination of the temporal and spatial versions of the BMCL model, the flame surface density (or the flame surface area per unit volume) can be estimated as $\Sigma = \left(\overline{t_F}/\delta_L\right)\nu$ (Bray, 1990; Deschamps *et al.*, 1992).

The temporal statistics of the instantaneous flame front dynamics as determined by the Rayleigh scattering technique give therefore a complete description of the turbulent premixed flame structure. The above relations and analyses are further developed in the next section in relation with the measurements corresponding to the flame configuration investigated here.

4. Results and Discussion

An example of the distributions of t_{rp} and t_{pr} is given on Fig. 4. As expected, $t_{rp} < t_{pr}$, due to the addition of the normal flame propagation velocity S_L to the convection velocity $U_r \cos\psi_{rp}$ (see Fig. 3 and Eq. 4). The asymmetric shape of these distributions can easily be explained as the flamelet transit times are bounded from below by the minimum times corresponding to flamelets crossing normally the flow lines

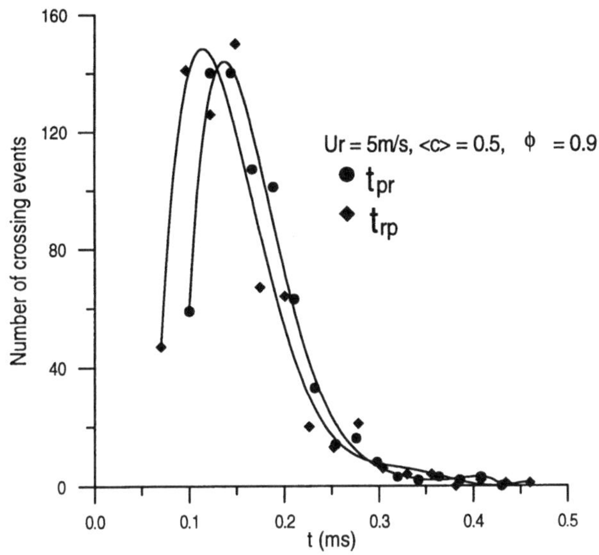

Fig. 4. Pdf distributions of t_{rp} and t_{pr} ($\overline{c} = 0.5$, $\phi = 0.9$, $U_r = 5$ m/s).

($\cos\psi = 1$) with the maximum instantaneous convection velocity, $U(t)_{max}$, of the reactants.

The instantaneous relations between the relevant quantities of flamelet dynamics can be further analyzed to infer an estimate of the laminar flamelet thickness δ_L normal to a flamelet element. δ_L is the shortest flamelet thickness and should therefore correspond to the shortest flamelet transit times, $(t_{rp})_{min}$ and $(t_{pr})_{min}$. The shortest crossings correspond furthermore to $\cos\psi = 1$. Assuming that $U(t)_{max}$ can be approximated as U_r $(u'_r \ll U_r)$, and that $S_L = S_L^0$ (weak strain), an estimate of the flamelet normal thickness δ_L can be inferred from the following relations:

$$(t_{pr})_{min} = \frac{\delta_L + d}{\overline{U_r} - S_L^0} \quad \text{and} \quad (t_{rp})_{min} = \frac{\delta_L + d}{\overline{U_r} + S_L^0} \tag{5}$$

Table 1. δ_L values for different ϕ.

ϕ	0.7	0.8	0.9	1
δ_L [mm] $U_r = 2$ m/s	0.31	0.16	0.12	0.14
δ_L [mm] $U_r = 5$ m/s	0.37	0.26	0.23	0.14
δ_L [mm] $=\nu/S_L^0$	0.067	0.0516	0.042	0.034

Table 1 summarizes the estimated values of δ_L for four equivalence ratios and for two cold flow velocities. For each pair of conditions (ϕ and U_r), eight locations within the flame brush are explored, by traversing it from $\bar{c} = 0$ to $\bar{c} = 1$ on the flame centerline. Also, the calculations are made separately by using $(t_{rp})_{min}$ and $(t_{pr})_{min}$. Therefore, the δ_L values on Table 1 are average values of 16 measurements for each pair of conditions. Average values of S_L^0 for methane/air mixtures from the literature are used in these estimations (S_L^0 = 0.23 m/s, 0.30 m/s, 0.37 m/s and 0.45 m/s for respectively ϕ = 0.7, 0.8, 0.9 and 1.0). Finally, the dimension d of the measurement volume is estimated as the diameter 0.15 mm of a spherical volume. Table 1 gives the values of δ_L after removal of the measurement volume dimension.

Table 1 shows interesting features. First, as expected, δ_L decreases with increasing strength of the mixture, and passes from 0.37 mm to 0.14 mm for U_r = 5 m/s, and from 0.31 mm to 0.14 mm for U_r = 2 m/s, when the mixture equivalence ratio is changed from 0.7 to its stoichiometric value. The determination of the same δ_L value for the two stoichiometric mixtures is noteworthy. The variations between the δ_L values of the two cases (U_r = 5 m/s and 2 m/s) can be attributed to different sources. First, some small variations in the setting of the mixture compositions are to be expected, as the relative accuracies of the flow metering system used here (sonic flow meters) depend on the flow rates. Second, δ_L values obtained here are those influenced by the flow; i.e. by mean and fluctuating strain rates, which are different for the two cases. Finally, and most importantly, the flame front statistics for the U_r = 2 m/s case are less rigorous due to the significantly reduced number of crossing events, which depends mainly on the convection velocity of the unburnt mixture (Boukhalfa and Gökalp, 1988).

Concentrating on the results for flames with U_r = 5 m/s, some more inferences can be made. First, it is interesting to note that the δ_L value obtained for the stoichiometric methane/air flame is very close to the characteristic preheat zone thickness computed for this flame by Echekki (1992) as equal to 0.155 mm. In fact, the definition adopted by Echekki, $(T_p - T_r)/(\mathrm{d}T/\mathrm{d}x)$, is close to the total laminar flame thickness. Table 1 indicates also the values of the Zeldovich thickness as estimated by the relation $\delta_L = \nu / S_L^0$, where ν (the kinematic viscosity) is taken equal to 0.155 cm^2/s for all mixtures. The Zeldovich thickness is accepted as representative of the reaction zone in a laminar flame. The average ratio between the experimentally estimated δ_L value and the Zeldovich thickness is found equal to 5.03, i.e. very close to that given by Jarosinski (1984) as the ratio between the total laminar flame thickness and the reaction zone thickness (Gökalp, 1987). The above discussions show that the laminar flame thickness values estimated from Rayleigh scattering measurements are plausible and that they are representative of the total flame thickness, as implied by the settings of the two thresholds used to determine them.

Also, the above comparisons seem to indicate that for the present turbulent flame conditions, the normal total laminar flame thickness is not different from that of an unstrained laminar flame. As will be shown below, the distributions of t_{rp} and t_{pr}, thus of the effective laminar flame thickness, are not due to a thickening of the laminar flame front, but to the variations of the orientation angle between the flamelet elements and the velocity vector of the reactants. This also implies that the distribution of the orientation angle is indicative of the distribution of flamelet curvatures.

The relationships between the relevant quantities of flamelet dynamics as depicted on Fig. 3 can be rewritten in the following form:

$$\frac{1}{t_{\rm rp}} = \frac{U_{\rm r}\cos\psi_{\rm rp} + S_{\rm L}}{\delta_{\rm L} + d} \quad \text{and} \quad \frac{1}{t_{\rm pr}} = \frac{U_{\rm r}\cos\psi_{\rm pr} - S_{\rm L}}{\delta_{\rm L} + d}\,. \tag{6}$$

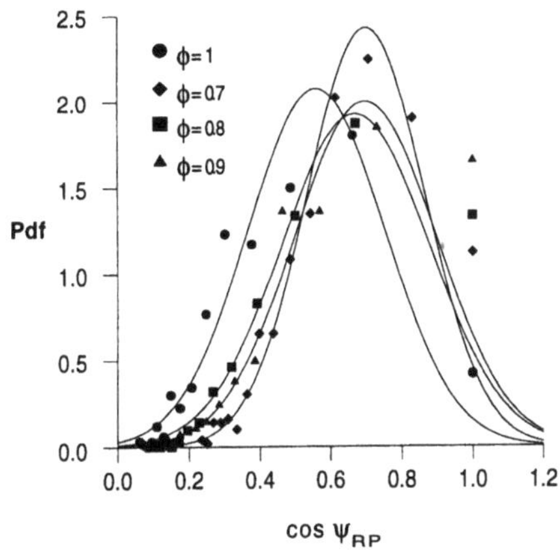

Fig. 5a. Pdf distributions of $\cos\psi_{\rm rp}$ ($\bar{c} = 0.5$, $U_{\rm r} = 5$ m/s).

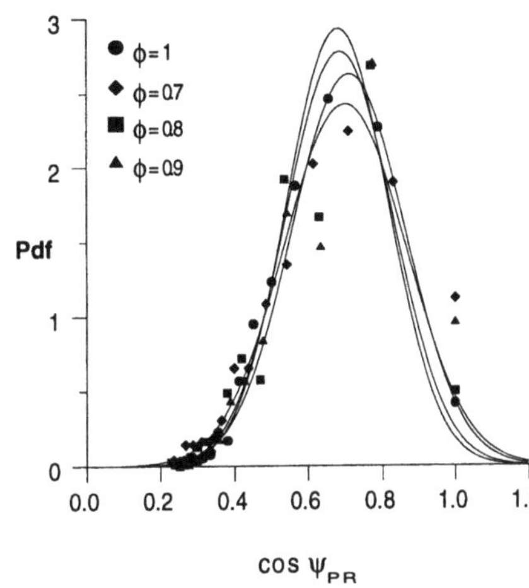

Fig. 5b. Pdf distributions of $\cos\psi_{\rm pr}$ ($\bar{c} = 0.5$, $U_{\rm r} = 5$ m/s).

The distributions of $\cos\psi_{\rm rp}$ and $\cos\psi_{\rm pr}$ for four flames with $U_{\rm r} = 5$ m/s are shown on Figs. 5. They correspond to a location in the flame brush where $\bar{c}$ is approximately 0.5. The pdfs are normalized so that the area under each of them is equal to 1. The curves on these figures correspond to Gaußian distributions.

Several observations can be made. First, the experimental distributions of the direction cosines are very close to clipped-Gaußian distributions. Second, there is no appreciable influence of the equivalence ratio. And, third, the distributions of $\cos\psi_{\rm rp}$ are broader, so that the corresponding rms values are higher. A plausible explanation has not yet been found for this last observation.

The mean values of the direction cosines are plotted on Figs. 6 versus $\bar{c}$, for the two reactants' velocities. For $U_{\rm r} = 5$ m/s, $\cos\psi_{\rm rp}$ and $\cos\psi_{\rm pr}$ are approximatively independent of $\bar{c}$ and respectively scattered around average values of 0.58 and 0.64. It should however be noted that measurements corresponding to both edges of the flame brush are not achievable with the technique developed here, because of the insufficient number of crossing events ; this is why the behaviour of the direction cosines near $\bar{c} = 0$ and 1 is not shown. It is also interesting to note that these results confirm those of Chew *et al.* (1990) and Lee *et al.* (1992) obtained by using flame imaging techniques. The variations with $\bar{c}$ of the direction cosines for the flames with

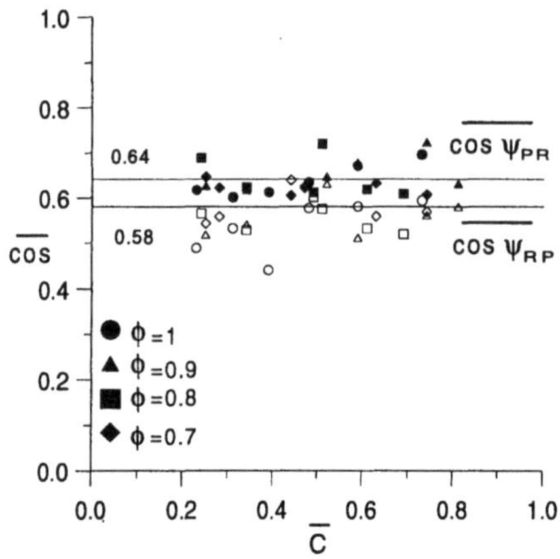

Fig. 6a. Mean values of $\cos\psi_{rp}$ and $\cos\psi_{pr}$ versus $\bar{c}$ ($U_r = 5$ m/s).

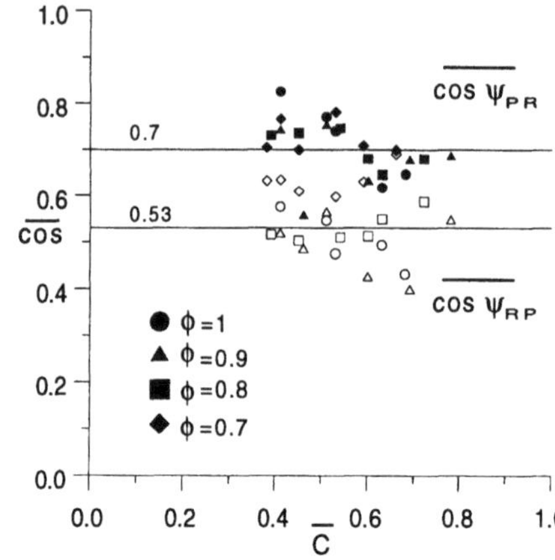

Fig. 6b. Mean values of $\cos\psi_{rp}$ and $\cos\psi_{pr}$ versus $\bar{c}$ ($U_r = 2$ m/s).

$U_r = 2$ m/s are shown on figure 6b. Two main differences are observed compared to the previous case. First, the distribution of the mean cosines values show significant variations with $\bar{c}$; this is again due to the reduced number of crossing events for these flames with a lower cold flow velocity, which reduces the statistical robustness of the analysis. This is also shown by the fact that the presented results for these flames are confined to a narrower $\bar{c}$ region. Second, the difference between the two averages of the cosines is more important than previously; a closer look at the figure shows that this difference increases with ϕ. This is due to the higher $\overline{S_L}/U_r$ ratio for these flames, which, according to the above relationships, introduces a significant difference between the two direction cosines.

The equations (1) to (4) above can be combined to obtain the expression (7) for the average mass of products created per unit volume of reactants during a flamelet crossing.

$$\frac{\overline{w}_F}{\varrho_r} = \overline{S}_L \overline{\frac{1}{V_n}} = \overline{S}_L \frac{\left(\overline{t_{rp}} + \overline{t_{pr}}\right)}{2\delta_L} \tag{7}$$

In this relation, $\overline{t_{rp}}$ and $\overline{t_{pr}}$ are determined from the experimental distributions of flamelet transit times; δ_L is the normal flamelet thickness determined as previously from the shortest flamelet transit times; $S_L = I_0 S_L^0$ is the effective laminar propagation velocity, which is in principle different from its unstrained planar laminar flame value S_L^0. As a first approximation, we determined $\overline{w}_F/\varrho_r$ with $I_0 = 1$. Its variation versus $\bar{c}$ is shown on Figs. 7 for $U_r = 5$ m/s and 2 m/s. Table 2 presents its averaged values through the flame brush as a percentage of ϱ_r. $\overline{w}_F/\varrho_r$ is almost constant through the flame brush, but depends strongly on both the mixture composition and its convection velocity. For both cases, $\overline{w}_F/\varrho_r$ increases with increasing equivalence ratio. For example, Fig. 7a indicates that for $U_r = 5$ m/s, a unit volume of unburnt gases with a stoichiometric composition needs to encounter approximately 5 flamelets to

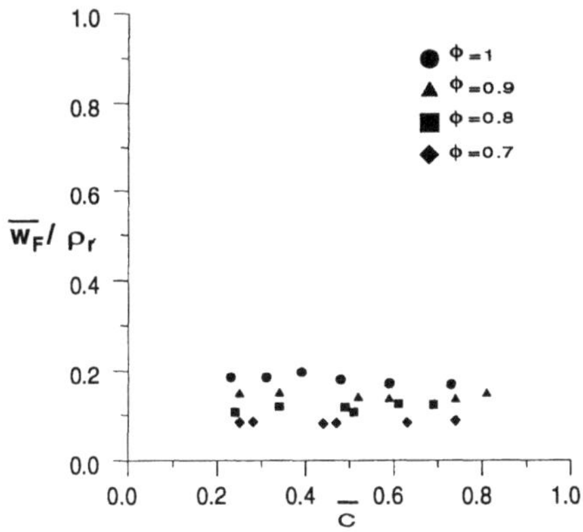

Fig. 7a. Variation of $\overline{w}_F/\varrho_r$. (U_r = 5 m/s)

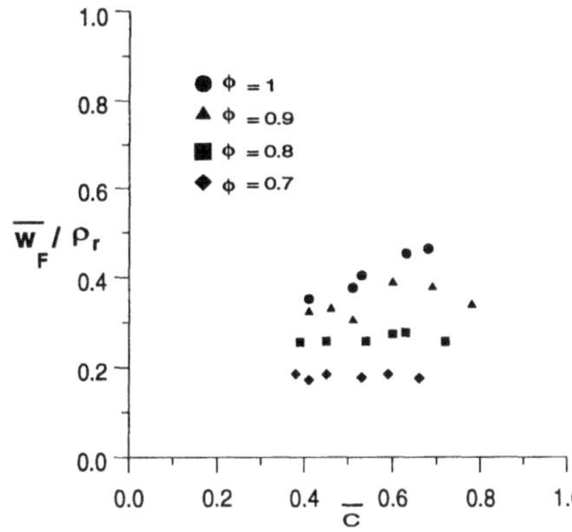

Fig. 7b. Variation of $\overline{w}_F/\varrho_r$. (U_r = 2 m/s)

be consumed entirely; this number is approximately doubled for a mixture with $\phi = 0.7$. As the flame transit times are increased when the convection velocity of the reactants is reduced, $\overline{w}_F/\varrho_r$ is higher for the flames with U_r = 2 m/s.

Table 2. $\overline{w}_F$ values as percentages of ϱ_r for different ϕ.

ϕ	0.7	0.8	0.9	1
(2 m/s)	18	31	34	41
(5 m/s)	8.5	12	14	18

To obtain the mean reaction rate at a given point within the turbulent flame brush, the average mass of products created per unit volume of reactants during a flamelet crossing should be multiplied by the flamelet crossing frequency ν. In terms of the characteristic times determined by the Rayleigh light scattering technique, $\overline{w}$ can be written as follows:

$$\frac{\overline{w}}{\varrho_r} = \frac{\overline{S}_L\left(\bar{t}_{rp} + \bar{t}_{pr}\right)}{2\delta_L}\,\frac{2}{\left(\bar{t}_r + \bar{t}_p + 2\bar{t}_F\right)} \tag{8}$$

The last term of this relation is simply the crossing frequency ν which takes into account the mean passage times of reactants and products (or the mean time interval between two consecutive flamelet crossings) and the flamelet transit time t_F. This more complete definition of the crossing frequency, which no longer assumes infinitely rapid flamelet crossings, generalizes the initial BML model (Bray, 1993) and is taken into account in our experimental estimate by using a single threshold set at $I^\star = 0.5$ for the determination of ν. The variation of $\overline{w}/\varrho_r$ versus $\bar{c}$ is presented on Fig. 8 for the flames with U_r = 5 m/s. As explained previously, the present technique is not able to gather sufficient statistical information on the edges of the flame brush where flamelet crossing events are rather scarce. But, it may be argued that this is not in itself a limitation as long as the central part of a turbulent premixed flame brush, where the bulk of the chemical transformation and therefore thermal energy production occurs, is correctly represented.

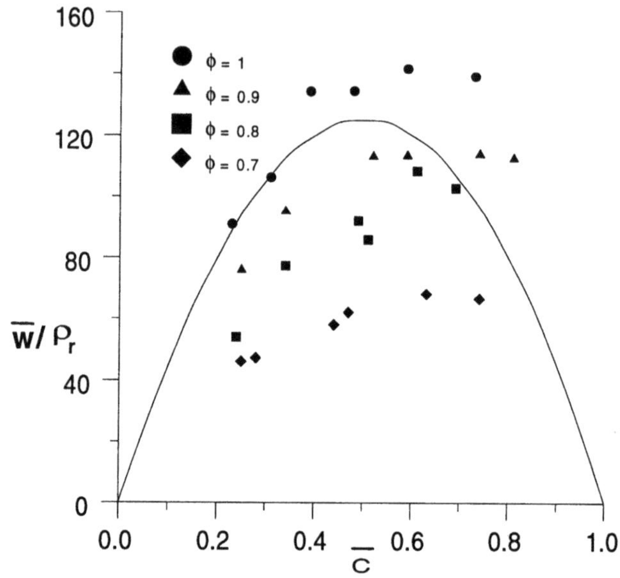

Fig. 8. Variation of $\overline{w}/\varrho_r$ versus $\overline{c}$ (U_r = 5 m/s).

As for $\overline{w}_F/\varrho_r$, the mean reaction rate $\overline{w}/\varrho_r$ depends strongly on the mixture equivalence ratio. It is important to recall here that our previous studies have clearly demonstrated that the flamelet crossing frequency ν is almost independent of the mixture composition (Boukhalfa and Gökalp, 1988; Deschamps, 1990). The dependence of $\overline{w}$ on ϕ is then solely related to the variation of $\overline{w}_F$ with ϕ. In order to compare this experimental estimate of the mean reaction rate with the BMCL model, we used the following algebraic model given in Bray (1990):

$$\frac{\overline{w}}{\varrho_r} = 3I_0\overline{c}(1-\overline{c})\left(\frac{u'}{l}\right)_r \tag{9}$$

where $\left(\frac{u'}{l}\right)_r$ is a characteristic turbulence time scale assumed constant through the flame brush and evaluated at the leading flame edge. The prediction calculated by using the experimental value of $\left(\frac{u'}{l}\right)_r$ and $I_0 = 1$ is also shown on Fig. 8.

5. Conclusion

An experimental technique and data analysis method have been developed to provide reliable data on instantaneous flame front statistics in turbulent premixed flames. The data can be used to estimate the mean local reaction rate in such flames, and provides a useful data base for modelling purposes. The methods developed are illustrated in the case of open conical turbulent premixed flames. The comparison with the Bray-Moss-Champion-Libby model is satisfactory and indicates possible ways of improving the model. Also, the methodology outlined here could be used in technologically relevant turbulent flames, either by hardening the Rayleigh light scattering system or by replacing it with an intrusive technique such as miniaturized ionization probes. These avenues are presently explored in our research program.

Acknowledgements

The support of the French Army Research Office (DRET groupe 7) and the Conseil Régional de la Région Centre, and fruitful discussions with Ken Bray, Ian Shepherd and Fred Gouldin are greatly appreciated.

References

Bray, K.N.C., Champion, M., and Libby, P.A. (1988) *Twenty-Second Symposium Int. on Combustion*, The Combustion Institute, p. 763.

Bray, K.N.C. (1990) Proc. R. Soc. Lond. A, **431**, p. 315.

Bray, K.N.C. (1993) Personal communication.

Boukhalfa, A. and Gökalp, I. (1988) *Twenty-Second Symposium Int. on Combustion*, The Combustion Institute, p. 755.

Cant, R.S., Pope S.P. and Bray K.N.C. (1990) *Twenty-Third Symposium Int. on Combustion* p. 809.

Chew, T.C., Bray, K.N.C., and Britter, R.E. (1990) Comb. Flame **80**, p. 65.

Deschamps, B. (1990) *PhD Thesis*, University of Orléans.

Deshamps, B. and Gökalp, I. (1990) Spectroscopy in Combustion studies, ed. G.T. Kalghatgi, IOP Short Meeting № 27, p. 41.

Deschamps, B., Boukhalfa, A., Chauveau C., Gökalp, I., Shepherd, I.G., and Cheng, R.K. (1992) *Twenty-Fourth Symposium Int. on Combustion*, The Combustion Institute, p. 469.

Echekki, T. (1992) *PhD Thesis*, Stanford University.

Gökalp, I. (1987) Comb. Flame **67**, p. 111.

Gouldin, F.C. (1987) Comb. Flame **68**, p. 269.

Jarosinski, J. (1984) Comb. Flame **56**, p. 337.

Lee, T.-W., North, G.L., and Santavicca, D.A. (1992) Combust. Sci. and Tech. **84**, p. 121.

Mounaïm-Rousselle, Ch., Deschamps, B., Boukhalfa, A., Chauveau, C. and Gökalp, I. (1993) *Proceedings of the Joint Meeting of the English and German sections of the Combustion Institute*, Cambridge, England, p. 44.

Poinsot, T., Veynante, D., and Candel, S. (1991) J. Fluid Mech. **228**, p. 561.

Said, R. and Borghi, R. (1988) *Twenty-Second Symposium Int. on Combustion*, The Combustion Institute, p. 569.

Shepherd, I.G. and Ashurt W., T. (1992) *Twenty-Fourth Symposium Int. on Combustion*, The Combustion Institute, p. 485.

Index of Contributors

Andersson, H. I. 53
Antonia, R. A. 165
Arad, E. 7

Bilger, R. W. 411
Binder, G. 301
Blackwelder, R. F. 221
Boivin, M. 85
Borghi, R. 439
Bradley, D. 385
Brown, R. J. 411

Chen, J. H. 123
Comte, P. 283

Deutsch, E. 85
Ducros, F. 283
Durst, F. 225

Feng, M. Q. 301
Ferrão, P. 427

Gökalp, I. 459
Gosman, A. D. 187
Gutmark, E. 207

Hadžić, I. 323
Hallbäck, M. 21
Hanjalić, K. 323
Heitor, M. V. 427
Hirata, M. 241

Inaoka, K. 365

Jakirlić, S. 323
Johansson, A. V. 21
Jovanović, J. 225

Kasagi, N. 241
Kawamura, H. 259
Kollmann, W. 123
Kristoffersen, R. 53
Kuroda, A. 241

Launder, B. E. 3
Lesieur, M. 283
Li, J. D. 411
Lienau, J. J. 123
Lindstedt, R. P. 389
Liu, R. 187

Mantel, T. 439
Mastorakos, E. 147
McGuirk, J. J. 187
Mi, J. 165
Mounaïm-Rousselle, Ch. 459

Nagano, Y. 343

Oberlack, M. 33

Picart, A. 439
Piomelli, U. 71

Sakthitharan, V. 389
Satake, S. 259
Sato, H. 343
Schadow, K. C. 207
Sender, J. 225
Simonin, O. 85
Sjögren, T. 21
Sokolov, M. 119
Squires, K. D. 71
Suzuki, K. 365

Tagawa, M. 343
Tardu, S. 301
Taylor, A. M. K. P. 147

Whitelaw, J. H. 147
Wolfshtein, M. 7

Yu, K. 207

MIX
Papier aus verantwortungsvollen Quellen
Paper from responsible sources
FSC® C105338
FSC
www.fsc.org

If you have any concerns about our products,
you can contact us on
ProductSafety@springernature.com

In case Publisher is established outside the EU,
the EU authorized representative is:
Springer Nature Customer Service Center GmbH
Europaplatz 3, 69115 Heidelberg, Germany

Printed by Libri Plureos GmbH
in Hamburg, Germany